Drilling Engineering

Drilling Engineering

J.J. Azar
Professor Emeritus
and former Director of Drilling Research (TUDRP)
Petroleum Engineering Department
University of Tulsa, Tulsa

G. Robello Samuel
Senior Technical Advisor (Drilling and Completions)
Halliburton Company (Drilling, Evaluation and Digital Solutions)
Houston

PennWell®

1421 South Sheridan Road
Tulsa, Oklahoma 74112-6600 USA

800.752.9764
+1.918.831.9421
sales@pennwell.com
www.pennwellbooks.com
www.pennwell.com

Director: Mary McGee
Managing Editor: Stephen Hill
Production / Operations Manager: Traci Huntsman
Production Editor: Tony Quinn
Senior Design Editor: Robin Remaley
Book Designer: Susan E. Ormston
Cover Designer: Sheila Brock

Library of Congress Cataloging-in-Publication Data

Azar, Jamal J., 1937-
Drilling engineering / J.J. Azar, G. Robello Samuel.
p. cm.
ISBN-13: 978-1-59370-072-0 (hardcover)
ISBN-10: 1-59370-072-5 (hardcover)
1. Oil well drilling. 2. Gas well drilling. I. Samuel, G. Robello. II. Title.
TN871.2.A92 2007
622'.3381--dc22

2006034984

Printed in the United States of America

1 2 3 4 5 12 11 10 09 08 07

Dedication

This book is dedicated to all of those who have contributed
to my achievements during my professional career: my former students,
colleagues from academia and oil/gas & related industries,
friends, and last but not least, my loving family.

JJ Azar, Professor Emeritus, PE
Member: SPE, NAE

Contents

Preface

Drilling engineering is a challenging discipline in the oil patch. It goes beyond what is found in textbooks. The technological advances in the past two decades have been very significant. These advances have allowed the oil industry worldwide to economically and successfully exploit oil and gas fields that may have not been possible before.

The fundamentals of fluid mechanics and solid mechanics, along with the basic scientific concepts of chemistry, form the basis of drilling engineering. The rewards and successes of drilling projects are predicated on the ability of the drilling engineer who fully understands all the engineering aspects and equipment required to drill a usable hole at the lowest dollar per foot ($/ft), in vertical well drilling, or at the highest equivalent barrel of oil per foot (bbl/ft) in horizontal/multilateral well drilling.

In the oil industry, it is common practice to assume that problems may be encountered during the drilling process of a well and can differ from one well to the other, even in the same area. The drilling engineer has the responsibility to be able to think beyond simply the engineering equations in combating some of these problems. He or she must anticipate drilling problems, devise solutions, and choose the best alternatives. They must be intuitive, must have common sense, and be able to think quickly and decisively. In this book, the authors have prepared illustrative examples and supplementary problems that hopefully will encourage the students to cultivate their creative thinking, develop confidence in reaching independently for what is needed to solve problems, and above all, energize their common sense to make wise decisions.

The book is written for educating students in drilling engineering in a two-semester course series: Drilling I and Drilling II. It is also intended to serve as a reference to engineers in field operations. It is suggested that Drilling I cover the following chapters: chapter 1—Rotary Drilling for Oil and Natural Gas, Chapter 2—Drilling Fluids, chapter 3—Fluid Flow and Associated Pressures in the Rotary Rig Circulating System, chapter 4—Drill Bit Hydraulics, chapter 5—Transport of Drilled Cuttings, and chapter 6—Prevention and Control Mechanics of Well Blowouts; where Drilling II covers: chapter 7—Directional and Horizontal Well Drilling, chapter 8—Drill Bit Mechanics, chapter 9—Drill String Design, chapter 10—Drilling Problems and Solutions, chapter 11—Casing and Cementing Design, and Chapter 12—Well Planning in Drilling Engineering.

It is recommended that non-conventional drilling techniques, which is not included in this text, such as under-balance drilling, slim hole drilling, coiled-tubing drilling, casing drilling, etc., be given to students in Drilling II as special topics assignments.

Acknowledgments

The Authors would like to thank and acknowledge the following contributors:

Tommy Warren
George King
James Lummus
David Chen

We also thank Peter Bern for his valuable review and contribution. Acknowledgements and thanks are also given to the following companies who have given permission to use different images of the tools: Halliburton, Baker Hughes, Smith International Inc., International Association of Drilling Contractors, Atwood Oceanics, and Fann Instrument Company.

Finally, Professor Azar would like to thank Dr. Adam T.Bourgoynae Jr., who was instrumental in starting him on the road to success in drilling engineering.

1

Rotary Drilling for Oil and Natural Gas

Introduction

Drilling for oil and natural gas requires two major constituents: manpower and hardware systems. The manpower encompasses a drilling-engineering group and a rig-operations group. The first group provides engineering support for optimum drilling operations, including rig selection and design of the mud program, casing and cementing programs, the hydraulic program, the drill bit program, the drill string program, and the well control program. After drilling begins, the daily operations are handled by the second group, which consists of a tool pusher and several drilling crews (derrick and motor personnel, drillers, etc.). The hardware systems that make up a rotary drilling rig are

1. A power generation system
2. A hoisting system
3. A drilling fluid circulating system
4. A rotary system
5. Well blowout control systems
6. A drilling data acquisition and monitoring system

This chapter will primarily address the form and functions of these systems and their various components.

The Process of Rotary Drilling

Whether drilling a *vertical hole* or a *directional hole* (fig. 1–1) for the purpose of producing oil and/or natural gas, several elements are needed to drill the hole successfully and economically (illustrated in fig. 1–2), including

- A force acting downward on a drill bit
- Rotation of the drill bit
- Circulation of fluid, called drilling fluid (liquid, gas, or gasified liquid), from the surface through the tubular, called the drill string, and back to the surface through the annular space, the area between the hole wall and the outside wall of the drill string

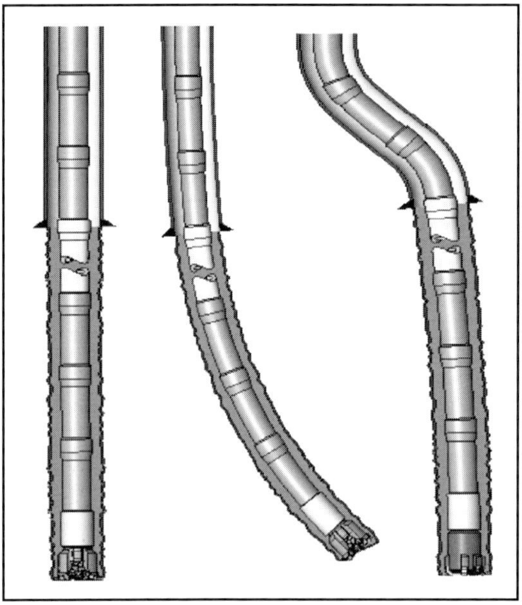

Fig. 1–1 Vertical and directional wells

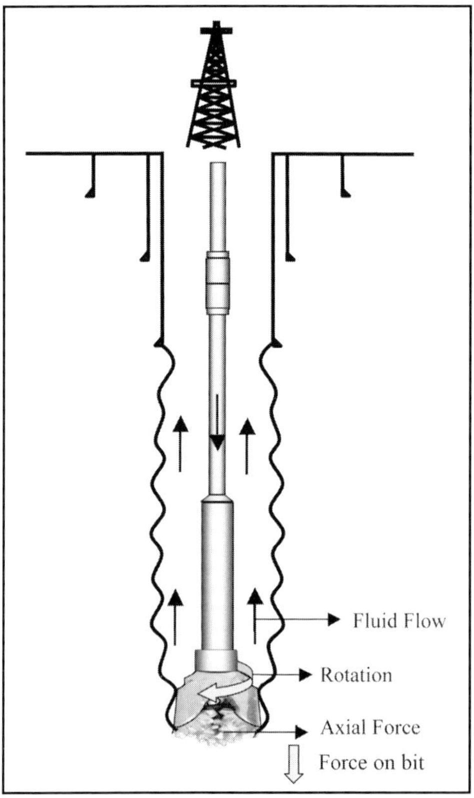

Fig. 1–2. Basic elements to drill a well

Force

To drill through a rock, a force (*weight on bit)* of a certain magnitude has to be applied downward on a drill bit. This force is provided by slacking off tension on the drilling line that supports the total weight of the drill string. The amount of force varies, depending on the hardness of rock and the type of drill bit in use, and is provided by the portion of the tubular (drill string) above the bit. In vertical well drilling, the amount of force is provided by the weight of the length of heavy pipe (drill collars) placed right above the bit. This is accomplished by allowing a part or all of the drill collar's length to go into compression. In directional wells, including horizontal wells, heavy pipes (drill collars and heavy-weight drill pipe) are no longer placed in the highly inclined portion of the hole; rather, they are placed in sections of lesser inclination or near vertical, to avoid excessive frictional forces that can significantly reduce the transfer of desired force (weight on bit) to the drill bit. (See fig. 1–3 for illustration.)

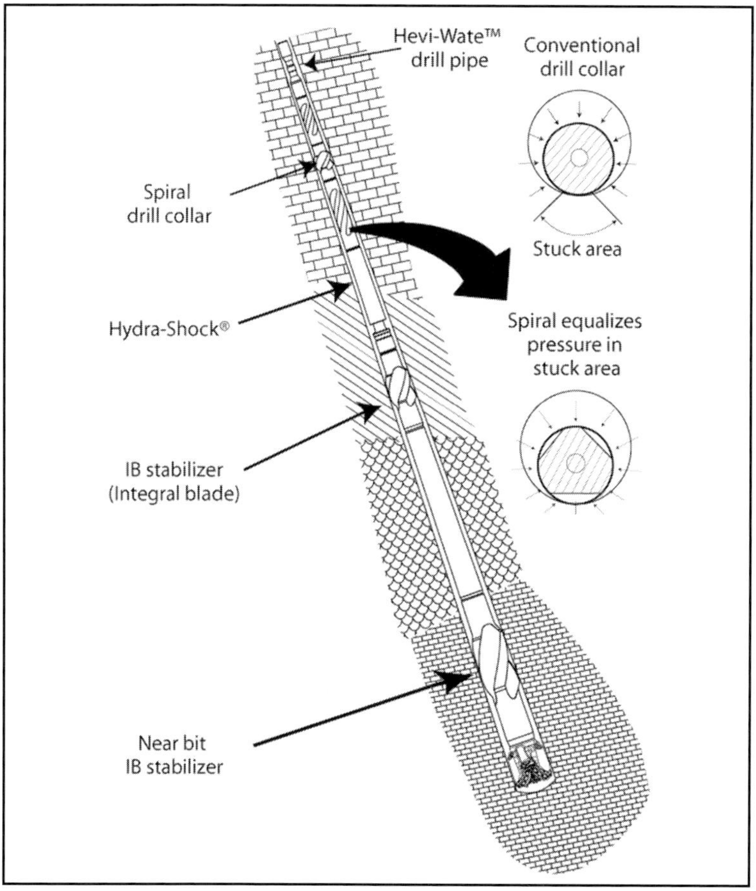

Fig. 1–3. Packed hole assembly (Courtesy Smith International Inc.)

Drill bit rotation

In rotary drilling, in contrast to the early cable tool drilling, the drill bit has to be rotated at a certain speed to cut and advance through the rock. Rotation of the drill bit may be accomplished from the surface or down hole. Surface rotation is done through the use of a conventional rotary table or top drive motor. Bottom-hole rotation is achieved by using downhole motors (positive-displacement mud motors or mud turbines or electric motors) that directly rotate the drill bit without drill string rotation.

Drilling-fluid circulation

During drilling, heat and rock cuttings are being generated continuously. If the process of making a hole is to be met successfully, a fluid has to be continuously circulated from the surface to the bottom and back to the surface again to dissipate the heat and remove cuttings.

Rotary Drilling Rigs

The function of a rotary drilling rig is to allow the drilling and completion of a usable hole at the lowest overall well cost. A schematic of a typical rig and its components is provided in figure 1–4. Rotary drilling rigs are classified as onshore (land) or offshore (marine) rigs. Their main design features are mobility, flexibility, and maximum depth of operation. Rig components, in general, are common to both onshore and offshore rigs. The only major difference is in the utilization of an extension pipe (drilling riser) between rig floor and seabed when drilling in a marine environment.

Modern land rigs are built-in units and are skid mounted so that they can be transported from one drilling site to another. Once on location, the various rig components are easily assembled to drill the well. The derrick is designed such that it can be assembled and disassembled with ease; its functions are to carry loads that are suspended in the wellbore and to provide space between the rig floor working area and the crown block, such that a certain length of the drill string can be made. A rig that can handle singles, doubles, or triples allows lengths of 30 ft, 60 ft, or 90 ft sections of drill string to be made, respectively, as drilling progresses; these sections are called stands. For moderate-depth wells, carrier-mounted rigs are used; these portable masts are mounted on flatbed vehicles as self-contained single units.

Offshore drilling is performed from the space above water on structures, called *platforms,* that may be classified as (1) mobile or (2) stationary or fixed. The selection of a particular rig carrier depends on the operations to be performed and the water depth of the location where such operations are to be conducted. Figure 1–5 shows a self-elevating platform rig, and figure 1–6 shows a fixed platform.

The drilling of wells in the oil patch may be for the purpose of exploration, development, infill, injection, or reentry into existing wells. Exploratory well drilling is conducted to find and evaluate potential reserves in various locations. In areas where reserves have not been proven, drilling is done most economically by the utilization of mobile rig carriers. These are defined as units that can be moved from one location to another as the need arises. Development well drilling is conducted to develop a field with proven reserves to its fullest production potential. Injection well drilling is conducted for the purpose of storing hydrocarbons, mainly natural gas; disposing of unwanted produced waters; or regaining the reservoir

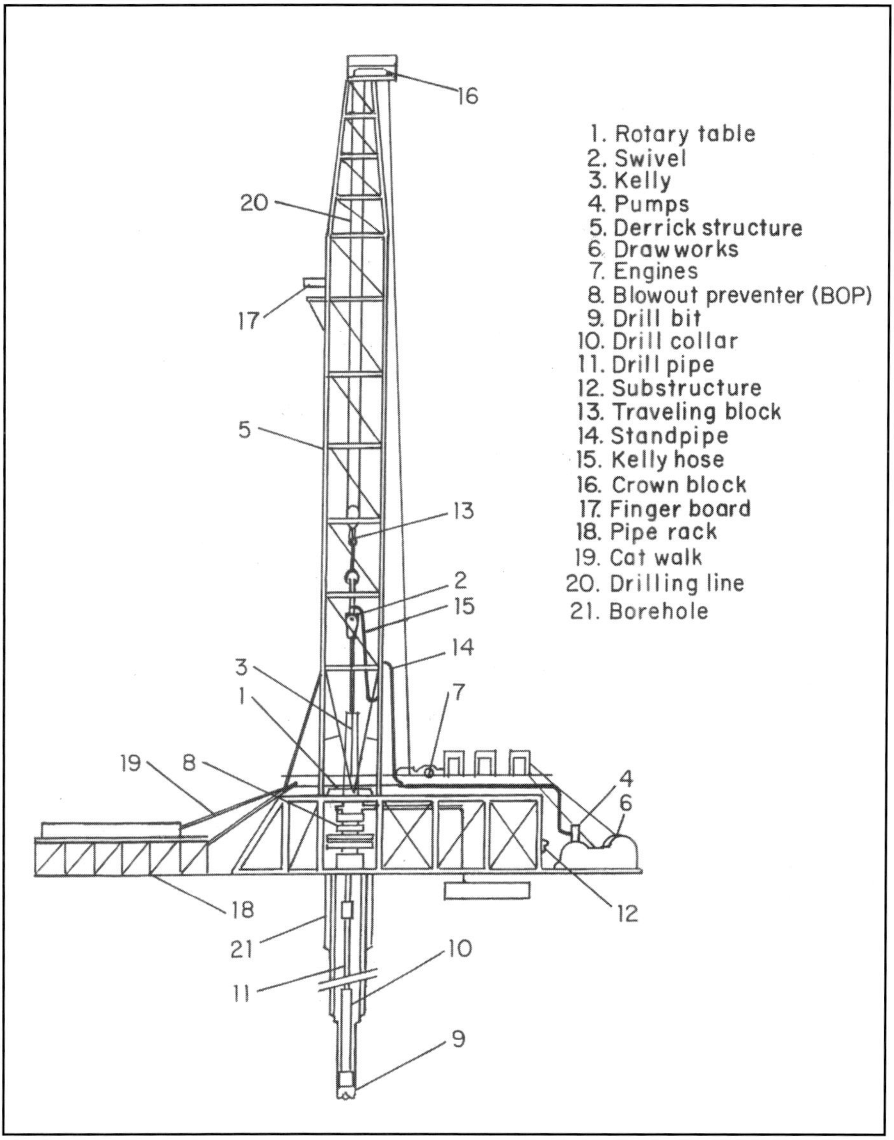

Fig. 1–4. Rotary drilling rig (Reprinted by permission; courtesy International Association of Drilling Contractors [IADC])

Fig. 1–5. Self-elevating platform (Courtesy Atwood Oceanics)

pressure by injecting water, steam, or gas into the reservoir. Reentry well drilling is conducted on existing wells for the purpose of tapping deeper producing zones, recompleting, or drilling laterals from existing main trunk (vertical, directional, or horizontal). Infill well drilling is conducted to replace depleted wells or drill additional wells to sustain production economics in certain fields.

Fig. 1–6. Fixed Platform (Courtesy Atwood Oceanics)

Offshore drilling utilizes rigs that are supported on man-made surfaces, in contrast to land rigs, which are supported on ground surface. The supporting surface for offshore rigs may be grouped as follows:

- Mobile units (jack-ups, semisubmersibles, and ships)
- Sea-bottom–supported fixed platforms

The power of rotary rigs is generally generated by diesel or gas-driven engines. The power is transmitted to the various rig systems (hoisting, rotary, circulating, etc.) by means of mechanical or electrical drives, which will be discussed later.

The electric drive system (DC/DC system) was introduced to rotary rigs in the mid '50s for use in offshore operations. Only recently, electric drives, particularly the AC/SCR (alternating current silicone-controlled rectifier), have gained prominence in application to land drilling rigs.

The DC/DC system, on the one hand, requires that one or more DC generators be specifically assigned to a DC motor to meet the desired load requirement for a specific unit (pumps, rotary motor, drawworks, etc.) at a controllable desired speed. In addition, AC generators must also be available, to operate other rig components that require alternating current. This specific assignment of generators and its driving engines always leads to operating with more engine capacity than the total rig power would justify.

The AC/SCR system, on the other hand, has the capability of passing alternating current in one direction only; therefore, it can produce DC power of controllable voltage, which in turn allows speed control of DC motors. Its advantage is that all of the generated AC power is fed to a common bus and converted to direct current only as needed.

The overall efficiency of power-generating systems may be defined as follows:

$$\text{Efficiency (\%)} = \frac{\text{output power}}{\text{input power}} \times 100 \quad (1.1a)$$

or

$$\eta = 100\frac{P_o}{P_i} \quad (1.1b)$$

The output power of an engine is generally expressed as a function of the engine rotary speed, N, and the output torque, T:

$$P_o = \frac{2\pi NT}{33{,}000} \quad (1.2)$$

where

T is output torque, in foot pounds (ft-lbs)
N is engine rotary speed, in revolutions per minute (rpm)
P_o is output power, in horsepower (hp)

The input power is expressed in terms of the rate of fuel consumption, Q_f, and the fuel heating value, H:

$$P_i = \frac{Q_f H}{2{,}545} \quad (1.3)$$

where

Q_f is rate of fuel consumption, in pounds per hour (lbs/h)
H is fuel heating value, in British thermal units per pound (BTU/lb)
P_i is input power, in hp

Example 1–1

A drilling rig has three diesel engines for generating the rig power requirement. Determine the total daily fuel consumption for an average engine running speed of 900 rpm, average output torque of 1,610 ft-lbs, and engine efficiency of 40%. The heating value of diesel oil is 19,000 BTU/lb.

From equations (1.1)–(1.3), the fuel consumption per engine can be solved for and is given by

$$P_i = \frac{48.5}{2{,}545}\frac{NT}{\eta} \text{ or } 48.5\frac{NT}{\eta H}$$

where

$N = 900$ rpm (given)
$\eta = 40\%$ (given)
$H = 19{,}000$ BTU/lb (given)
$T = 1{,}610$ ft-lbs (given)

Hence,

$$Q_f = 48.5\left(\frac{1{,}610 \times 900}{40 \times 19{,}000}\right) = 92.5\text{lbs/h}$$

or

$$Q_f = \frac{92.5\text{ lbs}}{\text{h}} \times \frac{24\text{ h}}{\text{d}} \times \frac{\text{gal}}{7.2\text{ lbs}} = 308\frac{\text{gal}}{\text{d}}$$

For the three engines, total $Q_f = 308 \times 3 = 924$ gal/d.

Rig Selection

The main objective of rig selection is to choose from the available rigs the one that will closely meet the criterion for drilling a usable hole at the lowest overall cost of the proposed well. Directly involved in the selection process is the evaluation of all of the following rig systems, which will be discussed later:

- Power generation system
- Hoisting system
- Fluid circulation system
- Rotary system
- Well control system
- Drilling data acquisition and monitoring system

In addition, quality of supervision and rig personnel must be given top priority. Other considerations in the selection process of rigs include

- Safety records
- Rig mobility and ease of handling
- Contractor dependability
- Contractual rates (footage, day, turnkey)
- Condition of all rig equipment

Minimum rig size specifications should be determined after completing the design requirements of all the drilling programs of the well and before soliciting bids for the job. Drilling programs that must be completed prior to rig selection include drilling fluid, drill string, drill bit, drilling-fluid hydraulics, well control, data acquisition and monitoring, and casing/cementing.

Rotary Drilling Systems

Hoisting system

The primary function of the hoisting system in rotary drilling is to hoist the drill and casing strings during drilling and casing operations, respectively. Its major components are the drawworks, the crown block, the traveling block, the hook, the drilling (wire) line, and the elevator. Figure 1–7 gives a schematic diagram of one such system. Following is a brief description of each component.

Drawworks. The *drawworks* is an assemblage of components that provides the hoisting and breaking power to enable the down- and uphole movements of heavy loads of the drill string and casings or other equipment. Its components include the drum, the brakes, the transmission, and the cathead.

Drawworks are rated by input horsepower. To determine the drawworks horsepower required for drilling a given well, the hook horsepower is first calculated using the weight of the heaviest drill string in air, along with a reasonable hoisting speed. A minimum of 100 ft/min is recommended for rating a rig. An overall hoisting efficiency of 65% is recommended to calculate the required drawworks horsepower rating.

Blocks and drilling line. The principle function of the *blocks and drilling line* is to provide a mechanical advantage while raising and lowering the extremely heavy loads into the wellbore. Their *efficiency* is measured by

$$\eta = \frac{\text{power output}}{\text{power input}} = \frac{P_o}{P_i} \tag{1.4}$$

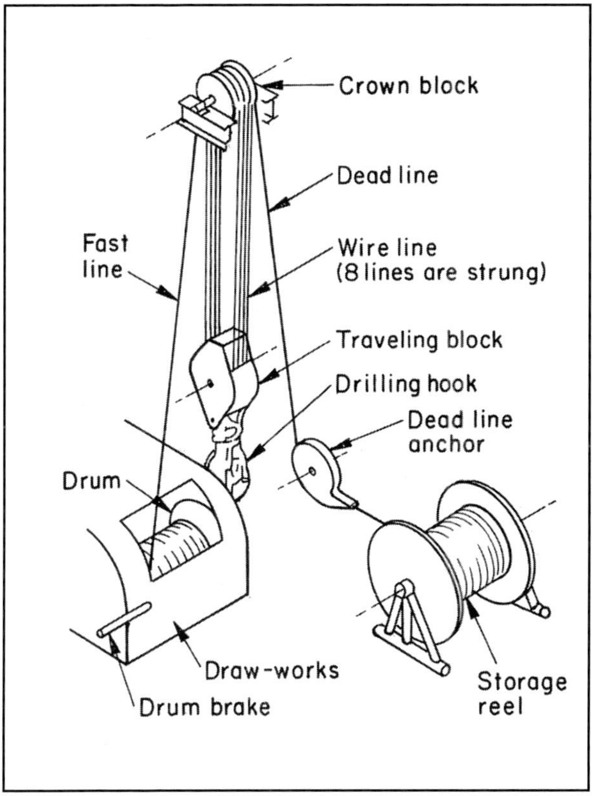

Fig. 1–7. Block and tackle (Reprinted by permission; courtesy IADC)

The output power is defined as follows:

$$P_o = F_h v_{tb} \tag{1.5}$$

where

F_h = hook load
v_{tb} = traveling-block velocity

Similarly, the input power from the drawworks to the fast line is given by

$$P_i = F_{fp} v_{fp} \tag{1.6}$$

where

F_{fp} = load in fast line
v_{fp} = fast line speed

The magnitude of F_{fp} depends on the number of lines strung between the crown block and the traveling block and the frictional forces induced between the contact surfaces of the lines and block sheaves.

Utilizing equations (1.5) and (1.6), equation (1.4) becomes

$$\eta = \frac{F_h v_{tb}}{F_{fp} v_{fp}} \tag{1.7}$$

It can be easily established that

$$v_{tb} = \frac{v_{fp}}{n} \tag{1.8}$$

Therefore,

$$\eta = \frac{F_h}{nF_{fp}} \tag{1.9}$$

where n is the number of lines strung between the crown block and the traveling block.

Example 1–2

A rotary rig that can handle triples is equipped with 1,200 hp drawworks. The efficiency of the hoisting system is 81%. Determine the time it takes to pull one stand at a hook load of 300,000 lbs.

Solution:

$$\eta = \frac{\text{power output}}{\text{power input}} = \frac{P_o}{P_i} = 0.81 = \frac{P_o}{1{,}200}$$

$P_o = 927$ hp, but $P_o = v_{tb}F_h$. Therefore,

$$v_{tb} = \frac{P_o}{F_h} = \frac{927\text{hp} \times 33{,}000 \ \frac{\text{ft-lb}}{\text{min}} \div \text{hp}}{300{,}000} = 107 \text{ ft/min}$$

and

$$t = \frac{L_s}{v_{tb}} = \frac{90 \text{ ft}}{107 \text{ ft/min}} = 0.84 \text{ min} = 50.5 \text{ s}$$

Drilling line. The *drilling line* is a wire rope that is spooled on the drum of the drawworks and strung between the crown and traveling blocks. Its maintenance in good working condition is of the utmost importance. Failure of the drilling line can lead to costly problems of fishing operations, injury to rig personnel, and damage to rig equipment.

The American Petroleum Institute (API) recommends the use of the *ton-mile* concept to evaluate drilling line service, wherein the traveling block moves a one-ton load a distance of one mile. To maintain even wear on the drilling line, it is recommended that the lift-off points be changed regularly at a set number of ton-miles based on drilling conditions. (See API-RP-9B.)

Traveling block. The *traveling block* and hook are rated in terms of tons of capacity. Ratings include a suitable factor of safety. These elements will safely suspend their full rated loads if they are in good condition. Minimum required ratings of the traveling block and hook are determined by the hook load required in order to handle the heaviest string of casing in air. However, if large drag forces are anticipated, these should also be considered in the hook load.

Crown block capacity. The load on the *crown block* is greater than the hook load. In the block and tackle system, the tension in all lines is equal except for friction. The hook load and the weight of the traveling block, hook, and other suspension equipment are equally distributed among the lines extending to the traveling block. All of these lines pull down on the crown block. In addition, the deadline to the anchor and the fast line to the drum also pull down on the crown block. The crown block capacity required in order to handle the net static hook load capacity can be calculated using the following formula:

$$R_C = \frac{(H_L + S)(n + 2)}{n} \tag{1.10}$$

where

R_C is the required crown block rating, in lbs
H_L is the net static hook load capacity, in lbs
S is the effective weight of suspended equipment, in lbs
n is the number of lines strung to the traveling block

Example 1–3 (Minimum drawworks horsepower)

What minimum drawworks horsepower is required in order to drill a well when using 10,000 ft of 16.6 lb drill pipe with an outside diameter (OD) of 4½ inches (in.) and 50,000 lbs of drill collars?

Solution:

$$\text{Drillstring weight} = 50{,}000 \text{ lbs} + (10{,}000 \text{ ft})\ (16.6 \text{ lb/ft})$$
$$= 216{,}000 \text{ lbs}$$

$$\text{Hook horsepower} = \frac{(216{,}000 \text{ lbs})\ (100 \text{ ft/min})}{33{,}000} = 655 \text{ hp}$$

$$\text{Required minimum drawworks horsepower rating} = \frac{655}{0.65} = 1{,}010 \text{ hp}$$

This method neglects the effect of buoyancy and the weight of the block and hook. In straight holes, buoyancy can be assumed to be offset by pipe drag. In directional holes, drag has to be considered.

Hoisting engines should have a horsepower rating for intermittent service equal to the required drawworks horsepower rating divided by 85% efficiency. Drawworks also have a rating of line pull efficiency; this factor depends on the number of lines strung between the crown block and the traveling block. (See table 1–1.)

Table 1–1. Line pull efficiency factor

Number of Lines	Efficiency Factor
6	0.874
8	0.841
10	0.81
12	0.77
14	0.74

Example 1–4 (Line pull calculation)

What line pull is required to handle a 500,000 lb casing load with 10 lines strung?

Solution:

$$\frac{500{,}000}{(10)\ (0.810)} = 61{,}800 \text{ lbs}$$

Fluid circulating system

The function of the fluid circulating system in rotary drilling is to allow the movement of a drilling fluid from the surface to the hole bottom and back to surface again. The main components of the system include

- Mud pumps/air compressors
- High-pressure surface connections
- Drill string
- Drill bit
- Return annulus
- Mud pits
- Mud treatment equipment

Figure 1–8 shows a schematic of the system.

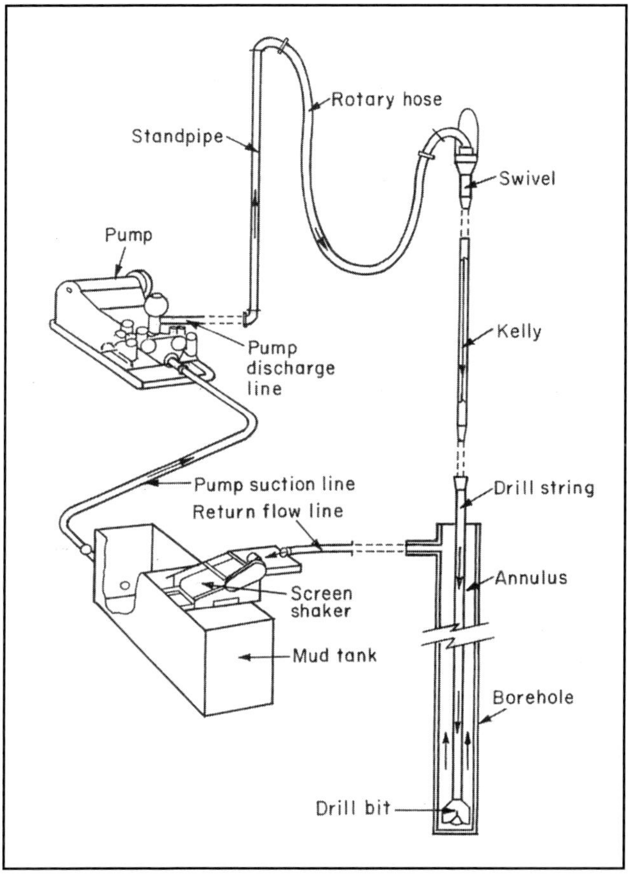

Fig. 1–8. Fluid circulation system (Reprinted by permission; courtesy IADC)

Mud System Evaluation

The configuration of tanks, shakers, stirring devices, and solids control equipment is unique to each rig. Rigs built simultaneously or in series at the same construction yard will reflect the individual taste and patterns of the different craftspersons and supervisors who worked on each tank. Unfortunately, aesthetics sometimes takes precedence over functionality.

Functionality and flexibility are required for cost-effective muds and good drilling operations. If the specific solids control equipment system does not provide adequate control of the mud, then the equipment should be corrected or upgraded to acceptable standards before a well is spudded.

Volume requirements

The volume of mud that must be maintained in the surface tank depends on hole volumes at given depths, the speed at which additional volumes are mixed, and the mud-processing ability. In most drilling operations, the surface volume of usable mud should be that required to fill the hole on trips, plus 100 barrels (bbl). Usable mud is mud that can be pumped without change of aerating pumps or introducing sediments to pump suctions. A typical 7 ⅞ in. hole being drilled with 4 ½ in. drill pipe and 6 ¼ in. drill collars requires about 60 bbl to fill the hole on a trip from 8,000 ft. Maintaining an active surface volume in excess of 160 bbl can result in unnecessarily high treatment cost.

With simple mud systems, (i.e., nonweighted systems that are run primarily through water dilution), the surface volume should be larger and can occupy several sequential pits to allow for settling. Each compartment will require frequent jetting to remove the settled solids. If the mud is more complex or weighted, then individual stirring compartments should facilitate mud treatment.

Mixing systems

Properly designed mixing systems should consist of high-speed centrifugal pumps running through a hopper system, plus a slug tank and/or a premix tank. The slug tank should consist of a compartment of 20–50 bbl capacity that can be directed to the pump suctions. The mixing or premix tank should be of 100–200 bbl capacity and also accessible directly by the mud pumps.

The hopper arrangement should be serviced by a centrifugal pump supplying 800–1,200 gallons per minute (gpm) and should preferably have accessible a redundant mixing pump. Few rigs offer mixing facilities that are available only by use of the standby mud pump. This type of arrangement will increase cost and leads to very slow volume additions.

Solids control needs

The need for solids control equipment varies greatly between areas and depends on the mud system in use. Solids control by dilution is effective for unweighted muds where adequate water is available, disposal is neither costly nor likely to cause environmental problems, and amount of liquid required is not excessive. The best way to control solids in drilling muds is through the use of effective solids removal equipment, including shale shakers, hydrocyclones, and centrifuges.

Shale shakers. *Shale shakers* are the most important part of the solids control equipment group. Ideally, the shaker should accommodate full-hole circulating rates over 120 mesh screens. Lesser performance will increase the cost of a sophisticated mud. Low-speed shakers using 40–60 mesh screens are adequate only for very simple mud systems.

Various screen arrangements are available on common shakers. Multiple-screen shakers are most widely used and offer better solids removal. The particle separation on the multideck screen is determined by the finest screen, which is usually the bottom screen.

Hydrocyclones. *Hydrocyclones (desanders* and *desilters)* are low-efficiency devices that must process large volumes of mud to be effective. However, many rigs have poorly built cyclone systems that move mud from place to place without removing many solids.

Cyclone devices should receive mud from one compartment and discharge into a successive compartment. Occasionally, back-flushing is desirable. Devices that receive and discharge into the same tank will offer only marginal to adequate solids removal.

Centrifuges. *Centrifuges* operate at high efficiency by processing small volumes. Most units process 60–80 gpm. Models with the capacity to handle up to 150 gpm may be available in some areas. Normally, centrifuges operate last in line, behind other solids control devices. Unlike screens, cyclones, and mud cleaners, which all operate continuously on the full mud circulating volume, centrifuges operate intermittently on a small fraction of the circulating volume, usually 5–10%.

Mud pumps

The *mud pumps* in rotary drilling are considered major components in the circulating system and are referred to as the heart of drilling-fluid hydraulics. Rotary rig pumps are reciprocating positive-displacement units and are either duplex or triplex pumps. Duplex pumps have two liners and move fluids on the forward and backward stokes (double acting). Triplex pumps have three liners and move fluids only on the forward stoke (single acting). (For a simplified schematic of these pump types, see fig. 1–9).

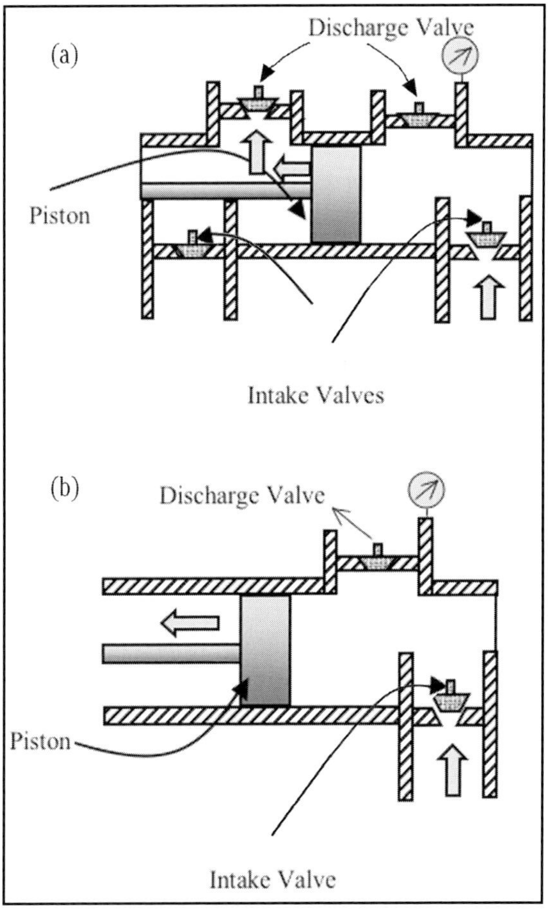

Fig. 1–9. Single- and double-acting reciprocating pumps

The theoretical volume of fluid being displaced by a pump with N liners can easily be formulated from fig. 1–9a and b. For a single-acting pump ($N_c = 3$),

$$V_t = \left(\frac{\pi}{4} D_L^2 L_S\right) N_c \tag{1.11}$$

and for a double-acting pump ($N_c = 2$),

$$V_t = \frac{\pi}{4} L_S N_c \left(2D_L^2 - D_r^2 \right) \tag{1.12}$$

where

D_L is the liner or piston diameter
L_S is the piston stroke or length stroke
N_c is the number of cylinders
D_r is the rod diameter

The actual volume, V_a, of fluid being pumped out is always less than the theoretical and depends on the volumetric *efficiency* of the pump—that is,

$$V_a = V_t E_v \tag{1.13}$$

Or in terms of pump flow rate, Q,

$$Q_a = Q_t E_v \tag{1.14}$$

where E_v = pump volumetric efficiency.

Entrained gas in the mud generally reduces the pump volumetric efficiency. Thus, it is important that the mud is degassed before its return to the suction pit.

Pump rating. Pumps are generally rated for maximum delivery pressures and flow rates. For a given pump delivery pressure, P, at a corresponding flow rate, Q, the hydraulic power of the pump is given by

$$\text{Hydraulic power} = PQ \tag{1.15a}$$

In field units,

$$H_{hp} = \frac{PQ}{1{,}714} \tag{1.15b}$$

where

Q is the pump flow rate in gpm
P is the corresponding pump delivery pressure in pounds per square inch (psi)
H_{hp} is the pump hydraulic power in hp

Equation (1.15) can be used to convert any pressure to hydraulic horsepower for any given flow rate.

Example 1–5

A double-acting duplex pump, with a 2.5 in. rod and a 20 in. stroke, is to be operated at 60 strokes/min (spm) for drilling down to 10,000 ft. The maximum available pump hydraulic horsepower is 1,360 hp. The pump delivery pressure recommended for optimum hydraulics is 3,423 psi. Determine the liner size to be used.

Solution: From equation (1.15), the required flow rate is,

$$Q = 1{,}714\ \mathrm{H_{hp}}/P = 1.714 \times 1{,}360/3{,}424 = 681\ \text{gpm}$$

The theoretical pump displacement flow from equation (1.12) is

$$\upsilon_t = \left(\frac{\pi}{4} N_c L_s\right)(2D_L^2 - D_r^2)$$

$$= \left(\frac{\pi}{4} 2 \times 20\right)(2D_L^2 - D_r^2)$$

$$= \frac{(40)\ (2D_L^2 - 6.25)}{294}\ \text{gal/stroke}$$

The theoretical flow rate of the pump operating at 60 spm is

$$Q_t = \frac{(40)(2D_L^2 - 6.25)60}{294}$$

$$= (16.33D_L^2 - 51.02)\ \text{gpm}$$

Therefore, from equation (1.14), $Q_a = Q_t E_v$. Or, $681 = (16.33D_L{}^2 - 51.02)(0.90)$. Solving for liner size, D_L, yields $D_L = 6.67$ in. (In practice, use the next available size, e.g., a 7 in. liner).

Multiplication factor 0.9 used is the efficiency—in this problem it is assumed.

Example 1–6

A single-acting triplex pump with 6 in. liner, 2 in. rods, and 20 in. strokes is operated at a delivery pressure of 2,250 psi and 60 spm for 4 min. The total volume of mud pumped during the 4 min period was measured to be 200 ft^3. Determine the volumetric pump efficiency and the hydraulic power developed.

Solution: The theoretical pump flow rate per stroke is given by

$$Q_t = \pi D_L^2 L_s N_c / 4 = \pi \times 6^2 \times 20 \times 3/4 = 1{,}697\ \text{in}^3/\text{stroke}$$

Or,

$$Q_t = (1{,}697 \times 7.48\ \text{gal/ft}^3)/(12\ \text{in.})^3 \times 60 = 440\ \text{gpm}$$

The actual measured flow rate is

$$Q_a = 200\ \text{ft}^3\ (7.48\ \text{gal/ft}^3) = 374\ \text{gpm}$$

Therefore, the volumetric efficiency is

$$E_v = \frac{Q_a}{Q_t} \times 100\% = \frac{374}{440} \times 100\% = 84.7\%$$

The hydraulic horsepower developed is

$$H_{hp} = \frac{QP}{1{,}714} = \frac{374 \times 2{,}250}{1{,}714} = 491\ \text{hp}$$

Example 1–7 (Volumetric efficiency, pump horsepower, shaft input horsepower)

A mud pump is required to deliver 300 gpm at 2,500 psi. What is the actual hydraulic horsepower? What is the required pump displacement based on 90% volumetric efficiency? What is the required pump horsepower rating? What is the shaft input horsepower?

Solution:

$$\text{Actual hydraulic power } H_{hp} = \frac{(300)(2{,}500)}{1{,}714} = 438\ \text{hp}$$

$$\text{Required pump displacement} = \frac{300}{0.9} = 333\ \text{gpm}$$

$$\text{Required pump horsepower rating} = \frac{(333)(2{,}500)}{(0.85)(1{,}714)} = 572\ \text{hp}$$

$$\text{Actual shaft input horsepower} = \frac{438}{0.85} = 515\ \text{hp}$$

Prime movers for mud pumps should have a continuous service rating (after derating for altitude) that is at least equal to the required pump horsepower rating calculated by the previous method.

Underpowered pumps. It is not uncommon for rigs to be equipped with mud pumps that are underpowered. With underpowered mud pumps, the manufacturer's discharge pressure ratings require correction as follows:

- If the engine delivers maximum continuous horsepower when the pump is operating at maximum spm, the manufacturer's discharge pressure ratings should be multiplied by the ratios of available horsepower to rated horsepower. For example, if 400 hp is available for a mud pump that is rated at 500 hp, discharge pressure ratings should be multiplied by a factor of 400/500, or 0.8.
- If engines deliver their maximum continuous horsepower at some speed below the maximum rated pump speed, the manufacturer's discharge pressure ratings should be multiplied by the following factor:

$$\frac{(\text{Avail. hp}) \times (\text{max. recommended SPM})}{(\text{rated input hp}) \times (\text{SPM at max hp})}$$

For example, if 400 hp is available at 60 spm for a pump that is rated at 500 hp at 65 spm, the rated discharge pressures should be corrected by a factor of

$$\frac{400 \times 65}{500 \times 60} = 0.866$$

Air compressors

Air compressors are used in rotary drilling when the drilling fluid in use is air or gas. Their design is beyond the scope of this book.

High-pressure surface connections

The high-pressure surface connections include swivel, standpipe, rotary hose, kelly, and a pressure surge chamber. The rotary hose, standpipe, and swivel are used to allow drill string travel during drilling and tripping operations. The surge chamber serves as a damper for any pressure surges which may be induced by the pump units. The kelly is generally a hexagonally shaped pipe that allows drill string rotation.

Drill string. The *drill string* includes the *kelly,* the *drill pipe,* and *drill collars.* The drill pipe is a seamless tubing made from various steel grades. The drill collars are thick wall, seamless tubings which are primarily to add stiffness at the lower end of the drill stem and to provide weight on the bit needed to penetrate the formation. The drill pipe and drill collar lengths are connected by tool joints. A thorough discussion on the drill string mechanics is given in chapter 9.

Return annulus. The *annulus* is the passageway for the drilling fluid returning to the surface and the generated cuttings which are being transported by fluid out of the hole. In offshore drilling a large-diameter pipe, referred to as a marine riser, is used as an extension of the wellbore at the ocean floor to the deck floor of the drilling platform.

Mud pits

Mud pits are generally steel tanks or containers whose main functions are (1) to contain circulated mud out of the wellbore, (2) to provide mud to the mud pumps for circulation down hole, and (3) to store reserve mud for hole fill up during tripping operations.

Drill bit

The *drill bit* is the cutting tool in rotary drilling. The application of weight through the drill collars, heavy-weight drill pipe, the rotation of the drill bit, and the circulation of the fluid are the primary factors that make the bit an effective cutting tool. The most commonly used bits in drilling today may be classified into two categories: (1) roller-cone bits and (2) drag bits. Their design and selection will be studied in chapter 8.

Rotary System

The rotary system (fig. 1–10) includes all of the rig components that are used to achieve bit rotation. These are drill pipe, drill collar, swivel, rotary table, kelly bushing and drive, and kelly. Some rigs are equipped with a top-drive motor to induce bit rotation instead of the conventional rotary table technique. Also, in directional well drilling, downhole motors are used to rotate the drill bit.

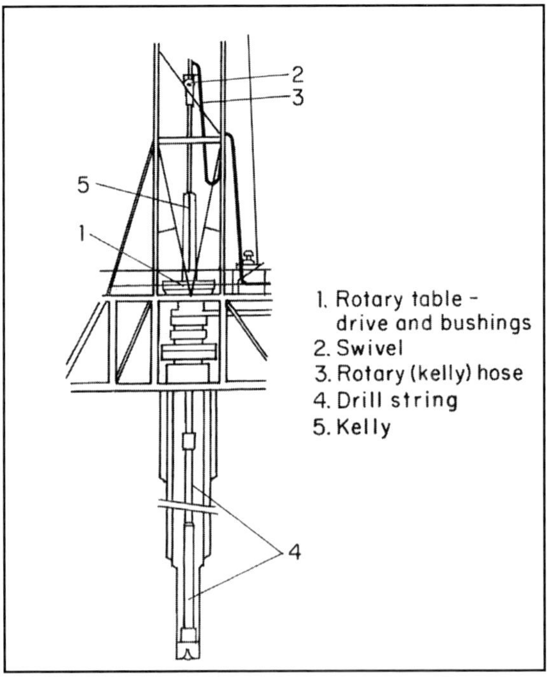

Fig. 1–10. Rotary system (Reprinted by permission; courtesy IADC)

The *swivel* supports the drill stem, allows rotation, and serves as a passageway for drilling fluids. The kelly is the first section of the drill stem just below the swivel. In conventional rotary table technique, it allows drilling ahead, at a total depth equal to its length, before a new section of drill pipe has to be added. Recently, top drives have become popular for inducing rotation to the drill bit—particularly in offshore drilling. They are more efficient and safer than the conventional rotary table system. They allow rotation and circulation during tripping in addition to full-stand-length connections during drilling.

Rotary power requirements

The rotary power requirement is generally small in vertical hole drilling, where rotational torque is approximately less then 15,000 ft-lbs. In directional drilling, high drill pipe rotational torque may exceed 80,000 ft-lbs and often requires high input rotary horsepower.

A computer model can be used to estimate the drill string torque; thus, the minimum required rotary power can be established. Less than 100 horsepower is usually required in vertical hole drilling. In directional drilling, the rotary power required may amount to several hundred horsepower.

Rotary horsepower can be calculated as follows:

$$H_{\text{rp}} = \frac{2\pi NT}{33{,}000} \tag{1.16}$$

where

N is the rotary table speed, in rpm
T is the torque, in ft-lbs

The difficulty of predicting rotary horsepower results from the unpredictable nature of the drill pipe torque. A number of factors influence the drill pipe torque, such as hole size, depth, type of bit, drill collar size, drill pipe size, bit weight, rpm, mud properties, hole inclination and severity and location of doglegs, use of reamers and stabilizers, and formation characteristics.

The drill pipe must transmit rotating power to the *bottomhole assembly* (BHA) and the bit. The following example illustrates calculation of the horsepower that the drill pipe can transmit without torsion failure. For example, if a drill pipe has a maximum recommended makeup torque of 20,000 ft-lbs, the rotary horsepower that can be transmitted at 100 rpm is

$$H_{\text{rp}} = \frac{2\pi \times 100 \times 20{,}000}{33{,}000} = 381 \text{ hp} \tag{1.16}$$

An empirical approach has been developed for estimating rotary horsepower requirements:

$$H_{\text{rp}} = FN \tag{1.17}$$

where

F is the torque factor
N is the rotary speed, in rpm

The torque factor, F, is generally estimated as follows:

F =	1.5 ± 1.75	for shallow holes less than 10,000 ft with light drill string
	1.75 ± 2.0	for 10,000–15,000 ft wells with average conditions
	2.0 ± 2.25	for deep holes with heavy drill string
	2.0 ± 3.0	for high torque

These empirical estimates are subject to many variables, but have proved to be reasonable estimates of rotary requirements. However, for highly deviated wells, torque/H_{rp} requirements must be closely calculated using available computer software programs.

Top-drive rotation

It has been reported that use of *top-drive* systems (electric or hydraulic motors) can result in total drilling cost savings of 10–20%. Even though the day rate of rigs equipped with top drive is higher than with the conventional rotary table system, this can be justified by consideration of overall cost savings. Below is a description of benefits and some disadvantages in the use of top drives.

Benefits. Operational time savings lead to actual cost savings, as follows:

- Connection. Two kelly connections are eliminated, assuming the stands are already made up on the rig floor.
- Laying down drill pipe. The top drive eliminates the need to lay down drill pipe when moving to next well.
- Trip time. Top drive eliminates the use of the kelly; therefore, less time is involved in handling it during trips.
- Washing and reaming. When washing and reaming is being conducted with conventional rotary table, two or three singles of pipe must be laid down and then picked up again to wash and ream down, with several minutes of circulation between connections. The top-drive system is made up on the last stand, with immediate rotation and circulation available to wash and ream. It also allows back-reaming in tight hole during tripping out.
- Underreaming and hole opening. If a hole interval must be underreamed or opened, the entire interval length in drill pipe must be laid down. When top drive is used, the time of laying down drill pipe and two-thirds of all connections are eliminated.
- Improved safety. Because there are fewer connections during drilling and rotary equipment has been eliminated on the rig floor, there is less exposure of the rig crew to potential safety hazards.
- Well control improvement. A hydraulic safety valve above the saver sub enables the driller to shut in the drill pipe faster during drilling and tripping. This eliminates the need to stab a safety valve manually if a kick is taken.
- Stuck pipe. The ability of reaming and back-reaming in full stands while tripping in or out minimizes the potential of pipe sticking.

Disadvantages. There are some operational drawbacks associated with top drive:

- More wear on drilling line
- More maintenance
- Derrick capacity reduction
- Lack of experience of rig crew
- Major top-drive failure can lead to considerable expense in down time to set aside equipment to allow conventional drilling

Well Control System

The primary function of the well control system is to prevent any further flow of formation fluid into the wellbore when a kick (an intrusion of formation fluids into the wellbore) is taken. The system requirements are to safely permit (1) shutting in the well at the surface, (2) controlling the removal of formation fluids from the wellbore, (3) pumping higher-density mud into the hole, and (4) stripping the drill pipe into or out of the hole. The mechanics of well control will be presented in chapter 6.

The basic components of the well control system are described in the following sections.

Blowout preventer stack

The blowout preventer (BOP) stack comprises

- Annular preventer
- Ram preventers
- Spools
- Internal preventers
- Casing head
- Flow and choke lines and fittings
- Kill lines and connections
- Mud- and gas-handling facilities
- Accumulators

Blowout prevention equipment is of primary importance in any rig selection. The BOP stack and system must have a pressure rating equal to or greater than the burst of the casing string and the wellhead. It is also necessary to check whether government and insurance regulations require that particular equipment be installed.

The following items should be specified, as a minimum, to comply with well control procedures and guidelines (fig. 1–11):

- Diverter
- BOP stack and arrangement
- Choke manifold
- Choke, kill line, and valves
- Minimum accumulator
- Kick control/detection equipment:
  - Pit volume indicator
  - Flow indicator
  - Trip tank
  - Internal drill string BOP valve

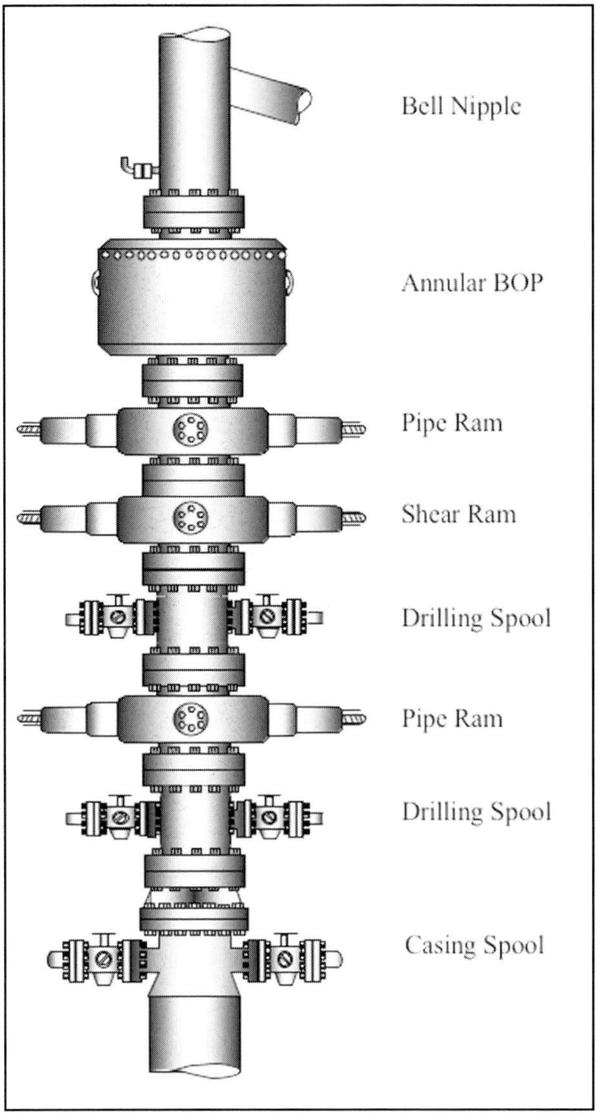

Fig. 1–11. Typical BOP arrangement

Ram preventers. There are three types of ram preventers:

- *pipe ram,* or casing ram
- *blind ram*
- *shear ram*

Pipe ram preventers are mechanical devices that close in the annular space around the drill pipe or casing when they are hydraulically activated. Pipe rams have semicircular openings matching the OD of the drill pipe or the casing. Rams that are designed to close in the annulus when no pipe is present are called blind rams. These rams, if activated while the drill string is in the hole, will not shut in the annulus. Rams that are designed to shear off drill pipe while string is in hole are referred to as shear rams. These kind of rams are activated only if all other preventers fail to shut the well.

Annular preventers. *Annular preventers* are well control devices that employ a ring of reinforced synthetic rubber as a packing unit that surround the drill string to cause the shutoff. In the full open position, the inside diameter (ID) of the packing unit is equal to the diameter of the preventer bore. These preventers will close and shut in the well regardless of the shape or diameter of the conduit that might be in the hole.

Drilling spools. *Drilling spools* are drill-through–type fittings that are placed in the BOP stack assembly to provide space between two consecutive pipe rams for temporary storage of tool joints during stripping operations and to allow attachment of the kill and choke lines.

Data Acquisition and Monitoring System

The data acquisition and monitoring system of a rotary drilling rig consists of all the devices used to monitor, analyze, display, record, and retrieve information regarding drilling operations. The parameters that are of prime concern are the following:

- Drilling rate
- Hook load
- Hole depth
- Pump pressure
- Flow rate
- Torque
- Rotary speed
- Mud density, temperature, salinity, and flow properties
- Mud tank level
- Pump strokes
- Weight on bit
- Hoisting speed

Drilling problems, such as lost circulation, well kicks, and pipe sticking, can be easily detected by the monitoring equipment. Drilling breaks, which are easily seen on the drilling rate chart, can provide information on changes of lithology and formation pressures. Excessive torques may indicate a bit bearing failure or an extremely high concentration of drilled cuttings in the wellbore annulus. A drastic increase in hook load or decrease in mud returning to the surface could indicate that a lost circulation zone has been encountered. A sudden increase in pit level indicates that formation fluids are entering the wellbore and, hence, that blowout is eminent. The proper maintenance of rotary speed, weight on bit, mud properties, and flow rates is of utmost importance for achieving optimum drilling conditions.

In conclusion, the monitoring system is the functional pulse for the entire drilling process. With the advent of personal computers and downhole measuring devices, the monitoring, recording, analysis, storage, and retrieval of drilling data has become a routine part of the process of drilling operations.

Special Systems for Offshore Drilling

Drilling at sea requires two special systems: (1) a motion compensating system for floating units, and (2) a marine riser system (fig. 1–12).

Riser system and components

The marine *riser* is an extension of the wellbore from the ocean floor to the derrick floor above the water surface. Its purpose is to guide and protect the drill string and to serve as a conduit for the return of drilling fluid and cuttings to the surface.

The riser pipe is the principal component of the marine riser system. It is a structure that is made of 40–50-ft-long joints, which are connected by means of rigid, pressure-tight couplings, called riser connectors. Choke and kill lines and control lines to the BOP stack are attached to the riser pipe. They are used to circulate mud to restore pressure balance in the hole when kicks occur. These lines are permanently attached to the riser joints but have their own couplings.

Large fixed-end moments are avoided by using ball joints at the riser ends. Ball joints are normally designed for rotations of up to 8–10° in any direction However, the drill string has to pass freely through the joints, which limits the drilling operation to bail joint angles of, at maximum, 4–5°.

At the upper end of the riser, a telescopic joint is used to compensate for the floating vessel's heave motions. The telescopic joint, also called a slip joint, is usually designed for stroke range of 15–45 ft.

The top tension on the riser is maintained approximately constant by means of tensioners. The capacity of each individual tensioner is between 60 and 80 thousand pounds (kips). The number of tensioners is usually four, six, or eight, resulting in a total capacity ranging from 240 to 640 kips or higher.

In deep water, added buoyancy is required in order to stay within the practical limits of the tensioner system. Several types of flotation systems are on the market, and the principal ingredient of most of them is air. The most widely used flotation materials are from the family of plastic composites called synthetic foam, composed of hollow spherical particles dispersed within some form of a binder.

Motion compensation system

The function of a motion compensating system is to limit the excursion of drill ships and other floating drilling units. The mooring system, which consists of a line, an anchor, a swivel, and a float, is one of the oldest means for motion compensation in ocean environments. Its important requirements are to be statically and dynamically stable under hold-down forces and to have a long life expectancy.

The line material may be of either the metallic type or the nonmetallic (synthetic) type, depending on the environmental, strength, and/or fatigue requirements. Metallic cables, although having a high strength-to-weight ratio, are susceptible to fatigue failures in a corrosive environment. This fatigue phenomenon becomes even more critical in the connection areas. Synthetic cables, by contrast, are corrosion resistant but are susceptible to biological damages. Recent developments in new synthetic fibers have circumvented this problem to an extent; hence, its application in offshore usages is steadily increasing.

In deep water, where a mooring system generally becomes ineffective, the concept of dynamic positioning is used. This consists of outboard thrusters that are capable of holding the drill ship on location without utilization of the anchor.

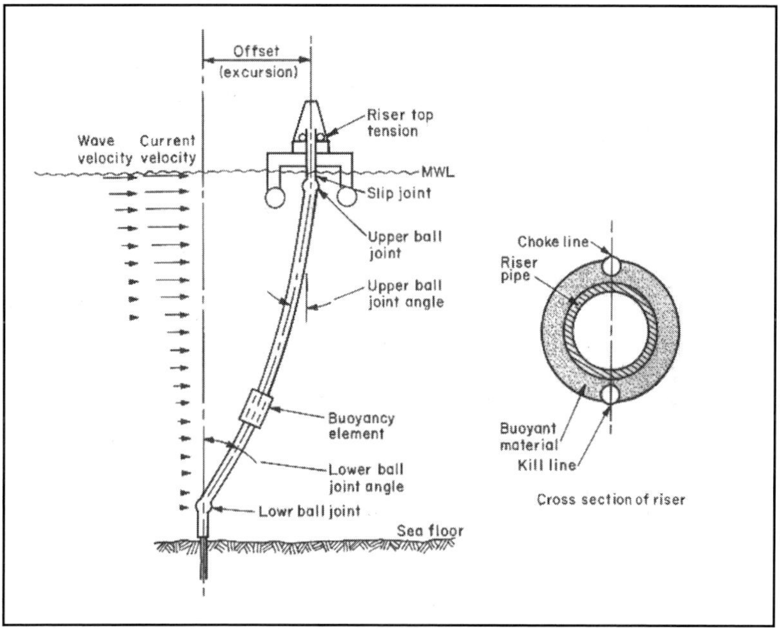

Fig. 1–12. A marine riser system.

Supplementary Problems

Problem 1–1. You are given the following data:

- Target well depth = 26,000 ft
- Drill pipe effective weight = 20 lbs/ft
- Drill collar effective weight = 100 lbs/ft
- Maximum expected weight on bit = 25,000 lbs
- Additional suspended loads to drill string = 60,000 lbs
- Number of lines strung between crown block and traveling block = 10 lines
- Block and tackle system efficiency = 81%
- Drawworks maximum input power = 2,000 hp
- Drawworks efficiency = 75%
- Rig can handle triples
- Time to handle a stand once out of hole = 1 min
- Drilling cost = $7,200/d

(a) Determine the cost of tripping out from 26,000 ft. Assume that the maximum allowed speed of pulling pipe out is that of pulling the first stand.

(b) If the rig can handle singles, what would be the cost in part (a)?

Problem 1–2. Consider a single-acting triplex pump, with stroke length of 12 in., rod diameter of 1.5 in., and pump factor of 6.0 at 100% volumetric efficiency. The pump was operated at 100 spm and 3,428 psi for 4 min. The amount of mud collected at the flowline was 2,040 gal.

(a) Determine the pump input hydraulic horsepower.

(b) Determine the liner diameter.

Problem 1–3. Select a liner size for a single-acting triplex mud pump to drill a well according to the following data:

- Hole sizes: 14 in. with 9 in. casing and 8 in. with 4 in. casing
- Minimum allowed annular fluid velocity = 90 ft/min
- Pump:
  - Rated at 700 hydraulic horsepower
  - Volumetric efficiency = 90% at 40 spm or less and 80% at spm greater than 40
  - Stroke length = 20 in.
  - Rod diameter = 2 ¼ in.
  - $(spm)_{max}$ and $(spm)_{min}$ = 70 and 25, respectively
  - Maximum pressure required for the circulating system = 3,616 psi
- Drill string:
  - Drill pipe: O.D. = 4 in.; weight = 11.85 lbs/ft
  - Drill collar: O.D. = 6 in.; weight = 90 lbs/ft
  - Flow rate = fluid velocity × conduit area of flow.

Problem 1–4. Determine the total time it takes to make a trip at 11,700 ft depth, given the following information:

- Rig can handle triples
- Drill pipe:
  - Length = 10,700 ft
  - Effective weight = 15 lbs/ft
- Drill collar: Effective weight = 80 lbs/ft
- Maximum available drawworks horsepower = 1,200 hp
- Drawline consists of 12 lines strung between crown block and traveling block
- Block and tackle efficiency = 85%
- Drawworks efficiency = 75%
- Rig crew can break a joint or connect a joint in average time of 18 s

Problem 1–5. Determine the engine fuel cost (in $/day), given the following information:

- Total well depth: 20,000 ft
- Drill collar weight = 150 lbs/ft
- Drill collar total length = 1,000 ft
- Drill pipe weight = 20 lbs/ft
- Average drag force on drill string while being tripped out at 60 ft/min = 3.5 lbs/ft
- Mechanical efficiency between power generating engine output and drawworks output = 70%
- Drawline consists of 10 lines strung between crown block and traveling block
- Diesel engine mechanical efficiency = 50%
- Diesel oil:
  - Heating value = 20,000 BTU/lb
  - Weight = 7.2 lbs/gal
  - Cost = $1.00/gal

Problem 1–6. A duplex pump (liner diameter = 6.5 in.; rod diameter = 35%; stroke length = 18 in.) is being used while drilling a well at 15,000 ft. At 50 spm, the pump is delivering 350 gpm of mud at a pressure of 4,000 psi. The pump mechanical efficiency is 70%.

(a) What must be the input horsepower to the pump?

(b) If the dynamic pressure in the circulating system is 2,200 psi, what percentage of the horsepower calculated in part (a) is wasted owing to friction pressure losses?

Problem 1–7. Determine the daily cost of running a diesel engine to power a duplex pump under the following conditions:

- Pump:
  - Single-acting duplex
  - Stroke length = 18 in.
- Rod diameter = 2 1/2 in.
- Liner size = 8 in.
- Delivery pressure = 1,000 psi at 40 spm
- Volumetric efficiency = 90%
- Mechanical efficiency = 85%
- Diesel engine: Efficiency = 50%
- Diesel oil:
  - Weight = 7.2 lbs/gal
  - Heating value = 19,000 BTU/lb
  - Cost = $1.15/gal

Problem 1–8. A trip is to be made from 20,000 ft depth. Determine the minimum time at which the first stand can be tripped out of the hole, given the following data:

- Rig can handle triples
- Drawworks: 1,000 hp; efficiency = 75%

- Drilling line: 12 lines
- Hoist efficiency = 80%
- Drill pipe: Effective weight = 14 lbs/ft
- Drill collar: Effective weight = 90 lbs/ft
- Drill collar length = 1,000 ft
- Other suspended loads = 30,000 lbs

Problem 1–9. You are given the following friction pressure loss-flow rate relationship:

$$P_f = cQ^m$$

where

P is the friction pressure loss
C is a constant
m is the flow exponent
Q is the flow rate
Find an expression to determine m if the following measurements are known:

$$P_f = P_{fi} \text{ at } Q = Q_i$$

$$P_f = P_{fij} \text{ at } Q = Q_j$$

Problem 1–10. Consider a single-acting triplex mud pump powered by a diesel engine. The pump data are as follows:

- Rod size = 2 ½ in.
- Liner size = 7 in.
- Stroke length = 12 in.
- Operating delivery pressure = 3,400 psi at 60 spm
- Volumetric efficiency = 90%
- Mechanical efficiency = 85%

For an engine efficiency of 55%, how many gallons of diesel fuel are being consumed per day? Heating value of diesel oil = 19,000 BTU/lb; density = 7.2 lbs/gal.

Problem 1–11. Consider a hoisting system with 12 lines. Determine the required drawworks horsepower for a maximum hook load of 400,000 lbs and a traveling block velocity of 2 ft/s. Assume hoist efficiency of 80% and drawworks efficiency of 70%.

Problem 1–12. A 10 lbs/gal liquid is being circulated at 540 gpm through the system shown in figure 1–13. A triplex-single-acting pump having a volumetric efficiency of 85% is being used. What must be the pump hydraulic horsepower requirement for the above operating conditions? Assume the friction pressure loss gradient in the circulating system is 0.06 psi/ft. The hydrostatic pressure of a liquid column is given by $P = 0.052h\gamma$, where h is the column height in feet and γ is the liquid density in lbs/gal.

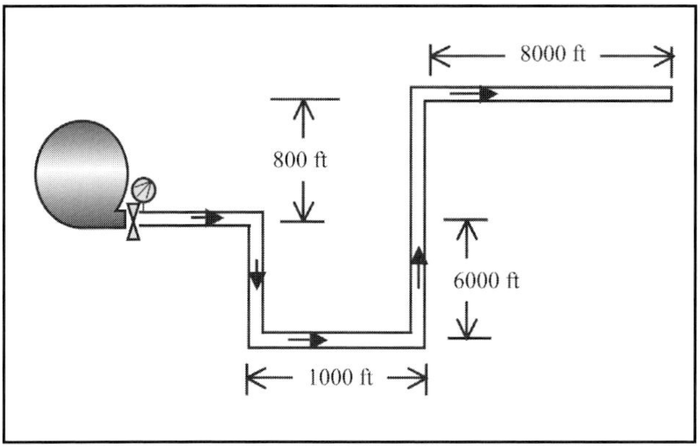

Fig. 1–13. Configuration of the fluid circulating system in problem 1–12

Problem 1–13. Determine the fuel cost (in \$/d) to run an engine at 1,800 rpm with 3,000 ft-lbs output torque. The engine efficiency at the aforementioned rotary speed is 30%. You are also given the following data:

- Diesel oil:
  - Cost = \$1.05/gal
  - Weight = 7.14 lbs/gal
- Heating value = 19,000 BTU/lb

Problem 1–14. Given one annular preventer, one shear ram preventer, and one pipe ram preventer, arrange the BOP stack, giving reason(s) why you have selected that stacking arrangement.

Problem 1–15. Find the cost (in \$/ft) to drill between 9,000 ft and 10,000 ft for a prospect well fitting the following description:

- Well depth = 17,000 ft
- Casing settings = 500 ft, 4,000 ft, 9,000 ft, 14,000 ft, 17,000 ft
- Rig can handle triples; rental cost = \$12,000/d
- Crew can handle a stand in average time of 1.5 min

Bit and control well data are given in tables 1–2 and 1–3, respectively.

Table 1–2. Bit data

Size (inch)	Average Life (hours)	Cost(\$)
22	40	3,000
17 ½	35	2,000
13 ½	20	1,400
8 ⅜	20	1,000
5 ¾	10	500

Table 1–3. Control well data

Well Number	k	a
1	180	0.00021
2	200	0.00024
3	300	0.00025

Problem 1–16. A 1,000 hp single-acting duplex pump has a volumetric efficiency of 80% when operated at 60 cycles/min. A delivery pump pressure of 3,000 psi is recommended to drill the last one thousand feet of hole. What liner size would you specify if the pump has a 20 in. stroke, 2.5 in. rod diameter, and is operating at 60 cycles/min?

Problem 1–17. In problem 1–16, what is the minimum rod diameter that can be used if the stress in the rod is not to exceed 40% of the material's yield stress and 50% of the critical buckling stress?

Problem 1–18. You are given the following data:

- Drawworks output horsepower = 800 hp
- Trip to be made from 20,000 ft well depth
- Effective drill pipe weight = 15 lbs/ft
- Effective drill collar weight = 80 lbs/ft
- Maximum required weight on drill bit is 80,000 lbs, between 15,000 and 20,000 ft well depth
- Hoist efficiency = 85%
- Drawworks efficiency = 80%

(a) Determine the amount of time it takes to trip out of hole, if the speed is maintained at a value equal to the maximum speed that can be obtained for pulling the first stand of pipe. Assume rig crew can handle a stand in average of 20 seconds. Rig can handle doubles.

(b) Determine the total cost of diesel oil used in tripping out. Assume diesel engine efficiency = 40% and drawworks efficiency = 80%; diesel oil heating value = 18,000 BTU/lb, weight = 7.2 lbs/gal, and cost = $1/gal.

Problem 1–19. You are given the following data:

- Drill string:
  - Drill collar OD = 7 in., and ID = 3 in.
  - Weight 100 lbs/ft
  - Length = 1,000 ft
- Drill pipe:
  - Drill pipe OD = 4 in., ID = 3.5 in.
  - Weight = 15 lbs/ft
- Drilling-fluid annular velocity requirement: 120 ft/min (minimum)

- Mud pump:
  - One single-acting triplex pump
  - Stroke length = 18 in.
  - Rod diameter = 2 in.
  - Volumetric efficiency = (100 – 0.25 spm)%
  - Mechanical efficiency = 60%
  - Maximum allowed surface pressure = 4,000 psi
  - Maximum spin = 100
  - Minimum spm = 25
- Hole sizes:
  - Upper interval: 18 in. hole (will be cased with 14 in. casing)
  - Lower interval: 10 in. hole (will be cased with 7 in. casing)
- Power-diesel engine: Output horsepower = 800 hp

Select a liner size to drill the upper and lower intervals of the well in accordance with these specifications.

Problem 1–20. Describe explicitly how a hole is made using the rotary drilling technique. Name the systems and the components that make up the rotary rig. Also, state briefly the function(s) of each system.

Problem 1–21. A well is to be drilled to a depth of 20,000 ft. The effective weight of drill pipe is 20 lbs/ft, and that of the drill collar is 120 lbs/ft. If the traveling blocks and swivel are rated at a maximum capacity of 200 tons, determine the required drawworks horsepower to trip at 10 ft/s from a depth of 19,000 ft. The drill collar length is 1,000 ft, and the hoisting system has 10 lines strung between traveling block and crown block (efficiency = 80%).

Problem 1–22. A portion of a well is to be drilled from depth A to depth B with the following requirements:

- Minimum flow rate = 240 gpm
- Maximum flow rate = 660 gpm
- The pump is a single-acting triplex pump with the following specifications:
- Maximum delivery pressure = 3,400 psi
- Hydraulic horsepower = 800 hp
- Volumetric efficiency = 80% for spm less than 50 and 70% for spm greater than 50

Specify the liner size that should be used to drill that portion of hole if the stroke length is 20 inches.

Nomenclature

D_L: Liner or piston diameter
D_r: Rod diameter
E_v: Pump volumetric efficiency
F_{fp}: Load in fast line
F_h: Hook load
H: Fuel heating value, in BTU/lb
H_L: Net static hook load capacity, in lbs
H_{hp}: Pump hydraulic power, in hp
L_s: Piston stroke or length stroke
n: Number of lines strung between the crown block and traveling block
N: Engine rotary speed, in rpm
N_c: Number of cylinders
P: Corresponding pump delivery pressure, in psi
P_i: Input power, in hp
P_o: Output power, in hp
Q: Flow rate, in gpm
Q_f: Rate of fuel consumption, in lbs/h
R_c: Required crown block rating, in lbs
S: Effective weight of suspended equipment, in lbs
T: Output torque, in ft-lbs
v_{fp}: Fast line speed
v_{tb}: Traveling-block velocity

2

Drilling Fluids

Characteristics of a Drilling Fluid

Definition

The term *drilling fluid* refers to a liquid, gas, or gasified liquid circulating continuum substance used in the rotary drilling process to perform any or all of the various functions required in order to successfully drill a usable wellbore at the lowest overall well cost.

Functions

Major functions. The major functions that a drilling fluid must perform vary with the type of fluid selected. In general, these functions are as follows:

- Drilled-cuttings removal—beneath the drill bit and from the annular space around the drill string
- Containment of subsurface formation fluid pressures
- Hole stabilization—chemical and/or mechanical—prior to casing/cementing

Minor functions. The minor functions of drilling fluids include

- Cooling and lubrication of drill string and drill bit
- Suspension of desired solids and ease of removal of undesired solids
- Reduction in weight of casing string and drill string while suspended in wellbore
- Aid in formation evaluation
- Cleaning of drill bit

Additional benefits

In addition to the drilling functions described previously, there are important benefits to be gained. If these fluids are properly selected, it is possible to achieve

- Minimum formation damage
- Minimum corrosion effects on drilling equipment
- Enhancement in rate of penetration (ROP)
- Minimum environmental impact
- Enhancement in safety issues
- Reduction in friction pressure losses

The failure of a drilling fluid to perform any or all of its required functions may lead to very costly drilling problems. In fact, most drilling problems are related, directly or indirectly, to the drilling fluid being used. Thus, the success of a drilling program and the ability to reduce the overall drilling cost are predicated on the proper selection and maintenance of a drilling fluid and on the understanding and application of its functions. The following sections describe each of these functions in brief.

Drilled-cuttings removal—beneath the drill bit. Removal of drilled solids from underneath the bit as soon as they are generated is essential if drilling cost is to be minimized for a given combination of bit type, weight on bit, rotary speed, and hydraulics. Drilling is generally conducted while maintaining a positive differential pressure across the wall of the wellbore. That is, drilling fluid is selected to have a pressure gradient higher than the formation fluid pressure gradient. Because of this differential pressure, the drilled solids are held on bottom beneath the bit, thus impeding the cutting of a new rock surface. In addition to the drilling rate impediment, these cuttings, if not removed immediately, will be reground, thus making their removal from the mud difficult and expensive. Therefore, a drilling fluid must provide the proper hydraulics at the bottom to create cross-flow fluid velocities sufficient to sweep efficiently all generated cuttings.

Drilled-cuttings removal—from the annular space. To avoid very costly problems, such as mechanical pipe sticking, excessive torque and drag, and casing and logging problems, the efficient transport of drilled cuttings to the surface through the annular space is a vital function of a drilling fluid. The factors that affect cuttings removal during drilling include fluid annular velocity, mud rheology, hole inclination, drill string rotation, annulus eccentricity, drilling rate, and the size and geometry of the cuttings. This will be discussed in more detail in chapter 5.

Formation fluid pressure containment. In drilling operations, if formation fluid is to be prevented from entering the wellbore, thus preventing kick occurrence, then a drilling fluid of sufficient density must be used to contain pore pressures. The pressures that may be encountered during drilling are classified as follows:

- Normal pressure if the pressure gradient is approximately 0.433 psi/ft of depth.
- Abnormal pressure if the pressure gradient is greater than 0.433 psi/ft of depth.
- Subnormal pressure if the pressure gradient is less than 0.433 psi/ft of depth.

Therefore—and as a safety consideration—in most drilling operations, a drilling fluid is selected that will provide a pressure gradient greater than the anticipated formation pressure gradient. This drilling condition is referred to as overbalance drilling; that is,

$$P_{df} - P_{ff} = \Delta P \quad (2.1)$$

where P_{df}, P_{ff} are the drilling-fluid and formation fluid pressures, respectively, and ΔP is the *differential pressure.*

Depending on whether ΔP is greater than zero, equal to zero, or less than zero, drilling is referred to as overbalanced drilling, balanced drilling, or underbalanced drilling, respectively. Most drilling operations are overbalanced drilling with the differential pressure, ΔP, generally maintained between 100 and about 500 psi.

Wellbore stabilization. Wellbore instability is an undesirable condition of a drilled hole interval that does not maintain its gauge size, shape, and/or structural integrity. The causes may be grouped as follows:

- Mechanical failure due to the unbalance of *in situ* stresses at the borehole walls
- Erosion due to drilling-fluid circulation
- Instabilities due to the chemical interaction between drilling fluid and formation

The types of instabilities may be grouped as follows:

- *Hole closure.* Hole closure is a narrowing process of the borehole. It generally occurs in plastic shale and salt sections. The problems associated with hole closure include an increase in torque and drag on drill string, an increase in potential pipe sticking, and difficulty in casings landing. Mud weight is the major controlling factor in mitigating this problem.
- *Hole enlargement.* Enlargement of borehole, commonly referred to as washout, is caused by hydraulic erosion, mechanical abrasion due to the drill string, and inherent sloughing of shale formations. The associated problems include difficulty in cementing, potential hole deviation, an increase in hydraulic requirement for effective hole cleaning, and difficulty in some logging-tool operations. In general, proper mud design, lower hydraulics, and elimination of severe drill string vibrations can minimize the problem.
- *Hole fracturing.* Borehole fracturing occurs when the drilling-fluid pressure gradient exceeds that of the formation fracture pressure gradient. The associated problems include lost circulation and possible kick occurrence. Accurate knowledge of formation fracture gradient and maintenance of the proper equivalent circulating-mud density (ECD) during drilling are the only means of combating the problem.
- *Hole collapse.* Borehole collapse occurs when the drilling-fluid pressure gradient is too low to maintain the structural integrity of the drilled section. The associated problems include pipe sticking and potential loss of well. It is important to know the formation fracture gradient and hence be able to select the proper mud weight, to minimize the occurrence of hole collapses.

The following formations, when drilled, can lead to a mechanically unstable wellbore:

- Unconsolidated formations
- High-pressure shale—caving and sloughing
- Water-sensitive shale—heaving and swelling
- Formations susceptible to hydraulic erosion

The first two problems are inherent mechanical problems due to the resulting unbalanced mechanical forces on the wall of the wellbore. Thus, for stabilization, mud weight is the essential mud property.

In water-sensitive shale, the problem is that a chemical interaction, between the shale and the drilling-fluid filtrate (the liquid part of the fluid), may cause hydration of the clays and therefore lead to swelling and heaving. Minimizing the chemical interaction between the drilling fluid and the formation is the only solution to this problem. This may be accomplished through the use of inhibitive mud systems.

Hydraulic *erosion* is caused by high annular fluid velocity and generally occurs in soft formations. Therefore, to achieve wellbore stability, the density, the chemical activity, and the flow properties of the drilling fluid must be properly designed and maintained for the particular situation at hand.

Cooling. Considerable heat is generated from the frictional contact of the rotating drill string and drill bit against the wall of the wellbore. This heat, if not dissipated, can cause premature failures of components of the drill bit and the drill string. Therefore, the presence of a circulating fluid is essential for heat dissipation.

Suspension and ease of removal of solids. There are several factors that have the tendency of preventing the settling of drilled solids and added solids such as weighting materials, when fluid circulation is stopped. For *nonthixotropic* (time independent) fluid, the density and size of solids and the density and viscosity of the fluid are the controlling parameters. However, in a *thixotropic* (time dependent) mud, the fluid develops a gel strength with time that holds the cuttings in suspension after fluid circulation is stopped.

In weighted drilling-fluid systems, the weighting solids, such as barite, must be suspended in the drilling fluid at all times. Hence, solids suspension is an important function of these fluids.

Drill bit cleaning. During drilling, certain formations and, depending on the type of drill bit that is being used, clogging and balling of the bit may occur. The type of fluid (its lubricity) and the amount of hydraulic horsepower expanded across the bit may prevent these problems.

Lubrication. Excessive torque and drag are common problems encountered in drilling directional wells in general and extended reach wells in particular. The mud system reduces the torque and drag by lubricating the hole and the drill string, thus allowing maximum rotational energy to be used at the bit, as well as allowing the bit to advance. Lubricants include bentonite, oil, detergents, graphite, asphalts, special surfactants, and walnut hulls. Bentonite acts as a lubricant only because it is slick when wet and reduces friction between the mud cake and drill string. A well-lubricated mud system will also minimize the wear on downhole equipment and casing.

Typical coefficients of friction for casing and open hole sections are given below:

	Cased hole	Open hole
Air	0.35–0.55	0.40–0.60
Foam	0.30–0.40	0.35–0.55
Lignosulfate	0.20–0.25	0.20–0.30
Polymer	0.15–0.22	0.20–0.30
Oil base	0.10–0.20	0.15–0.20

Certain lubricant additives can reduce the coefficient of friction, and a higher friction coeficient may be used for casing/liner running operation.

Corrosion prevention. The elements and compounds that can cause corrosion of drilling equipment include oxygen (O_2), carbon dioxide (CO_2), hydrogen sulfide (H_2S), and salts. Oxygen scavengers and amine-based inhibitors are generally effective at a reasonable cost. Oxygen corrosion is further minimized by using effective deaeration surface equipment. Wells that are drilled into formations that contain toxic and corrosive gases such as H_2S and CO_2 pose serious concerns during the drilling operations. Not only do these gases pose health and safety problems to personnel, but they also present dangers to metal components from hydrogen embrittlement, blistering, and stress cracking. The mud system can be treated with scavengers to remove these contaminants. Mud density should be increased whenever practical to prevent further influx of gases from the formation as it is being drilled.

Aid in formation evaluation. Drilling mud provides a conductive medium between the logging tools and the formation. The physical and chemical properties of drilling fluids affect many well logs. Mud properties must be compatible with the logs selected; conversely, the logs selected must be compatible with the mud properties. For example, a high-salt-content mud may prevent the use of a spontaneous-potential tool. If the salt concentrations of the mud are identical to those of the formation fluids, there is no potential to record. Oil muds inhibit the use of resistivity logs since the oil acts as an insulator to minimize current flow. The selection of appropriate logs must be closely coordinated with the mud program to allow proper formation evaluation. The mud must also allow collection of information from cuttings (mud logging) and cores.

Minimizing formation damage. The protection of potential pay zones is important for any drilling fluid. The deposition of a filter cake that allows a drilling operation to continue and protects a productive zone is an important consideration in mud system design. Formation damage is generally a reduction in permeability near the wellbore. The problem can be severe in low-permeability reservoirs or in reservoirs with high clay content. A good mud cake helps stabilize the wall of the hole and will reduce mud filtrate loss into the formation. This reduces the interstitial clay swelling and can cause permeability reductions. Colloidal solids, polymers, or clay can be lost into the formation and can cause plugging. All the properties and the chemical composition of the mud play a role in formation damage.

Drilling-Fluid Selection

Criteria

The main criterion used for the design of a drilling fluid program is generally minimum overall well cost. Other criteria that are considered include: production concerns, exploration concerns, environmental impact, safety, application performance, and logistics.

Further considerations

The considerations that must be taken into account when selecting drilling fluids to drill a well are

- Well type
- Problem formations
- Drilling rig
- Producing formations and kind of production

- Casing program
- Makeup water
- Potential corrosion
- Environmental impact
- Availability of products in international operations

A discussion of each of these considerations follows.

Well type. Drilling may be classified as development or wildcat drilling: In areas where drilling has been conducted previously and the geological conditions are known, additional wells to be drilled are referred to as development wells. Otherwise, it is a wildcat well, commonly referred to as an exploratory well. In the latter, the drilling fluid must be chosen so that all geological information can be obtained easily and safely. Also, the drilling fluid must be of such composition to allow quick changes to yield a fluid system suited for whatever unexpected problems are encountered. By contrast, in development wells, the drilling-fluid program is selected and the mud systems are designed along with other programs (hydraulic, drill bit, casing, etc.) to meet the criterion of minimum overall drilling cost.

Problem formations. If either the formation that is being drilled or the fluid that is being used to drill it is affected, directly or indirectly, by the other, then the formation is classified as a problem formation. Examples are

- Shale formations—water sensitive (heaving and swelling) or pressured (caving and sloughing)
- Anhydrite formations (lime or gypsum)
- Salt formation
- High-temperature formation
- Abnormal pressure formation
- Inherently fractured formation (loss circulation zones)

Shale intervals. The problems associated with the drilling of shale intervals (depending on the type of shale) are

- Mechanical pipe sticking
- High torque/drag
- Annular hole–cleaning difficulties
- Logging difficulties
- Mud contamination

Depending on the severity of the problem, adjustments to the properties of the drilling fluid may control the shale. Otherwise, very careful and specific design requirements of the drilling fluid may have to be implemented.

Anhydrite intervals. When bentonite-treated freshwater muds are used to drill anhydrite formations, the calcium ions released into the mud system will retard bentonite hydration and flocculate the mud that has been hydrated, thus affecting mud viscosity and fluid loss. Depending on the massiveness of the intervals, either the calcium ions are treated out of the system or the mud is converted to an inhibitive one (e.g., a gypsum- or lime-treated mud) that tolerates the contaminating ions. In general, economics dictates whether to treat out the ions or to convert to an inhibitive mud system.

Salt intervals. The contamination of bentonite-treated freshwater fluids from the drilling of salt sections has effects similar to those of the anhydrite formations. The contaminating ions can be magnesium (Mg^{2+}), calcium (Ca^{2+}), and/or chloride (Cl^{-}) ions. As in the anhydrite intervals, economics dictates whether these ions are to be treated out or an inhibitive fluid should be used. Often, if thick salt sections are expected to be encountered, then either salt-saturated or oil-based muds may have to be used.

High-temperature formation. Wellbore temperatures in excess of 250°F generally reduce the effectiveness of drilling-fluid chemical additives and can thus result in changes to such fluid properties as viscosity and fluid loss. Therefore, when high-temperature formations are expected to be encountered, certain drilling fluids that are tolerant to high temperatures have to be selected, such as oil-based mud systems.

Abnormal pressure formation. When encountered unexpectedly, abnormal pressures will cause the intrusion of formation fluids into the drilling fluids, resulting in mud contamination and undesirable kicks. A good knowledge of expected formation pressures and the proper selection of mud weight can alleviate this problem.

Loss circulation zones. A *loss circulation zone* is a formation interval that allows whole drilling fluid to be lost into the zone. If only part of the whole mud is lost, then the interval is referred to as partial-loss circulation zone. If no mud return to the surface is occurring, then the loss is a total loss of circulation. In the first case, cost is the governing factor, whereas in the second case, both the mud cost and the loss of desired geological information govern the fluid selection process.

Drilling rig. The two major components in a rotary drilling rig that figure in drilling fluid selection are the drilling-fluid control system (solids control and chemical treatment) and the circulating hydraulic horsepower available in the circulating system. The success of a mud program in achieving optimum drilling is predicated on the proper selection of the rig and its layout.

Producing formation. In formations that are intended to be zones for oil/gas production, the fluid filtrate (the liquid phase of the drilling fluid) must have minimum or no adverse effects on the producing formations. For example, in sand formations that may contain hydratable clays, a freshwater filtrate intrusion will result in swelling of the clay particles, thus impairing the reservoir production rates.

Casing program. Some of the controlling factors in the selection of casing-setting points are similar to those that are considered in drilling fluids. For instance, changes in lithology and isolation of troublesome formations are typical requirements for setting casing at designated depths. These same requirements enter into the selection of drilling fluids. Hence, a well-designed drilling-fluid program will minimize casing-setting requirements and thus reduce overall well costs.

Makeup water and availability. The source and the chemical composition of the makeup water are two of the primary considerations in the selection of the mud programs. The amount of chlorides, calcium, and magnesium must be determined so that a mud system compatible with the makeup water chemical composition can be selected. The source and the availability of the makeup water must be considered so that mud treatment cost can be minimized. For instance, if, on the one hand, freshwater is abundant on location, then mud dilution may be the most economical treatment. If, on the other hand, water is being transported by trucks, then solids control equipment can be more economically utilized. As a further example, in offshore drilling, it is more economical to utilize seawater; therefore, the mud program must be selected accordingly.

Corrosion. *Corrosion,* combined with mechanical cyclic loading, is a major cause of drill string premature failure. It has been shown that the presence of dissolved gases, such as O_2, CO_2, H_2S, and chlorides, in drilling muds decreases the life expectancy of drill pipe significantly. Endurance limit, as used in noncorrosive fatigue analysis, no longer characterizes metal fatigue behavior when various corrosives are present. In a corrosive environment, drill pipe failure occurs at any applied cyclic stress if the number of fatigue cycles becomes sufficiently large.

Environmental impact. Although drilling fluids and drilled cuttings are properly treated for any hazardous materials prior to their disposal, environmental impact is still an important consideration in the selection and safe implementation of such fluids. For example, in offshore drilling, mineral oil–based or synthetic oil–based mud systems are universally selected over the conventional diesel oil–based system.

Classification of Drilling Fluids

Drilling fluids may be classified as pneumatic or liquid. Depending on the liquid phase, liquid drilling fluids are of two types: water based and oil based. Pneumatic drilling fluids include dry gas (air, natural gas, and nitrogen), mist, foam, and gasified muds. A brief description of each is given in the following sections.

Pneumatic drilling fluids

Pneumatic drilling fluids are used in potential loss circulation zones, low-reservoir-pressure zones and in underbalanced drilling operations. The advantages of pneumatic drilling are improved rate of penetration, longer drill bit life, better control of loss circulation zones, and less damage to formations. The disadvantages in case of dry air/natural gas are possible downhole fire, corrosion, hole deviation, and hole erosion. Following is a breakdown of pneumatic drilling fluids.

Dry air. It has been well established that *dry-air* drilling yields the highest ROP among all fluids. The two major functions of air drilling fluid are removal of drilled cuttings and cooling of downhole equipment. The limitations of dry-air drilling are water-oil producing formations, mechanically unstable wellbore, and in some cases, economics.

Mist. *Mist* drilling occurs when water is injected into the airstream. Mist is used instead of dry air when the encountered formation produces small amounts of water. The limitations of unstable wellbore and sometimes economics still apply, as in dry air. The functions of mist are the same as those of dry air.

Foam. *Foam* drilling occurs when water and foaming surfactants are injected into the airstream, resulting in a viscous foam that is similar to aerosol shaving cream. Its functions are similar to those of dry air. Its application is where loss circulation zones are anticipated and cannot be economically sealed.

Gasified mud. *Gasified mud* occurs when air or nitrogen is injected simultaneously with the circulated mud slurry into the drill string. The application of gasified mud is in weak formations that have fracture gradients less than the very minimum mud weight gradient that can be prepared from the conventional makeup of liquid muds. It is also used in underbalanced drilling applications.

Liquid drilling fluids

When pneumatic drilling fluids cannot be used, because of economics or required functions that may not be achieved, then liquid drilling fluids have to be selected. These fluids can be classified as follows:

- Clear water drilling fluids (freshwater or seawater)
- Water-based drilling muds (noninhibitive or inhibitive)
- Oil-based drilling muds

Clear water drilling fluids. In trouble-free formations, clear water drilling is the most economical of all liquid drilling fluids. Higher ROPs and longer bit life are two major benefits. Limitations of clear water drilling are abnormal pressure zones, water-sensitive formations, and mechanically unstable wellbore.

Drilling mud. It is universally understood that when a liquid (water or oil or a combination of water and oil) is treated with a clay substance, such as bentonite or a polymer, the result is a drilling fluid that is termed drilling mud. Following are the primary phases of a drilling mud and its primary components:

- Liquid phase:
  - Water—freshwater, seawater, or specially prepared saltwater
  - Oil—diesel oil, mineral oil, or synthetic oil
- Solid phase:
  - Inert solids, such as barite (added for weighting)
  - Active solids, such as clay (added to build a viscosity) or polymers and other chemicals

When the continuous liquid phase (external phase) is water, the mud system is called water based; otherwise, it is an oil-based mud. The reactive solids, which are the clays, are the bodybuilding part of the mud system, along with polymers. Sodium montmorillonite (bentonite) and attapulgite are typical active solids. The inert (nonreactive) solids, such as barite, are the part of the mud system that is added at the surface, to increase the weight to a desired density. Chemicals are added to the mud system to react with the active clays, to achieve certain mud properties, and will be discussed later. The following sections provide a brief description of the various types of muds and their specific applications.

Water-based drilling muds—noninhibitive. Noninhibitive water-based drilling muds are the least expensive and are easy to make and maintain. However, their application ceases when it is expected that high-temperature formations, dispersive formations, or formations that may contain certain contaminants, such as H_2S, will be encountered. Noninhibitive drilling fluids include spud mud (native mud), clear water mud, bentonite-treated mud, and lignite/lignosulfonate (dispersed) mud.

Spud mud. In some areas, surface hole drilling into a clay/shale-bearing formation while using clear water results in a viscous native mud (spud mud). Native mud is generally discarded after conductor pipe is set.

Bentonite-treated mud. To provide viscosity for improved cuttings-carrying capacity in vertical and near-vertical holes, to suspend barite in weighted mud systems, and to prevent minor hole sloughing problems, bentonite clay is added to clear water, resulting in what are simply called bentonite-treated mud systems. These muds are generally utilized in drilling trouble-free shallow wells.

Phosphate mud. When bentonite muds become contaminated with calcium or magnesium ions, phosphates are added to control their viscosity. The resulting mud is called a phosphate-treated mud system. The phosphates are ineffective in controlling fluid loss and become unstable at bottom-hole temperatures in excess of 150°F. Therefore, their efficient utilization is limited to the top portion of a hole or in shallow wells, where temperature and fluid loss are not critical.

Gel-chemical mud. Gel-chemical mud consists of bentonite and small concentrations of thinners, such as quebracho or one of the lignosulfonates. Their application is similar to that of the phosphates; however, they can be utilized to greater well depths.

Lignite/lignosulfonate mud. Muds heavily treated with lignosulfonates, a more effective thinner, are called lignosulfonate mud systems. When lignite, a more effective fluid loss agent, is utilized along with lignosulfonate, the resulting mud system is referred to as lignite/lignosulfonate mud. Some of these muds are formulated to be stable up to 375°F. Although lignosulfonates are used to control yield and gel strengths, the treated mud is more resistant to calcium and chloride contaminations.

Water-based drilling muds—inhibitive. Inhibitive drilling fluids include calcium-based muds, salt-based muds, potassium-based muds, and polymer drilling muds.

Calcium-based muds. When lime or gypsum is added to bentonite-treated mud, the sodium clays (bentonite) are converted to calcium-based clays, resulting in what is called a calcium-based mud. Lime mud, on the one hand, is formulated by adding caustic soda, organic thinners, lime, and a fluid-loss control agent to bentonite-based muds. Gyp mud, on the other hand, is formulated by adding gypsum (calcium sulfate) and lignosulfonates to bentonite-treated mud. Calcium-based mud is utilized in the drilling of thick anhydrite sections and also where minor sloughing shale and saltwater flows are common.

Salt-based muds. The salinity of salt muds can range from 10,000 ppm (parts per million) up to complete saturation of 315,000 ppm salt. The salinity may be due to inherently saline makeup water, addition of salt at the surface, or salt dissolved from salt stringers. The conventional salt muds are those whose base is either prehydrated bentonite or attapulgite clay and are utilized in drilling salt sections, anhydrite, gypsum, and minor troublesome shale sections. Certain polymers are used to viscosify saltwater, resulting in special-application salt muds, as used in workover operations and offshore drilling.

Potassium-based muds. Potassium-based muds are the most common types used in the drilling of water-sensitive shale. Their performance is based on the cationic exchange of potassium for sodium or calcium ions on smectites and interlayer clays.

Potassium-based muds are superior to calcium muds in terms of their shale-inhibition properties. The shale stabilization phenomenon is known to take place as follows. Potassium is exchanged for sodium and calcium when drilling in shale that contains montmorillonite, resulting in a more stable mud system, less susceptible to hydration. There are four types of potassium-based muds:

- KCl—polymer (KCl-Partial Hydrolytic Polyacrylamide [PHPA])
- KCl—cationic polymer
- KOH—lignite
- KOH—lime

Polymer drilling mud. Polymer drilling fluids are those that have been treated with a certain type of polymer. Advances in polymer chemistry have had a significant impact, diversifying the applications of common drilling fluids. For example, polymers are added to viscosify, to control fluid loss, to flocculate or deflocculate certain solids, to encapsulate wellbore walls, to provide high-temperature mud stabilization, and to extend the yield of bentonite. Polymer fluids generally contain less than 5% by volume of total low-gravity solids and therefore enhance rate of penetration. These fluids are referred to as true low-solids fluids and are grouped as follows:

- Nondispersed low solids polymer muds (PHPA, PAC/CMC, and Ben-Ex)
- High-temperature deflocculated polymer muds (Polytemp, Pyro-Drill, and Therma-Drill)

Oil-based drilling muds. Drilling muds are called oil-based muds if the continuous liquid phase (external phase) is oil (diesel, mineral, or synthetic). They are inert to contamination, as by H_2S, CO_2, salt, anhydrite, and active shales. There are two types of oil-based mud: invert emulsion and oil muds. Oil muds contain less than 5% of dispersed water, whereas invert emulsion mud contains dispersed water greater than 5%. In either case, oil is the continuous phase, and water is the dispersed phase (internal phase). These muds are highly temperature-stable fluids and can be weighted to any weight after adjusting the oil/water ratio. The use of oil-based muds requires safeguards for environmental protection and safety. Their main applications are in high-temperature formations, water-sensitive shales, thick salt sections, areas where lubricity is critical (as in extended-reach drilling), low-pore-pressure formations, and formations that contain corrosive elements (H_2S, CO_2, etc.).

Water-Based Drilling-Mud Contaminants

A contaminant is any material intrusion into a mud system that causes undesirable changes in the preselected properties. Solids of all kinds—and specifically those of micron and submicron size—are the most prevalent contaminant. They affect all mud properties and slow the drilling rate. Effective solids removal equipment is the primary means of removing undesirable solids. Most other contaminants are chemical in nature and require chemical treatment for their removal:

- Calcium—treat with soda ash or sodium bicarbonate
- Magnesium—treat with caustic soda
- Carbonate—treat with gypsum or lime
- Bicarbonate—treat with lime
- Sulfide—treat with zinc carbonate or zinc oxide

The following sections describe, in brief, the sources of these contaminants and their effect on mud properties.

Salt and saltwater

Contamination of the mud by salt or saltwater is indicated by an increase in the chloride concentration, high viscosity, increased gel strength, and an increase in fluid loss, a rapid decrease in pH, or possibly foaming. This contamination can be treated by increasing the mud weight, in the case of a saltwater flow, and increasing caustic, sulfomethylated tannin and lignosulfonate treatment. In the case of bedded salt, pull up three or four stands above the salt to prevent saturating the system. Use gel, Desco, lignosulfonate, and CMC or Drispac to aid in filtrate control and replace bentonitic solids diluted by water influx.

High solids

This condition is indicated by high funnel viscosity, high PV, high *methylene blue test* (MBT), reduced response to chemical treatment, and high volume of solids on mud retort. It is treated by centrifuging and discarding of low-gravity solids and dilution with water or by dilution alone (dilution may be the most expensive step). The centrifuge should be run on the active system while water is added and the pit volume and weight are kept relatively constant. A small-mesh shaker screen should be used, and makeup water should be added downstream from the shaker. The size of the screen will depend on the type of shale shaker. In general, the smaller the mesh screen is, the better.

Anhydrite (gypsum)

This form of contamination is indicated by high viscosity, high flat gels, and increased fluid loss. An analysis of the filtrate will show increased concentrations of both calcium and sulfates. Note that in most cases, tests for sulfates are not run at the rig site, and special chemicals will be required. When drilling small thin anhydrite stringers, the mud may be treated with soda ash or, in some cases, phosphates or sodium carbonate. If massive anhydrite is encountered, the mud can either be converted to a gyp mud or to a high pH (where calcium solubility is limited).

Cement

Cement contamination is indicated by an increase in viscosity, gel strength, fluid loss, and calcium content in the filtrate. The pH will dramatically rise owing to the hydroxyl ion in the cement. Depending on the pH, this condition can be treated with phosphates, sulfomethylated tannins, bicarbonate of soda, and soda ash. The mud may be pretreated with bicarbonate of soda before drilling cement. Organic thinners will restore the desired rheological properties to the mud. Pretreatment with small amounts of sodium bicarbonate prior to drilling cement will substantially reduce the effect of cement contamination.

Drilling-Mud Properties, Field Tests, and Control

It is mandatory to specify not only the type of drilling mud for each hole interval to be drilled but also the properties of such muds. These are the density, flow properties or rheology, filtration, and solids content, as well as chemical properties. To avoid costly drilling problems, these properties must be field controlled and properly maintained at their preselected values. For this reason, it is essential to monitor any changes by conducting field tests and thereby determining the cause of any problem and finding a solution. Here, emphasis is placed on the definition and functions of mud properties, on their field test and control, and on common practices and errors that are observed.

Viscosity

Viscosity is the internal resistance of a fluid to flow. This phenomenon is attributed to the attraction between molecules of a liquid and is a measure of the combined effects of adhesion and cohesion on suspended particles and the liquid environment. The greater this resistance is, the higher will be the viscosity. In other words, it is the resistance offered by a liquid to the forces trying to make it flow. Viscosity will be discussed further later.

The most familiar device for the checking of viscosity is the Marsh funnel, which measures viscosity qualitatively in seconds per quart (see fig. 2–11). Another instrument, the rotational direct-indicating viscometer, measures viscosity quantitatively and will be discussed later. The procedure for the use of these has been standardized by the American Petroleum Institute (API) in API-RP-13B ("Standard Procedure for Testing Drilling Mud").

To determine Marsh funnel viscosity, the funnel is filled to the bottom of the screen. The funnel holds 1,500 cubic centimeters (cc). The time in seconds required for a quart (946 cc) to flow out the bottom of the funnel is recorded as the Marsh funnel viscosity. The funnel must be calibrated so that the viscosity of water is 25.5–26.5 s/quart at 65–75°F. It should be kept in mind that the funnel viscosity is only a relative value and cannot be used for correlations or calculations of flow regimes. Its primary function is to monitor the viscosity trend. A definite recorded continuous increase or decrease in funnel viscosity means that there are mud problems and should be checked with an instrument such as the rotational viscometer to determine the exact problem and its cause(s).

Rotating *viscometers* (see fig. 2–12) are driven with an electrical motor. The instrument constants are adjusted so that *plastic viscosity* (PV) and *yield point* (YP), which will be discussed shortly, are obtained by using rotor-sleeve speeds of 300 and 600 rpm. The apparent viscosity in centipoise is one-half the 600 rpm reading. Since it is one-speed viscosity, the apparent viscosity is only a general indicator of mud thickness.

The flow properties of the mud are determined with the rotating viscometer by the following procedure. Place a freshly stirred sample in a suitable container and immerse the rotor-sleeve exactly to the scribed line. When the 600 rpm dial reading has reached a steady value, it is recorded. The rotating speed is shifted to 300 rpm, and when this dial reading comes to a steady value, it is recorded.

The PV in centipoise equals the 600 rpm reading minus the 300 rpm reading. The YP in pounds per hundred square feet (lbs/100 ft^2) equals the 300 rpm reading minus the PV.

The *gel strengths* are also measured with the rotating viscometer. The sample should be stirred at high speed for 10 s (or constant reading), then allowed to stand undisturbed for 10 s. With the gears in neutral, a slow (about three rpm), steady motion on the hand wheel is applied. The maximum reading is the initial gel in pounds per hundred square feet. The mud is restirred for 10 s and allowed to stand for 10 min. The measurement is repeated as before, and the maximum reading is recorded as the 10 min gel strength in lbs/100 ft^2.

When an electrically driven instrument is used, the gel readings are the maximum value obtained at the low-speed (three rpm) setting. The PV and the YP are important values that can be used to determine the pressure drops in the circulating system. They are also key parameters in determining how to manage a mud system.

The PV is a function of the concentration, size, and shape of the solids particles in the mud. It can be controlled by dilution or by mechanical means. The YP is a function of the electrostatic forces between colloidal particles in the mud. It can be controlled by chemical thinners, dispersants, and so forth.

Viscosity control. It is most common to add bentonite and/or polymers to increase viscosity. To decrease the viscosity, dilution is generally considered. However, if a large amount of liquid is required, this can lead to changes in other properties and, therefore, a higher mud cost. An alternative to dilution is the addition of thinners or dispersants, such as quebracho, tannins, lignosulfonates, lignites, and polyphosphates. The addition of any of these thinners to a mud that has an unusually high concentration of low-gravity solids may actually increase viscosity. In fact, with the exception of water, anything added to a mud in such condition may increase the viscosity.

Density (mud weight)

The term weight is used in connection with mud more often than *density,* even though density is the more correct term. This is a consequence of the first attempts to measure properties of muds. These first tests included hanging one-gallon buckets on spring scales and reporting the weight of the mud in *pounds per gallon* (ppg, or lbs/gal). Most tests of mud density are still reported in pounds per gallon. Mud weight can also be reported in terms of pounds per cubic foot (lbs/ft^3), grams per cubic centimeter (g/cc), specific gravity (spg), or pressure gradient (psi/ft).

Ideally, a mud weight as low as the weight of water is desired, for optimum drilling rates and for minimizing the chances of fracturing the formation. However, in practice, mud weights in excess of two times the weight of water may be necessary, to contain abnormal pressures or to mechanically stabilize unstable formations.

Mud hydrostatic pressure. The hydrostatic pressure at a depth D for a column of mud having a density γ_m can be easily derived and is given by

$$P_h = k\gamma_m D \tag{2.2}$$

where

D is the depth of mud column
γ_m is the mud weight
k is the conversion unit factor

For example, if P_h is in psi, γ_m is in lbs/gal, and D is in ft, then the value of k in equation (2.2) is 0.052.

Mud weight field test. The most commonly used device to measure mud weight in the field is the conventional mud balance (fig. 2–1). However, unless the mud sample is completely free of entrapped air/gas bubbles, this device will not give the true mud weight. A new device called the Tru-Wate cup has been developed as an alternative for weight measurement. It is a pressurized mud balance (fig. 2–2) that has proved to be effective in eliminating the effects of entrapped gases in mud samples, thus giving the true density.

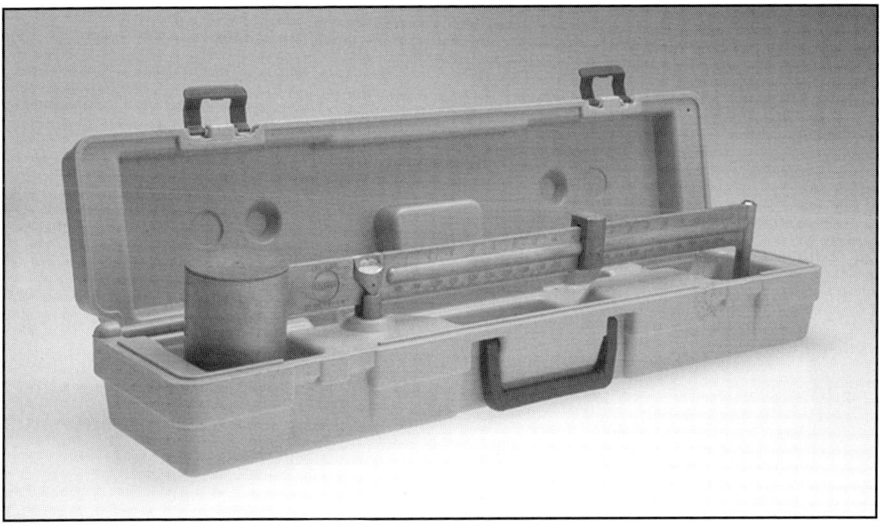

Fig. 2–1. Conventional mud balance (Courtesy Fann Instrument Company)

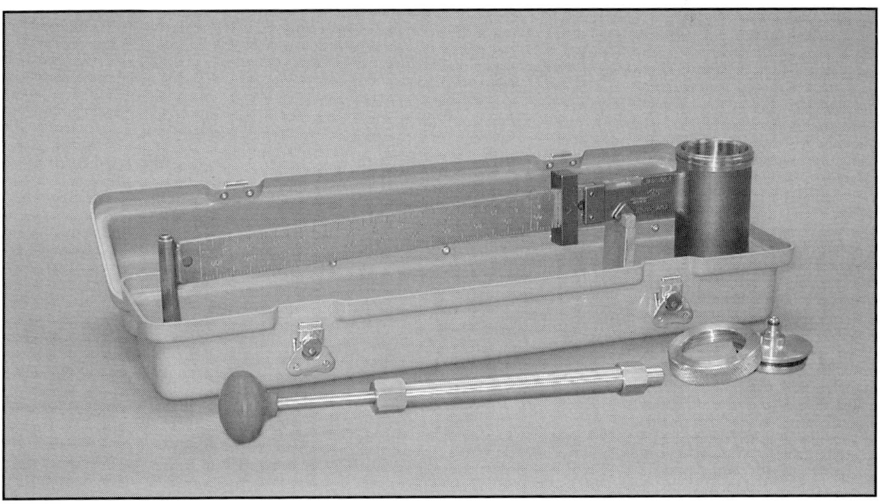

Fig. 2–2. Pressurized mud balance (Courtesy Fann Instrument Company)

Misreported mud weights may erroneously call for mud weight adjustments, which can lead to well kicks, the fracturing of a formation, and higher mud drilling costs. The most common causes of such errors are

- Lack of calibration of the mud balance
- Air/gas cut mud
- Improper procedures followed in measuring mud weight

Mud weight control. Control of mud weight within a specified range in the field is one of the major concerns in the implementation of an optimum drilling-mud program. Drilled solids, particularly those of micron and submicron size, are the most detrimental to the rate of penetration (drilling rate). If these solids are not removed from the mud system, they will adversely affect the weight and other important properties of the mud. The control of mud weight in the field is accomplished by

- Utilization of effective solids removal equipment
- Addition of chemical flocculants at the flowline
- Dilution with water

Solids removal by mechanical means utilizes high-speed screen shakers with 50 mesh size or finer, followed by desanders, desilters, and mud cleaners for unweighted muds. Centrifuges, which can discriminate among solids on the basis of their specific gravities, are used for weighted muds; centrifuges are also used in unweighted muds if a true low-solids mud system is to be maintained. Flow-line flocculation through the addition of chemicals is also used to enable the solids removal equipment to extract submicron-size drilled solids and is an essential means of controlling the mud weight.

A less effective supplementary technique in mud weight control is dilution by water. A continuous but small regulated stream of water should be added at the flow line, as excessive dilution will result in excessive treatment costs. In oil-based muds, dilution for the purpose of reducing the mud weight is costly and detrimental to the drilling rate and therefore must be avoided.

The mud weight measurement at the flow line reflects the effective hydrostatic head in the annulus and will depend on the concentration of drilled cuttings, which in turn depends on the drilling rate. The mud weight measurement at the suction pit is representative of the effectiveness of the solids removal equipment. Failure to maintain the proper mud weight at the suction tank will result in poor drilling performance and, therefore, a higher drilling cost

Mud-weighting mathematical relations. All the materials present in a mud contribute to its density. The main components are the liquid, the clay, and the barite. The chemicals used to control viscosity and gel properties are usually present only in small amounts and are neglected in weighting calculations. Freshwater has a density of about 8.33 lbs/gal, the bentonitic clay commonly added to a mud has a specific gravity of about 2.5, and finely ground barite has a specific gravity of about 4.3.

Four fundamental equations are used in the development of mud-weighting mathematical relations.

- Material balance equation:

$$W_f = W_o + W_a \tag{2.3}$$

where W_f, W_o, W_a are the final mixture weight, the original liquid weight, and the added material weight, respectively.

- Volume balance equation:

$$V_f = V_o + V_a \tag{2.4}$$

where V_f, V_o, V_a are the final mixture volume, the original liquid volume, and the added material weight, respectively.

- Relationship between weight and volume:

$$\gamma = \frac{W}{V} \tag{2.5}$$

where γ, W, V are the weight density, the weight, and the volume, respectively.

- Volume balance in low specific gravity solids:

$$V_f f_{vf} = V_o f_{vo} \tag{2.6}$$

where f_{vf}, f_{vo} are the volume fractions of low-specific-gravity solids in the final mixture and in the original liquid, respectively.

Equations (2.3)–(2.6) may be combined and rearranged algebraically to suit the particular calculations desired. From equations (2.3) and (2.4), for example, the following equation can be obtained to describe a general relationship:

$$V_1D_1 + V_2D_2 + V_3D_3 + V_4D_4 = V_fD_f \tag{2.7a}$$

$$V_1 + V_2 + V_3 + V_4 = V_f \tag{2.7b}$$

where

V_1, V_2, V_3, V_4 are the respective volumes of materials 1, 2, 3, and 4.
D_1, D_2, D_3, D_4 are the respective densities of materials 1, 2, 3, and 4.

Simultaneous solution of equations (2.7a) and (2.7b) results in any two sought unknowns with the rest of the parameters as knowns.

The following are example formulations for commonly encountered field problems:

- Required amount of weighting materials to increase original mud density, γ_o, to a final density γ_f.

$$W_{wm} = \frac{42(\gamma_f - \gamma_o)}{1 - (\gamma_f / \gamma_{wm})} \tag{2.8}$$

where W_{wm} is the required amount of weighting material (in lbs/bbl or original mud).

- Average weight density, γ_{av}, of two added materials i and j of weights γ_i, and γ_j, respectively:

$$\gamma_{av} = \frac{\gamma_i \gamma_j}{f_w \gamma_i + (1 + f_w) \gamma_j} \tag{2.9}$$

where $f_w = W_j/(w_i + w_j)$ is the weight fraction of material j with respect to added weights of materials i and j.

- Amount of liquid volume, V_l, required to make up a mud of total volume V_f having a weight γ_f:

$$V_l = V_f \frac{1-(\gamma_f/\gamma_a)}{1-(\gamma_l/\gamma_a)} \tag{2.10}$$

where γ_a is the weight density of added material or average weight density γ_{av}, of material added to the liquid.

- Amount of solids *i* and *j*, such as clay and barite, required to make up a specified final mud volume V_f and density γ_f:

$$w_i = \frac{42 f_w (\gamma_f - \gamma_o)}{1 - f_w(\gamma_o/\gamma_j) - (1-f_w)(\gamma_o/\gamma_i)} \tag{2.11}$$

$$w_i = \frac{42(1-f_w)(\gamma_f - \gamma_o)}{1 - f_w(\gamma_o/\gamma_j) - (1-f_w)(\gamma_o/\gamma_i)} \tag{2.12}$$

where w_i, w_j are the solid materials *i* and *j* (in lbs/bbl of final mud).

- Final mud density, γ_f, when a certain liquid volume, V_l, of density γ_l is added to a mud system of original density (γ_o) and volume (V_o):

$$\gamma_f = \frac{\gamma_o + \alpha\gamma_l}{1+\alpha} \tag{2.13}$$

where $\alpha = V_l/V_o$.

Mud weight increase due to fast drilling. When large-surface holes are drilled at high ROPs, the mud weight will increase rapidly unless mechanical solids removal equipment is used and sufficient water is added to the system. Controlled drilling may be required in order to control mud weight in the annulus and to avoid lost circulation.

An estimate of the average annulus mud weight ($\gamma_{m,av}$) can be formulated from the previous equations and is given by

$$\gamma_{m,av} = \frac{\gamma_{ps}Q + 0.85D_h^2 R}{Q + 0.0408D_h^2 R} \tag{2.14}$$

where

$\gamma_{m,av}$ is the average annular mud weight, in lbs/gal
Q is the flow rate, in gpm
γ_{ps} is the measured mud weight at pump suction, in lbs/gal
D_h is the diameter of hole, in in.
R is the ROP, based on time the pump is on—before, during, and after the joint is drilled down (in ft/min)

For example, if the pump is on for 1.5 min while a 30 ft joint is drilled down, R will equal 20.0 ft/min. Suppose the hole diameter is 17½ in., the flow rate is 800 gpm, and γ_{ps} is 9.0 lbs/gal; then,

$$\gamma_{m,av} = \frac{(800 \times 9) + (0.85 \times 20 \times 17.5 \times 17.5)}{800 + (0.0408 \times 17.5 \times 17.5 \times 20)} = 11.8 \text{ lbs/gal} \tag{2.15}$$

Equivalent circulating density. In small-diameter holes, those of 9⅞ in. or less, friction pressure loss in the annulus becomes an important factor that must be taken into consideration. This pressure can cause a further, significant increase in mud weight as compared to the increase due to cuttings accumulations under low drilling rates.

The effect of friction loss is expressed in terms of mud weight and is called the *equivalent circulating density* (ECD) or the equivalent mud weight (EMW) and can be expressed as follows:

$$\text{ECD} = \gamma_m + \frac{P_f}{0.052D} \tag{2.16}$$

where

γ_m is the mud weight, in lbs/gal, as measured on surface
P_f is the annulus friction loss, in psi
D is the depth, in ft, at which the ECD is being calculated

Fluid loss

Fluid loss is defined as the loss of a mud filtrate (liquid phase) into a permeable formation that is being drilled. Because of positive differential pressure (i.e., the pressure difference between the mud pressure in the wellbore and the formation pore pressure), the mud filtrate tends to flow into the formation; this results in the buildup of mud solids deposited on the wellbore walls, thus forming what is commonly referred to as mud cake (filter cake). The term *spurt loss* is used to characterize the initial loss of filtrate to formation at time practically equal to zero. After a mud cake is formed, the loss of filtrate from then on is referred to as the continuous fluid loss.

It has been pointed out in the literature that high spurt loss is advantageous, while high continuous loss is detrimental. It is the general consensus that the rate and amount of high continuous fluid loss into a formation can adversely affect the rate of penetration, water-sensitive and sloughing shale, and the producing formation and can cause differential-pressure pipe sticking. Any or all of these can occur in the field. Therefore, emphasis is placed on specifying and controlling the mud fluid loss property. However, the general tendency is to specify fluid losses much lower than necessary. This can result in drilling problems and, therefore, higher well cost. The best solution, so as not to over- or underspecify fluid loss properties, is to obtain information from offset wells in the same area.

Fluid loss tests. There are two types of tests for fluid loss measurement:

- API static filtration test (standard low pressure and temperature and high pressure and temperature)
- Dynamic filtration test

The static filtration test includes the standard API test at room temperature and 100 psi differential pressure and is the *field test* for fluid loss measurement. The high-pressure, high-temperature test is a laboratory test and is conducted at a differential pressure of 500 psi and a temperature of 300°F. The static tests are indicative of the loss of fluid and the buildup of the filter cake when the fluid is not moving.

The dynamic test represents the loss of fluid and the filter cake buildup while the drilling mud is being circulated. Dynamic tests are strictly laboratory conducted tests.

Fluid loss measurement. The static fluid loss and filter cake characteristics of a mud are determined using an API-approved filter press. In the standard API test, the sample is placed in the filter press cell, and a pressure of 100 psi is applied to the top of the cell with a pressure regulator. At the end of 30 min, the volume of filtrate in cubic centimeters is recorded as the standard API filtrate loss. The cake thickness in 32nds of an inch is also recorded. Although an exact description of the cake characteristics is virtually impossible, reported observations (whether the cake is hard, soft, tough, firm, rubbery, etc.) may be useful in conveying some idea of the cake consistency.

The high-temperature, high-pressure (HTHP) water loss is basically the same, except that more elaborate equipment is required in order to hold the temperature at 300°F and to apply 500 psi differential pressure across the filtration medium, which is one-half the API standard filtration area. The collected filtrate is multiplied by, to account for the difference in area between the two filter papers used. This test is designed to give some idea of the performance of the mud under downhole conditions. As with the standard API fluid-loss test, the HTHP API test determines only the relative amount of filtrate that may be lost. Actual downhole filtration will vary according to the formation porosity, temperature, differential pressure and fluid dynamics. The test equipments used for this purpose are filter presses described in figures 2–3 and 2–4.

Factors that affect fluid loss. To recognize the parameters that affect the fluid loss, consider Darcy's law of flow through porous media, which is given by

$$\frac{dQ_f}{dt} = \frac{kA\Delta P}{\mu h} \tag{2.17}$$

where

Q_f is the amount of filtrate volume
k is the cake permeability
A is the filtration area
ΔP is the differential pressure
μ is the filtrate viscosity
h is the cake thickness
t is the filtration time

Under the assumption that a volume of mud, Q_m, is totally filtered, a volume material balance can be easily written, as

$$Q_m = Q_s + hA \tag{2.18}$$

and

$$Q_s = f_{vc} hA \tag{2.19}$$

where

Q_m is the total volume of mud filtered
Q_s is the volume of solids deposited in mud cake
f_{vc} is the volume fraction of solids in cake

Fig. 2–3. Standard filter press (Courtesy Fann Instrument Company)

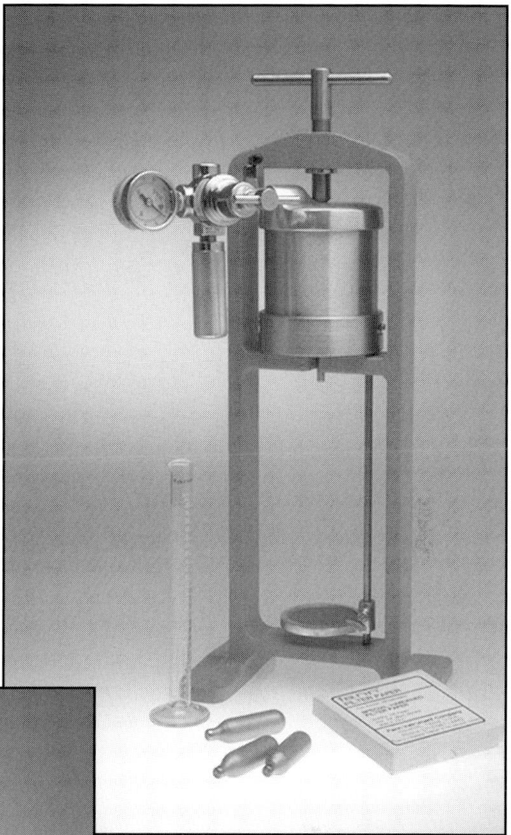

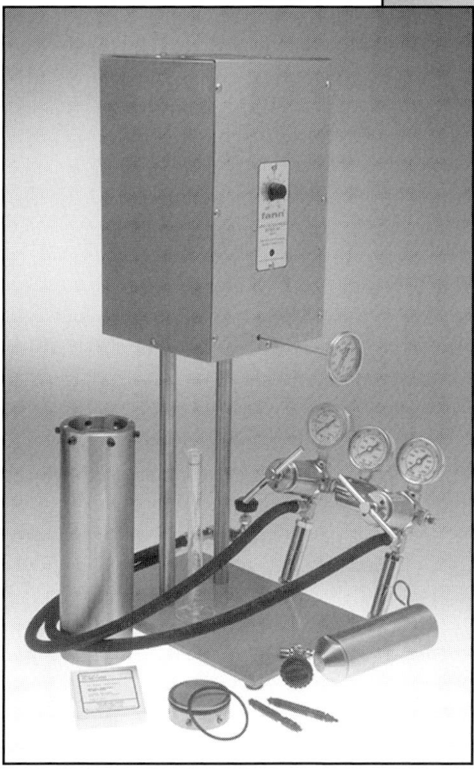

Fig. 2–4. HTHP filter press (Courtesy Fann Instrument Company)

The volume fraction of solids in mud, f_{vm}, can be written as

$$f_{vm} = \frac{Q_s}{Q_m} \tag{2.20}$$

Substituting equations (2.18) and (2.19) into equation (2.20) yields

$$f_{vm} = \frac{f_{vc} hA}{Q_f + hA} \tag{2.21}$$

Solving for the cake thickness, h, the following is obtained:

$$h = \frac{Q_f}{A[(f_{vc}/f_{vm}) - 1]} \tag{2.22}$$

Utilizing equation (2.22) in equation (2.17), assuming that Q is the only time-dependent parameter, and then integrating yields

$$Q_f = A\left[\frac{2k[(f_{vc}/f_{vm}) - 1]\Delta P_t}{\mu}\right]^{1/2} + Q_s \tag{2.23}$$

where Q_s is the constant of integration, equal to spurt loss at $t = 0$.

On the basis of equation (2.23), it can be concluded that the filtrate volume per unit area, Q_f/A, under static conditions is affected by the following factors:

- Time
- Differential pressure
- Solids (amount, type, and size and size distribution)
- Permeability of mud cake
- Filtrate viscosity

Following is a brief discussion of each of these factors.

Time. The filtrate volume (fluid loss) increases in direct proportion to the square root of time, as shown in figure 2–5. In general, the line does not pass through the origin, as indicated in the figure. Theoretically, this means that at $t = 0$, there is a sudden loss of filtrate (referred to as spurt loss) to the formation before a mud cake is formed and begins to restrict the flow of the filtrate into the formation.

A common error in reporting fluid test results is to measure the filtrate for only 7.5 min and multiply that by 2 and call it the standard 30 min API test. This practice is based on the reasoning that the square root of the ratio of 30 over 7.5 is 2. This would be true if the spurt loss, Q_s, is zero or if the line in figure 2–5 passes through the origin. Because this practice can result sometimes in large errors, depending on how far off the true intercept is from the origin, it should be avoided. If, however, it must be used, then the following equation should be used to convert the 7.5-min test to the 30 min test:

$$Q_f(\text{API}) = 2(Q_t - Q_s) + Q_s \tag{2.24}$$

where Q_t is the filtrate volume collected in $t = 7.5$ min and Q_s is the spurt loss.

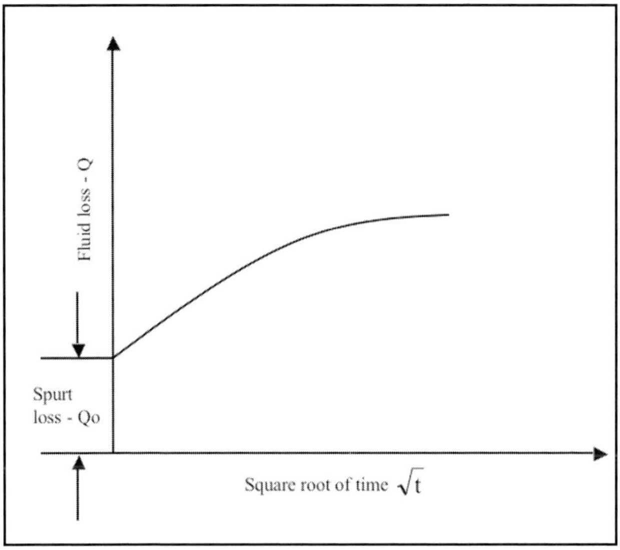

Fig. 2–5. Effect of time on fluid loss

Pressure differential. The effect that pressure differential has on filtration is dictated by the compressibility of the mud cake. For example, the permeability of a compressible mud cake will decrease for an increase in differential pressure, resulting in negligible increase in fluid loss. However, for an incompressible mud cake, the permeability will not change for an increase in differential pressure, thus leading to a more pronounced increase in fluid loss.

Solids. It has been shown in equation (2.24) that the fluid is proportional to the quantity $\sqrt{(f_{vc}/f_{vm}) - 1}$. This should be true if the volume fraction of the solids in the mud cake is not a function of time. It has also been shown that an increase in mud solids will result in a decrease of fluid loss. However, this practice is not recommended; instead polymers in addition to bentonite, can be used to reduce fluid loss.

Cake permeability. It is obvious on the basis of equation (2.24) that the higher the mud cake permeability is, the higher will be the fluid loss. The most effective and desirable means to reduce fluid loss, therefore, is to have low mud cake permeability. The size, the shape, and the deformability of the solids in the mud system are important factors in controlling the permeability. Particles of one micron or smaller size will contribute to the best mud cake composition. In addition, thin and flat particles are more effective than other, irregular shapes.

Filtrate viscosity. Equation (2.24) shows that fluid loss in porous media is inversely proportional to the filtrate viscosity—that is, the higher the viscosity is, the lower will be the fluid loss. However, increasing the filtrate viscosity is not a recommended practice, since it will result in an increase in mud viscosity that could cause drilling problems.

Many of the materials that are used to increase viscosity also serve to reduce the filtration rate. Among these materials are bentonite, sodium polyacrylate, CMC and other polymers, and starch. Some thinners, such as lignite and the lignosulfonates, also reduce filtration rate. Attapulgite, lime, gypsum, and salt will each increase viscosity but also tend to increase filtration rate. Oil emulsified in water mud lowers the API filtration rate but greatly increases dynamic filtration. Oil has unpredictable effects on the HTHP filter loss. Table 2–1 lists common additives that control filtration rates.

Table 2–1. Common fluid-loss additives

Bentonite	Provides a good basic filter cake.
CMC	Reduces filtration rate by coating solids and by minimizing flocculation. Starts breaking down at 300°F.
Starch	Reduces filtration rate by coating solids; will spoil in fresh-water environment where pH is less than 11.5. A salt concentration of 250,000 ppm will break down at temperatures of 275°F.
Lignosulfonates	Reduce filtration by deflocculating the mud. Also, increase the viscosity of the filtrate, start breaking down at 300°F.
Lignites	Reduce filtration by deflocculating and plugging open spaces in the filter cake. Start breaking down at 300°F.

As the temperature of water is increased, its viscosity decreases. Therefore, as the temperature of the mud sample is increased, one can expect the fluid loss to increase, because the lower viscosity allows the filtrate to flow through the filter cake more easily. As the pressure on a mud sample is increased, the filter cake generally becomes more compacted and more impervious to the flow of the filtrate. This tends to offset some of the effect of the increase in mud temperature.

In bentonite muds, filtration rates are similar at different pressures. However, with polymers and other additives in the mud, the filtration rate is pressure related. The composition of the mud has a significant effect on the filtration rate. If the solids in a mud consist of coarse particles, such as native clays, shales, barite, and fine silt, the filter cake will be less compacted than with a mud whose solids are finely divided particles of high-grade bentonite, PAC, and barite. Most muds carry a certain amount of drilled solids (native clays and fine silt). The objective in good mud control is to minimize drilled solids content. By the judicious addition of bentonite and other fluid-loss reducers, a better particle-size distribution can be attained. A balanced particle-size distribution will provide the optimum filter-cake compactability.

The HTHP fluid-loss test is more rigorous than the standard API test and is more appropriate when drilling at high temperature and pressure. It gives a better idea of the quality of a mud with regard to composition and particle-size distribution. In fact, the standard API test can sometimes be misleading: cases are known where the API test showed a fluid loss of 3.5–5.0 cc when an HTHP test on the same sample showed 30.0–35.0 cc; after the mud was conditioned with bentonite and other fluid-loss reducers and thinners, the HTHP test showed a fluid loss of 10.5–12.0 cc and the standard API test showed 3.0–4.0 cc.

Alkalinity (M_f, P_f, P_m, M_m, P_1, P_2)

The *alkalinity* measurements on the mud and the mud filtrate consist of titrating a small sample with a standardized sulfuric (or nitric) acid solution to the end point of an indicator solution (when the indicator changes color). The two commonly used indicator solutions are phenolphthalein and methyl orange.

Phenolphthalein has an end point of 8.3 pH and will have a pink color in solutions with pH of higher 8.3. As the acid is added drop by drop, the pH of the sample will be lowered and at the pH of 8.3 the pink color will disappear. The color change is the end point of the titration.

The end point of the methyl orange indicator is pH of 4.3 and the titration is carried out of the sample titrated to the phenolphthalein end point, using the same standardized acid solution. Methyl orange will have yellow color above a pH of 4.3 which will turn to pink at the end point of 4.3 pH.

The titration with the phenolphthalein indicator is referred to as the P end point depending on whether a sample of filtrate or a sample of mud is titrated, abbreviated as the P_f or the P_m value. The methyl orange titration is similarly abbreviated as the M_f or M_m value depending on the sample titrated.

All alkalinities are reported simply as the volume of the standardized acid (in cc) needed to reach the indicator end point for a 1 cc sample. As an example, if 1.4 cc of standardized acid was used to titrate the P end point of a 2cc filtrate sample, the P_f is reported as 0.7 cc on the mud report. The M end point is reported as the total volume of acid required to reach the M end point in a given sample, including the acid used to titrate the P end point.

A point often overlooked in the field is that if a filtrate or mud sample has a P_f or P_m value of greater than zero (even 0.01), the pH of that sample cannot be less than 8.3 (the end point of the indicator). Commonly the mud reports contain pH and P_f measurements that are commonly incongruous.

The P_f and P_m values are utilized to determine the free lime content in lime base drilling fluids. The P_f and P_m values were also utilized to estimate the hydroxyl, bicarbonate, and carbonate ions in the filtrate. More recently, API has standardized alternate filtrate alkalinity analysis as based on P_f and M_f, and two newly defined alkalinity measurements, P_1 and P_2 and the calculations for determining the hydroxyl, bicarbonate and carbonate ions in the filtrate in the API-RP-13B, pages 18–19.

Treatment of alkalinity. If there are bicarbonates present, they must be converted to carbonates before they can be removed from solution. To do this, calcium hydroxide (in the form of slaked lime) may be added. The soluble calcium combines with carbonates to form limestone, which precipitates. However, both lime and gypsum are not very soluble; thus, it may take numerous circulations to precipitate the carbonate ions.

pH

The pH, or hydrogen ion concentration, is a measure of the relative acidity or alkalinity. The pH value ranges from 0 to 14, with 0–6 being acid, 7 being neutral, and 8–14 being alkaline. Except for salt muds, the pH of mud is seldom below 7. Generally, the pH of mud falls between 8 and 11, depending on the mud type.

The pH may be determined by a pH meter, which measures the potential difference and gives a direct dial reading of the pH. It is accurate, but because of the delicate nature of the instrument, it has limited field use. pH papers are more commonly used. These strips change color in accordance with the acidity or alkalinity of the filtrate. There are a variety of pH papers available to cover the entire pH range. When the chloride content exceeds 10,000 ppm, the results become less accurate.

The pH of mud plays a major role in controlling the solubility of calcium. At high pH values, calcium solubility is very limited; this makes high-pH mud suitable for use in the drilling of carbonate formations, which normally are susceptible to erosion and dissolution by freshwater mud. The pH value is also an important indicator for the control of corrosion. A minimum value of 9.5 should always be maintained to prevent oxygen corrosion of casing, drill pipe, and so forth. A high pH tends to disperse (disintegrate) the active clays in the mud.

Methylene blue test (cation exchange capacity)

This test gives an estimate of the cation exchange capacity of mud solids. A standardized solution of *methylene blue* dye is added to 1 ml of mud that has been treated with hydrogen peroxide and sulfuric acid and is then boiled to decompose the polymers and organics (which have a very high exchange capacity and would otherwise interfere with the test). The methylene blue is added in increments until the mud solids no longer absorb the dye. This end point is determined by putting a drop of the solution on a hard filter paper (standard API filter paper: Whatman No. 50 or equivalent). When the dye is in excess, a halo of free dye will form around the blue dot. The halo that forms is turquoise in color and will be very distinct from the blue color of the dye. The results are usually reported as equivalent lbs/bbl bentonite. It would be more accurate to report the results as milliequivalents per hundred grams (mEq/100 g) of mud solids.

$$\text{Methelyne blue capacity} = \frac{\text{cc of dye}}{\text{cc of mud}} \tag{2.25}$$

Reactive solids (expressed in lbs/bbl), including bentonite, are equal to five times the methylene blue capacity.

Sand content

API sand content is not restricted to sand; rather, it is a convenient measure of all particles in the mud that are coarser than 200 mesh (74 microns). The results are reported as a percent by volume. Sand content should be kept at 1/4% to trace.

Chlorides

Chloride content is determined by titrating a 1 ml sample of filtrate, to which is added a potassium chromate indicator, with a silver nitrate solution of known strength. Although chloride content is reported in ppm, in reality, it is milligrams per liter (mg/l). The value for chloride can be multiplied by 1.65 to give the salinity.

Total hardness

Total hardness (Ca^{2+} + Mg^{2+}) is found by titrating a 1 ml sample of filtrate, to which is added a versenate hardness indicator and a buffer. The titrating is done with a standard versenate filtrating solution. The results are reported as ppm calcium, even though the magnesium hardness is included.

The actual calcium hardness may be determined by using a calcium indicator solution and 2 ml of one normal solution of sodium hydroxide (NaOH) (A different strength solution may be used, but the pH of the sample must be above 9.5 to precipitate the magnesium chloride). The sample is titrated with standard versenate solution, and the results are reported as ppm calcium.

The magnesium hardness can be determined as the difference between the total hardness and the calcium hardness.

Solids content

Today, much effort is expended to maintain *low-solids* muds. It has been proved that low-solids muds are more easily controlled and provide faster ROPs in all areas of operations. Low-solids muds have a near-minimum amount of bentonite, clay, or other low-gravity solids. Total solids content is determined, to a great extent, by the density needed to control formation pressures. Solids content is determined by distilling the liquid (water and sometimes oil) from a known volume of mud.

A *retort* is used for this test. The mud is placed in a steel container and is heated until the liquid is vaporized. The vapors are passed through a condenser and are collected in a graduated cylinder. The volume of the liquid is measured and converted to percent. The solids, both suspended and dissolved, are equal to 100% minus the liquid percent. This procedure also gives the percent oil, if any, in the mud. There are charts and formulas to determine the specific gravity of the solids. From the specific gravity of the solids, it is possible to determine the approximate percent clay and barite present in the mud. Provided that retorts are in good condition and care is exercised in running the test, the accuracy attained is generally ±5%. This is one of the least costly mud tests that can be run in the field. Therefore, as in oil mud tests, trends should be noted, rather than using a single retort test as an absolute.

Aniline point

The *aniline point* of oil in oil-based drilling fluids is the guide to whether damage can be expected to occur to rubber parts of a drilling-rig circulating system. Oils with a high aromatics content are more detrimental to rubber products than those having low aromatics content. The relative aromatics content of an oil is indicated by its aniline point. Inasmuch as oils having low aromatics content have high aniline points and vice versa, the oils with high aniline points are the most desirable for use in drilling fluids. The aniline point of the oil-based muds should be 150°F or above to minimize damage to the rubber parts.

Mud viscosity

Mud viscosity, as defined previously, is the internal resistance of a fluid to its forced flow. On the one hand, only Newtonian fluids, such as water, have true viscosity that can be described by one term (constant). Non-Newtonian fluids, on the other hand, require more than one component (constant) of viscosity in order to describe their flow behavior, as will be shown later, in the section "Mud Rheology."

Viscosity measurement. Drilling-fluid *viscosity* measurement may be qualitative (relative measurement) or quantitative (absolute measurement). The qualitative measurement is conducted through the use of a Marsh funnel, as discussed previously. The absolute measurement is conducted through the use of a rotational viscometer and allows the calculation of PV, yield stress, gel stress, power low index (n), and consistency index (K); these rheological constants will be defined later.

Throughout the circulating system, the desired viscous properties are those that will lead to effective bottom cleaning of generated cuttings (low viscosity), effective annular hole cleaning (sufficient viscosity in vertical and near-vertical holes and low viscosity in inclined holes), and suspension of weighting materials and, in some cases, suspension of drilled cuttings when circulation is stopped. In any type of mud, excessive viscous properties will result in excessive friction pressure losses and, therefore, excessive surge/swab pressures, excessive bottom-hole circulating mud densities, which in turn can lead to very costly drilling problems.

Mud Rheology

Definitions

Rheology. Rheology is the study of the deformation of fluids. Its importance is recognized in the analysis of fluid-flow velocity profiles, fluid viscosity, friction pressure losses, ECD, and annular hole cleaning. It is the basis for all analyses of wellbore hydraulics.

Fluid-flow velocity. Figure 2–6 shows the velocity profile of a fluid between two parallel plates, the lower of which is stationary while the upper is moving at a velocity v. The layer of fluid adjacent to the bottom plate has zero velocity, while the top layer has a velocity equal to that of the moving upper plate. The rate at which each layer moves past the other is termed the *shear rate*.

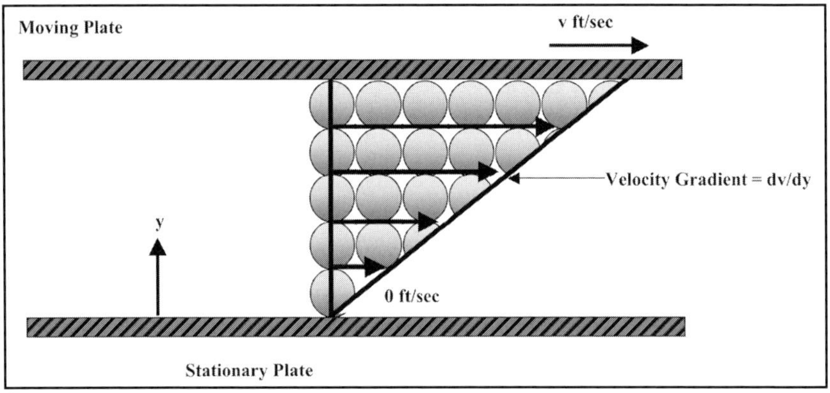

Fig. 2–6. Velocity gradient between two parallel plates

Shear stress–shear rate. The shear stress, τ, is defined as an applied force, F, acting along a unit surface area, A, tending to deform the fluid element as shown. The shear rate, γ, is the velocity gradient—that is, the change in velocity of a fluid layer moving along the x-axis with respect to another layer a unit distance away along a mutually perpendicular axis, such as the y-axis in Cartesian coordinates or the r-axis in polar coordinates. Mathematically, these are expressed as follows:

Shear rate:

$$\gamma = \frac{dv}{dy} = \frac{dv}{dr} \tag{2.26}$$

Shear stress:

$$\tau = \frac{F}{A} \tag{2.27}$$

The field units are inverse seconds (s^{-1}) for the shear rate and lbs/100 ft^2 for the shear stress. It has been shown experimentally that the force per unit area applied to a fluid is proportional to the velocity change of two fluid layers a unit distance apart—that is,

$$\frac{F}{A} \approx \left(\frac{dv}{dy}\right) \approx \left(\frac{dv}{dr}\right) \tag{2.28}$$

Newtonian and non-Newtonian fluids. Drilling fluids are classified into two major groups: Newtonian and non-Newtonian fluids.

Newtonian fluids are those whose flow behavior can be fully described by a single term called the Newtonian viscosity, μ. For these fluids, examples of which include water and light oil, the shear stress is related to shear rate linearly (i.e., $n = 1$ in eq. [2.28]), with the proportionality constant being the Newtonian constant viscosity, μ. (See fig. 2–7.)

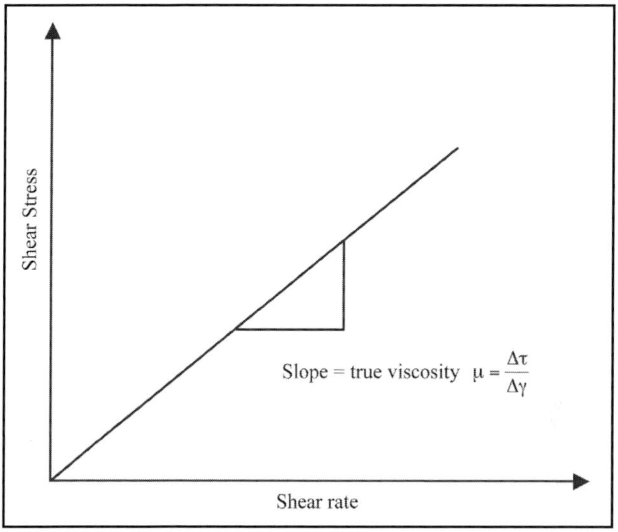

Fig. 2–7. Shear rate–shear stress relationship of Newtonian fluids

For a Newtonian fluid,

$$\tau = \mu\gamma \tag{2.29}$$

In engineering units,

τ is in dynes/cm^2 = 4.79 lbs/100 ft^2
γ is in s^{-1}
μ is in poise = dyne × s/cm^2

The field unit of viscosity is the centipoise (1 poise = 100 centipoise). The field unit of shear stress is lbs/100 ft^2.

A non-Newtonian fluid (figs. 2–8 to 2–10) is one whose viscous property cannot be described by a single term, as had been done in equation (2.19) for a Newtonian fluid. Rather, the viscous property is approximated to behave in accordance with one of the following assumed models:

- Bingham plastic:

$$\tau = \tau_y + \mu_p \gamma \tag{2.30}$$

- Power law:

$$\tau = K\gamma^n \tag{2.31}$$

- Power law yield:

$$\tau = \tau_y + K\gamma^n \tag{2.32}$$

where

τ_y is the yield value or yield stress
μ_p is the Bingham PV
K is the consistency index
n is the power law index

Note that in each of the previous shear stress–shear rate relationship expressions, more than one viscous term is needed to fully describe the flow behavior of the fluid.

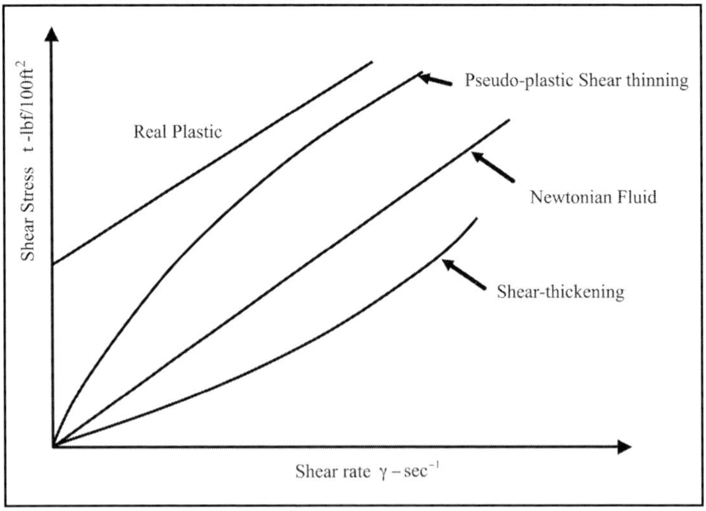

Fig. 2–8. Shear rate–shear stress relationship of time-independent non-Newtonian fluids

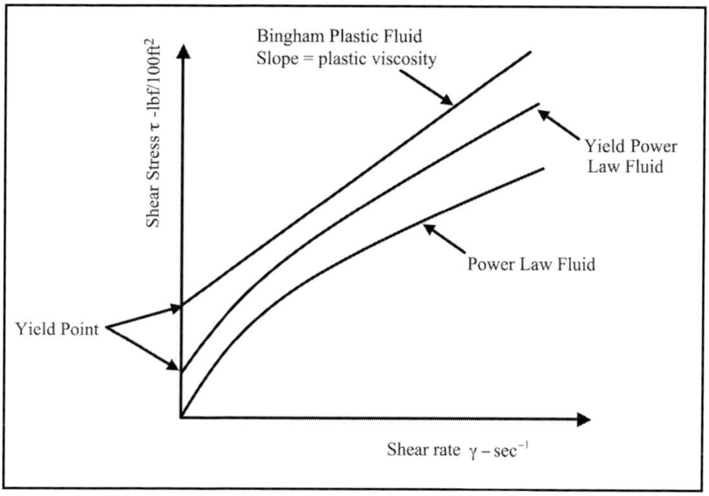

Fig. 2–9. Shear rate–shear stress relationship of non-Newtonian fluids

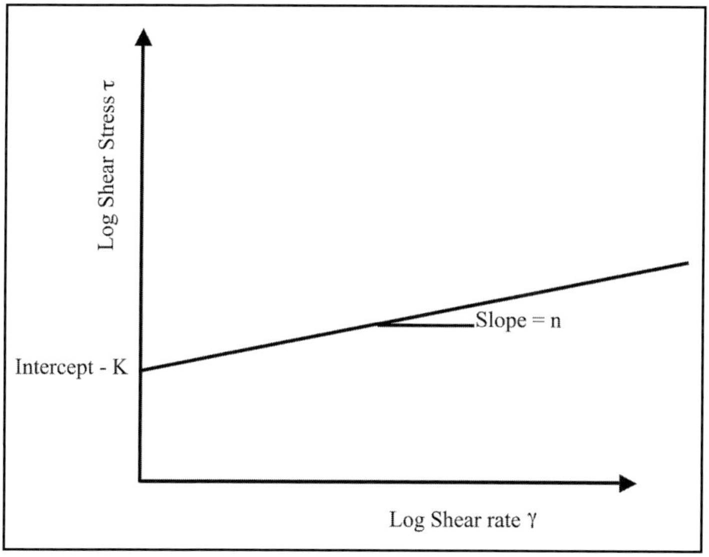

Fig. 2–10. Shear rate–shear stress relationship of a power law fluid on logarithmic scale

Plastic viscosity **(μ_p).** The PV is the part of the flow resistance of the fluid caused by mechanical friction within the fluid. This mechanical friction is due to the interaction of individual solid particles, the interaction between solid and liquid particles, and the deformation (shearing) of the liquid particles under shear stress. The amount of solid particles, as well as their size, distribution, and shape, has direct bearing on the PV. Although there is no quantitative relationship between the PV and the solid particles in a mud system, the measured value of the PV is a qualitative indication of mud solids content.

Yield value (τ_y**)**. The yield value, or *yield stress*, is the part of the flow resistance of the fluid caused by electrochemical forces within the fluid. These electrochemical forces are due to the electrical charges on the surface of reactive particles, the electrical charges on the submicron particles, and in water-based muds, the presence of the electrolytes. The field unit for the yield value is lbs/100 ft^2.

Gel strength (τ_g**)**. The *gel strength* is a measurement of the electrochemical forces within the fluid under static conditions. Its field unit is the same as that of the yield strength.

Thixotropic fluid. Most non-Newtonian drilling muds exhibit a flow behavior that is time dependent. A thixotropic fluid is one whose internal structure builds up gel strength while at rest, or a fluid for which shear stress decreases while it is being sheared at a constant shear rate.

As thixotropic fluids are sheared at constant shear rate, their gel structure breaks down in a short time period; hence, the Bingham plastic and power law equations (eq. [2.30] and [2.31], respectively) may be utilized to approximate their flow behavior after the gel structure has been broken.

Viscoelastic fluids. Certain polymers have a certain degree of elasticity and under high shear rates tend to elongate, although they revert back to their common chain lengths when shear rate decreases to a nominal level. These fluids are referred to as viscoelastic fluids or muds with *shear-thinning* characteristics. That is, the mud becomes thin while flowing at high velocity through the bit nozzles and then thickens again while flowing in the annulus at considerably lower velocity.

Field measurements of viscosity

There are two routinely conducted field measurements of viscosity: qualitative measurement and quantitative measurement. The qualitative viscosity measurement, or the funnel viscosity, is a measurement used only to detect any variation in the viscous properties of the fluid. The Marsh funnel (fig. 2–11) is the instrument used for this purpose. The funnel viscosity is the time in seconds it takes for one quart of mud to flow out of the Marsh funnel. Note that the funnel viscosity of a mud system is dependent on the flow properties (e.g., PV and YP) of that mud, and not vice versa. If the flow properties of a given mud system are at the proper values, then the funnel viscosity will likewise be at its proper value. However, the corrective measures for adjusting the mud flow properties must not be made solely on the basis of the funnel viscosity measurement, because it is not an independent mud property.

The quantitative field measurements of a mud system are accomplished by utilizing a Fann V-G meter (fig. 2–12; for a detailed description and the recommended procedure, see API-RP-13B). The numerical values of the mud properties obtained using a V-G meter are the PV, the YP, and the 10 s to 10 min gel strengths. In terms of the viscometer dial reading, θ_N, and the rotational speed, N (in rpm), the mud properties are calculated using the following expressions:

For a Newtonian fluid ($\tau = \mu\gamma$):

$$\mu = 300\,\frac{\theta_N}{N} \tag{2.33}$$

where

μ is the Newtonian viscosity in centipoise
θ_N is the viscometer dial reading at any rotational speed N.

Fig. 2–11. Marsh funnel (Courtesy Fann Instrument Company)

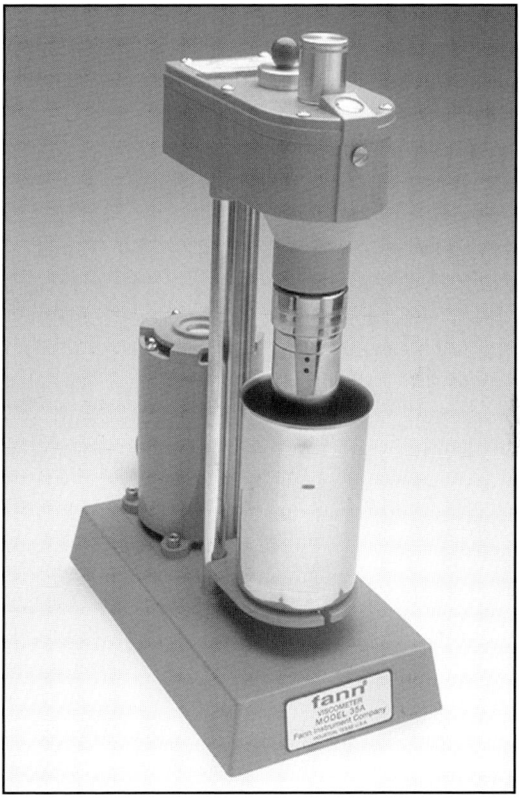

Fig. 2–12. Fann V-G Meter (Courtesy Fann Instrument Company)

For a non-Newtonian fluid,
– Using the Bingham plastic model ($\tau = \tau_y + \mu_p \gamma$),

$$\mu_p = 300 \frac{\theta_N}{N} - \frac{300\tau_y}{N} \tag{2.34}$$

For $N = 300, 600$ rpm,

$$\mu_p = \theta_{600} - \theta_{300} \tag{2.35a}$$

$$\tau_y = 2\theta_{300} - \theta_{600} \tag{2.35b}$$

where
μ_p = is the PV
θ_{600}, θ_{300} are the viscometer dial readings at $N = 600$ rpm and $N = 300$ rpm, respectively
τ_y is the YP in lbs/100 ft
τ_g is the gel strength in lbs/100 ft taken after mud has been thoroughly stirred and then kept undisturbed for 10 s and 10 min time at a rotational speed of three rpm.

– Using the power law fluid model ($\tau = K\gamma^n$),

$$n = 3.32\log_{10} \frac{\theta_{2N}}{\theta_N}$$

$$K = \frac{\theta_N}{\gamma_N^n} \tag{2.36}$$

where
n is the power law index
K is the consistency index

– Using the yield power law fluid model ($\tau = \tau_y + K\gamma^n$),

$$n = 3.32\log_{10} \frac{(\theta_{2N} - \theta_0)}{(\theta_N - \theta_0)} \tag{2.37}$$

$$K = \frac{(\theta_N - \theta_0}{\gamma_N^n} \tag{2.38}$$

where θ_0 denotes zero gel (lbs/100 ft).

The shear rate, γ (in s^{-1}), imparted to a sample of drilling mud by a variable speed viscometer is commonly expressed in terms of the rotational speed, N (in rpm), of the rotating sleeve (rotor). Based on the geometry of these viscometers, the rotational speed can be converted to shear rate by the following equation:

$$\gamma = 1.7N \tag{2.39}$$

Mud viscosity control. The control of the viscous properties (PV, YP, and gel strengths) of drilling muds within specified ranges is an essential prerequisite to the overall drilling cost control of a well. The control of mud viscosity is important so as to achieve the following benefits:

- Maintains an optimum hole cleaning
- Prevents settling of weight material
- Releases generated cuttings at the surface for easy removal
- Controls friction pressure losses in the circulating system
- Minimizes swab/surge pressures during tripping and casing operations

The viscous properties of mud systems are affected by:

- Size, shape, and amount of concentrations of total solids in the mud at a given temperature and pressure
- Flocculants or deflocculants added to the mud system
- Contaminants (cement, salt, anhydrite, gypsum, etc.) entering the mud during drilling

The continual monitoring of the funnel viscosity, whereby any abrupt change or gradual trend of change in viscosity can be detected, is a must. A comprehensive mud test can be conducted afterward to identify the cause and the required treatments that can be established.

Plastic viscosity control. As discussed previously, the PV is affected by the concentration, shape, and size of total solids in the mud system and by the presence of chemical contaminants. Therefore, the control of the PV can be accomplished by

- Dilution (not recommended if large amount of liquid is required)
- Effective solids removal equipment
- Addition of chemical flocculants at the flow line

Yield strength control. The magnitude of the electrochemical forces within the mud system under flow conditions is reflective of the mud system yield strength. These forces are due to the electrical charges on the submicron-size solid particles and the presence of the electrolytes in the mud water phase. Thus, control of the yield strength at the optimum desired level depends directly on the effective control of the drilled solids in the mud system and the proper chemical treatments.

In mud systems where drilled solids may be tolerated to some extent, the yield strength can be properly maintained by

- Solids removal
- Additions of chemical dispersants (lignites and lignosulfonates)
- Additions of chemical flocculants at flow line
- Additions of bentonite or a polymer

Whereas the first three of these mud treatments decrease the mud yield value, the fourth increases it.

In low-solids mud systems, where drilled solids buildup is not tolerated, the key to controlling the yield strength at the proper value is flow-line flocculation in conjunction with the use of the proper solids control equipment.

It may be concluded that to maintain the optimum flow properties, a mud system's yield value must be properly maintained within an exact range specified in accordance with the hole conditions and the type of mud being used. Effective solids control is the most significant factor in controlling the yield value of a mud system, minimizing treatment requirements and enhancing the efficiency of the drilling operation.

Gel strength control. The gel strength is the measurement of attractive (electrochemical) forces within the mud system under no-flow conditions. It differs from the yield value because of its time dependence and is broken up after flow is limited. As with other properties, the ability to maintain the proper value of gel strength depends on effective solids control. Although the type of treatment used to control the gel strength will depend on the mud system in use, the primary means of control is flow-line flocculation and minimal treatments with viscosifiers.

In summary, the following generalizations can be made about the economical control of mud flow properties at optimum conditions:

- An increase in yield value accompanied by small or no change in PV can be controlled by the addition of chemicals only.
- An increase in PV accompanied by small or no change in yield value can be controlled by solids removal or addition of water.
- A simultaneous increase in yield value and PV can be controlled by addition of a combination of chemicals, water dilution, and solids removal.

Effective circulating viscosity. Most drilling muds commonly used in the field exhibit viscous properties which are shear rate dependent. During normal drilling, the mud being circulated experiences different velocities in the various sections of the circulating system, ranging from practically 0 ft/s in the pits to over 300 ft/s across the jets of the drill bit. These wide-ranging velocities give rise to mud shear rates of less than 5 s in mud pits to over 100,000 s across the bit jets. The effective circulating viscosity, therefore, can be defined as the viscosity of the mud at a given shear rate in a particular section of the circulating system under given conditions of pressure and temperature.

From the previous discussion, it is clear that the standard mud viscosity measurements do not define the exact mud viscosity at any of the shear rate ranges that are most important in efficient drilling. For example, knowing and controlling mud viscosities at the bit jets allows the improvement of the drilling rates and annular hole cleaning and, in the mud pits, the achievement of efficient solids removal. The effective mud viscosities in the shear rate ranges encountered in the annulus can be determined using the following equation:

$$\mu_e = 47{,}913K\left(\frac{v_a}{D_h - D_p}\right)^{n-1} \tag{2.40}$$

where

μ_e is the effective viscosity (in centipoise)
K is the consistency index
n is the power law index
D_h, D_p are the hole and outer pipe diameters, respectively
v_a is the annular fluid velocity

Measurement of solids

The purpose of solids measurement is to determine the amount of overall solids content, the sand content, and the reactive bentonite content in a mud system. The sand content is a measurement of the volume fraction of particles that are retained in a 200-mesh screen by using the sample mud and must be taken at the flow line. (E.g., see fig. 2–13.)

The mud retort is used to determine the volume fraction of oil, water, and total solids that make up the drilling mud. (E.g., see fig. 2–14; the procedure for operating the instruments is given in API-RP-13b, pp. 12–14.)

The methylene blue test (MBT) is conducted to determine the amount of reactive bentonite-type solids in the drilling mud. It is used in conjunction with retort data, mud weight, and filtrate analysis in evaluating the respective amounts of each type of solids in the drilling mud. As a diagnostic tool, it can isolate the cause of viscosity problems and identify the causes of high fluid loss. The cation exchange capacity of solids particles forms the basis of the test. (E.g., see fig. 2–15; the standard API procedure for the MBT test is given in API-RP-13b, pp. 21–22.)

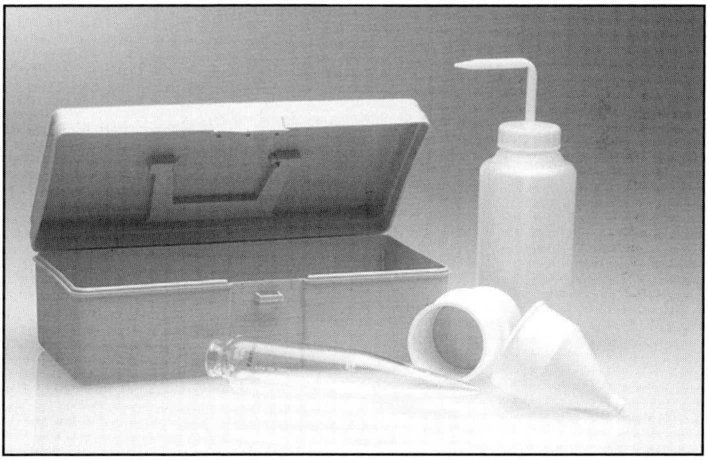

Fig. 2–13. Sand content kit (Courtesy Fann Instrument Company)

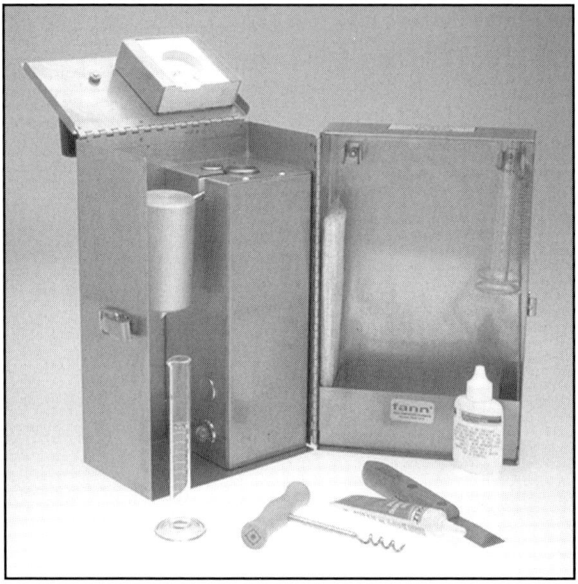

Fig. 2–14. Oil and water retort kit (Courtesy Fann Instrument Company)

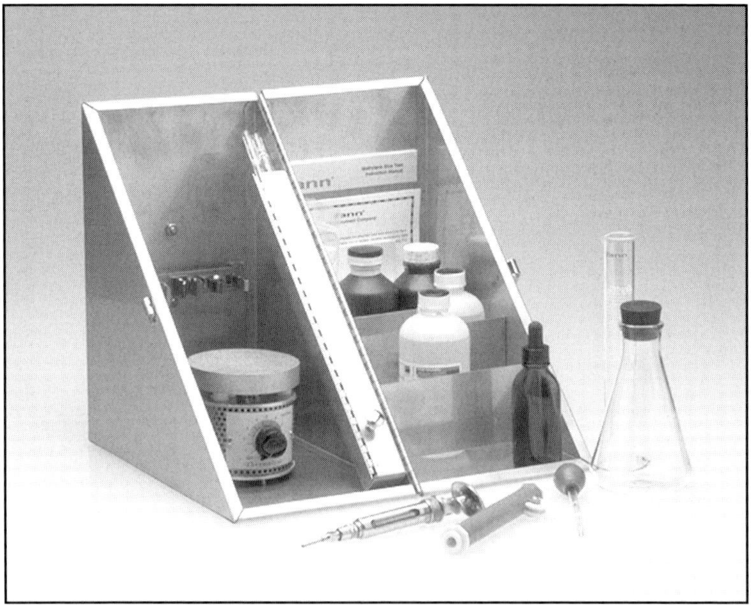

Fig. 2–15. MBT kit (Courtesy Fann Instrument Company)

Solids control. The need for solids control in mud systems has been well accepted in drilling operations. Some of the major proven advantages of low-drilled-solids content in muds are lower viscosity, improved mud cakes, longer bit and pump life, and higher drilling rates. The ability to control drilled solids in muds is governed by the selection of the mud system itself and the selection of the rig, by the control of drilling rates in various sections of hole, and by the solids removal equipment and techniques employed in the field.

The following mechanical solids removal equipment can be used in the field:

- Shale shakers
- Desanders
- Desilters
- Centrifuges
- Mud cleaners

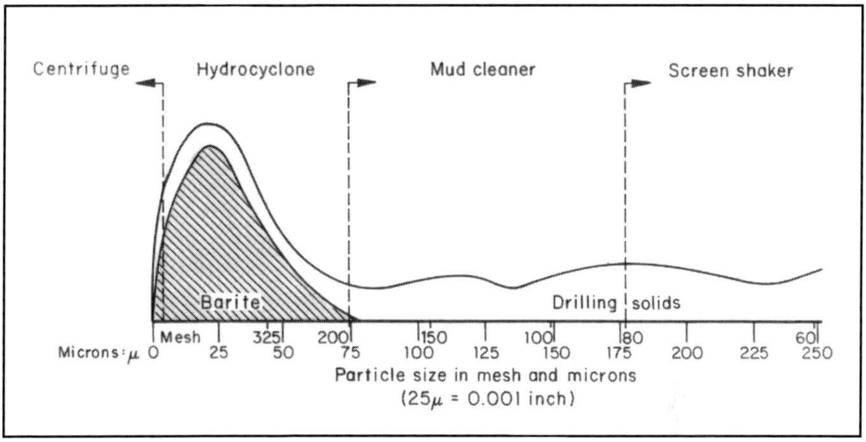

Fig. 2–16. Effectiveness of separation equipment in the removal of solids from a drilling fluid

These pieces of equipments are designed to remove solids in various ranges of the particle size spectrum, as shown in figure 2–16. Except for the screen shakers, all mechanical solids removal is based on the physical law of settling of solids—that is,

$$v_{sp} = \frac{(\gamma_p - \gamma_m)d_p^2}{\mu} \tag{2.44}$$

where

v_{sp} is the particle settling velocity
γ_p, γ_m are the particle and mud densities, respectively
d_p is the particle diameter of largest dimension
μ is the mud viscosity
γ, W, V are the weight density, weight, and volume, respectively

Drilling-Mud Additives

Drilling-mud additives may be grouped into seven categories:

1. Viscosifiers:
   - Bentonite
   - Attapulgite
   - Polymers
2. Viscosity reducers:
   - Phosphates
   - Tannates
   - Lignites
   - Lignosulfonates
   - Sodium polyacrylates
3. Weighting materials:
   - Barite
   - Iron oxide
   - Calcium carbonate
   - Dissolved salts
   - Galena
4. Fluid-loss reducers:
   - Bentonite
   - Starch
   - Polymers
5. Emulsifiers
6. Lost circulation materials:
   - Walnut shells
   - Fibrous
   - Cellophane flakes
   - Diesel/bentonite
7. Special additives:
   - Flocculants
   - Corrosion control chemicals
   - Defoamers
   - pH control additives

The following sections provide a brief description of some of these additives and their practical limitations in drilling muds.

Bentonite

Bentonite, a sodium montmorillonite, is primarily used as a viscosifier and as a fluid-loss reducer in freshwater-based muds. Viscosity is imparted to the freshwater by the hydration of bentonite particles. In saltwater-based muds, bentonite can be first hydrated in freshwater and then added to the saltwater, to control viscosity and fluid loss.

Attapulgite

Attapulgite, a hydrous magnesium aluminum silicate, is used as a viscosifier in saltwater-based muds. Because of its needlelike particle structure, attapulgite does not offer adequate fluid-loss control.

Polymers

Polymers are recent advances in drilling-mud additives. Some commonly used polymers are listed in table 2–2.

Phosphates

Listed below are some of the commonly used phosphates for viscosity-reducing agents, all of which are limited to a temperature environment of 150°F:

- SAPP (sodium acid pyrophosphate)
- STP (sodium tetraphosphate)
- TSPP (tetrasodium pyrophosphate)
- SHMP (sodium hexametaphosphate)

Table 2.2. Polymers commonly used in drilling mud

Polymer	Primary Function	Temperature Limitation (°F)
X-C (Xanthum Gum)	Viscosifier	250°
Drispac	Fluid loss reducer	300°
CMC	Fluid loss reducer	250°
HEC (hydroxyethylcellulose)	Viscosifier	250°
(X–C + HEC)	Viscosifier	250°
Ben-Ex (copolymer)	Bentonite extender & Flocculants	250°
Lo-Sol (polyacrylamide)	Bentonite extender	250°
PHPA	Viscosifier	250°

Tannates

Quebracho and hemlock trees are two extract sources of tannin. Both additives are effective in the reduction of mud viscosity in a temperature environment not exceeding 250°F.

Lignites

Lignites are humic acids, a degradation product of decayed vegetation. They are used as thinners, emulsifiers, and fluid-loss reducers and are temperature stable up to 375°F.

Lignosulfonates

Lignosulfonates (calcium, modified sodium, etc.) act as dispersants and are used to reduce viscosity and gel strength and to provide some fluid-loss control.

Sodium polyacrylates

These additives are used in low-solids nondispersed muds to control yield strength, gel strength, and fluid loss. They are temperature stable up to 350°F. These additives become ineffective for viscosity reduction if drilled-solids concentration is too high or if the calcium content is greater than 200 ppm.

Barite

Barite, a barium sulfate, occurs as a natural mineral. It has a specific gravity range of 4.2–4.6. Mud weights up to 20 lbs/gal can be obtained with barite. It is the most commonly used weighting material.

Iron oxide

Iron oxide has a specific gravity of 4.9–5.3, and mud weights up to 24 lbs/gal may be obtained. It is known to have a detrimental effect on fluid loss and is quite erosive. It is less expensive and more abundant than barite.

Calcium carbonate

Calcium carbonate is a limestone that has a specific gravity of 2.7. It is primarily used to obtain mud weights up to 11 lbs/gal in oil-based muds and workover fluids.

Dissolved salts

Sodium chloride, calcium chloride, and calcium chloride/calcium bromide are some of the salts used to weight up drilling and workover fluids. Sodium chloride is used to drill salt sections, and mud weights up to 10 lbs/gal can be obtained. Calcium chloride is used mainly for workover fluids, and mud weights up to 12 lbs/gal can be obtained. Calcium chloride/calcium bromide is used for workover fluids with mud weights ranging between 12 lbs/gal and 15 lbs/gal.

Supplementary Problems

Problem 2–1. Using viscometer-related equations, determine the PV and yield values for a Bingham mud. The following data are given:

Fann V-G Meter Dial Reading	Corresponding Shear Rate (s)
45	340
65	510

Problem 2–2. A kick was taken as indicated in figure 2–17 while drilling with 12 lbs/gal mud. The shut-in drill pipe and casing pressures were recorded at 800 psi and 1,150 psi, respectively. Determine the cost of additional barite required in order to drill the encountered high-pressure zone at a differential pressure of 500 psi. Barite weight: 36 lbs/gal; barite cost: $8.50/100 lbs.

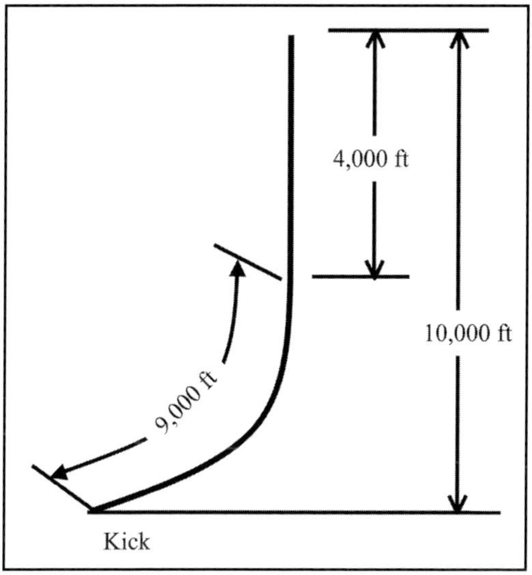

Fig. 2–17. Kick encountered in problem 2–2

Problem 2.3 You are given the following well data:

- Intermediate casing set at depth = 10,000 ft
- Drill collar length = 1,200 ft
- Maximum anticipated induced friction pressure losses in annulus during normal drilling:
  - Δp_{fadc}= 0.05 psi/ft (behind drill collar)
  - Δp_{fadp} = 0.012 psi/ft (behind drill pipe)
- Maximum anticipated surge and swab pressures during tripping:
  - Δp_{swab} = 0.02 psi/ft
  - Δp_{surge} = 0.02 psi/ft
- Required differential pressure = 400 psi

Select the required mud weight to drill the interval, from 10,000 to 14,000 ft. Under the anticipated conditions (see fig. 2–18), show whether there will or will not be problems that could cause possible well blowouts.

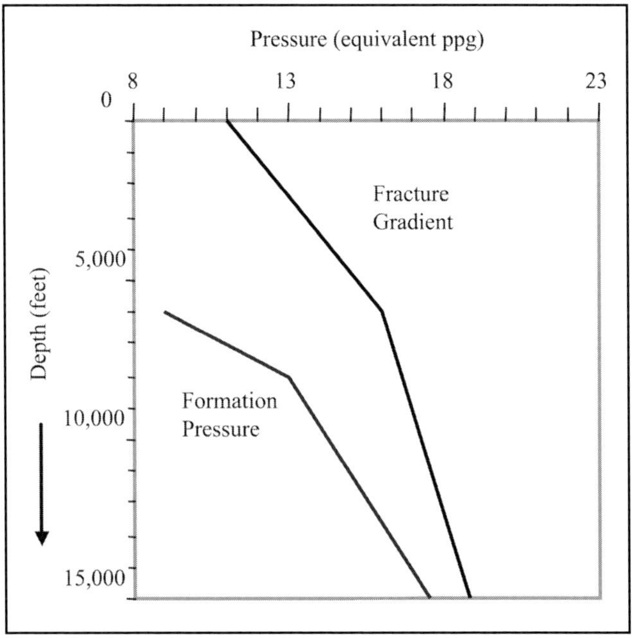

Fig. 2–18. Drilling conditions in problem 2–3

Problem 2–4. Find the value for k in the equation $P = k\gamma_m D$ if P is given in psi, γ_m is the specific gravity, and D is given in ft.

Problem 2–5.

(a) List the functions of mist drilling fluid.
(b) List at least three types of inhibitive mud systems.
(c) List drilling-fluids selection criteria.
(d) When can dry-air drilling be used?

Problem 2–6. It is desired to mix a volume V_f of mud with weight density γ_m. The mud composition is as follows: liquid (V, γ), barite (V_b, γ_b), and bentonite (V_c, γ_c). Weight fraction of barite in added solids is denoted a. Develop (derive) the equations that will allow you to make up the mud system.

Problem 2–7. Kick was taken in the well shown in figure 2–19. The BOP stack was closed. Shut-in casing pressure is 900 psi, and shut-in drill pipe pressure is zero. Explain why.

Problem 2–8. Determine the amount (in lbs/bbl) of weighing material with specific gravity of 5.2 that must be added to a mud system to increase its pressure gradient from 0.52 psi/ft to 0.624 psi/ft.

Problem 2–9. It is desired to mix X barrels of γ_m lbs/gal water-based mud. Bentonite (weight: γ_a, in lbs/gal), and barite (weight: γ_b, in lbs/gal) are to be added in the ratio of a lbs of bentonite to b lbs of barite. Determine the amounts of water (V_w), bentonite (W_c), and barite (W_b) needed to make up the mud system.

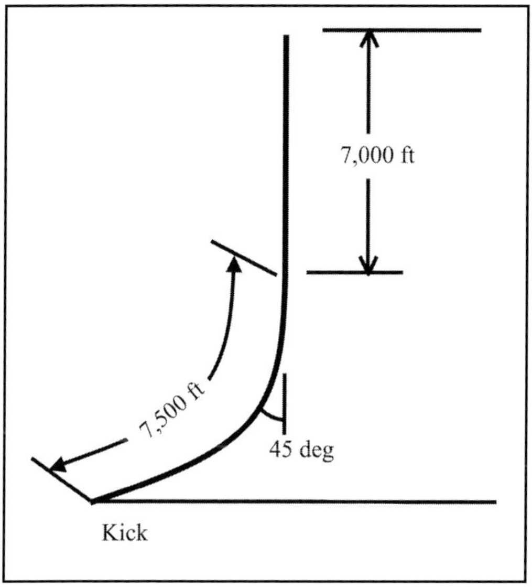

Fig. 2–19. Kick encountered in problems 2–7 and 2–15

Problem 2–10. In the dilution of mud with a liquid, it is necessary to know the resulting mud density for a given amount of liquid added. On the basis of the weight and volume balance concepts of materials, derive an expression for the resulting density of the diluted mud. Show the derivations.

Problem 2–11. You are given the following data:

rpm	Fann V-G Meter Dial Readings
N_i	θ_i
N_j	θ_j

Determine expressions for the flow properties of a Newtonian fluid and a Bingham fluid in terms of N and θ.

Problem 2–12. The spurt loss of a mud is known to be 2 cc. If a filtrate of 10 cc is collected in 15 min using a filter press, what is the standard API filtrate for this mud?

Problem 2–13. It is desired to increase a mud weight from 12 lbs/gal to 0.78 psi/ft using barite having a specific gravity of 4.4. Determine the cost owing to mud weight increase, assuming barite cost is $101/100 lbs.

Problem 2–14. List the following:

(a) Seven criteria used in the selection of drilling fluids.
(b) The three basic types of drilling fluids.
(c) Functions of dry-air drilling fluids.
(d) The composition of drilling muds.

Problem 2–15. A kick was taken at the indicated depth for the well shown in figure 2–19 while drilling with 12 lbs/gal mud The recorded shut-in casing and shut-in drill pipe pressures were 600 psi and 800 psi, respectively. How much barite must be added to bring the well under control and maintain a 400 psi differential pressure while drilling the interval?

Nomenclature

A Filtration area
D Depth
D_h Hole diameter
d_p Particle diameter of largest dimension
D_p Outer pipe diameter
F Force
f_{vc} Volume fraction of solids in cake
f_{vf} Volume fraction of low specific gravity solids in final mixture
f_{vo} Volume fraction of low specific gravity solids in original liquid
f_w Volume percent of water, expressed as a fraction
h Cake thickness
k Cake permeability
K Consistency index
n Power law index
P_{df} Drilling-fluid pressure
P_f Annulus friction loss, in psi
P_{ff} Formation fluid pressure
P_h Hydrostatic pressure
Q Flow rate, in gpm
Q_f Amount of filtrate volume
Q_m Total volume of mud filtered
Q_s Volume of solids deposited in mud cake
ΔP Differential pressure
R Penetration rate (ROP)
t Filtration time
v_a Annular fluid velocity
V_a Added material weight
V_f Final mixture volume
V_o Original liquid volume
v_{sp} Particle settling velocity
W_a Added material weight

W_f Final mixture weight
W_o Original liquid weight
W_{wm} Required amount of weighting material, in lbs/bbl, of original mud
γ Shear rate
γ_a Added weight
γ_f Final mud weight
$\gamma_{m,av.}$ Average annular mud weight, in lbs/gal
γ_p Particle density
γ_{ps} Measured mud weight at pump suction, in lbs/gal
θ_n Viscometer dial reading at any rotational speed N
μ Filtrate viscosity; also Newtonian viscosity, in centipoise
μ_e Effective viscosity
μ_p Bingham PV
τ_y Yield value or yield stress
τ_g Gel strength

3

Fluid Flow and Associated Pressures in the Rotary Rig Circulating System

Introduction

In contrast to static fluid columns, the determination of pressures while the fluid is moving is more complex. Because most drilling fluids are non-Newtonian, the mathematical treatment of the problem is even more difficult to formulate. However, in spite of all the complexities, the calculations of friction pressure losses in the rotary rig circulating system are important and essential because of their direct bearing on

- Drill bit hydraulic program design
- ECD during tripping in and out of wellbore
- ECD during drilling and well control operations

Drill bit hydraulic program design

Pump operating requirements, optimum flow rates, and corresponding optimum drill bit nozzle sizes are necessary calculations to ensure optimum drilling conditions.

In drilling, pump pressure (P_p) required in order to drill to a certain depth for a given set of operating conditions is equal to the total friction pressure (P_f) losses in the circulating system plus the dynamic pressure changes (generally those across jet bits [P_B])—that is,

$$P_p = P_f + P_B \tag{3.1}$$

where

P_p is the required pump delivery pressure
P_f is the circulating-system friction pressure loss
P_B is the dynamic pressure across bit nozzles

Pressure changes during tripping and casing operations

During *tripping out* (upward movement of pipe in a fluid-filled wellbore), a decrease of pressure in the annulus, referred to as the *swab* pressure (P_{asw}), is experienced. This effectively reduces the hydrostatic head of the mud in use to a value that can be less than what the formation fluid pressure is, thus leading to a kick.

Mathematically,

$$P_{aei} = P_{ahi} - D_i \Delta P_{a\,swab} \tag{3.2}$$

where

P_{aei} is the equivalent mud pressure at some well depth D_i in the annulus
P_{ahi} is the mud hydrostatic pressure at D_i
$\Delta P_{a\ swab}$ is the swab pressure gradient in the annulus (in psi/ft)

or in terms of mud weight,

$$\gamma_{me} = \gamma_{mh} - \Delta\gamma_{a\ swab} \tag{3.3}$$

where

γ_{me} is referred to as the equivalent circulating mud weight
γ_{mh} is the actual mud weight while not moving
$\Delta\gamma_{a\ swab}$ is the change in mud weight due to the swab pressure

For safe drilling, $\gamma_{me} > \gamma_{ff}$, where γ_{ff} is the formation pressure gradient.

Likewise, during *tripping in* (downward movement of pipe in a fluid-filled wellbore), an increase in annular pressure, referred to as the *surge* pressure (P_{surge}), is experienced. This increases the mud hydrostatic head to a value that may exceed the fracture gradient of the formation at some depth D_i.

Mathematically,

$$P_{aei} = P_{ahi} + D_i\,\Delta P_{a\ surge} \tag{3.4}$$

or

$$\gamma_{me} = \gamma_{mh} + \Delta\gamma_{a\ surge} \tag{3.5}$$

Again, for safe drilling, $\gamma_{me} < \gamma_{frac}$, where γ_{frac} is the formation fracture gradient.

Pressure changes during drilling

During drilling, the friction pressure losses in the annulus will effectively increase the mud weight, resulting in an ECD that, as in the case of surge pressures, may cause fracturing of formation: $\gamma_{me} = \gamma_{mh} + \Delta\gamma_{af}$, where $\Delta\gamma_{af}$ is the change in mud weight in annulus owing to the friction pressure loss gradient in the annulus. For safe drilling, $\gamma_{me} < \gamma_{frac}$. A similar situation may be encountered during well control operations.

The purpose of this chapter is to present developments and equations in regard to the calculation of pressure changes throughout the circulating system for noncompressible drilling fluids only. More specifically, the pressure losses due to friction in the surface connections, inside the drill string and in the annular space, and the dynamic pressure drop across the jets of the drill bit will be treated.

Example 3–1

You are given the following well data:

- Last casing setting = 14,000 ft total measured depth (TMD)
- Target depth (TD) = 13,500 ft true vertical depth (TVD)
- Hole size to TD: 7 $^7/_8$ in.
- Drill pipe:
  - OD = 4$^1/_4$ in.
  - ID = 3$^1/_2$ in.
  - Air weight = 26.88 lbs/ft
- Drill collar:
  - OD = 6 in.
  - ID = 3$^1/_2$ in.
  - Air weight = 100.8 lbs/ft
- Pumps: 2,000 hp; E_v = 85%; P_{max} = 3,800 psi
- Minimum annular fluid velocity for hole cleaning = 120 ft/min
- Coefficients of friction at walls: 0.1, cased sections; 0.2, open hole section
- Mud:
  - Bingham plastic; specific gravity = 2.0 for next depth interval
  - $\theta_{600} = 55$; $\theta_{300} = 35$; 10 s/10 min; gel = 18/30 lbs/100 ft^2 (Fann V-G meter)
- Formation below 14,000 ft TMD:
  - Fracture gradient = 18.0 lbs/gal
  - Pore pressure gradient = 0.83 psi/ft
- Friction pressure losses while drilling at 14,000 ft TMD:
  - $\Delta P_{fdp} = \Delta_{fdc}$ = 0.075 psi/ft
  - ΔP_{fadp} = 0.01 psi/ft
  - ΔP_{fadc} = 0.08 psi/ft;
- During tripping from 14,000 ft TMD:

  $\Delta P_{surge} = \Delta P_{swab}$ = 0.025 psi/ft (annulus)
- Maximum 56 kips of Weight on bit is allowed. If not, the L_{dc} can be assumed to be 1500 ft.

  (a) Explain, in detail, why any drilling problem may or may not occur while drilling at or tripping from 14,000 ft TMD.

  (b) How much pressure is available to be utilized across the bit while drilling at 14,000 ft TMD, under the specified conditions (also see fig. 3–1)? Assume a surface connections length of 800 ft of drill pipe.

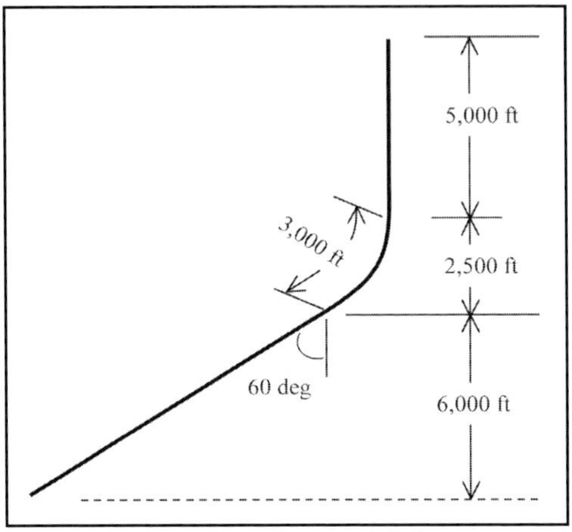

Fig. 3–1. Drilling conditions in example 3–1

Solution:

(a) During drilling, the concern is formation fracture at the casing shoe (14,000 ft TMD):

$$\gamma_e = \gamma_m + \Delta\gamma_{af}$$

for

$$\gamma_m = 2 \text{ sp.gr} \approx 16.66 \text{ lbs/gal}$$

Length of the drillcollar can be calculated as below:

$$L_{fCD} = \frac{\text{Design Factor} \times \text{WOB}}{W_{dc} \times \text{BF} \times \cos\theta} = \frac{1 \times 56}{100.8 \times 0.746 \times \cos 60} \approx 1500\text{ft}$$

$$\Delta\gamma_{af} = \frac{L_{dp}\Delta P_{gadp} + L_{dc}\Delta P_{fadc}}{k(\text{TVD})}$$

$$= \frac{(0.01 \quad 12{,}500) + (0.08 \times 1{,}500)}{0.052 \times 10{,}500}$$

$$= 0.449 \text{ lbs/gal}$$

$$\gamma_e = 16.66 - 0.641 = 16.02 \text{ lbs/gal}$$
$$= 0.833 \text{ psi/ft}$$
$$> 0.83 \text{ psi}$$

During tripping, there are three concerns:

1. Kick due to swab pressure:

$$\gamma_e = \gamma_m - \Delta\gamma_{swab} \text{ (annulus)}$$

for

$$\gamma_m = 16.66 \text{ lbs/gal}$$

$$\Delta\gamma_{swab} = \frac{\Delta P_{fa\ swab}\,(TMD)}{k(TVD)}$$

$$= \frac{0.025 \times 14{,}000}{0.052 \times 10{,}500}$$

$$= 0.641 \text{ lbs/gal}$$

$$\gamma_e = 16.66 - 0.641 \approx 16.02 \text{ lbs/gal}$$
$$= 0.833 \text{ psi/ft}$$
$$> 0.83 \text{ psi}$$

2. Fracturing due to surge pressure:

$$\gamma_e = \gamma_m + \Delta\gamma_{surge}$$

$$= 16.66 + \frac{0.025 \times 14{,}000}{0.52 \times 10{,}500}$$

$$= 17.3 < 18.0 \text{ lbs/gal} \therefore \text{ No fracture occurs}$$

3. Breaking circulation owing to fracture (explained later):

$$\Delta P_{frac} = \frac{\tau_g}{300(D_h - D_{op})}$$

$$= \frac{30}{300(7.875 - 6)}$$

$$= 0.053 \text{ psi/ft}$$

(b) The total pump pressure required in order to drill at a given depth for the specified set of operating conditions is

$$P_p = P_f + P_b = P_{fdb} + P_{fdc} + P_{fadp} + P_{fadc} + P_{fs} + P_b$$

$$P_f = L_{dp}\Delta P_{fdp} + L_{dc}\Delta P_{fdc} + L_{dp}\Delta P_{fadp} + L_{dc}\Delta P_{fadc} + L_s\Delta P_{fbp}$$

where

$$\Delta P_{fdp} = \Delta P_{fdc} = 0.075 \text{ psi/ft (given)}$$

$$\Delta P_{fadc} = 0.08 \text{ psi/ft, and } \Delta P_{fadp} = 0.01 \text{ psi/ft}$$

$$L_s = 12{,}500 \text{ ft, psi/ft}, L_{dc} = 1{,}500 \text{ ft, and } L_s = 800 \text{ ft (given)}$$

$$P_f = (0.75 \times 12{,}500) + (0.075 \times 1{,}500) + (0.01 \times 12{,}500) + (0.081 \times 1{,}500) + (0.075 \times 800)$$
$$= 1{,}355 \text{ psi}$$

$$P_{p,\,max} = 3{,}800 \text{ psi (given)}$$

P_B is the bit pressure drop that can be made available
$P_{p,max-P_f} = 3{,}800 - 1{,}355 = 2{,}445$ psi

Mechanical Energy and Pressure Balance

The study of fluid dynamics is based on three physical laws:

- Conservation of energy
- Conservation of momentum
- Conservation of mass

These laws, when combined with the fluid rheological models (Newtonian, Bingham, power law, and yield power law), the fluid state (compressible or incompressible), fluid-flow regime (laminar or turbulent), and the conduit type (pipe, annular or slot flow), constitute all the conditions required in order to formulate a fluid dynamic problem.

The mechanical energy balance equation for an incompressible fluid entering a physical system at point *i* and leaving it at some point *j* (fig. 3–2) can be written as

$$\gamma(D_j - D_i) + \frac{\gamma}{2g}(\bar{v}_j - \bar{v}_i) + P_j - P_i = W_p - W_f \tag{3.6}$$

where

D is the elevation
γ is the fluid weight density
$\bar{v}$ is the average fluid velocity
P is the pressure
W_p is the work done by pump
W_f is the friction energy loss

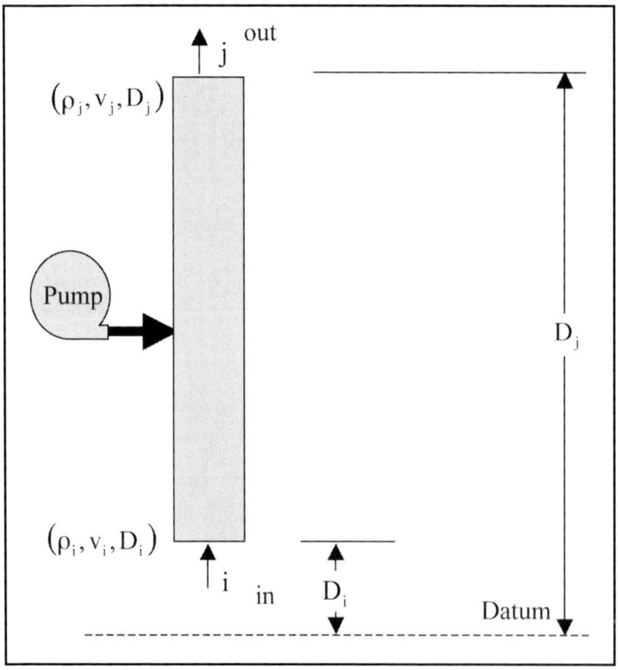

Fig. 3–2. Physical system

All the terms in equation (3.1) have the dimensions of energy per unit volume (i.e., FL/L^3 [where F denotes force and L denotes length], the units in which pressure is expressed).

Hence, equation (3.6) can be written as a pressure balance equation:

$$P_h + P_d + P_j - P_i = P_p - P_f \tag{3.7}$$

where

P_h is the the hydrostatic head (pressure)
P_d is the dynamic pressure
P_p is the pump pressure
P_f is the pressure loss due to friction

In drilling, the circulation system can be visualized as a U-tube (fig. 3–3) in which $D_i = D_j$, $P_i = P_j$, and the only dynamic pressure of great significance is that across the jets of the bit, P_b. Therefore, equation (3.7) is simplified as

$$P_b + P_f = P_p \tag{3.8}$$

Pressure due to friction in the circulation system, P_f, can be written as follows:

$$P_f = P_{fdp} + P_{fdc} + P_{fadp} + P_{fadc} + P_{fs} \tag{3.9}$$

where

P_{fdp}, P_{fdc} are the respective pressure friction losses inside drill pipe and drill collar
P_{fadp}, P_{fadc} are the respective pressure friction losses in the annulus around drill pipe and drill collar, respectively
P_{fs} is the pressure friction loss inside surface connections
P_b is the dynamic pressure change across the bit

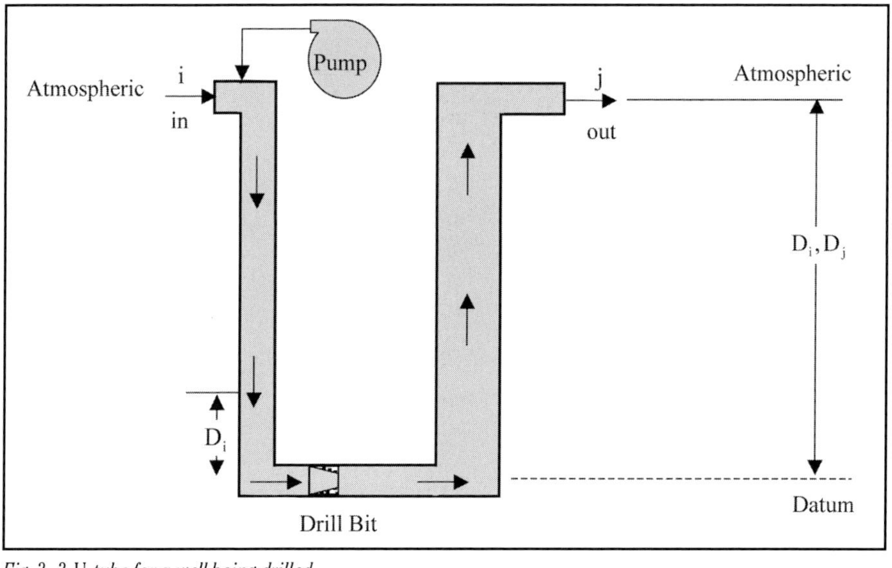

Fig. 3–3. U-tube for a well being drilled

On the basis of equation (3.8), it becomes clear that the pump pressure required is that which is needed to overcome the frictional forces throughout the circulating system and to accelerate the fluid across the drill bit nozzles (*jets*) from a comparatively negligible nozzle entrance velocity to an extremely high exit velocity, in excess of 300 ft/s.

Pressure Drop across the Bit Nozzles (Jets)

If two reference sections, *i* and *j*, are located just at the entrance and the exit of the bit jets, as shown in figure 3–4, then equation (3.7) can be reduced to

$$P_h + P_b + P_j - P_i = -P_f \tag{3.10}$$

where P_p was set to zero, since there is no pump between the two stations considered (*i* and *j*).

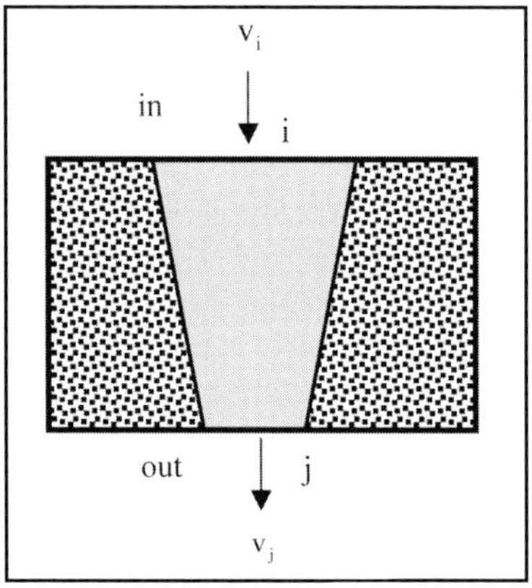

Fig. 3–4. Bit nozzle

Also, it can be assumed that the pressure loss due to friction, P_f, and the pressure due to elevation can be ignored in comparison to the dynamic pressure, P_b. Therefore, equation (3.10) can be rewritten as

$$P_i - P_j = \Delta P = P_b = \frac{\gamma}{2g}\left(\bar{v}_j^2 - \bar{v}_i^2\right) \tag{3.11}$$

However, v_i is very small compared to v_j and is neglected; therefore, the pressure drop across the jets becomes

$$P_b = \frac{\gamma}{2g}\bar{v}_j^2 = \frac{\gamma}{2g}\bar{v}_{exit}^2 \tag{3.12}$$

Solving for the jet exit velocity,

$$\bar{v}_{exit} = \sqrt{\frac{2g}{\gamma} P_b} \tag{3.13a}$$

The actual exit velocity of the jets is always smaller than will be predicted by equation (3.14), owing to the assumption made regarding the frictionless state of the jets. To compensate for this difference, a modifying factor called the nozzle (jet) discharge coefficient, C_d, which will be implemented in equation (3.14), is used to give

$$\bar{v}_{exit} = C_d \sqrt{\frac{2g}{\gamma} P_b} \tag{3.13b}$$

The value of C_d, is assumed to be 0.95 unless otherwise specified.

If the total flow rate across the jets is assumed to be Q and the total area of jets open to flow is A_t, then $\bar{v}_{exit} = Q/A_t$. Substituting into equation (3.13b) and solving for the pressure drop across the bit yields

$$P_b = \frac{\gamma Q^2}{2gA_t^2 C_d^2} \tag{3.14}$$

Equation (3.14) becomes

$$P_b = \frac{(8.3 \times 10^{-5})\gamma Q^2}{C_d^2 A_t^2} \tag{3.15}$$

when the following field units are used:

Q is in gpm
γ is in lbs/gal
A_t is in in.^{2}
C_d is a nondimensional term
P_b is in psi

Example 3–2

A 10 lbs/gal drilling mud is flowing at 500 gpm through a jet bit. Compare the hydraulic horsepower expended across the bit for each of the following nozzle sets:

- 9-9-9
- 11-11-11
- 13-13-13

Note that nozzle sizes (diameters) are expressed in 32nds of an inch (e.g., 9-9-9 means three nozzles each with a diameter of 9⁄32 of an inch). Assume $C_d = 0.95$ for all nozzle sets.

From equation (1.15), bit hydraulic horsepower (BHH) can be expressed as follows:

$$\text{BHH} = \frac{P_b Q}{1{,}714}$$

The following values are substituted into equation (3.15):

- $Q = 500$ gpm
- $\gamma = 10$ lbs/gal
- $C_d = 0.95$
- $A_t = 3\left(\frac{\pi d^2}{4}\right) = 3\left(\frac{\pi}{4}\right)\left(\frac{13}{32}\right)^2 = 0.389 \text{ in.}^2$

This gives

$$P_b = \frac{(8.3 \times 10^{-5})(500)^2(10)}{(0.95)^2(0.389)^2} = 1{,}519 \text{ psi}$$

Therefore,

$$\text{BHH} = \begin{array}{ll} (1{,}519 \times 500)/1{,}714 = 443 \text{ hp} & \text{for 13-13-13 nozzles} \\ (2{,}975 \times 500)/1{,}714 = 859 \text{ hp} & \text{for 11-11-11 nozzles} \\ (6{,}645 \times 500)/1{,}714 = 1{,}931 \text{ hp} & \text{for 9-9-9 nozzles} \end{array}$$

Friction Pressure Losses in the Rotary Rig Circulating System

Excluding the drill bit, which has been discussed already, the flow through the circulating system can be divided into (1) pipe flow (flow inside the drill string and surface connections) and (2) annular flow (flow in the annular space between the wellbore walls and the outer walls of the drill string). Depending on the fluid velocity, fluid flow is characterized as laminar or turbulent. In the former, the fluid particle velocity (fluid velocity profile) can be mathematically defined; therefore, mathematical relationships for friction pressure losses can be formulated, as will be demonstrated in this section. By contrast, in the latter, the velocity profile cannot be mathematically defined; therefore, the friction pressure loss formulations are strictly empirical.

Laminar flow is a streamline flow in which all fluid particles move along lines parallel to the axis of the conduit and adjacent fluid layers slip past each other with no mixing of particles. In steady-state conditions, inside circular conduits, laminar flow can be visualized as a series of concentric cylinders slipping past one another, as shown in figure 3–5. The layer adjacent to the walls is stationary. As the fluid velocity becomes higher, the fluid particles move in a very disorderly, tumbling motion, in what is termed turbulent flow.

The *Reynolds* criterion is generally used to determine whether a flow regime is laminar or turbulent. For Newtonian pipe flow under ideal conditions, it has been established experimentally that the critical Reynolds number—that is, the number above which flow is turbulent—has a value approximately equal to 2,100. Mathematically, the Reynolds number of a pipe, N_{Rp}, is given by

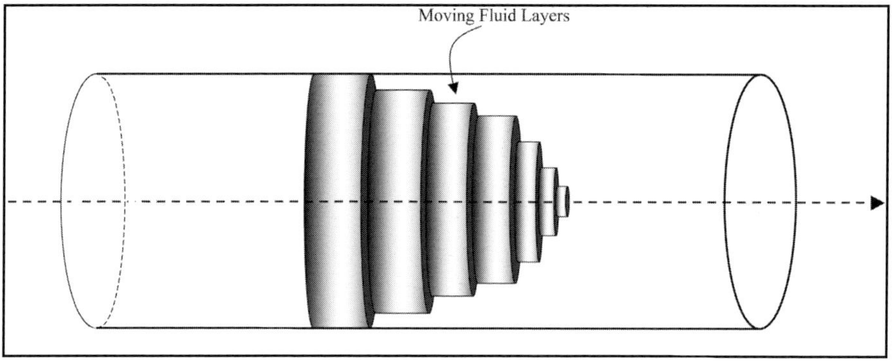

Fig. 3–5. Concentric layers of fluid slipping past one another

$$N_{Rp} = \frac{\rho \bar{v}_p D_{ip}}{\mu} \quad (3.16)$$

where

ρ is the fluid mass density
$\bar{v}_p$ is the average fluid velocity
D_{ip} is the ID of a pipe
μ is the Newtonian fluid viscosity

Therefore, if the Reynolds number for pipe flow calculated by equation (3.16) is less than 2,100, the flow is laminar flow; otherwise it is turbulent.

If field units are used, equation (3.16) becomes

$$N_{Rp} = \frac{928 \gamma \bar{v}_p D_{ip}}{\mu} \quad (3.17)$$

where

γ is in lbs/gal
$\bar{v}_p$ is in ft/s
D_{ip} is in in.
μ is in centipoise

For annular flow, equation (3.17) is rewritten as

$$N_{Ra} = \frac{928 \gamma \bar{v}_a D_{ep}}{\mu} \quad (3.18)$$

where

$\bar{v}_a$ is the average annular fluid velocity
D_{ep} is the annulus equivalent pipe diameter

The following correlations for calculating the equivalent diameter, D_{ep}, have been presented in the literature:[1]

$$D_{ep} = \sqrt{\frac{D_h^2 + D_{op}^2 - \left(D_h^2 - D_{op}^2\right)}{\ln\left(D_h / D_{op}\right)}} \qquad (3.19a)$$

$$D_{ep} = 0.816(D_h - D_{op}) \qquad (3.19b)$$

$$D_{ep} = \frac{1}{2}\sqrt{D_h^2 - D_{op}^2}\sqrt[4]{D_h^4 - D_{op}^4 - \left[\frac{D_h^2 - D_{op}^2}{\ln\left(D_h / D_{op}\right)}\right]^2} \qquad (3.19c)$$

When using equation (3.19c), the fluid average velocity in the annulus must be calculated on the basis of equivalent pipe diameter, D_{ep}; that is,

$$\bar{v}_a = \frac{Q}{A_a} = \frac{Q}{(1/4)\pi D_{ep}^2} \qquad (3.20)$$

where Q is the flow rate.

For non-Newtonian fluids, equations (3.17) and (3.18) can be expressed as

$$N_{Rp} = \frac{928\gamma \bar{v}_p D_{ip}}{\mu_e} \qquad (3.21)$$

and

$$N_{Ra} = \frac{928\gamma \bar{v}_a D_{ep}}{\mu_e} \qquad (3.22)$$

where μ_e is the equivalent Newtonian viscosity for a non-Newtonian fluid, as will be discussed later.

Example 3–3

An oil well is expected to produce at a rate of 480 standard barrels per day (SBPD). If the viscosity of the oil is 5 centipoise and its specific gravity is 0.8, determine the minimum size of production tubing to be installed in the well such that laminar flow is maintained. Assume that the fluid is Newtonian and that the critical Reynolds number, N_{Ra} is 2100.

From equation (3.17),

$$N_{Ra} = \frac{928\gamma \bar{v} D_p}{\mu} < 2{,}100$$

For pipe flow, the average fluid velocity is

$$\bar{v} = \frac{Q}{2.45} D_{ip}^2$$

where

Q is in gpm
D_{ip} is the ID of a pipe, in in.
$\bar{v}$ is in ft/s

Thus,

$$Q = 480\frac{\text{bbl}}{\text{d}} \times \frac{42\text{ gal}}{\text{bbl}} \times \frac{\text{d}}{24\text{ h}} \times \frac{\text{h}}{60\text{ min}}$$

$$= 14\text{ gpm}$$

$$\bar{v} = \frac{14}{2.45}D_{ip}^2$$

$$2{,}100 > \frac{(928)(0.8)(8.34)(14)}{(2.45)(5)^2 D_{ip}}$$

$$D_{ip} < 3.37\text{ in.} \approx 3\frac{1}{4}\text{ in.}$$

Friction Pressure Losses in Pipes and Annuli—Laminar Flow

The application of fluid dynamics to drilling is concerned with the fluid flow inside the drill string (pipe flow) and in the annular space between the drill string outer wall and the wellbore wall (annular flow). If the average fluid velocity is low enough, as discussed previously, then the laminar-flow regime prevails, allowing the development of mathematical relationships between fluid-flow rate, fluid-flow properties (rheological properties), flow conduit geometry, and friction pressure losses, subject to the following assumptions:

- Isothermal flow
- Incompressible fluid
- Concentric circular annulus
- Nonrotating drill string
- Layers of fluid adjacent to pipe and wellbore walls are stationary
- Shear rate produced at any point in the flow stream is a function only of the shear stress at that point.
- Steady-state flow (time independent)

Extensive research has shown that these assumptions are quite valid for the liquid muds used in drilling operations. Their inclusion will hardly justify the computational complexity for practical applications.

To derive the relationship between the friction pressure loss, P_f, the fluid shear stress, τ, and the conduit geometry radius, r, consider the forces that are acting along the fluid layer.

These forces are located at r and $r + \Delta r$ in the radial direction and at L and $L + \Delta L$ along the wellbore axis, as shown in figure 3–6.

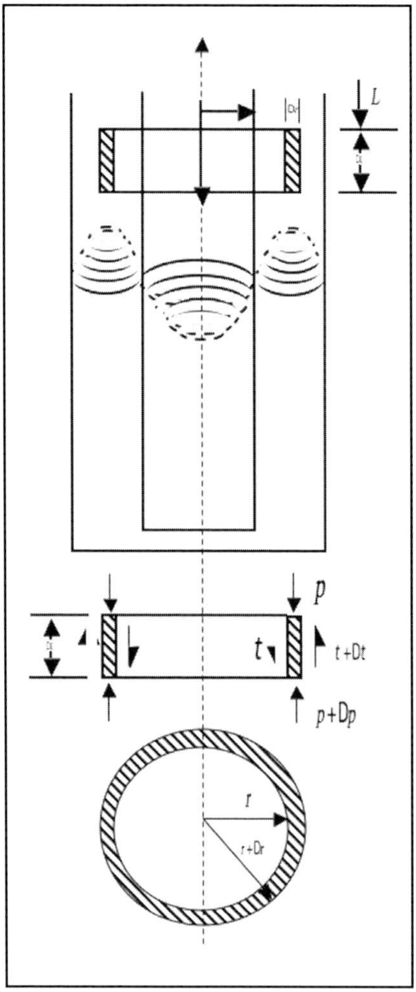

Fig. 3–6. Elements of flowing fluid

Using Newton's law of motion, $\sum \vec{F} = m\vec{a}$, where m denotes mass and a denotes acceleration. Alternatively, for a nonaccelerating fluid, as is the case here, $\sum \vec{F} = 0$.

Thus, summation of forces induced on the isolated layer of fluid owing to the fluid motion yields

$$(P_f + \Delta P_f)(2\pi r)\Delta r - P_f(2\pi r)\Delta r + (\tau + \Delta\tau)[2\pi(r + \Delta r)\Delta L] - \tau(2\pi r)\Delta L = 0 \qquad (3.23)$$

After simplification and by ignoring the higher-order terms of differentials,

$$\frac{\Delta P_f}{\Delta L} + \frac{\Delta \tau}{\Delta r} + \frac{\tau}{r} = 0 \tag{3.24}$$

As $r \to 0$ and $L \to 0$, equation (3.24) becomes

$$\frac{d\tau}{dr} + \frac{\tau}{r} + \frac{dP_f}{dL} = 0 \tag{3.25}$$

where dP_f/dL is the friction pressure loss gradient and is independent of r. Solving for τ by integrating equation (3.25) yields

$$\tau = -\frac{r}{2}\left(\frac{dP_f}{dL}\right) + \frac{c_1}{r} \tag{3.26}$$

where c_1 is a constant of integration and must be determined from known boundary conditions.

Equation (3.26) is valid regardless of the drilling-mud rheology. For pipe flow, the variable r in equation (3.26) can be zero; hence, $\tau \to \infty$ at the center. This leads to an unrealistic solution, contrary to the fact that the shear stress is maximum at the pipe walls and approaches zero at the center. Therefore, for pipe flow, c_1 must be chosen equal to zero; thus,

$$\tau = -\frac{r}{2}\left(\frac{dP_f}{dL}\right) \tag{3.27}$$

The shear rate, γ, of a fluid flowing through a circular conduit is related to the fluid particle velocity:

$$\gamma = -\frac{dv}{dr} \tag{3.28}$$

where v is the instantaneous fluid particle velocity and the minus sign indicates a decrease in fluid velocity as r increases from the center of the conduit. The shear rate of a drilling mud is related to the shear stress through the various rheological models discussed previously. The most commonly used models are

Newtonian:

$$\tau = \mu\gamma \tag{3.29a}$$

Bingham:

$$\tau = \tau_y + \mu_p\gamma \tag{3.29b}$$

Power law:

$$\tau = k\gamma^n \tag{3.29c}$$

Yield power law:

$$\tau = \tau_y + k\gamma^n \tag{3.29d}$$

The total flow rate in a circular conduit is related to conduit area open to flow and fluid velocity by the following integral:

$$Q = \int_A v\,dA = \int_r v(2\pi r)\,dr \tag{3.30}$$

The manipulation of equations solving for the Reynolds number (eqq. [3.16]–[3.18], [3.21], and [3.22]) will yield the sought relationships for friction pressure losses in pipe and annular flow under the laminar-flow regime.

Newtonian fluid

Figure 3–7 shows the velocity profile and shear stress distribution of a Newtonian fluid. Consider model (3.29a) and equation (3.26), as well as the following relationships:

$$\gamma = \frac{dv}{dr} \tag{3.31a}$$

$$Q = \int_v (2\pi r)\,dr \tag{3.31b}$$

Substituting equation (3.31a) into model (3.29a) and then substituting the result into equation (3.26) and integrating with respect to r yields

$$v(r) = \frac{r^2}{4\mu}\left(\frac{dP_f}{dL}\right) - \frac{c_1}{\mu}(\ln r) + c_2 \tag{3.31c}$$

For annular flow, the boundary conditions are:

$$v(R_{op}) = 0$$
$$v(R_h) = 0 \tag{3.31d}$$

where

R_{op} is the pipe outer radius
R_h is the hole radius

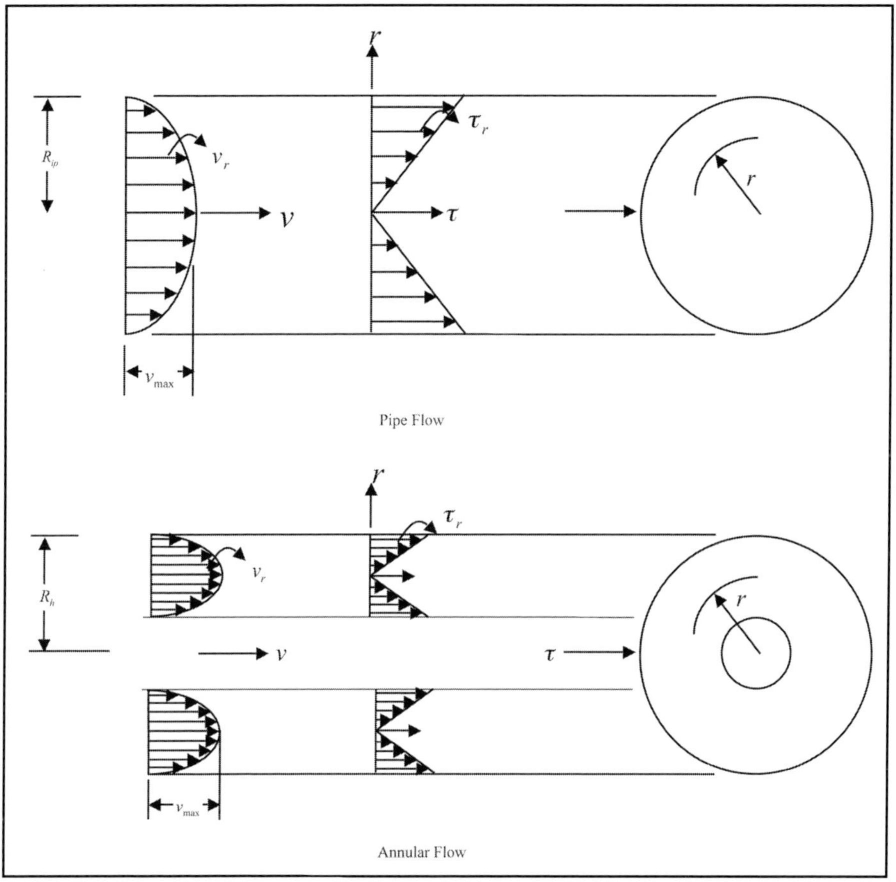

Fig. 3–7. Velocity profile and shear stress distribution of a Newtonian fluid. Top, *Pipe flow.* Bottom, *Annular flow.*

Substituting equation (3.31d) into equation (3.31c) and solving for the constants c_1 and c_2 yields

$$c_1 = \left[\frac{R_h^2 - R_{op}^2}{4\ln\left(R_h / R_{op}\right)}\right]\left(\frac{dP_f}{dL}\right) \tag{3.31e}$$

$$c_2 = \frac{1}{4\mu}\left(\frac{dP_f}{dL}\right)\left[\frac{\ln R_{op}\left(R_h^2 - R_{op}^2\right)}{\ln\left(R_h / R_{op}\right)} - R_{op}^2\right] \tag{3.31f}$$

Substituting equations (3.31e) and (3.31f) back into equation (3.31c),

$$v(r) = -\frac{1}{4\mu}\left(\frac{dP_f}{dL}\right)\left[\left(R_{op}^2 - r^2\right) - \left(R_h^2 - R_{op}^2\right)\frac{\ln\left(r / R_{op}\right)}{R_h / R_{op}}\right] \tag{3.31g}$$

Substituting equation (3.31g) into equation (3.31b) and integrating with respect to r yields

$$Q = -\frac{\pi}{8\mu}\left(\frac{dP_f}{dL}\right)\left[\left(R_h^4 - R_{op}^4\right) - \frac{\left(R_h^2 - R_{op}^2\right)^2}{\ln\left(R_h / R_{op}\right)}\right] \tag{3.31h}$$

The total flow rate is related to the average fluid velocity by the following relation:

$$Q = A\bar{v} \tag{3.31i}$$

where A is the area open to flow—that is,

$$A = \begin{array}{ll} \pi\left(R_h^2 - R_{op}^2\right) & \text{for annular flow} \\ \pi R_{ip}^2 & \text{for pipe flow} \end{array} \tag{3.31j}$$

where R_{ip} is the internal pipe radius. Utilizing equations (3.31i) and (3.31j) in equation (3.31h) and integrating with respect to L yields

$$\frac{P_f}{L} = \Delta P_{fa} = \frac{\mu\bar{v}_a}{1{,}500} \Bigg/ \left[D_h^2 + D_{op}^2 - \frac{D_h^2 - D_{op}^2}{\ln\left(D_h / D_{op}\right)}\right] \quad \text{(NLAF)} \tag{3.32}$$

where

μ is the Newtonian viscosity, in centipoise
$\bar{v}_a$ is the average fluid annular velocity, in ft/s
D_h is the hole diameter, in in.
D_{op} is the pipe OD, in in.
P_f / L is the friction pressure loss gradient, in psi/ft
NLAF denotes Newtonian fluid and laminar annular flow

For pipe flow, $D_{op} \rightarrow 0$, thereby reducing equation (3.32) to

$$\frac{P_f}{L} = \Delta P_{fp} = \frac{\mu\bar{v}_p}{1{,}500 D_{ip}^2} \quad \text{(NLPF)} \tag{3.33}$$

where

$\bar{v}_p$ is the average fluid velocity inside pipe, in ft/s
NLPF denotes Newtonian fluid and laminar pipe flow
D_{ip} is the pipe ID, in in.

Equation (3.33) is the well-known Hagen-Poiseuille equation for Newtonian fluids flowing in circular pipes under a laminar-flow regime. Note that the minus sign befor dP_f/dL means loss of pressure and need not be carried in the final equation.

Example 3–4

You are given the following well data:

- Well depth = 10,000 ft TMD
- Hole size = 8⅞ in.
- Drill pipe:
  - OD = 5 in.
  - ID = 4 in.
- Drill collar:
  - OD = 7 in.
  - ID = 4 in.
  - Length = 1,000 ft
- Mud is considered Newtonian, with $\mu = 40$ centipoise and $\gamma = 8.5$ lbs/gal
- Minimum allowed annular average fluid velocity = 100 ft/min

Determine the percentage of pump hydraulic horsepower that will be lost owing to friction while drilling at 10,000 ft and a flow rate that corresponds to the minimum allowed annular velocity. Assume laminar flow throughout the circulation system. Also assume surface connections to be equivalent to 500 ft of drill pipe.

Solution: The hydraulic horsepower that will be lost owing to friction (HHF) is given by

$$\text{HHF} = \frac{P_f Q}{1{,}714}$$

Further,

$$Q = \bar{v}_{a\,min} A_{a\,max} = \bar{v}_{a\,min} \frac{\pi}{4}\left(D_h^2 - D_{odp}^2\right)$$

$$= 0.041(100)\frac{\pi}{4}\left(\overline{8.875}^2 - \bar{5}^2\right) \approx 173 \text{ gpm}$$

$$P_f = P_{fdp} + P_{fdc} + P_{fadp} + P_{fadc} + P_{fs}$$

$$= \bar{L}_{dp}\Delta P_{fdp} + L_{dc}\Delta P_{fdc} + L_{dp}\Delta P_{adp} + L_{dc}\Delta P_{fadc}$$

Since $D_{idp} = D_{idc}$ and $\Delta P_{fdp} = \Delta P_{fdc}$ or

$$\Delta P_{fp} = \frac{\mu \bar{v}_p}{1{,}500 D_{ip}^2}$$

it follows that

$$\bar{v}_p = \frac{Q}{A_{ip}} = \frac{Q}{(\pi/4)D_{ip}^2}$$

Thus,

$$\bar{v}_p = \frac{173}{2.45(4)^2} = 4.4 \text{ ft/s}$$

$$\Delta P_{fp} = \frac{40(4.4)}{1{,}500(4)^2} = 7.32 \times 10^{-3} \text{psi/ft}$$

$$\Delta P_{fadp} = \frac{\mu \bar{a}_{adp}}{1{,}500} \Bigg/ \left[D_h^2 + D_{odp}^2 - \frac{D_h^2 - D_{opd}^2}{\ln\left(D_h / D_{odp}\right)} \right]$$

$$= \frac{(40)(100/60)}{1{,}500} \Bigg/ \left[8.875^2 + 5^2 - \frac{8.875^2 - 5^2}{\ln(8.875/5)} \right]$$

$$= 4.44 \times 10^{-3} \text{ psi/ft}$$

Similarly, for ΔP_{fadc},

$$\bar{v}_{adc} = \frac{Q}{\pi/4}\left(D_h^2 - D_{odc}^2\right)$$

$$= \frac{173}{2.45\left(8.875^2 - 7^2\right)} = 2.4 \text{ ft/s}$$

$$\Delta P_{fadc} = \frac{(40)(2.4)}{1{,}500} \Bigg/ \left[8.875^2 + 7^2 - \frac{8.875^2 - 7^2}{\ln(8.875/7)} \right]$$

$$= 0.272 \text{ psi/ft}$$

$$P_f = \left(7.82 \times 15^3\right)(10{,}000 + 500) + \left(4.44 \times 10^{-3}\right)(9{,}000)$$

$$+ (0.278)(1{,}000)$$

$$= 388 \text{ psi}$$

$$\text{HHF} = \frac{388 \times 173}{1{,}714} = 39 \text{ hp}$$

or, expressed as a percentage,

$$\frac{39}{2{,}000} \times 100\% = 1.85\%$$

Example 3–5

Referring back to example 3–4, determine whether formation fracturing will occur if the flow rate is 400 gpm. Assume that the formation fracture gradient is 9 lbs/gal.

ECD is given by

$$\gamma_{em} = \gamma_{hm} + \Delta\gamma_{af}$$

At 10,000 ft,

$$\Delta\gamma_{af} = \frac{L_{dp}\Delta P_{fadp} + L_{dc}\Delta P_{fadc}}{0.052 \times 10{,}000}$$

$$\Delta P_{fadc} = \Big|_{Q=10{,}000\text{ gpm}} = \frac{400}{173}\Delta P_{fadp}\Big|_{Q=173} = 0.01 \text{ psi/ft}$$

Likewise,

$$\Delta P_{fadc} = \frac{400}{173}(0.272) = 0.629 \text{ psi/ft}$$

$$\Delta\gamma_{af} = \frac{(0.01 \times 9{,}000) + (0.629 \times 1{,}000)}{0.052 \times 10{,}000} = 1.38 \text{ lbs/gal}$$

$$\gamma_{em}\Big|_{D=10{,}000\text{ ft}} = 8.5 + 1.38 = 9.88 \text{ lbs/gal} > 9 \text{ lbs/gal}$$

Therefore, fracture will occur.

Bingham fluid

The appropriate equations for the Bingham model are equations (3.26), (3.28), and (3.29b), and

$$Q = \int_r \nu(2\pi r)dr \tag{3.34a}$$

A drilling mud whose flow behavior is governed by the Bingham model has the velocity profile and shear stress distribution shown in figure 3–8.

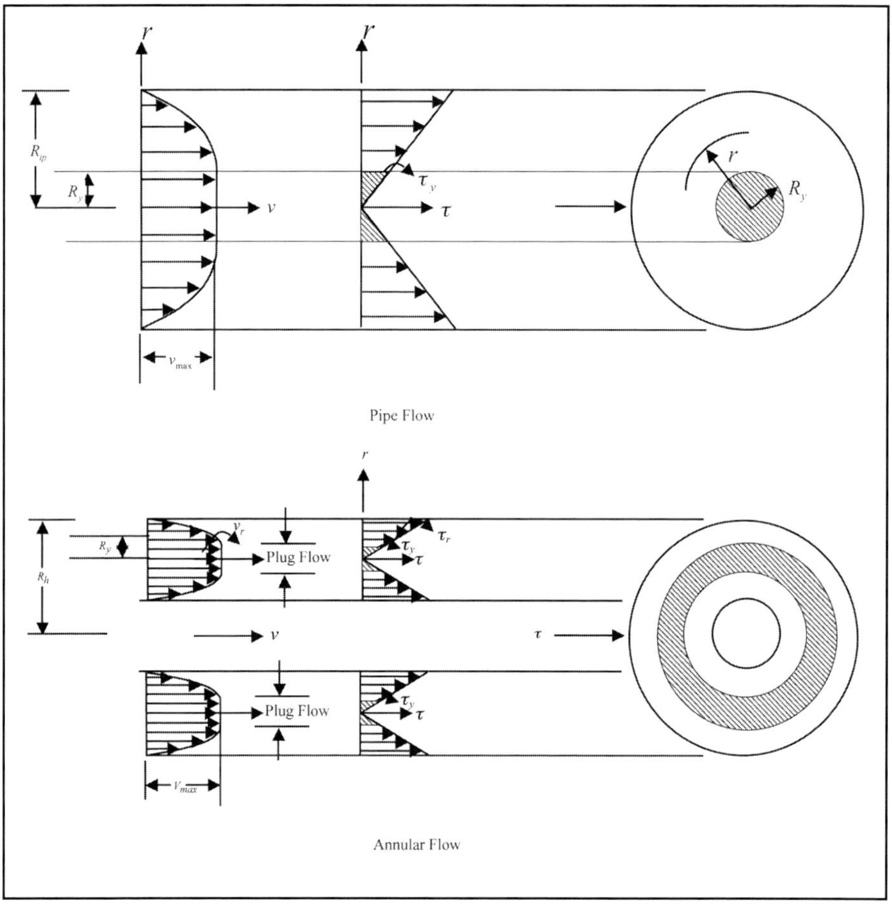

Fig. 3–8. Velocity profile and shear stress distribution of a non-Newtonian fluid with yield strength. Top, *Pipe flow.* Bottom, *Annular flow.*

Considering pipe flow, equation (3.26) becomes

$$\tau = \frac{r}{2}\left(\frac{dP_f}{dL}\right) \tag{3.34b}$$

Substituting equation (3.28) into equation (3.29b) and then equating the result with equation (3.34b) yields

$$\tau_y = \mu_p \frac{d\nu}{dr} = -\frac{r}{2}\left(\frac{dP_f}{dL}\right) \quad r > R_y \tag{3.34c}$$

Integrating yields

$$v(r) = \frac{r^2}{4\mu_p}\left(\frac{dP_f}{dL}\right) + \frac{\tau_y r}{\mu_p} + c_1 \tag{3.34d}$$

where, from the boundary conditions $v(R_{ip}) = 0$,

$$c_1 = \frac{R_{ip}^2}{4\mu_p}\frac{dP_f}{dL} - \frac{\tau_y}{\mu_p}R_{ip} \tag{3.34e}$$

Hence,

$$v(r) = \frac{1}{4\mu_p}\frac{dP_f}{dL}\left(r^2 - R_{ip}\right) + \frac{\tau_y}{\mu_p}\left(r - R_{ip}\right) \quad r > R_y \tag{3.35}$$

From equation (3.35), the velocity from the plug region is

$$v_y = v\left(R_y\right) = \frac{1}{4\mu_p}\left(\frac{dP_f}{dL}\right)\left(R_y^2 - R_{ip}^2\right) + \frac{\tau_y}{\mu_p}\left(R_y - R_{ip}\right) \tag{3.36}$$

but from equation (3.34b),

$$\tau\left(R_y\right) = \tau_y = -\frac{R_y}{2}\left(\frac{dP_f}{dL}\right) \tag{3.37}$$

Further,

$$\begin{aligned} v_y &= \frac{1}{4}\left(\frac{dP_f}{dL}\right)\left(R_y^2 - R_{ip}^2\right) - \frac{1}{2\mu_p}\left(\frac{dP_f}{dL}\right)R_y\left(R_y - R_{ip}\right) \\ &= -\frac{1}{4\mu_p}\left(\frac{dP_f}{dL}\right)\left(R_y - R_{ip}\right)^2 \quad r \le R_y \end{aligned} \tag{3.38}$$

On substituting equations (3.35) and (3.38) into equation (3.34a) and carrying out the appropriate integrations, the following is obtained:

$$Q = -\frac{\pi R_{ip}^4}{8\mu_p}\left(\frac{dP_f}{dL}\right)\left[1 - \frac{4}{3}\left(\frac{\tau_y}{\tau_{ip}}\right) + \frac{1}{3}\left(\frac{\tau_y}{\tau_{ip}}\right)^4\right] \tag{3.39}$$

On the basis of equation (3.39), the shear stress at the pipe wall ($r = R_{ip}$) must exceed the yield stress before flow can be initiated. In practice, the shear stress at the walls, τ_{ip}, is more than double the yield stress, τ_y. Therefore, the last term in equation (3.39) can be dropped with no appreciable error introduced. Hence, equation (3.39) reduces to

$$Q = -\frac{\pi R_{ip}^4}{8\mu_p}\left(\frac{dP_f}{dL}\right)\left[1 - \frac{4}{3}\left(\frac{\tau_y}{\tau_{ip}}\right)\right] \tag{3.40}$$

However, $\tau_{ip} = r(R_{ip}) = (R_{ip}/2)dP_f/dL)$, and $Q = \bar{v}_p A = \pi R_{ip}^2 \bar{v}_p$. Therefore,

$$\pi R_{ip}^2 \bar{v}_p = -\frac{\pi R_{ip}^2}{8\mu_p}\left(\frac{dP_f}{dL}\right)\left[1 + \left(\frac{8}{3}\right)\frac{\tau_y}{R_{ip}\left(dP_f/dL\right)}\right] \tag{3.41}$$

Solving for dP_f/dL and converting to field units yields

$$\frac{P_f}{L} = \Delta P_{fp} = \frac{\mu_p \bar{v}_p}{1,500 D_{ip}^2} + \frac{\tau_y}{225 D_{ip}} \quad \text{(BLPE)} \tag{3.42}$$

where

P_f is in psi

L is in ft

τ_y is the yield value, in lbs/100 ft[2]

BLPE denotes Bingham fluid and laminar pipe flow

μ_p is the plastic viscosity, in centipoise

$\bar{v}_p$ is in ft/s

D_{ip} is in in.

The exact mathematical derivations of equations for friction pressure loss in annular flow for non-Newtonian fluids having a yield value are quite complex. However, Melrose et al.[2] have shown that annular flow can be accurately approximated by slot flow (i.e., flow between two parallel surfaces) as long as the ratio of pipe diameter to hole diameter is greater than 0.3, a condition that is generally met in drilling applications.

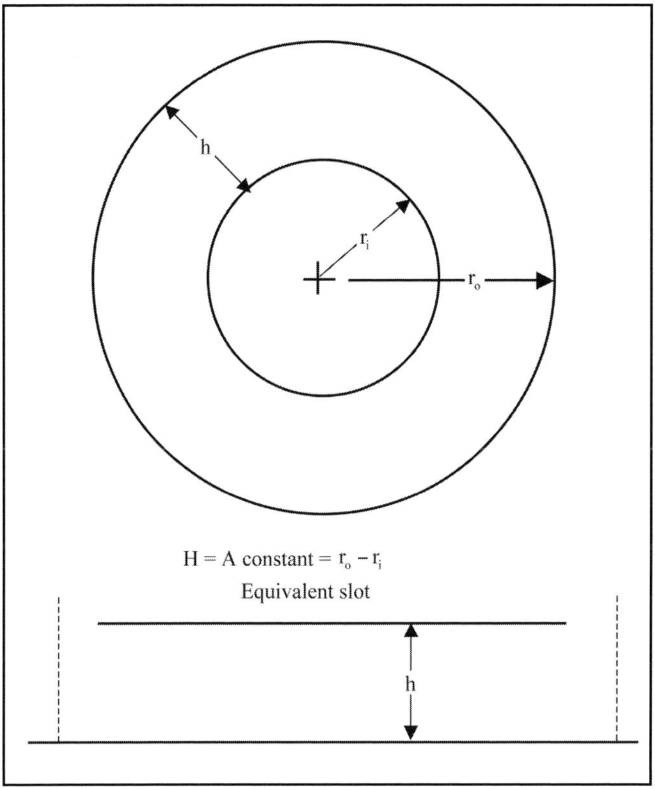

Fig. 3–9. Slot equivalents of concentric and eccentric annuli

As indicated in figure 3–9, an annulus can be represented by a slot whose geometric height is defined as follows (shown in fig. 3–10):[3]

$$h = \sqrt{R_h^2 - \varepsilon^2 c^2 \sin^2\theta} - R_{op} + \varepsilon c \cos\theta \tag{3.43}$$

where

$c = R_h - R_{op}$

ε is the annulus eccentricity: where $\varepsilon = 100e / (R_h - R_{op})$, where e is the distance between the center of the hole and the center of the pipe

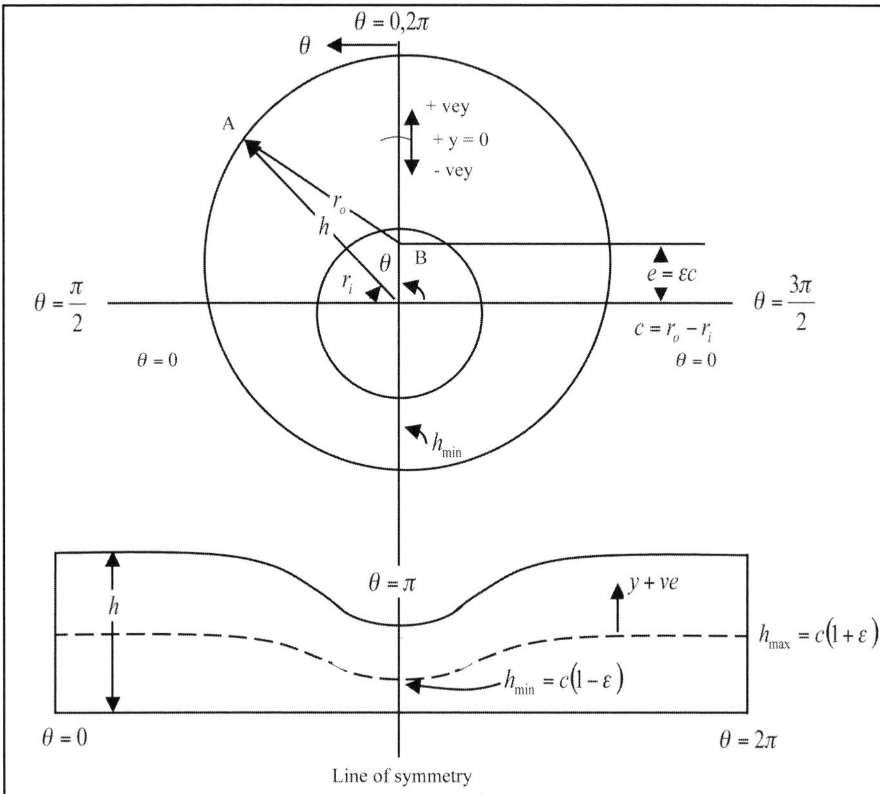

$$h = \left(r_o^2 - \varepsilon^2 c^2 \sin^2\theta\right)^{\frac{1}{2}} - r_i + \varepsilon c \cos\theta$$

Fig. 3–10. Nomenclature for eccentric annulus and equivalent slot

For concentric annuli, where $\varepsilon = 0$, equation (3.43) reduces to

$$h = R_{\text{h}} - R_{\text{op}} \tag{3.44}$$

The mean width of the slot is given by

$$w = \frac{\left(2\pi R_{op} + 2\pi R_h\right)}{2} = \pi\left(R_h + R_{op}\right) \tag{3.45}$$

Hence, the slot area open to flow can be expressed as the product of slot height and width, or

$$A = hw = (R_{\text{h}} - R_{\text{op}})\pi(R_{\text{h}} + R_{\text{op}}) = \pi(R_{\text{h}}^2 - R_{\text{op}}^2) \tag{3.46}$$

For a Bingham fluid and considering the Cartesian coordinates of figure 3–9, it is evident that

$$\tau = -y\frac{dP_{\rm f}}{\rm dL} + c_1 \tag{3.47a}$$

$$\gamma = -\frac{d\nu}{dy} \tag{3.47b}$$

$$Q = \int_A \nu dA \tag{3.47c}$$

Thus, by utilization of equations (3.43)–(3.47) and model (3.29b), the following relation can be derived:

$$\tau_w = \tau\big|_{h=0} = \frac{h}{2}\left(\frac{dP_{\rm f}}{dL}\right) \tag{3.48}$$

$$Q = -\frac{wh^3}{12\mu_{\rm p}}\left(\frac{dP_{\rm f}}{dL}\right)\left[1 + \frac{3}{2}\left(\frac{\tau_{\rm y}}{\tau_w}\right) - \frac{1}{2}\left(\frac{\tau_{\rm y}}{\tau_w}\right)^3\right] \tag{3.49}$$

For most practical cases, the term $(\tau_y/\tau_w)^3$ in equation (3.49) can be dropped without any significant error. Expressing Q in terms of the average fluid velocity, $\bar{\nu}$, and the conduit area, wh, and using equations (3.44) and (3.48) yields

$$-\frac{dP_{\rm f}}{\rm dL} = \frac{12\mu_{\rm p}\bar{\nu}_{\rm a}}{\left(R_{\rm h} - R_{\rm op}\right)^2} + \frac{3\tau_{\rm y}}{R_{\rm h} - R_{\rm op}} \tag{3.50}$$

Integrating with respect to L and using field units,

$$\frac{P_{\rm f}}{L} = \Delta P_{\rm fa} = \frac{\mu_{\rm p}\bar{\nu}_{\rm a}}{1{,}000\left(D_{\rm h} - D_{\rm op}\right)^2} + \frac{\tau_{\rm y}}{200\left(D_{\rm h} - D_{\rm op}\right)} \quad \text{(BLSF)} \tag{3.51}$$

where BLSF denotes Bingham fluid and laminar slot flow. For a Newtonian fluid, $\tau_y = 0$, and $\mu_p = \mu$; hence, equation (3.50) becomes

$$\Delta P_{\rm fa} = \frac{\mu\bar{\nu}_{\rm a}}{1{,}000\left(D_{\rm h} - D_{\rm op}\right)^2}\,{\rm L} \quad \text{(NLSF)} \tag{3.52}$$

Example 3–6

Solve example 3–4 assuming the fluid is a Bingham fluid having the following rheological properties: $\theta_{600} = 55$; $\theta_{300} = 35$; 10 s/10 min; gel = 8/17 (Fann V-G meter readings).

Solution:

$$\Delta P_{fp} = \Delta P_{fdp} = \Delta P_{fdc} = \frac{\mu_p \bar{v}_p}{1{,}500 D_{ip}^2} + \frac{\tau_y}{225 D_{ip}}$$

$$\mu_p = \theta_{600} - \theta_{300} = 55 - 35 - 20 \text{ centipoise}$$

$$\tau_y = \theta_{300} - \mu_p = 35 - 20 = 15 \text{ lbs/100 ft}^2$$

$$\bar{v}_p = 4.4 \text{ ft/s (from ex. 3–4)}$$

$$\Delta P_{fp} = \frac{20(4.4)}{1{,}500(4)^2} + \frac{15}{225(4)} = 0.02 \text{ psi/ft}$$

$$\Delta P_{fadp} = \frac{\mu_p \bar{v}_a}{1{,}000\left(D_h - D_{odp}\right)^2} + \frac{\tau_y}{200\left(D_h - D_{odp}\right)}$$

$$\bar{v}_{adp} + \frac{100 \text{ ft/min}}{60 \text{ s/min}} = 1.67 \text{ ft/s (from ex. 3–4)}$$

$$\Delta P_{fadp} + \frac{20(1.67)}{1{,}000(8.875-5)^2} + \frac{15}{200(8.875-5)} = 0.022 \text{ psi/ft}$$

$$P_f = (0.02 \times 10{,}500) + (0.022 \times 9{,}000) + (0.054 \times 1{,}000) = 462 \text{ psi}$$

$$\text{HHF} = \frac{P_f Q}{1{,}714} = \frac{462 \times 173}{1{,}714} = 46.6 \text{ hp}$$

$$\% \text{ Loss} = \frac{46.6}{2{,}000} \times 100\% = 2.33\%$$

Example 3–7

Solve example 3–5 for the fluid in example 3–6.

Solution:

From example 3–6,

- $\mu_p = 20$ centipoise
- $\tau_y = 15$ lbs/100 ft^2
- $\tau_g =$ gel $= 8/17$ lbs/100 ft^2

From example 3–5,

- $Q = 400$ gpm
- $\gamma_{frac} = 9$ lbs/gal

While drilling at 1,000 ft, the ECD is

$\gamma_{em} = \gamma_{hm} + \Delta\gamma_{af}$

for

$$\Delta\gamma_{af} = \frac{L_{dp}\Delta P_{fadp} + L_{dc}\Delta P_{fadc}}{0.052(1,000)}$$

Since laminar flow is assumed and friction pressure loss is directly proportional to flow rate,

$$\Delta P_{fadp} = \frac{400}{173}(0.022) = 0.051 \text{ psi/ft}$$

$$\Delta P_{fadc} = \frac{400}{173}(0.054) = 0.125 \text{ psi/ft}$$

or

$$\Delta\gamma_{af} = \frac{(0.051 \times 9,000) + (0.125 \times 1,000)}{0.052 \times 10,000} = 1.13 \text{ lbs/gal}$$

$$\gamma_{em} = 8.5 + 1.13 = 9.62 > 9 \text{ lbs/gal (fracture)}$$

Power law fluid

Recall equations (3.26), (3.28), (3.29c), and (3.30). For pipe flow, c_1 must be selected equal to zero. Thus,

$$-\frac{dv}{dr} = \left[-\frac{dP_f}{dL}\left(\frac{r}{2k}\right)\right]^n \tag{3.53}$$

Separating variables and integrating yields

$$v = -k^{-\frac{1}{n}}\left[-\frac{1}{2}\left(\frac{dP_f}{dL}\right)\right]^{\frac{1}{n}}\left(\frac{n}{n+1}\right)r^{\frac{n+1}{n}} + c \tag{3.54}$$

From the boundary condition, $v = 0$ at $r = R_{ip}$, the constant c is obtained:

$$c = k^{-\frac{1}{2}}\left[-\frac{1}{2}\left(\frac{dP_f}{dL}\right)\right]^{\frac{1}{n}}\left(\frac{n}{n+1}\right)\left(R_{ip}^{\frac{n+1}{n}} - r^{\frac{n+1}{n}}\right) \tag{3.55}$$

Substituting equation (3.55) into equation (3.30), integrating, and recognizing that $Q = \pi R_{ip}^2 \bar{v}_p$ yields

$$\frac{dP_f}{dL} = \frac{2k\bar{v}^n (3n+1)^n}{R_{ip}^{n+1} n^n} \tag{3.56}$$

In field units and after integrating with respect to L, equation (3.56) becomes:

$$\frac{P_f}{L} = \Delta P_f = \frac{k\bar{v}_p^n}{143,700 D_{ip}^{n+1}}\left(\frac{3n+1}{0.0416}\right)^n \quad \text{(PLPF)} \tag{3.57}$$

where

P_f/L is in psi/ft
v_p is in ft/s
k is in equivalent centipoise
PLPF denotes power law fluid and laminar pipe flow

The friction pressure loss equation in annular laminar flow for a power law fluid can be derived similarly:

$$\Delta P_{fa} = \frac{k\bar{v}_a^n}{143,700\left(D_h - D_{op}\right)^{n+1}}\left(\frac{2n+1}{0.0208}\right)^n \quad \text{(PLAF)} \tag{3.58}$$

where PLAF denotes a power law fluid and laminar annular flow.

Example 3–8

Solve examples 3–6 and 3–7 assuming a power law fluid with the following Fann V-G meter readings: $\theta_{600} = 55$; $\theta_{300} = 35$.

Solution—example 3–6:

$$n = 3.32 \log \frac{\theta_{600}}{\theta_{300}} = 3.32 \log \frac{55}{35} = 0.652$$

$$k \cong \frac{510\theta_{300}}{511^n}$$

$$k = \frac{35 \times 510}{511^{.652}} = 306 \text{ equivalent centipoise}$$

For Q = 173 gpm, $\bar{v}_{dp} = \bar{v}_{dc} = \bar{v}_p = 4.4$ ft/s (from previous exx.), $\bar{v}_{adp} = 1.67$ ft/s and $\bar{v}_{adc} = 2.4$ ft/s:

$$\Delta P_{\text{fp}} = \frac{306(4.4)^{0.652}}{143{,}700(4)^{1.652}}\left[\frac{(3\times 0.652)+1}{0.0461}\right]^{0.652}$$
$$= 9.1\times 15^{3}\ \text{psi/ft}$$

$$\Delta P_{\text{fadp}} = \frac{306(1.67)^{0.652}}{143{,}700(8.875-7)^{1.652}}\left[\frac{(2\times 0.652)+1}{0.0208}\right]^{0.652}$$
$$= 6\times 15^{5}\ \text{psi/ft}$$

$$\Delta P_{\text{fadc}} = \frac{306(2.4)^{0.652}}{143{,}700(8.875-7)^{1.652}}\left[\frac{(2\times 0.652)+1}{0.0208}\right]^{0.652}$$
$$= 0.019\ \text{psi/ft}$$

Therefore,

$$P_{\text{f}} = \left(9.1\times 15^{3}\right)(10{,}500) + \left(6\times 15^{3}\right)(9{,}000) + 0.019(1{,}000)$$
$$= 169\ \text{psi}$$

$$\text{HHF} = \frac{169\times 173}{1{,}714} = 17\ \text{hp}$$

$$\%\ \text{Loss} = \frac{17}{2{,}000}\times 100\% = 0.85\%$$

Solution—example 3–7:

$$\Delta P_{\text{fadp}} = \frac{400}{173}\left(6\times 15^{3}\right) = 0.014\ \text{psi/ft}$$

$$\Delta P_{\text{fadc}} = \frac{400}{173}(0.019) = 0.044\ \text{psi/ft}$$

Solving for ECD,

$$\gamma_{\text{em}} = \gamma_{\text{hm}} + \Delta\gamma_{\text{af}} = 8.5 + \frac{(0.014\times 9{,}000)+(0.044\times 1{,}000)}{0.052\times 10{,}000}$$
$$= 8.83 < 9\ \text{ppg}\ (\text{no fracture})$$

Yield power law (Herschel-Bulkley) fluid[4]

Similar treatments can be carried out for yield power law fluid in laminar flow. The results are summarized in this section.

For pipe flow, the friction pressure loss may be expressed as

$$\Delta P_{\mathrm{fp}} = \frac{1}{281\left(D_{\mathrm{h}} - D_{\mathrm{p}}\right)}\left\{\theta_0 + k\left[\frac{G\bar{v}_{\mathrm{p}}}{63.9\left(D_{\mathrm{h}} - D_{\mathrm{p}}\right)}\right]^n\right\} \text{ (PYLPF)} \tag{3.59}$$

where

θ_0 denotes "zero" gel
PYLPE denotes power yield fluid and laminar pipe flow
$G = [(3n + 1) / 4n][(1 - 0.877) / 0.123]^{1/n}$

For annular flow,

$$\Delta P_{\mathrm{fa}} = \frac{1}{281\left(D_{\mathrm{h}} - D_{\mathrm{p}}\right)}\left\{\theta_0 + k\left[\frac{G\bar{v}_{\mathrm{a}}}{63.9\left(D_{\mathrm{h}} - D_{\mathrm{p}}\right)}\right]^n\right\} \text{ (PYLAF)} \tag{3.60}$$

where PYLAF denotes power yield fluid and laminar annular flow; further,

$$\alpha = 1 - \left[1 - \left(\frac{D_{\mathrm{p}}}{D_{\mathrm{h}}}\right)^{0.37n^{-0.14}}\right]^{2.7n^{0.14}} \tag{3.61}$$

where

$$G = \left[\frac{(3-\alpha)n+1}{(4-\alpha)n}\right]\left[1+\frac{\alpha}{2}\right]\left[n\frac{(-10.877)^{\frac{1}{n}}}{0.123}\right]$$

Friction Pressure Losses in Pipes and Annuli—Turbulent Flow

As discussed previously, there are no mathematical developments that lead to the determination of friction pressure losses in pipes and annuli when the flow is turbulent. However, there is a vast amount of experimental data that explains the main factors that influence such pressure drops. Through dimensional analysis, these factors have been grouped in nondimensional forms, such as the Fanning friction factor and the Reynolds number, that allow the empirical determination of those friction pressure losses.

The Fanning friction factor in pipe flow for Newtonian fluids is given by

$$f_{\mathrm{p}} = \frac{D_{\mathrm{ip}}\Delta P_{\mathrm{fp}}}{2\rho\bar{v}_{\mathrm{p}}^2} \tag{3.62a}$$

where

f_p is the Fanning friction factor, which is also given in equation (3.62b)
D_{ip} is the pipe ID
ΔP_{fp} is the pipe friction pressure loss gradient
ρ is the fluid mass density
$\bar{v}_p$ is the fluid average velocity in pipe flow

In field units,

$$f_p = \frac{25.8 D_{ip} \Delta P_{fp}}{\gamma_m \bar{v}_p^2} \tag{3.62b}$$

Likewise, the Reynolds number is given by,

$$N_{Rp} = \frac{\rho \bar{v}_p D_{ip}}{\mu} \tag{3.63}$$

where

μ is the Newtonian viscosity
N_{rp} is the Reynolds number in pipe flow

Equations (3.62) and (3.63) are valid only for Newtonian fluids in pipe flow. The critical Reynolds number above which pipe flow is turbulent is determined experimentally and is approximately 2,100. The friction factor is given by the Colebrook equation:[5]

$$\frac{1}{\sqrt{f}} = 3.48 - 4 \log \left(\frac{2\varepsilon}{D_{ip}} + \frac{9.35}{N_{Rp}\sqrt{f}} \right) \tag{3.64}$$

Alternatively,[6]

$$\frac{1}{\sqrt{f}} = 2.28 - 4 \log \left(\frac{\varepsilon}{D_{ip}} + \frac{21.25}{N_{Rp}^{0.9}} \right) \tag{3.65}$$

where ε is the pipe roughness with the same dimensions as D_{ip}.

For annular flow, equations (3.62) and (3.63) are rewritten as

$$f_a = \frac{1}{2}\left(D_e \frac{\Delta P_{fa}}{\rho \bar{v}_a^2} \right) \tag{3.66a}$$

or

$$f_a = \frac{25.8 \Delta P_{fa} D_e}{\gamma_m \bar{v}_a^2} \tag{3.66b}$$

$$N_{Ra} = \frac{\rho \bar{v}_a D_e}{\mu} \tag{3.67}$$

where

f_a, N_{Ra} are the Fanning friction factor and the Reynolds number, respectively, for annular flow

$\bar{v}_a$ is the fluid average annular velocity

D_e is the equivalent pipe diameter

The correlations for calculating the equivalent diameter, D_e, have been presented already, in equations (3.19a–c).

For non-Newtonian fluids, equations (3.63) and (3.67) can be expressed as

$$N_{Rp} = \frac{\rho \bar{v}_p D_{ip}}{\mu_e} \tag{3.68}$$

$$N_{Ra} = \frac{\rho \bar{v}_a D_a}{\mu_e} \tag{3.69}$$

where μ_e is the equivalent Newtonian viscosity for a non-Newtonian fluid, as will be discussed in later sections.

Equivalent Newtonian Viscosity

Several criteria have been proposed to determine the transition from laminar to turbulent flow for non-Newtonian fluids. Some of these are based on the critical Reynolds number of 2,100, which was established experimentally in pipe flow for Newtonian fluids.

The value of μ_e in equations (3.68) and (3.69) may be found by assuming that the friction pressure loss gradient for Newtonian fluid in laminar flow is approximately equal to the friction pressure loss gradient for non-Newtonian fluids; that is,

$$\Delta P_{FpN} = \Delta P_{fpNN} \tag{3.70a}$$

$$\Delta P_{faN} = \Delta P_{faNN} \tag{3.70b}$$

where

ΔP_{fpN}, ΔP_{faNN} are the pipe and annulus friction pressure loss gradients for Newtonian fluids, respectively

ΔP_{fpNN}, ΔP_{faNN} are the pipe and annulus friction pressure loss gradients for non-Newtonian fluids, respectively

Thus, for a Bingham fluid in a pipe (μ_{eBp}) or an annulus (μ_{eBa}),

$$\mu_{eBp} = \mu_p + \frac{20 D_{ip} \tau_y}{3 \bar{v}_p} \tag{3.71a}$$

$$\mu_{eBa} = \mu_p + \frac{5(D_h - D_{op})\tau_y}{\bar{v}_a} \tag{3.71b}$$

Similarly, for a power law fluid in a pipe (μ_{epp}) or an annulus (μ_{epa}),

$$\mu_{epp} = \frac{k\overline{v}_p^{\,n-1}}{96D_{ip}^{\,n-1}}\left[\frac{3+(1/n)}{0.0416}\right]^n \tag{3.72a}$$

$$\mu_{epa} = \frac{k\overline{v}_a^{\,n-1}}{144\left(D_h - D_{ip}\right)^{n-1}}\left[\frac{2+(1/n)}{0.0208}\right]^n \tag{3.72b}$$

Turbulent Pipe Flow of Non-Newtonian Fluids

Many correlations related to turbulent pipe flow analysis of non-Newtonian fluids have appeared in the literature, the work of Metzner and associates being the most noted.[7] Their work is based on a concept referred to as the *generalized Reynolds number,* which initially was developed for power law fluids and was later extended to other time-independent non-Newtonian fluids.[8]

It was stated in preceding sections that the Reynolds number used in the analysis of Newtonian fluid flow can be applied to a non-Newtonian fluid if an appropriate equivalent Newtonian viscosity, μ_e, can be calculated for the non-Newtonian fluid. The Reynolds number is given by equation (3.68), where the equivalent Newtonian viscosity is defined by

$$\mu_e = \frac{\tau_w}{8\overline{v}_p / D_{ip}} \tag{3.73}$$

and τ_w is the laminar wall shear stress corresponding to the shear rate, $8\overline{v}_p/D_{ip}$.

Substituting equation (3.73) into equation (3.67) yields

$$N_{Rp} = \frac{\rho \overline{v}_p D_{ip}}{\tau_w}\left(\frac{8\overline{v}_p}{D_{ip}}\right) \tag{3.74}$$

It can be shown from Skelland[9] that

$$\tau_w = k'\left(\frac{8\overline{v}_p}{D_{ip}}\right)^{n'} \tag{3.75}$$

where k' and n' vary with $8\overline{v}_p/D_{ip}$. On a logarithmic plot of τ_w versus $8\overline{v}_p/D_{ip}$, n' is the slope of the tangent and k' is the intercept on the ordinate at $8\overline{v}_p/D_{ip}$ equal to unity. For power law fluids, n' and k' are constants over any range of τ_w—that is, $n' = n$ and $k' = k$.

The generalized Reynolds number can be obtained by substituting equation (3.75) into equation (3.74):

$$N_{\text{RpGe}} = \frac{\rho \bar{v}_{\text{p}}^{2-n'} D_{\text{ip}}^{n'}}{k' 8^{n'-1}} \tag{3.76}$$

For power law fluids, equation (3.76) becomes

$$N_{\text{RpGe}} = \frac{\rho \bar{v}_{\text{p}}^{2-n'} D_{\text{ip}}^{n'}}{K\left[(3n+1)/4n\right]^{n} 8^{n-1}} \tag{3.77}$$

where n and k are the power law constants.

The Fanning friction factor, which was discussed earlier, can be defined here as follows:

$$f = \frac{D_{\text{ip}} P_{\text{f}} / 4L}{\rho \bar{v}_{\text{p}}^{2} / 2} = \frac{\tau_{\text{w}}}{\rho \bar{v}_{\text{p}}^{2} / 2} \tag{3.78}$$

where

P_{f} is the friction pressure loss
L is the pipe length

By utilization of equation (3.75), equation (3.78) becomes

$$f = 16 \bigg/ \frac{\rho \bar{v}_{\text{p}}^{2-n'} D_{\text{ip}}^{n'}}{k' 8^{n-1}} \tag{3.79}$$

or

$$f = \frac{16}{N_{\text{RpGe}}} \tag{3.80}$$

Equation (3.80) is similar to that of Newtonian fluid with exception of the definition of the Reynolds number.

Dodge and Metzner[10] have developed the following empirical correlation for the friction factor in smooth pipe on the basis of power law fluids:

$$\sqrt{\frac{1}{f}} = \frac{4}{(n')^{0.75}} \log\left(N_{RpGe} f^{1-n^{\frac{1}{2}}} \right) - \frac{0.4}{(n')^{1.2}} \tag{3.81}$$

Dodge and Metzner have also shown that the correlation in equation (3.81) is valid for non–power law fluids provided that k' and n' are determined at the actual existing wall shear stress under the prevailing conditions. Metzner has advised the use of existing relationships developed for Newtonian fluids in rough pipes.

Although the generalized Reynolds number applies to all non-Newtonian fluids, for Bingham plastic fluid in smooth pipes, the Tomita[11] relationship is recommended:

$$\sqrt{\frac{1}{f_B}} = 4\log\left(N_{RpB}\sqrt{f_B}\right) - 0.40 \tag{3.82}$$

where

$N_{RpB} = (D_{ip}\bar{v}_p\rho/\mu_p)(1-x)(x^4 - 4x + 3)/3$

$f_B = D_{ip}P_f/[2\rho\bar{v}_p^2(1-x)]$

$x = \frac{\tau_y}{\tau_w} = \frac{r}{R_{ip}}$

μ_p is the Bingham plastic viscosity

$R_{ip} = D_{ip}/2$

Turbulent Annular Flow of Non-Newtonian Fluids

Work on turbulent annular flow of non-Newtonian fluids has been very limited. Frederickson and Bird[12] have recommended the following correlations for power law fluids:

$$f = \frac{\Delta P_f D_h\left[1-\left(d_{op}/D_h\right)^2\right]}{2\left[1+\left(D_{op}/D_h\right)\right]\rho\bar{v}_a^2} \tag{3.83}$$

$$N_{Re} = \frac{D_h^n\bar{v}_a^{2-n}\rho/\left(D_{op}/D_h\right)}{2^{n-3}\left[1-\left(D_{op}/D_h\right)^2\right]/\left[1+\left(D_{op}/D_h\right)\right]}\Theta \tag{3.84}$$

In equation (3.84),

$$\Theta = \frac{Q}{\pi\left(D_h/2\right)^2\left[\Delta P_f\left(D_h^2/D_{op}\right)^{\frac{1}{n}}\right]} \tag{3.85}$$

where Q is the flow rate.

Skelland[13] has suggested that turbulent annular flow can be represented by the Blasius equation:

$$f = \frac{0.0791}{\left(N_{Re}\right)^{0.25}} \quad 3\times10^3 < N_{Re} < 10^5 \tag{3.86}$$

where N_{Re} is as given in equation (3.84).

Example 3–9

You are given the following data for the directional S-shaped well shown in figure 3–11:

- Mud:
  - Bingham plastic; $\gamma = 10.2$ lbs/gal
  - $\theta_{600} = 65$; $\theta_{300} = 40$; 10 s/10 min; gel = 15/25 lbs/100 ft^2
- Bit: Tri-cone roller cone jet bit (12-12-12)
- Drill pipe:
  - OD = 4.5 in.
  - ID = 4 in.
  - Air weight = 14 lbs/ft
- Drill collar:
  - OD = 7.5 in.
  - ID = 4 in.
  - Air weight = 110 lbs/ft
  - $L_{DC} = 1{,}000$ ft
- Surface connections: 500 ft
- Hole: Last intermediate casing set (8.5 in. ID set at 16,000 ft TVD open hole washed out to 8.5 in. using a 7.5 in. bit).

Determine the pump hydraulic horsepower (PHHP) required in order to drill to TD. Assume pump volumetric efficiency of 85% and a maximum flow rate of 400 gpm.

Solution:

$$\text{PHHP} = \frac{P_p Q}{1{,}714}$$

$$P_p = P_{fdp} + P_{fdc} + P_{fadp} + F_{fadc} + P_{fs} + P_B$$

Velocities:

$$\bar{v}_{dp} = \bar{v}_{dc} = \frac{Q}{2.45 D_i^2} = \frac{400}{2.45(4)^2} = 10.2 \text{ ft/}$$

Flow regimes (find N_{Rdp}, N_{Rdc}, N_{Radp}, N_{Radc} given $N_{Rdp} = N_{Rdc}$):

$$N_{Rdp} = N_{Rdc} = N_{Rp} = \frac{928 \gamma_m \bar{v}_p D_{ip}}{\mu_{ep}}$$

$$\mu_{ep} = \mu_p + \frac{6.66 \tau_y D_{ip}}{\bar{v}_p}$$

$$N_{Radp} = \frac{928 \gamma_m \bar{v}_{adc} D_{edc}}{\mu_{eadp}}$$

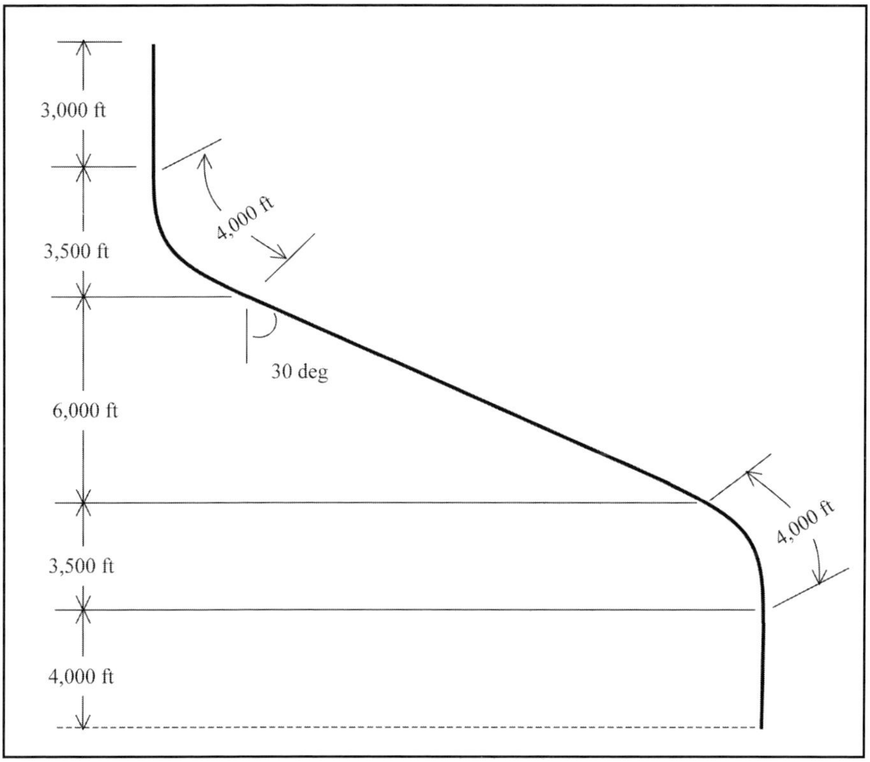

Fig. 3–11. Directional S-shaped well in example 3–9

$$D_{edp} = 0.816\left(D_h - D_{odp}\right)$$

$$N_{Radc} = \frac{928\gamma_m \bar{v}_{adc} D_{edc}}{\mu_{eadc}}$$

$$D_{edc} = 0.816\left(D_h - D_{odc}\right)$$

$$\mu_{eadp} = \mu_p + \frac{5\tau_y\left(D_h - D_{odp}\right)}{\bar{v}_{adp}}$$

$$\mu_{eadc} = \mu_p + \frac{5\tau_y\left(D_h - D_{odc}\right)}{\bar{v}_{adc}}$$

$\mu_p = \theta_{600} - \theta_{300} = 65 - 40 = 25$ centipoise

$\tau_y = \theta_{300} - \mu_p = 40 - 25 = 15$ lbs/100 ft^2

$\mu_p = 64.4$ centipoise

$\mu_{eadp} = 117.6$ centipoise

$\mu_{eadc} = 32.4$ centipoise

$D_{edp} = 3.26$ in.

$D_{edc} = 0.816$ in.

$$\mu_{eadp} = \mu_p + \frac{5\tau_y\left(D_h - D_{odp}\right)}{\bar{v}_{adp}}$$

$$\mu_{eadc} = \mu_p + \frac{5\tau_y\left(D_h - D_{odp}\right)}{\bar{v}_{adc}}$$

$\mu_p = \theta_{600} - \theta_{300} = 65 - 40 = 25$ centipoise

$\tau_y = \theta_{300} = \mu_p = 40 - 25 = 15$ lbs/100 ft^2

Thus,

$N_{RP} = 5{,}956 > 2{,}100$ (turbulent)

$N_{Radp} = 850 < 2{,}100$ (laminar)

$N_{Radc} = 2{,}414 > 2{,}100$ (turbulent)

To find ΔP_{fp} and ΔP_{fadc} for turbulent flow, the Fanning friction factor must be calculated. From equation (3.65), assuming $\varepsilon = 0$,

$$\frac{1}{\sqrt{f}} = 2.28 - 4\log\left(\frac{24.25}{N_R^{0.9}}\right)$$

$$N_{Ra} = \frac{928\gamma_m \bar{v}_a D_{ea}}{\mu_p}$$

$$\Delta P_{fp} = \frac{\gamma_m \bar{v}_p^2 f_P}{25.8 D_{ip}}$$

$$\Delta P_{fadp} = \frac{\mu_p \bar{v}_a}{1{,}000\left(D_h - D_{op}\right)^2} + \frac{\tau_y}{200\left(D_h - D_{op}\right)}$$

(Thus, slot flow obtains, because $D_{odp}/D_h = 0.53 > 0.3$.)

$$\Delta P_{\text{fadc}} = \frac{\gamma_{\text{m}} \bar{v}_{\text{sbadc}}^{2} f_{\text{adc}}}{25.8 D_{\text{edc}}}$$

$$N_{\text{Rp}} = \frac{928 \times 10.2 \times 10.2 \times 4}{25} = 15,448$$

$N_{\text{Radc}} = 3,151$

$f_{\text{p}} = 0.007$

$f_{\text{adc}} = 0.011$

$$\Delta P_{\text{fp}} = \frac{10.2 \times \overline{10.2}^{2} \times 0.007}{25.8 \times 4} = 0.072 \quad \text{psi/ft}$$

$\Delta P_{\text{fadp}} = 0.024$ psi/ft

$$\Delta P_{\text{fadc}} = \frac{10.2 \times \overline{10.2}^{2} \times 0.011}{25.8 \times 0.816} = 0.554$$

$$P_{\text{B}} = \frac{\left(8.31 \times 10^{-5}\right)\gamma_{\text{m}}\theta^{2}}{C_{\text{d}}^{2} A_{\text{t}}^{2}} = \frac{\left(8.31 \times 10^{-5}\right)(10.2)\overline{400}^{2}}{\overline{0.95}^{2}\left[3(\pi / 4)(12/32)^{2}\right]^{2}} = 1,367 \text{ psi}$$

$$\text{Required PHHP} = \frac{1,039}{0.85} = 1,222 \text{ hp}$$

Annular Friction Pressure Losses Due to Pipe Movement

When a pipe moves up and down along the axis of a wellbore filled with a nonmoving fluid, friction pressure losses are induced. Their treatments are identical to those when the fluid is being circulated while the pipe is stationary. The only difference in the analytical development of friction pressure loss equations is in the applied boundary conditions. (See fig. 3–12 for illustration.)

The main concern in friction pressure losses due to pipe movement is related to the annulus section. If the pipe is being pulled out of the wellbore at some pipe velocity, v_{p}, pressure reduction in the annulus (lower equivalent mud weight) will occur. This annular pressure reduction is referred to as swab pressure, P_{sw}. The danger is that induced swab pressures, if excessive, will cause kicks. Therefore, the following condition must be maintained while pulling pipe out of hole:

$$P_{\text{ec}} = P_{\text{m}} - P_{\text{sw}} > P_{\text{ff}} \tag{3.87a}$$

or, in terms of pressure gradient,

$$\gamma_{ec} = \gamma_m - \gamma_{sw} > \gamma_{ff} \tag{3.87b}$$

where
P_{ec} is the equivalent circulating mud pressure
P_m is the mud hydrostatic pressure
P_{sw} is the swab pressure
P_{ff} is the formation fluid pressure
γ is the pressure gradient (in lbs/gal, psi/ft, or spg)

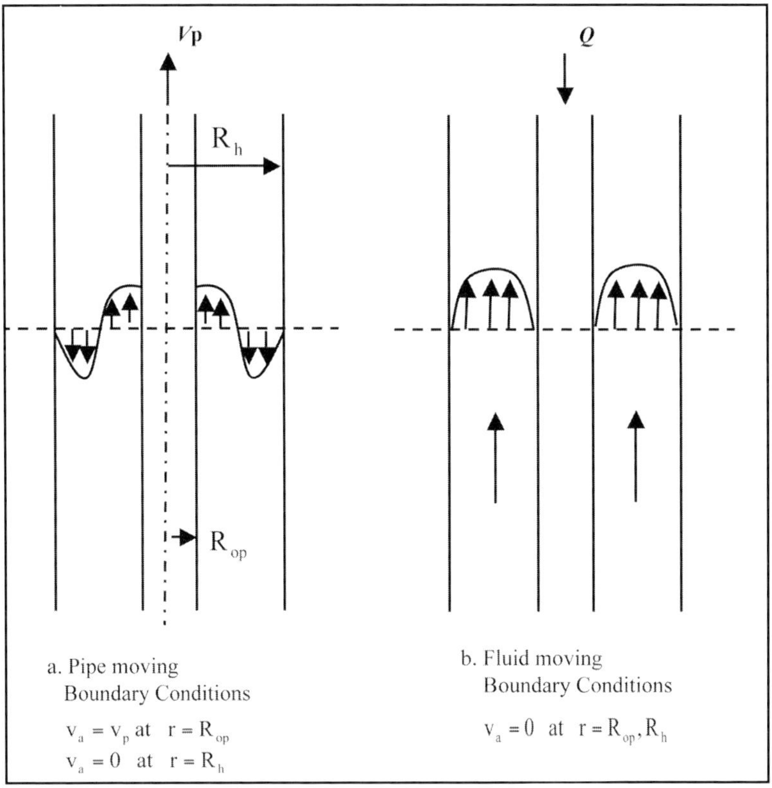

Fig. 3–12. Laminar-flow velocity profile. a, *Pipe movement.* b, *Fluid movement.*

Similarly, when lowering pipe into the wellbore at some pipe velocity, v_p, an increase in annulus pressure (higher equivalent mud weight) will occur. This increase is called surge pressure, P_{surge}. The danger is that surge pressures, if excessive, will cause formation fracturing and thus may lead to total loss circulation, which in turn will cause kicks. Therefore, the following condition must be maintained while lowering pipe into the hole:

$$P_{ec} = P_m + P_{surge} < P_{frac} \tag{3.88a}$$

or

$$\gamma_{ec} = \gamma_m + \gamma_{surge} < \gamma_{frac} \tag{3.88b}$$

where P_{frac} is the formation fracture pressure and P_{surge} is the surge pressure.

The two operations during which surge and swab pressures occur are tripping and casings. These operations are conducted many times during the drilling of a well, during tripping in particular; therefore, proper practice must be followed to avoid the costly problems of blowouts.

Swab/Surge Pressure Calculations

The difficulty in swab/surge pressure analytical development is determining the actual fluid-flow behavior and, therefore, the annular fluid velocity profiles resulting from pipe movement. For Newtonian fluids, by use of a slot flow approach to laminar flow, an expression can be easily derived to determine friction pressure losses during pipe movement:

$$\Delta P_{fsp} = \frac{\mu\left(\bar{v}_a - 0.5v_p\right)}{1{,}000\left(D_h - D_{op}\right)^2} \tag{3.89}$$

where ΔP_{fsp} is the friction pressure loss gradient for slot flow owing to pipe movement for a Newtonian fluid in laminar flow. Note that v_p has the opposite sign of $\bar{v}_a$. From equation (3.32) for Newtonian fluids,

$$\Delta P_{fs} = \frac{\mu\bar{v}_a}{1{,}000\left(D_h - D_{op}\right)^2} \tag{3.90}$$

On the basis of the resemblance between equations (3.89) and (3.90), it can be generalized that, if an equivalent average fluid annular fluid $\bar{v}_{ae}$ can be defined, the friction pressure loss equations previously derived for the case in which the fluid is moving are equally applicable to the case in which the pipe is moving. For example, equation (3.89) can be rewritten to be identical to equation (3.90):

$$\Delta P_{fs} = \frac{\mu\bar{v}_{ae}}{1{,}000\left(D_h - D_{op}\right)^2} \tag{3.91}$$

where $\bar{v}_{ae} = \bar{v}_a - 0.5v_p$ and is referred to as the effective fluid velocity in the annulus.

Burkhardt[14] has generalized equation (3.91) to include non-Newtonian fluids by introducing what he called a *clinging factor*, K_c:

$$\bar{v}_{ae} = \bar{v}_a - K_c v_p \tag{3.92}$$

This value can be calculated approximately using the following equations:

- For laminar flow,

$$K_c = \frac{1}{2 \ln \alpha} + \frac{\alpha^2}{1-\alpha^2} \tag{3.93}$$

- For turbulent flow,

$$K_c = \frac{-\sqrt{(\alpha^4+\alpha)/(1+\alpha)}+\alpha^2}{1-\alpha^2} \tag{3.94}$$

where $\alpha = D_{op}/D_h$. The clinging factor represents the contribution that the drill string viscous drag makes to the effective fluid annular velocity.

For the performance of swab/surge pressure calculations, the average annular fluid velocity, $\bar{v}_a$, owing to pipe movement needs to be determined. It is necessary to consider the following two cases:

1. Closed-end pipe
2. Open-ended pipe

For a closed-end pipe, the annular flow rate Q_a is equal to the rate of fluid being displaced by the pipe—that is, $Q_a = (\pi/4)D^2_{op}v_p$. Since the end of pipe is closed, this is the total flow rate in the annulus section. Therefore, the average fluid annular velocity can be written as $v_a = Q/A_a$, where A_a is the annulus cross section open to flow. Further,

$$\bar{v}_a = \frac{D_{op}^2 v_p}{D_h^2 - D_{op}^2} \tag{3.95}$$

For an open-ended pipe, the problem of finding the average annular velocity, $\bar{v}_a$, owing to pipe movement becomes complex. This is especially true in actual field application, where geometric variations of hole and pipe exist. Since friction pressure losses in the annulus owing to pipe movement are higher for assumed closed-end pipe than open-ended pipe, the latter case is not treated here; for a treatment of this case, see the work of Bourgoyne et al.[15]

Example 3–10

You are given the following well data:

- Current well depth = 10,000 ft TVD
- Mud:
  - Bingham mud; $\gamma_m = 10$ lbs/gal
  - $\theta_{600} = 65$; $\theta_{300} = 40$ (Fann V-G meter)

- Hole size = 7⅞ in.
- Drill pipe:
  - OD = 4 in.
  - ID = 3¼ in.
- Drill collar:
  - OD = 6 in.
  - ID = 3¼ in.
  - Length = 700 ft
- Formation:
  - Pressure gradient = 0.5 psi/ft
  - Fracture gradient = 0.56 psi/ft

Determine the maximum pipe speed at which tripping may be performed. Assume a closed-end pipe.

Solution: The concerns are swab pressures (tripping out) or surge pressures (tripping in) that may cause kicks or formation fracturing, respectively—that is, $P_{ff} \geq P_{hm} - P_{swab}$ and $P_{frac} \leq P_{hm} + P_{surge}$.

$$\Delta P_{surge}, \Delta P_{swab} = \frac{\mu_p \bar{v}_{ae}}{1{,}000\left(D_h - D_{op}\right)^2} + \frac{\tau_y}{200\left(D_h - D_{op}\right)}$$

For closed-end pipe, $\bar{v}_a = D_{op}^2 v_p / (D_h^2 - D_{op}^2)$, or

$$\bar{v}_{adp} = \frac{D_{op}^2 v_p}{D_h^2 - D_{odp}^2}$$

$$\bar{v}_{adc} = \frac{D_{odc}^2 v_p}{D_h^2 - D_{odc}^2}$$

$$K_{cdp} = \frac{1}{2 \ln \alpha_{dp}} + \frac{\alpha_{dp}^2}{1-\alpha_{dp}^2} \quad \text{(laminar flow)}$$

$$K_{cdp} = \frac{-\sqrt{\alpha_{dp}^4 / \left(1+\alpha_{dp}\right)} + \alpha_{dp}^2}{1-\alpha_{dp}^2} \quad \text{(turbulent flow)}$$

where $\alpha_{db} = D_{odp} / D_h = 0.51$.

For K_{cdc}, $\alpha_{dc} = D_{odc} / D_h = 0.762$; therefore,

$$\bar{v}_{adp} = 0.35 v_p$$

$$\bar{v}_{adc} = 1.38 v_p$$

$$K_{cdp} = \begin{matrix} -0.391 \text{ (laminar flow)} \\ -0.484 \text{ (turbulent flow)} \end{matrix}$$

$$K_{cdc} = \begin{matrix} -0.453 \text{ (laminar flow)} \\ -0.5 \quad \text{(turbulent flow)} \end{matrix}$$

Thus,

$$\bar{v}_{aedp} = \begin{matrix} 0.35v_p = 0.391v_p = 0.741v_p \text{ (laminar flow)} \\ 0.384v_p \text{ (turbulent flow)} \end{matrix}$$

$$\bar{v}_{adc} = \begin{matrix} 1.83v_p \text{ (laminar flow)} \\ 1.88v_p \text{ (turbulent flow)} \end{matrix}$$

Assuming laminar slow flow around drill pipe,

$$\Delta P_{swab}, \Delta P_{surge} = \frac{\mu_p \bar{v}_{aedp}}{1{,}000\left(D_h - D_{odp}\right)^2} + \frac{\tau_y}{200\left(D_h - D_{odp}\right)}$$

$$\mu_p = \theta_{600} - \theta_{300} = 65 - 40 = 25 \text{ centipoise}$$

$$\tau_y = \theta_{300} - \mu_p = 40 - 25 = 15 \text{ lbs/100 ft}^2$$

$$\Delta P_{swab}, \Delta P_{surge} = \frac{25\left(0.741v_p\right)}{1{,}000\left(\overline{7.875}^2 - \bar{4}^2\right)} + \frac{15}{200(7.875 - 4)}$$

$$\Delta P_{swab} \le \Delta P_{hm} - \Delta P_{ff} = 0.52 - 0.5 = 0.02$$

$$\Delta P_{surge} \ge \Delta P_{fac} - \Delta P_{hm} = 0.56 - 0.52 = 0.04$$

or

$v_p \le 1.5$ ft/s for swab
$v_p \le 51$ ft/s for surge

Supplementary Problems

Problem 3–1. You are given the following well data (see fig. 3–13):

- Last casing setting = 14,000 ft TMD
- TD = 13,500 ft TVD
- Hole size to TD = 7⅞ in.
- Drill pipe:
  - OD = 4¼ in.
  - ID = 3½ in.
  - Air weight = 26.88 lbs/ft
- Drill collar:
  - OD = 6 in.

  - ID = 3½ in.
  - Air weight = 100.8 lbs/ft
  - Length = 1,500 ft
- Pumps:
  - 2,000 hp
  - E_V = 85%
  - P_{max} 3,800 psi
- Minimum annular fluid velocity for hole cleaning = 120 ft/min
- Coefficients of friction at walls: 0.2, cased sections; 0.30, open hole section
- Mud:
  - Bingham plastic
  - Specific gravity = 2.0 for next depth interval
  - θ_{600} = 55; θ_{300} = 35; 10 s/10 min; gel = 18/35 lbs/100 ft^2 (Fann V-G meter)
- Formation below 14,000 ft TMD:
  - Fracture gradient = 18.0 lbs/gal
  - Pore pressure gradient = 0.83 psi/ft

Indicate whether any drilling problem will occur while drilling at or tripping from 14,000 ft TMD. Assume flow rate of 500 gpm and pipe trip velocity of 10 ft/s. Provide details.

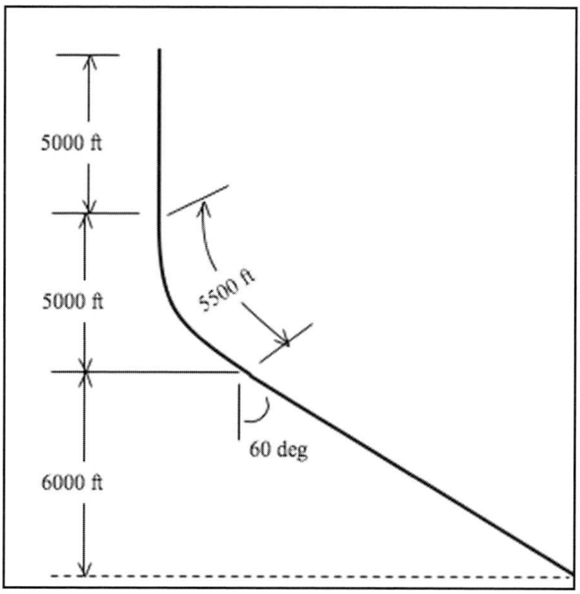

Fig. 3–13. Well schematic for problem 3–1

Problem 3–2. You are given the following well data:

- Casing-setting depths = 1,000 ft, 8,000 ft, 18,000 ft, and 20,000 ft TMD
- Last hole size = 7⅞ in. at 18,000 ft TMD
- Drill pipe:
  - OD = 4½ in.
  - ID = 3½ in.
  - Air weight = 26.88 lbs/ft
- Drill collar:
  - OD = 6½ in.
  - ID = 3½ in.
  - Air weight = 100.8 lbs/ft
  - Length = 1,500 ft
- Mud:
  - Bingham plastic
  - Specific gravity = 1.5 between 8,000 ft and 18,000 ft TMD
  - Specific gravity = 2.0 to TD
  - $\theta_{600} = 55$; $\theta_{300} = 35$; 10 s/10 min; gel = 18/30 lbs/100 ft^2 (Fann V-G meter)
- Formation below 18,000 ft TMD:
  - Fracture gradient = 17.2 lbs/gal
  - Pore pressure gradient = 0.83 psi/ft

Indicate whether any drilling problems will occur while (a) drilling at 18,000 ft and (b) tripping from 28,000 ft. Assume flow rate of 650 gpm and trip pipe velocity of 20 ft/s. Provide details.

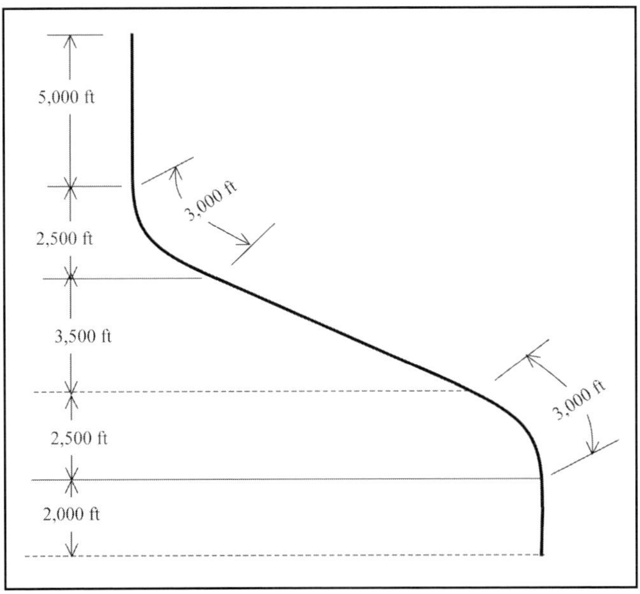

Fig. 3–14. Well schematic for problem 3–2

Problem 3–3. You are given the following well data:

- TD = 10,000 ft TVD
- Last intermediate casing set at depth = 5,000 ft TVD
  - OD = 9⅞ in.
  - Wall thickness = 0.3125 in.
- Next hole size = 8½ in. tri-cone roller cone jet bit (hole washed out to 9¼ in.)
- Minimum velocity required for annular hole cleaning = 120 ft/min
- Drill pipe:
  - OD = 4½ in.
  - ID = 3¾ in.
  - Air weight = 18.10 lbs/ft
- Drill collar:
  - OD = 7 in.
  - ID = 3¾ in.
  - L_{dc} = 1,200 ft
- Mud:
  - Bingham plastic
  - $\theta_{600} = 37$; $\theta_{300} = 25$; 10 s/10 min; gel = 7/16 lbs/100 ft^2
  - $\tau_y = 13$ lbs/100 ft^2
- Steel weight = 66.3 lbs/gal
- Coefficient of friction between formation and pipe = 0.2
- Formation fracture pressure gradient at casing shoe = 17 lbs/gal

  (a) What is the maximum weight on bit (WOB) that can be applied while drilling at 7,500 ft TVD?

  (b) At 15,000 ft TMD to TD, the following operating conditions are recommended: jet sizes 14-14-14 and flow rate corresponding to the required minimum annular velocity. Determine the required PHHP for a volumetric efficiency at 85%.

  (c) Determine whether a kick can occur under the conditions given in part (b).

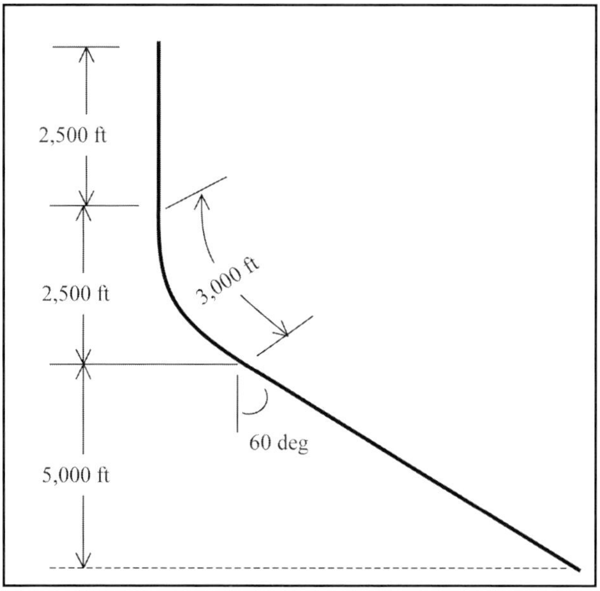

Fig. 3–15. Well schematic for problem 3–3

Problem 3–4. You are given the following well data:

- Current well depth = 16,000 ft
- Target well depth = 17,000 ft
- Drill pipe:
  - OD = 4.5 in.
  - Wall thickness = 0.43
  - Air weight = 18.7 lbs/ft
- Drill collar:
  - OD = 7 in.
  - ID = 2 in.
  - Air weight = 120 lbs/ft
- Maximum anticipated WOB = 80,000 lbs
- Maximum anticipated formation pressure = 0.83 psi/ft
- Total required differential pressure = 0.4 lbs/gal
- Mud:
  - Power law
  - $\theta_{600} = 37$; $\theta_{300} = 25$ (Fann V-G meter)

(a) Determine the hook load while drilling at 16,000 ft with WOB = 50,000 lbs.

(b) Determine the pump pressure required in order to drill to TD. Assume a flow rate of 250 gpm and an 8.5 ft drill bit with 9-11-11 jets. Neglect surface connections.

(c) Determine the equivalent circulating mud weight at 17,000 ft depth under the conditions given in part (b).

Problem 3–5. You are given the following well data (see fig. 3–16):

- Open hole size = 8⅞ in. (washed out to 9⅞ in.) to TMD (TD) of 20,250 ft
- Drill pipe:
  - OD = 5 in.
  - ID = 3 in.
- Drill collar:
  - OD = 7½ in.
  - ID = 3 in.
  - Length = 800 ft
- Mud:
  - Bingham; γ_m = 13 lbs/gal
  - θ_{600}= 40; θ_{300}= 25; gel = 30 lbs/100 ft^2 (Fann V-G meter)
- Last casing setting depth = 13,000 ft TMD
- Pump: P_{max} = 4,300 psi (allowed)
- Minimum annular fluid velocity = 90 ft/min

Determine the maximum flow rate while drilling at TMD of 13,000 ft and the maximum tripping in pipe velocity that will not cause any formation fracturing. Neglect surface connections.

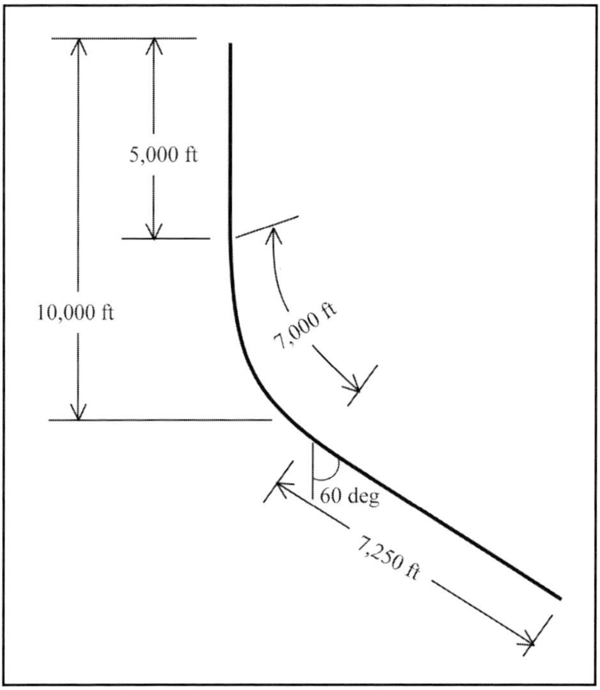

Fig. 3–16. Well schematic for problem 3–5

Problem 3–6. You are given the following well data:

- Last casing setting depth = 10,500 ft TMD
- Casing ID = 8½ in.
- TD = 16,000 ft TVD
- Drill bit: 7⅞ in. tri-cone roller cone
- Hole size washed out to 8½ in.
- Maximum required WOB to TD = 40,000 lbs
- Drill pipe:
  - OD = 4½ in.
  - ID = 3½ in.
  - Air weight = 27 lbs/ft
- Drill collar:
  - OD = 6½ in.
  - ID = 3½ in.
  - Air weight = 100 lbs/ft
- Coefficients of friction at walls: 0.25, casings; 0.3, open hole
- Steel weight = 65 lbs/gal
- Mud:
  - Power law
  - $\theta_{600} = 65$; $\theta_{300} = 40$ (Fann V-G meter)
- Rig surface connections: 600 ft of drill pipe
- Minimum allowed annular velocity = 120 ft/min
- Differential pressure required = 0.5 lbs/gal
- Casing shoe of 10,000 ft TVD to TD
- Formation:
  - Pressure gradient = 0.52 psi/ft
  - Fracture gradient = 13 lbs/gal
- Pump-duplex-pump factor = 3.7 gal/stroke

Determine the safe operating conditions, for $P_{p\ max\ =}$= 4,500 psi (drilling and tripping) at a depth of 16,000 ft TVD.

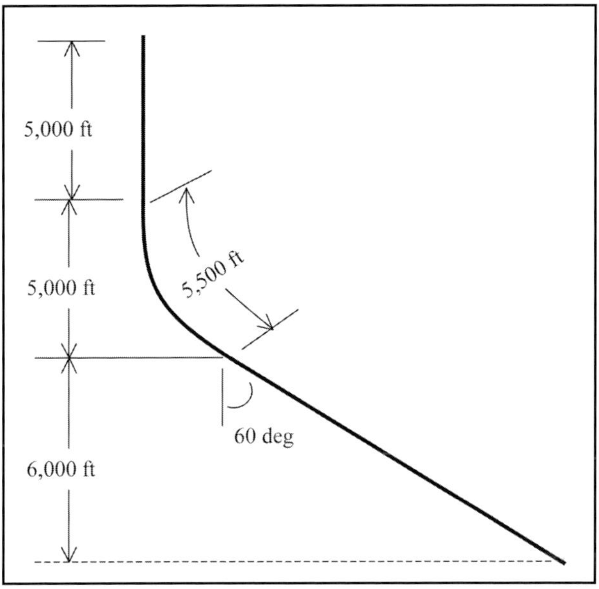

Fig. 3–17. Well schematic for problem 3.6

Problem 3–7. You are given the following well data:

- Drill pipe:
  - OD = 4.5 in.
  - ID = 4 in.
  - Weight = 12.75 lbs/ft
- Drill collar:
  - OD = 7.5 in.
  - ID = 4 in.
  - Weight = 107.3 lbs/ft
  - Length = 1,000 ft
- Intermediate casing:
  - OD = 10.75 in.
  - Wall thickness = 0.4 in.
- Current depth = 10,000 ft TVD
- TD = 15,000 ft
- Setting depth = 6,500 ft TVD
- Drill bit: 8.5 in. tri-cone roller cone jet with 9-9-9 nozzles
- Mud:
  - Bingham plastic; $\gamma = 10$ lbs/gal
  - $\theta_{600} = 95$; $\theta_{300} = 50$; $\tau_g = 30$ lbs/100 ft^2 (gel)
- Minimum annular velocity required for cuttings removal = 120 ft/min

- Pump:
  - 2,000 hp
  - Maximum allowed surface pressure = 4,800 psi
  - Volumetric efficiency = 90%
- Pump pressure at current depth:
  - At $Q = 250$ gpm, $P_p = 2{,}333$ psi
  - At $Q = 350$ gpm, $P_p = 4{,}377$ psi

(a) Determine whether formation fracturing will occur during breaking circulation. Assume that the formation fracture gradient at the casing shoe is 0.545 psi/ft.

(b) Find the bottom-hole equivalent circulating mud density at a flow rate of 450 gpm for (i) normal circulation and (ii) reverse circulation.

Problem 3–8. Derive the friction pressure loss equation in laminar flow for Newtonian fluids using the slot approximation for the annular flow.

Problem 3–9. Derive equation (3.87).

Nomenclature

a	Acceleration
A	Area open to flow
BLPE	Bingham fluid and laminar pipe flow
C_d	A nondimensional term
D	Elevation
D_{ep}	Equivalent pipe diameter; also, annulus equivalent pipe diameter
D_h	Hole diameter, in in.
D_{ip}	Pipe ID, in in.
D_{op}	Pipe OD, in in.
e	Distance between center of hole and center of pipe
f_a	Fanning friction factor for annular flow
f_p	Fanning friction factor for pipe flow
L	Length, in ft
N_{Ra}	Reynolds number for annular flow
N_{Rp}	Reynolds number for pipe flow
NLPF	Newtonian fluid and laminar pipe flow
P_b	Dynamic pressure change across the bit
P_B	Dynamic pressure across bit nozzles
P_d	Dynamic pressure
P_f	Friction pressure loss; also, circulating-system friction pressure loss
P_{fdp}, P_{fdc}	Friction pressure loss inside drill pipe and drill collar, respectively
P_{fadp}, P_{fadc}	Friction pressure loss in annulus around drill pipe and drill collar, respectively
ΔP_{fp}	Pipe friction pressure loss gradient

P_{fs}	Friction pressure loss inside surface connections
P_h	Hydrostatic head (pressure)
P_p	Required pump delivery pressure; also, pump pressure
Q	Flow rate, in gpm
R_h	Hole radius
R_{ip}	Internal pipe radius
R_{op}	Pipe outer radius
W_f	Friction energy loss
W_p	Work done by pump
γ	Fluid weight density
$\Delta\gamma_{a\ swab}$	Change in mud weight owing to the swab pressure
γ	Mud weight, in lbs/gal
γ_{me}	Equivalent circulating mud weight
γ_{mh}	Actual mud weight
ε	Annulus eccentricity; also, pipe roughness (with same dimensions as D_{ip})
θ_0	Zero-second gel
μ	Newtonian viscosity, in centipoise
μ_p	Plastic viscosity (PV), in centipoise
$\bar{v}$	Average fluid velocity
$\bar{v}_a$	Average annular fluid velocity, in ft/s
$\bar{v}_p$	Average fluid velocity inside pipe, in ft/s
ρ	Fluid mass density
NLPF τ_y	Yield value, in lbs/100 ft^2

References

1 Bourgoyne, A. T., M. E. Chenevert, K. K. Millheim, and F. S. Young. 1986. *Applied Drilling Engineering.* SPE Textbook Series, Vol. 2. Crittendon, B. C. 1959. "The Mechanics of Design and Interpretation of Hydraulic Fracture Treatment." *Journal of Petroleum Technology.* October, 21–29.

2 Melrose, J. C., J. G. Savins, W. R. Foster, and E. R. Parish. 1958. "A Practical Utilization of the Theory of Bingham Plastic Flow in Stationary Pipes and Annuli." *Transactions of the AIME.* 213, 316.

3 Iyoho, A. W., and J. J. Azar. 1980. "An Accurate Slot-Flow Model for Non-Newtonian Fluid Flow through Eccentric Annuli." Paper SPE 9447, presented at the SPE-AIME Annual Fall Meeting, Dallas.

4 Herschel, W. H., and R. Bulkley. 1926. *Kolloid Zeitschrift,* 39, 291–300. Herschel, W. H., and R. Bulkley. 1926. *Proceedings of the ASTM, Part II,* 26, 621–29.

5 Colebrook, C. F. 1938–39. "Turbulent Flow in Pipes, with Particular Reference to the Transition Region between the Smooth and Rough Pipe Laws." *Journal of the Institute of Civil Engineers,* London.

6 Tomita, Y. 1959. *Bulletin J.S.M.E.,* 2, 469. Tomita, Y. 1961. *Bulletin J.S.M.E.,* 4, 77.

7 Metzner, A. B. 1956. In *Advances in Chemical Engineering,* edited by T. B. Drew and J. W. Hoopes. Vol. 1, p. 77. New York: Academic Press. Metzner, A. B. 1957. *Ind. Eng. Chem.,* 49, 1429.

[8] Dodge, D. W., and A. B. Metzner. "Turbulent Flow of Non-Newtonian Systems." *Journal of the American Institute of Chemical Engineers*, Journal 5, 189, 1959.

[9] Metzner, A. B., and J. C. Reed. 1955. "Flow of Non-Newtonian Fluids—Correlation of the Laminar, Transition, and Turbulent-Flow Regions." *Journal of the American Institute of Chemical Engineers,* 1, 434.

[10] Skelland, A. H. P. 1967. *Non-Newtonian Flow and Heat Transfer.* New York: John Wiley & Sons.

[11] Tomita, *Bulletin J.S.M.E.,* 2, 469. Tomita, *Bulletin J.S.M.E.,* 4, 77.

[12] Fredrickson, A. G., and R. B. Bird. 1958. "Non-Newtonian Flow in Annuli." *Ind. and Eng. Chem.,* 347.

[13] Skelland. *Non-Newtonian Flow and Heat Transfer.*

[14] Burkhardt, J. A. 1961. "Wellbore Pressure Surges Produced by Pipe Movement." *Journal of Petroleum Technology,* June, 595–605. Burkhardt, *Transactions of the AIME,* 222.

[15] Bourgoyne et al. *Applied Drilling Engineering.*

4

Drill Bit Hydraulics

Introduction

The optimum utilization of mud pump horsepower is of significant importance in rotary drilling operations. An understanding and analytical evaluation of the power-consuming components of a rotary rig are essential to the optimum use of such available power.

Drill bit hydraulics is generally associated with the use of jet bits. The purpose of the jet nozzles is to improve the cleaning action of the drilling fluid at the bottom of the hole. Drilled cuttings of rocks that are not efficiently removed from underneath the drill bit will be reground and hence can cause wasteful bit wear and extend drilling time, thereby increasing drilling cost.

It is recognized throughout the petroleum industry that the fluid energy expended at the rock-bit interface has a significant effect on drilling performance. However, what is not recognized is how this hydraulic energy can be best utilized in achieving optimum removal of drilled cuttings from beneath the bit. Certain advances—such as the maximizing concepts of bit hydraulic power, jet impact force, and jet velocity—have provided general guidelines for bit hydraulic optimization. These criteria have served to prevent severe cuttings accumulations, but are far from being analytically reliable tools for achieving true optimum bottom-hole cleaning. The debate continues among investigators offering conflicting views; some contend that there is no limit to the benefit in penetration rate from the application of maximum hydraulic power, while others claim that for every formation type, there is an optimum power level above which those benefits will be offset by the cost of fuel and equipment maintenance.

Drilling hydraulics is the most complicated and least understood drilling variable because of the interactive complex relationships between the drilling fluid, the bit, and the formation. Although fluid-flow technology is very well developed, in drilling, the flow mechanism on the bottom of the hole is not yet understood. There is strong evidence, however, in the literature on bit hydraulics that the cross-flow of the impinging fluid beneath the bit is the governing factor in bottom-hole cuttings removal. Nevertheless, this belief is far from being conclusively proved. Some research has been and still is being conducted to add to the understanding of drilling-fluid momentum transfer mechanism, cross-flow velocities, fluid impact, and other concepts that are thought to be pivotal in bottom-hole cleaning.

Current field practice in bit hydraulic design involves the selection of an optimum flow rate and corresponding optimum bit nozzle sizes that maximize one of the following parameters:

- Drill bit hydraulic horsepower
- Jet impact force
- Jet nozzle velocity

Recently, the concept of jet impact pressure has been presented in the literature.[1] However, its practical field application has not yet been demonstrated. The first two criteria—namely, bit hydraulic horsepower and jet impact force—are the most commonly used and therefore will be the focus of the remainder of this chapter.

Pump Pressure Requirement in Rotary Drilling

The available pump surface pressure, normally referred to as the circulating *standpipe pressure,* is expanded throughout the circulating system in the following manner:

- Friction pressure loss in surface connections (P_{fs})
- Friction pressure loss inside drill pipe (P_{fdp})
- Friction pressure loss inside drill collar (P_{fdc})
- Friction pressure loss in annulus around drill pipe (P_{fadp})
- Friction pressure loss in annulus around drill collar (P_{fadp})
- Dynamic pressure change across bit (P_b)

Mathematically this can be written as follows:

$$P_s = P_{fs} + P_{fdp} + P_{fdc} + P_{fadp} + P_{fadc} + P_b \tag{4.1}$$

or

$$P_s = P_f + P_b \tag{4.2}$$

where P_f is the total friction pressure loss and P_s is the maximum allowed surface pressure, $P_{s\,max}$ (or some specified optimum operating surface pressure, $P_{s\,opt}$); P_b, the pressure drop ac ross the bit, can be determined using the equations developed in chapter 3.

Example 4–1

A 15,000 ft well is to be drilled. It has been determined that the following maximum pressure gradients in the circulating system will be encountered while drilling the last 1,000 ft hole interval:

(a) $\Delta P_{fdb} = 0.1$ psi/ft
(b) $\Delta P_{fdc} = 0.3$ psi/ft
(c) $\Delta P_{fadp} = 0.01$ psi/ft
(d) $\Delta P_{fadc} = 0.07$ psi/ft
(e) $P_b = 2{,}200$ psi

Determine the surface pressure requirement for each of these maximum pressure gradients. The drill collar length is 1,000 ft and the surface connections equivalent drill pipe length is 600 ft.

Solution: From equations (4.1) and (4.2),

$$\begin{aligned} P_s &= (L_{dp} + L_s)\Delta P_{fdp} + L_{dc}\Delta P_{fdc} + L_{dp}\Delta P_{fadp} + L_{dc}\Delta P_{fadc} + P_b \\ &= (14{,}600 \times 0.1) + (1{,}000 \times 0.3) + (14{,}000 \times 0.01) + (1{,}000 \times 0.07) + 2{,}200 \\ &= 4{,}170 \text{ psi} \end{aligned}$$

Hydraulic Power Requirement

Hydraulic power, H_h, is defined as the product of a flow rate Q and a corresponding pressure P—that is,

$$H_h = PQ \tag{4.3}$$

Multiplying both sides of equation (4.2) by Q and utilizing equation (4.3) gives the following:

$$P_s Q = P_f Q + P_b Q \tag{4.4}$$

or

$$H_{hs} = H_{hf} + H_{hb} \tag{4.5}$$

where

H_{hs} is the hydraulic pump power

H_{hf} is the hydraulic power due to friction pressure loss in circulating system.

H_{hb} is the hydraulic power due to dynamic pressure change across the drill bit.

In field units, equation (4.3) becomes,

$$H_h = \frac{PQ}{1{,}714} \tag{4.6}$$

where H_h is the hydraulic horsepower (in hp), P is in psi, and Q is in gpm.

Example 4–2

In example 4–1, determine the pump hydraulic horsepower required in order to achieve a pump volumetric efficiency of 80%. Assume a flow rate of 350 gpm.

Solution:

$$H_{hf} = \frac{P_f Q}{1{,}714} = \frac{1{,}970 \times 350}{1{,}714} = 402.28 \text{ hp}$$

$$H_{hb} = \frac{P_b Q}{1{,}714} = \frac{2{,}200 \times 350}{1{,}714} = 449.24 \text{ hp}$$

Thus, by use of equation (4.5), H_{hs} = 402.28 + 449.24 = 852 hp. For 80% volumetric efficiency, the pump hydraulic horsepower must be H_{hp} = 852/0.80 = 1,065 hp.

Flow Exponent (α)

From chapter 3, it can be deduced easily that friction pressure loss. P_f, is related to flow rate, Q, by the relation

$$P_f = CQ^{\alpha} \tag{4.7}$$

where C is a constant (a function of mud flow properties, hole geometry, pipe geometry, etc.) and α is a *flow exponent*, varying between 1 and 2, that either can be measured in the field, as will be shown subsequently, or can be assumed to have a value of 1.86.

Taking the logarithm of both sides of equation (4.7) yields

$$\log P_f = \log(CQ^{\alpha}) = \log C + \alpha \log Q \tag{4.8}$$

If equation (4.8) is plotted on a log-log paper, the resulting curve is a straight line whose slope is α. Therefore, if two points, $i(P_{fi}, Q_i)$ and $j(P_{fj}, Q_j)$ are known, then the slope of the line is determined by

$$\alpha = \frac{\log P_{fj} - \log P_{fi}}{\log Q_j - \log Q_i} \tag{4.9a}$$

or

$$\alpha = \frac{\log(P_{fj} / P_{fi})}{\log(Q_j / Q_i)} \tag{4.9b}$$

In the field, while drilling at a certain depth using a known mud weight and bit nozzle sizes, the circulating pump pressure (standpipe pressure) can be easily recorded at various flow rates. Thus, at flow rates Q_j and Q_i, the corresponding standpipe pressures of P_{sj} and P_{si} can be recorded, respectively.

From equation (4.2), $P_{sj} = P_{fj} + P_{bj}$, and $P_{si} = P_{fi} + P_{bi}$. Further, P_{bj} and P_{bi} can be calculated from equation (4.9b):

$$P_{bj,i} = \frac{\left(8.3 \times 10^{-5}\right)\gamma Q_{j,i}^{2}}{A_t^2 C_d^2} \tag{4.10}$$

which in turn allows the determination of P_{fj} and P_{fi}:

$$P_{fj} = P_{sj} - \frac{\left(8.3 \times 10^{-5}\right)\gamma Q_j^2}{A_t^2 C_d^2} \tag{4.11a}$$

$$P_{fi} = P_{si} - \frac{\left(8.3\times10^{-5}\right)\gamma Q_i^2}{A_t^2 C_d^2} \tag{4.11b}$$

With P_{fj} and P_{fi} and corresponding flow rates and given known Q_j and Q_i values, the flow exponent, α, can be determined using equation (4.9b).

Example 4–3

A well is being drilled at 12,000 ft while using a 15.5 lbs/gal mud and a drill bit having 14-14-14 nozzles. At flow rates of 400 gpm and 300 gpm, the circulating standpipe pressures recorded were 4,622 psi and 2,800 psi, respectively. Determine the flow exponent,α, for the well.

Solution: The flow exponent is determined using equation (4.9b), given $Q_j = 400$ gpm and $Q_i = 300$ gpm. From equation (4.11a),

$$\begin{aligned} P_{fj} &= P_{sj} - \frac{(8.3\times10^{-5})\gamma Q_j^2}{A_t^2 C_d^2} \\ &= 4,622 - \frac{\left(8.3\times10^{-5}\right)(15.5)(400)^2}{(\pi/4)4(14/32)^2(0.95)^2} \\ &= 4,622 - 3,150 = 1,472 \text{ psi} \end{aligned}$$

Similarly,

$$\begin{aligned} P_{fj} &= P_{sj} - \frac{\left(8.3\times10^{-5}\right)\gamma Q_i^2}{A_t^2 C_d^2} \\ &= 2,800 - 1,800 = 1,000 \text{ psi} \end{aligned}$$

Therefore,

$$\begin{aligned} \alpha &= \frac{\log(1,472/1,000)}{\log(400/300)} \\ &= 1.34 \end{aligned}$$

Maximum Drill Bit Hydraulic Horsepower Criterion

The *maximum bit hydraulic horsepower* criterion assumes that optimum bottom hole cleaning is achieved if hydraulic horsepower across the drill bit is maximized with respect to the flow rate Q. Thus, expressing this hydraulic horsepower explicitly in terms of Q and then using differential calculus leads to the desired flow rate to ensure optimum bottom-hole cleaning.

Drill bit hydraulic horsepower was defined previously as $H_{hb} = P_b Q$. However, P_b is a function of flow rate and can be eliminated by making the substitution $P_b = P_s - P_f = P_s - C_1 Q_1^{\alpha}$. Therefore,

$$\begin{aligned} H_{hb} &= (P_s - CQ^{\alpha})Q \\ &= P_s Q - CQ^{\alpha+1} \end{aligned} \tag{4.12}$$

where H_{hb} is the dependent variable and is a function of flow rate, Q, which is the independent variable.

Thus, by use of differential calculus, taking the first derivative of H_{hb} with respect to Q and setting the result equal to zero,

$$\frac{dH_{hb}}{dQ} = P_s - (\alpha + 1)CQ^{\alpha} = 0 \tag{4.13}$$

Since $P_f = CQ^{\alpha}$,

$$P_s - (\alpha + 1)P_f = 0 \tag{4.14a}$$

or

$$P_f = \frac{1}{\alpha + 1} P_s \tag{4.14b}$$

which is the root that makes H_{hb} a maximum. Note that $d^2H_{hb}/dQ^2 < 0$ at $P_f = [1/(\alpha+1)]P_s$, which proves that H_{hb} is maximum at this value of P_f.

Hence, on the basis of the maximum bit hydraulic horsepower criterion, the optimum bit hydraulics will be achieved if friction pressure loss in the circulating system is maintained at an optimum value of

$$P_{f\,opt} = \frac{1}{\alpha + 1} P_{s\,max} \tag{4.15}$$

or at the optimum pressure drop across the nozzles,

$$\begin{aligned} P_{b\,opt} &= P_{s\,max} - P_{f\,opt} \\ &= \frac{\alpha}{\alpha + 1} P_{s\,max} \end{aligned} \tag{4.16}$$

From equation (4.9b), the following can be written,

$$\alpha = \frac{\log\left(P_{fopt} / P_{fqa}\right)}{\log\left(Q_{opt} / Q_a\right)} \tag{4.17}$$

where $P_{f\ opt}$ is as defined in equation (4.15), P_{fqa} is the friction pressure loss calculated or measured at some flow rate Q_a, and Q_{opt} is the sought optimum flow rate. Solving for Q_{opt} yields

$$Q_{opt} = Q_a \text{ antilog}\left[\frac{1}{\alpha}\log\left(\frac{P_{f\ opt}}{P_{fqa}}\right)\right] \tag{4.18}$$

Thus, if α is known, then $P_{f\ opt}$ can be calculated using equation (4.18). At some assumed flow rate, Q_a, P_{fqa} can be calculated from the friction pressure loss equations developed in Ch. 3 or measured in the field, then Q_{opt} can be found, using equation (4.18).

With Q_{opt} known and P_{bopt} from equation (4.16), equation (4.10) can be rewritten as

$$P_{bopt} = \frac{\left(8.3\times10^{-5}\right)\gamma Q_{opt}}{A_{t\,opt}^2 C_d^2}\ , \tag{4.19}$$

or, solving for $A_{t\ opt}$,

$$A_{t\ opt} = \sqrt{\frac{\left(8.3\times10^{-5}\right)\gamma Q_{opt}}{C_d^2 P_{b\ opt}}} \tag{4.20}$$

With the total optimum area of the nozzles known, the corresponding optimum *nozzle diameters*, d_{nopt}, can be found:

$$A_{t\ opt} = \frac{\pi}{4}(d_{1\ opt}^2 + d_{2\ opt}^2 + d_{3\ opt}^2 + \cdots + d_{n\ opt}^2) \tag{4.21}$$

If the nozzles are all of equal size (i.e., $d_1 = d_2 = d_3 = L = d_n$), then

$$A_{t\ opt} = \frac{\pi}{4} n d_{n\ opt}^2 \tag{4.22}$$

Solving for $d_{n\ opt}$ yields

$$d_{n\ opt} = 2\sqrt{\frac{A_{t\ opt}}{n\pi}} \tag{4.23}$$

where n is the number of nozzles of the drill bit. For a tri-cone roller cone bit, n = 3. For drag bits, n can be more than 3.

Importantly, the *optimum flow rate,* Q_{opt}, to be used in drilling an interval of depth must be bounded by maximum and minimum flow rates, Q_{max} and Q_{min}, respectively; that is, $Q_{min} \leq Q_{opt} \leq Q_{max}$. The maximum flow rate is governed by the following:

- Available rig hydraulic horsepower and the allowed operating pressure of the pump; that is, $Q_{max} = H_{hp}\eta_v/P_{s\,max}$, where η_v is the pump volumetric efficiency
- Potential hole erosion problems arising from high flow rates
- Potential formation-fracturing problems arising from excessive annular friction pressure losses or the induced ECD: $P_{frac} > P_h + P_{af} = \text{ECD}$, where P_{frac} is the formation fracture pressure, P_h is the mud hydrostatic pressure, and P_{af} is the friction pressure loss in the annulus

The minimum flow rate is governed by the cuttings transport phenomenon. For acceptable annular hole cleaning, a minimum annular fluid average velocity, $v_{a\,min}$, must be maintained: $Q_{min} = v_{a\,min}A_{a\,max}$, where $A_{a\,max}$ is the largest annulus cross-sectional area open to flow. This is discussed in detail in chapter 3

Example 4–4

You are given the following well data:

- Drill pipe:
  - OD = 4.5 in.
  - ID = 3.64 in.
  - Weight = 20 lbs/ft
- Drill collar:
  - OD = 7 in.
  - ID = 2 in.
  - Weight = 120.3 lbs/ft
  - Length = 1,000 ft
- Mud:
  - Bingham
  - $\theta_{300} = 21$; $\theta_{600} = 29$ (Fann V-G meter)
  - Weight = 15.5 lbs/gal
- Pump:
  - National duplex (double acting)
  - Maximum allowed operating pressure = 5,440 psi
  - Hydraulic horsepower = 1,600 hp
  - Volumetric efficiency = 80%
- Minimum required annular fluid velocity: 85 ft/min average
- Well geometry:
  - Shape as shown in figure 4–1
  - Last intermediate casing is 9⅝ in. at 12,000 ft TVD
- Drill bit: 12⅞ in. tri-cone with 3-14 nozzles to 12,000 ft; next bit to be used is 8⅜ in. tri-cone (assume that the hole is washed out to 9⅝ in.)

- Field data: At 12,000 ft, while using the 8⅞ in. tri-cone 3-14 nozzles,
  - $Q_1 = 300$ gpm
  - $Q_2 = 400$ gpm
  - $P_{p1} = 2{,}966$ psi
  - $P_{p2} = 4{,}883$ psi

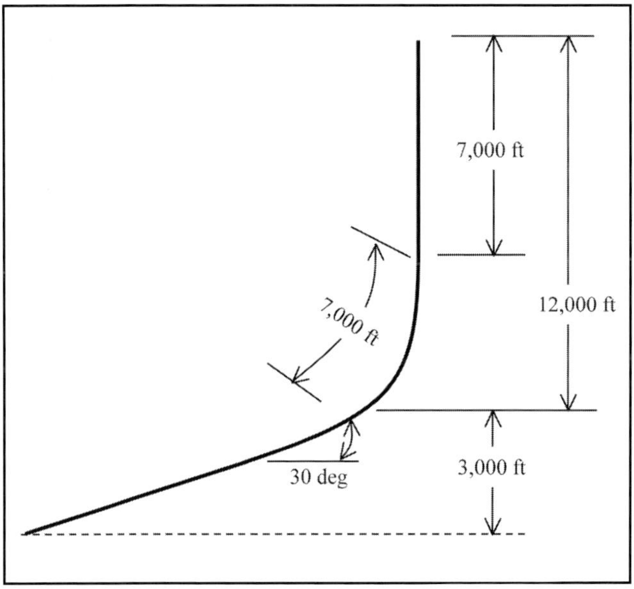

Fig. 4–1. Well schematic in example 4–4

Using the bit hydraulic horsepower criterion, determine the optimum nozzle sizes to be used to drill the next depth interval. Assume current depth is 12,000 ft.

Solution: The pressure drops through the bit at flow rates of 300 and 400 gpm can be determined as follows:

$$P_b = \frac{\left(8.311 \times 10^{-5}\right)\gamma Q^2}{A_t^2 C_d^2}$$

Assume that $C_d = 0.95$. The total nozzle area is

$$A_t = \frac{\pi}{4}(d_1^2 + d_2^2 + d_3^2) = \frac{3\pi}{4}\left(\frac{14}{32}\right)^2 = 0.45099 \text{ in.}^2$$

Further,

$$P_{b1} = \frac{\left(8.311\times10^{-5}\right)(15.5)(300)^2}{(0.95)^2(0.45099)^2} = 631.6 \text{ psi}$$

$$P_{b2} = \frac{\left(8.311\times10^{-5}\right)(15.5)(400)^2}{(0.95)^2(0.45099)^2} = 1{,}122.8 \text{ psi}$$

Then, the friction pressure losses at the above given flow rates are $P_f = P_p - P_b$, or

$$P_{f1} = 2{,}966 - 631.6 = 2{,}334.4 \text{ psi}$$

$$P_{f2} = 4{,}883 - 1{,}122.8 = 3{,}760.2 \text{ psi}$$

Therefore, the flow exponent for the given well geometry can be determined as follows:

$$\alpha = \frac{\log(P_{f2}/P_{f1})}{\log(Q_2/Q_1)} = \frac{\log(3{,}760.2/2{,}334.4)}{\log(400/300)} = 1.66$$

The maximum and minimum flow rates are

$$Q_{max} = 1{,}714(\eta_v)\left(\frac{H_{hp}}{P_{p\,max}}\right) = 1{,}714(0.8)\left(\frac{1{,}600}{5{,}440}\right) = 403.3 \approx 403 \text{ gpm}$$

$$Q_{min} = 2.448\left(D_h^2 - D_p^2\right)v_{min} = 2.448(9.875^2 - 4.5^2)\left(\frac{85}{60}\right) = 267.9 \approx 268 \text{ gpm}$$

The optimum friction pressure loss $P_{\text{f opt}}$, using the maximum hydraulic horsepower criterion, is

$$P_{\text{f opt}} = \left(\frac{1}{\alpha + 1}\right) P_{\text{p max}}$$
$$= \frac{1}{1.66 + 1}(5,440)$$
$$= 2,047.4 \approx 2,047 \text{ psi}$$

The optimum pressure drop at the bit can be calculated as

$$P_{\text{b opt}} = P_{\text{p max}} - P_{\text{f opt}}$$
$$= 5,440 - 2,047 = 3,393 \text{ psi}$$

The optimum flow rate is

$$Q_{\text{opt}} = Q_{\text{a}} \text{ antilog}\left[\frac{1}{\alpha}\log\left(\frac{P_{\text{f opt}}}{P_{\text{fqa}}}\right)\right]$$
$$= 300 \text{ antilog}\left[\frac{1}{1.66}\log\left(\frac{2,047}{2,334}\right)\right]$$
$$= 227 \text{ gpm}$$

This is greater than the Q_{min} of 268 gpm and less than the Q_{max} of 403 gpm; therefore, the Q_{opt} value of 277 gpm is acceptable. The optimum total area of nozzles is

$$A_{\text{opt}} = \sqrt{\frac{\left(8.311 \times 10^{-5}\right)\gamma\left(Q_{\text{opt}}\right)^2}{C_{\text{d}}^2 \Delta P_{B\,\text{opt}}}}$$
$$= \sqrt{\frac{\left(8.311 \times 10^{-5}\right)(15.5)(277.1)^2}{(0.95)^2(3,392.6)}}$$
$$= 0.18 \text{ in.}^2$$

For three equal-sized nozzles, the optimum (from eq. [4.23]) is,

$$d_{n\,\text{opt}} = 2\sqrt{\frac{0.18}{3\pi}} = 0.28 \text{ in.}$$

Therefore, three 9/32 in. nozzles should be used. The actual nozzle flow area is

$$A_t = \frac{3\pi}{4(32)^2}(9)^2 = 0.1863\,\text{in.}^2$$

which is acceptable.

Maximum Jet Impact Force Criterion

The *maximum jet impact force* criterion assumes that optimum bottom-hole cleaning is achieved by maximizing the jet impact force with respect to the flow rate. Thus, the desired value of Q that makes F_j maximum is found by expressing this impact force, F_j, in terms of the flow rate Q and using differential calculus.

The jet impact force (hydraulic impact) induced at the bottom of the hole by the mud can be derived from Newton's second law of motion and is given by

$$F_j = BQ\sqrt{P_b} \tag{4.24}$$

for $B = 0.01823C_d\sqrt{\gamma}$, where Q is in gpm, γ is in lbs/gal, and P_b is in psi. The pressure change across the drill bit is $P_b = p_s - p_f = p_s - CQ_\alpha$. Therefore, equation (4.24) becomes

$$F_j = BQ\sqrt{P_s - CQ^\alpha} \tag{4.25}$$

Prior to carrying out the differentiation there are two limitations to the hydraulic impact force that need to be considered: (1) limitations owing to the maximum available pump hydraulic horsepower, $H_{p\,max}$, and (2) limitations owing to the maximum allowed surface operating pressure, $P_{s\,max}$. During the drilling of the shallow portion of the well, where the system friction pressure loss requirement is small and the flow rate requirement is large, the hydraulic impact force is limited only by the limited pump hydraulic horsepower, $H_{p\,max}$; therefore, available surface pressure is expressed as

$$P_s = \frac{H_{p\,max}}{Q} \tag{4.26}$$

Substituting equation (4.26) into equation (4.25) yields

$$\begin{aligned} F_j &= BQ\sqrt{\frac{H_{p\,max}}{Q} - CQ^\alpha} \\ &= B\sqrt{H_{p\,max}Q - CQ^{\alpha+2}} \end{aligned} \tag{4.27}$$

Differentiating equation (4.27) and setting to zero yields

$$\frac{dF_j}{dQ} = \frac{(1/2)B\left[H_{p\max} - (\alpha + 2)CQ^{\alpha+1}\right]}{\sqrt{H_{p\max}Q - CQ^{\alpha+2}}} = 0 \tag{4.28}$$

For a valid solution, the numerator in equation (4.28) must be equal to zero: $(1/2)B[H_{p\max} - (\alpha + 2)CQ^{\alpha+1}] = 0$, or $H_{p\max} - (\alpha+2)CQ^{\alpha}Q = 0$, or $P_sQ - (\alpha + 2)CP_fQ = 0$. Further, solving for the optimum friction pressure that must be maintained in circulating system yields

$$P_{f\,opt} = \frac{1}{\alpha + 2}P_{s\,opt} \tag{4.29}$$

The optimum pressure that must be maintained across drill bit is

$$P_{opt} = P_{s\,opt} - P_{f\,opt} = \frac{\alpha + 1}{\alpha + 2}P_{s\,opt} \tag{4.30}$$

As the well is drilled deeper, the system friction pressure loss requirement increases, while the flow rate requirement decreases. Therefore, the hydraulic impact force will be limited by the limited maximum allowed pump pressure, $P_{s\max}$. Hence, equation (4.25) becomes $F_j = BQ\sqrt{P_{s\max} - CQ^{\alpha}}$. Differentiating and setting equal to zero yields

$$\frac{dF_j}{dQ} = \frac{(1/2)B\left[2P_{s\max}Q - (\alpha + 2)CQ^{\alpha+1}\right]}{\sqrt{P_{s\max}Q^2 - CQ^{\alpha+2}}} = 0 \tag{4.31}$$

For valid solution, the numerator in equation (4.31) must be zero: $(1/2)B[2P_{s\max}Q - (\alpha + 2)CQ^{\alpha+1}] = 0$, or $2P_{s\max} - (\alpha+2)P_f = 0$. Therefore,

$$P_{f\,opt} = \frac{2}{\alpha + 2}P_{s\max} \tag{4.32}$$

Hence,

$$P_{b\,opt} = P_{s\max} - P_{f\,opt} = \frac{\alpha}{\alpha + 2}P_{s\max} \tag{4.33}$$

Using equations (4.29), (4.30), (4.32), and (4.33), the optimum flow rate, Q_{opt}, and the optimum nozzles sizes, $d_{n\,opt}$, can be determined on the basis of the jet impact force maximization criterion.

Example 4–5

Refer to the well data given in example 4–4.

(a) Using the impact force criterion, determine the optimum nozzle sizes to be used to drill the next interval of depth, starting at 12,000 ft.

(b) Determine the maximum total hydraulic horsepower that will be utilized under the drilling conditions in part (a). Assume a bit life of 50 h and a drilling rate of 10 ft/h. Friction pressure losses at the flow rate found in part (a) are given as follows: $\Delta P_{fdc} = 1.587$ psi/ft; $\Delta P_{fadp} = 0.0126$ psi/ft; $\Delta P_{fadc} = 0.0254$ psi/ft.

Solution:

(a) From example 4–4, $\alpha = 1.66$, $Q_{min} = 268$ gpm, and $Q_{max} = 403$ gpm. Since the interval in question is at a deeper part of the well, pump surface pressure will be the limiting factor. Therefore, from equations (4.32) and (4.33),

$$P_{f\,opt} = \frac{2P_{s\,max}}{2+\alpha} = \frac{2 \times 5{,}440}{3.66} = 2{,}975 \text{ psi}$$

$$P_{b\,opt} = P_{s\,max} - P_{f\,opt} = 5{,}400 - 2{,}975 = 2{,}465 \text{ psi}$$

Further,

$$Q_{opt} = Q_a \text{ antilog}\left[\frac{\log\left(P_{f\,opt}/P_{fqa}\right)}{\alpha}\right] = 300 \text{ antilog}\left[\frac{\log(2{,}975/2{,}334)}{1.66}\right] = 347 \text{ gpm}$$

which is bound by the maximum and minimum flow rates and therefore is acceptable. Hence,

$$A_{opt} = \frac{\left(8.3 \times 10^{-5}\right) \times 15.5 \times 347^2}{0.95^2 \times 2{,}465} = 0.2639 \text{ in.}^2$$

For three equal nozzles, $d_{n\,opt} = 10.71/32$ in.; thus, use three 11/32 in. nozzles.

(b) The total footage drilled by the bit will be $A = RT = 50$ ft/h × 10 h = 500 ft.

Then, the well TMD and TVD are found: TMD = 14,000 + 500 = 14,500 ft, and TVD = 12,000 + 500 sin 30° = 12,250 ft.

Since the frictional pressure loss gradient inside the drill pipe is not given, it must be calculated at 347.0 gpm. The average velocity inside the drill pipe ($\overline{v}_{dp}$) is

$$\overline{v}_{dp} = \frac{24.51Q}{D_p^2} = \frac{24.5 \times 347.2}{(3.64)^2} = 642.3 \text{ ft/min}$$

The plastic viscosity (PV) and the yield point (YP), respectively, are

$$PV = \theta_{600} - \theta_{300} = 29 - 21 = 8 \text{ centipoise}$$

$$YP = \theta_{300} - PV = 21 - 8 = 13 \text{ lbs/100ft}^2$$

Then, the critical velocity for turbulent flow (v_c) is found:

$$v_c = \frac{64.57(PV) + (64.57)\sqrt{(PV)^2 + 9.9\rho_m D_p^2 (YP)}}{15.6 \times 3.64}$$

$$= 194.78 \text{ ft/min}$$

Since $\overline{v}_{dp} > v_c$ the flow regime is turbulent.

The Reynolds number is

$$N_R = \frac{928\overline{v}D_p\gamma_m}{PV}$$

$$= \frac{928(642.3/60)(3.64)(15.6)}{8} = 70{,}513.2$$

The friction factor (assuming smooth pipe) is

$$f = \frac{0.0791}{\sqrt[4]{N_R}}$$

$$= \frac{0.0791}{\sqrt[4]{70{,}513.2}} = 0.004854$$

The pressure loss gradient is

$$\Delta P_{fdp} = \frac{f\gamma_m v_{dp}}{93{,}000 D_p}$$

$$= \frac{0.004854(15.6)(642.3)^2}{(3.64)} = 0.09228 \text{ psi/ft}$$

At the new depth, the total friction pressure loss is

$$\begin{aligned} P_f &= (\Delta P_{dp} + \Delta P_{adp})L_{dp} + (\Delta P_{dc} + \Delta P_{adc})L_{dc} \\ &= (0.09228 + 0.0126)(13{,}500) + (1.587 + 0.0254)(1{,}000) \\ &= 3{,}028 \text{ psi} \end{aligned}$$

The pressure drop at the bit is

$$\begin{aligned} P_b &= \frac{(8.311 \times 10^{-5})\gamma Q^2}{C_d^2 A_t^2} \\ &= \frac{(8.311 \times 10^{-5})(15.6)(347.2)^2}{(0.95)^2(0.2784)^2} = 2{,}234 \text{ psi} \end{aligned}$$

Thus, the total pump pressure is

$$\begin{aligned} P_p &= P_f + P_b \\ &= 3{,}028 + 2{,}234 = 5{,}262 \text{ psi} \end{aligned}$$

and the required hydraulic horsepower (HHP) is

$$\begin{aligned} \text{HHP} &= \frac{QP_p}{1{,}714} \\ &= \frac{347.2 \times 5{,}262}{1{,}714 \times 0.80} = 1{,}333 \text{ hp} \end{aligned}$$

Supplementary Problems

Problem 4–1. You are given the following well data and well trajectory shown in fig. 4–2:

- Open hole size = 8⅞ in. (washed out to 9⅞ in.) to TD of 20,250 ft
- Drill pipe:
  - OD = 5 in.
  - ID = 3 in.
- Drill collar:
  - OD = 7½ in.
  - ID = 3 in.
  - Length = 800 ft

- Mud:
  - Bingham; $\gamma_m = 13$ lbs/gal
  - Gel = 30 lbs/100 ft^2
  - $\theta_{600} = 40$; $\theta_{300} = 25$ (Fann V-G meter)
- Last casing set at 13,000 ft TD
- Constant slope of hydraulics = 1.85
- Pump: $P_{max} = 4{,}300$ psi (allowed)
- Minimum annular fluid velocity = 90 ft/min

Neglect surface connections and assume that the minimum flow rate is allowed.

(a) Determine the optimum nozzle sizes to be used for optimum hydraulics (bottom-hole cleaning) at a TD of 13,000 ft. Use the bit hydraulic horsepower criterion.

(b) If the formation fracture gradient is 0.831 psi/ft, indicate whether fracture will occur (i) while drilling under the conditions specified in part (a), and (ii) when a trip is made at a TD of 20,000 ft.

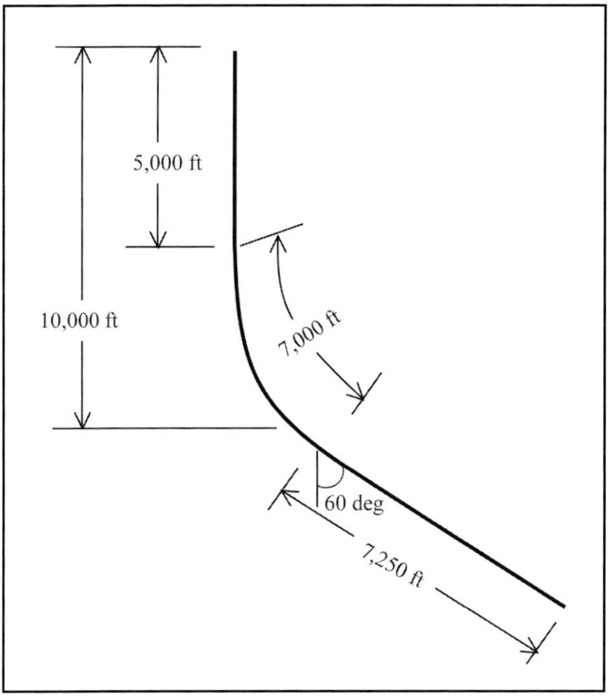

Fig. 4–2. Well trajectory in problems 4–1 and 4–11

Problem 4–2. You are given the following data (see also fig. 4–3):

- Last casing set at 10,500 TMD
- Casing ID = 8.5 in.
- TD = 16,000 ft TVD
- Drill bit: 7⅞ in. tri-cone roller cone (hole size washed out to 8½ in.)
- Maximum required WOB to TD = 40,000 lbs
- Drill pipe:
  - OD = 4½ in.
  - ID = 3½ in.
  - Air weight = 27 lbs/ft
- Drill collar:
  - OD = 6½ in.
  - ID = 3½ in.
  - Air weight = 100 lbs/ft
- Coefficients of friction at walls: 0.1, casings; 0.2, open hole
- Steel weight = 65 lbs/gal
- Mud:
  - Power law
  - $Q_{600} = 65$; $Q_{300} = 40$ (Fann V-G meter)
  - Gel strength = 15/25 lbs/100 ft^2
- Friction pressure losses while drilling at 16,000 ft TVD: $\Delta P_{adc} = 0.012$ psi/ft; $\Delta P_{adc} = 0.079$ psi/ft; $\Delta P_{fdp} = \Delta P_{fdc} = 0.065$ psi/ft.
- Surge/swab pressures in annulus while tripping from 16,000 ft TVD: $\Delta P_{surge} = \Delta P_{swab} = 0.02$ psi/ft
- Rig surface connections = 600 ft of drill pipe
- Minimum allowed annular velocity = 120 ft/min
- Differential pressure required = 0.5 lbs/gal
- Casing shoe of 10,000 ft TVD to TD
- Formation pressure gradient = 0.52 psi/ft
- Formation fracture gradient = 13 lbs/gal
- Pump: Duplex
  - Pump factor = 3.7 gal/stroke
  - $P_{max} = 4{,}500$ psi

(a) Indicate whether any drilling problems will occur under the above conditions. Show details.

(b) Using the jet impact force criterion, determine the optimum bit nozzle sizes to be used for drilling the next interval of hole starting at 10,500 ft TMD with 12-12-12 jets, given the following recordings: (i) at 805 gpm, pump pressure = 2,000 psi; (ii) at 905 gpm, pump pressure = 2,500 psi.

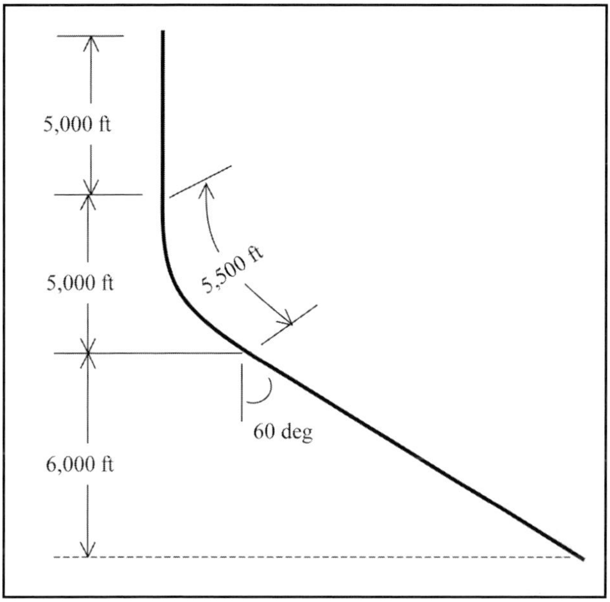

Fig. 4–3. Well conditions in problem 4–2

Problem 4–3. You are given the following well data:

- Drill pipe:
  - OD = 4.5 in.
  - ID = 4 in.
  - Air weight = 12.75 lbs/ft
- Drill collar:
  - OD = 7.5 in.
  - ID = 4 in.
  - Air weight = 107.3 lbs/ft
  - Length = 1,000 ft
- Intermediate casing:
  - OD = 10¾ in.
  - Wall thickness = 0.4 in.
- Current depth = 10,000 ft TVD
- TD = 15,000 ft
- Settings depth = 6,500 ft TVD
- Drill bit: Tri-cone roller cone jet bit; 8½ in. with 9-9-9 nozzles.
- Mud:
  - Bingham plastic; $\gamma = 10$ lbs/gal
  - $Q_{600} = 95$; $Q_{300} = 50$; $\tau_g = 30$ lbs/ft^2

- Minimum annular velocity required for cuttings removal = 120 ft/min
- Pump:
  - 2,000 hp
  - Maximum allowed surface pressure = 4,800 psi
  - Volumetric efficiency = 90%
- Pump pressure at current depth:
  - At Q_1 = 250 gpm, P_p = 2,333 psi
  - At Q_2 = 350 gpm, P_p = 4,377 psi

(a) Determine the optimum flow rate and nozzle sizes to be used for drilling the next interval of depth. Use maximum jet impact force and maximum bit hydraulic horsepower criteria.

(b) Determine if formation fracturing will occur during breaking circulation. Assume that the formation fracture gradient at the casing shoe is 0.545 psi/ft.

(c) Find the bottom hole equivalent circulating mud density at a flow rate of 450 gpm for (i) normal circulation and (ii) reverse circulation.

Problem 4–4. You are given the following well data (see also fig. 4–4):

- Last casing setting = 14,000 ft TMD
- TD = 13,500 ft TVD
- Hole size to TD = 7⅞ in.
- Drill pipe:
  - OD = 4¼ in.
  - ID = 3½ in.
  - Air weight = 26.88 lbs/ft
- Drill collar:
  - OD = 6 in.
  - ID = 3½ in.
  - Air weight = 100.8 lbs/ft
  - Length = 1,500 ft
- Pumps: 2,000 hp; $\eta_?$ = 85%; P_{max} = 3,800
- Minimum annular fluid velocity for hole cleaning: 120 ft/min
- Coefficients of friction at walls: 0.1, cased sections; 0.2, open hole section
- Mud:
  - Bingham plastic
  - Specific gravity = 2.0 for next depth interval
  - θ_{600} = 55; θ_{300} = 35; 10 s/10 min; gel = 18/30 lb/100 ft^2 (Fann V-G meter)
- Formation below 14,000 ft TMD:
  - Fracture gradient = 18.0 lbs/gal;
  - Pore pressure gradient = 0.83 psi/ft

- Friction pressure losses:
  - While drilling at 14,000 ft TMD: $\Delta P_{fdp} = \Delta P_{fdc} = 0.075$ psi/ft; $\Delta P_{fadp} = 0.01$ psi/ft; $\Delta P_{fadc} = 0.08$ psi/ft
  - During tripping from 14,000 ft TMD: $\Delta P_{surge} = \Delta P_{swab} = 0.025$ psi/ft (annulus)

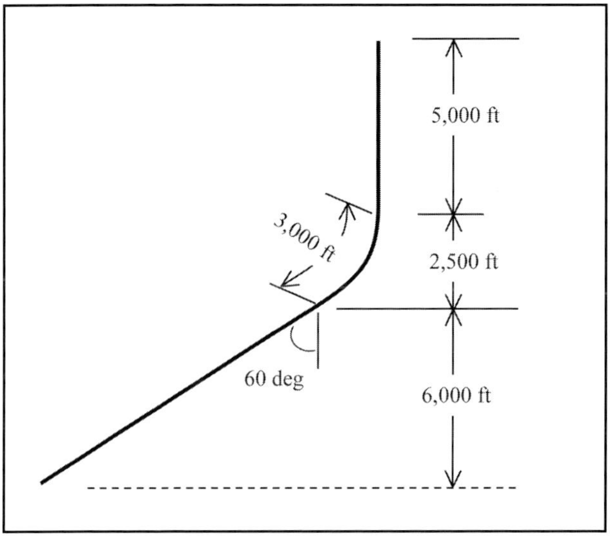

Fig. 4–4. Well conditions in problem 4–4

(a) Indicate whether any problem will occur while drilling at or tripping from 14,000 ft TMD. Show details.

(b) Determine the maximum WOB that can be applied while drilling below the last casing seat. (Steel weight = 64.6 lbs/gal.)

(c) Using the impact force criterion, determine the optimum bit hydraulics at 14,000 ft TMD. Assume a flow power index of $m = 1.86$; $\Delta P_{fadp} = 0.62$ psi/ft, and $\Delta P_{fadc} =$ 0.3 psi/ft at $Q = 564$ gpm.

Problem 4–5. You are given the following data for a directional S-shaped well (see also fig. 4–5):

- Mud:
  - Bingham plastic
  - $\gamma = 10.2$ lbs/gal
  - $\theta_{600} = 65$; $\theta_{300} = 40$; 10s/10 min gel = 15/25 lbs/100 ft^2
- Bit: Tri-cone roller cone jet bit
- Drill pipe:
  - OD = 4½ in.
  - ID = 4 in.
  - Air weight = 14 lbs/ft

- Drill collar:
  - OD = 7½ in.
  - ID = 4 in.
  - Air weight = 110 lbs/ft
  - Length = 1,000 ft
- Pumps: Duplex (double acting)
  - Rated hydraulic horsepower = 2,000 hp
  - Maximum allowed surface pump pressure = 4,800 psi
  - Minimum pump flow rate = 200 gpm
  - Volumetric efficiency = 90%
- Annular velocity = 70 ft/min (minimum)
- Hole: Last intermediate casing (ID = 8½ in.) set at 16,000 ft TVD (open hole washed out to 8½ in. using 7½ in. bit)

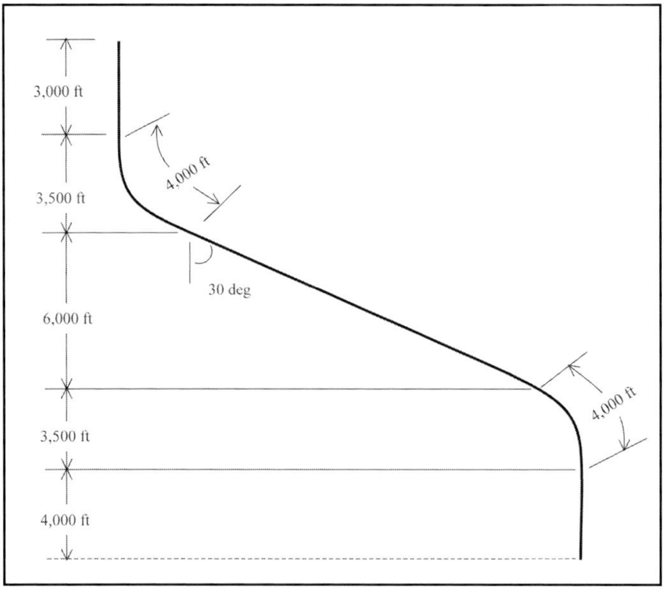

Fig. 4–5. Well conditions in problem 4–5

(a) Determine the optimum bit hydraulics to be used at TVD = 16,000 ft and TVD = 20,000 ft, using the hydraulic horsepower criterion. Assume a power flow index of $m = 1.8$.

(b) Determine the differential pressure while drilling at 18,200 ft TVD, assuming $\Delta P_{fadp} = 0.0158$ psi/ft, $\Delta P_{fadc} = 0.0491$ psi/ft, and $\Delta P_{fadp} = 0.073$ psi/ft.

Problem 4–6. You are given the following well data:

- Mud:
  - Plastic viscosity = 30 centipoise
  - Yield value = 8 lbs/100 ft^2
  - Gel strength = 15 lbs/100 ft^2
  - Weight = 12 lbs/gal
- Well depth = 10,000 ft
- Drill bit nozzle sizes: 13-13-13
- Pump:
  - Total hydraulic horsepower = 1,800 hp
  - Volumetric efficiency = 80%
  - Maximum allowed pump pressure = 4,500 psi
  - Minimum flow rate = 350 gpm

Pump pressures of 4,000 psi and 2,500 psi were recorded while circulating mud at 500 gpm and 390 gpm, respectively.

(a) Determine the optimum nozzle sizes to be used for the next bit run, using the maximum hydraulic horsepower criterion for optimum hydraulics.

(b) What percentage of the pump hydraulic horsepower is wasted while flowing at 500 gpm?

(c) What percentage of the pump hydraulic horsepower is utilized under the conditions in part (a)?

Problem 4–7. You are given the following well data:

- Last casing set at 5,000 ft
- Hole size: 8¾ in. at 5,000 ft to 10,000 ft TVD
- Drill pipe:
  - OD = 5 in.
  - ID = 4 in.
  - Air weight = 25.6 lbs/ft
- Drill collar:
  - OD = 7 in.
  - ID = 4 in.
  - Air weight = 84 lbs/ft
- Coefficients of friction at walls: 0.1, cased sections; 0.2 open hole section
- Mud:
  - Bingham plastic; between 5,000 ft and 10,000 ft
  - Fann V-G dial reading: 45; Corresponding Shear Rate (sec^{-1})340 s^{-1}
  - Fann V-G dial reading: 65; Corresponding Shear Rate (sec^{-1})510 s^{-1}
  - Gel = 18/30 lbs/100 ft^2
  - $\gamma = 1.7 \times N sec^{-1}$, where N is the Fann rotational speed: $\mu_p = (300\theta_N/N) - (300\tau_y/N)$

- Formation:
  - Fracture gradient = 18.2 lbs/gal
  - Pore pressure gradient = 0.78 psi/ft
- Pump: Triplex
  - L_S = 12 in.
  - D_L = 6½ in.
  - $(spm)_{max}$ = 120
  - $(spm)_{min}$ = 25
  - Mechanical efficiency = 85%
  - Volumetric efficiency = [100 – (0.15 × spm)]%
- Maximum input horsepower = 1,300 hp
- Maximum allowed surface pressure = 3,500 psi
- Minimum annular velocity allowed = 120 ft/min
- Maximum anticipated WOB below 5,000 ft = 60,000 lbs
- Differential pressure required for drilling safety = 0.1 lbs/gal
- A kick was taken at 5,000 ft
  - P_{sidp} = 400 psi
  - P_{sic} = 800 psi
- Bit discharge coefficient: 0.95
- Drilled rock cuttings: Weight density = 20.8 lbs/gal
- Steel weight density = 65 lbs/gal
- Surface connections: 800 ft of drill pipe
- Hoisting system: 12 lines; efficiency = 85%
- Drawwork efficiency = 90%
- Diesel engine efficiency = 60%
- Bit in use before tripping at 5,000 ft: three-jet roller cone 16-16-16
- At Q = 250 gpm, P_p = 708 psi
- At Q = 500 gpm, P_p = 2,137 psi

(a) Design the optimum hydraulic program between 5,000 ft and 10,000 ft in increments of 2,000 ft depth intervals. Use the maximum bit hydraulic horsepower criterion.

(b) While drilling at the conditions determined in part (a), the rock cuttings concentration in the mud occupying the annulus is 10% of the annulus volume. Determine whether formation fracturing will occur.

(c) Determine the engine horsepower requirement to trip out from 10,000 ft at a speed of 3 ft/s.

(d) Determine whether the available pump is adequate to break circulation at 9,000 ft.

Problem 4–8. You are given the following well data (see also fig. 4–6):

- Last casing setting = 18,000 ft TMD
- TD = 17,200 ft TVD
- Hole size to TD = 8⅞ in.
- Drill pipe:
  - OD = 4¼ in.
  - ID = 3½ in.
  - Air weight = 26.88 lbs/ft
- Drill collar:
  - OD = 6 in.
  - ID = 3½ in.
  - Air weight = 100.8 lbs/ft
- Pumps:
  - 2,000 hp
  - $\eta_v = 90\%$
  - $P_{max} = 4{,}500$ psi
- Minimum annular fluid velocity for hole cleaning = 120 ft/min
- Coefficients of friction at walls: 0.15, cased sections; 0.25, open hole section
- Mud:
  - Bingham plastic
  - $\theta_{600} = 55$; $\theta_{300} = 35$(Fann V-G meter)
  - 10 s/min; gel = 15/35 lbs/100 ft^2
- Formation below 18,000 ft TMD:
  - Fracture gradient = 17.2 lbs/gal;
  - Pore pressure gradient = 0.78 psi/ft
- Pressure changes in circulating system while drilling at 18,000 ft TMD:
  - $\Delta P_{fdp} = \Delta P_{fdc} = 0.080$ psi/ft
  - $\Delta P_{fadp} = 0.075$ psi/ft
  - $P_B = 1{,}904$ psi
- Circulating standpipe pressure (pump pressure) = 3,800 psi
- During tripping from 18,000 ft TMD, $\Delta P_{surge} = \Delta P_{swab} = 0.03$ psi/ft (annulus)
- Steel weight = 65 lbs/gal
- Maximum WOB below 18,000 ft TMD = 33,000 lbs

(a) Indicate whether any drilling problem will occur while drilling at or tripping from 18,000 ft TMD. Show details.

(b) Using the bit hydraulic horsepower criterion, determine the optimum bit hydraulics at 20,000 ft TMD. Assume a flow power index of $m = 1.86$; also assume that total friction pressure loss in the annulus is 25% of total friction pressure loss in the drill string.

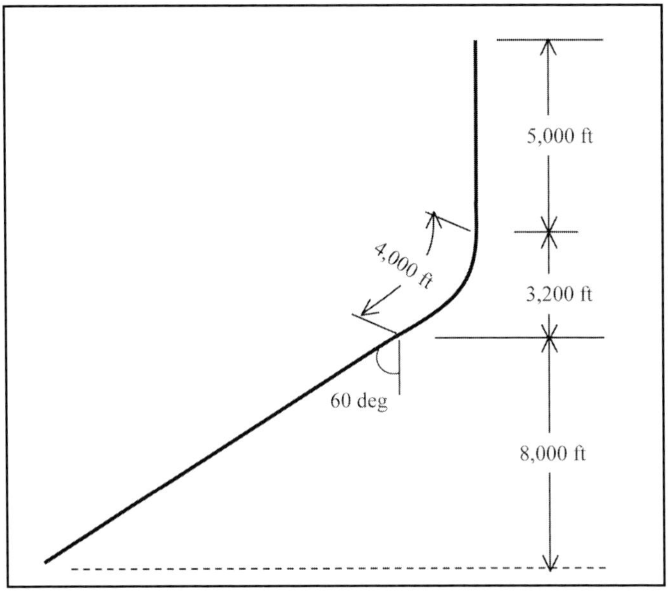

Fig. 4–6. Well conditions in problem 4–8

Problem 4–9. Determine the optimum bit hydraulics using the maximum jet impact force criterion at 1,000 ft increments for an interval of the well between the intermediate casing at 8,000 ft and 13,000 ft. Mud data for the well are given in table 4–1, and the well data plan is as follows:

- Pump horsepower = 1,400 hp
- Pump efficiency = 0.85
- Maximum allowed standpipe pressure = 4,500 psi
- Minimum annular velocity = 120 ft/min
- Drill pipe: 4.5 in. by 3.875 in.
- Drill collars: 600 ft of 8.0 in. by 3.00 in.
- Hole size = 10.5 in.
- Surface equipment equivalent to 400 ft of drill pipe

Table 4–1. Mud data for problem 4–9

Depth (ft)	Mud Density (lb/gal)	Plastic Viscosity (cp)	Yield Point (lb/100 ft ^{2})
8,000–11,000	11	25	5.0
11,000–13,000	13	35	8

Problem 4–10. You are given the following well data (see also fig. 4–7):

- TD = 11,500 ft TVD
- Casing (7⅞ in.) set at 8,000 ft
- Differential pressure = 300 psi
- Drill pipe: OD = 4.25 in.; ID = 3.5 in.
- Drill collar: OD = 6 in.; ID = 3.5 in.; length = 1,000 ft
- Drill bit: Tri-cone three-jet (6 7/8 in.); at 8,000 ft to 11,500 ft (assume hole will be washed out to 7⅞ in.)
- Pump: Duplex
  - Total hydraulic horsepower = 1,200 hp
  - Volumetric efficiency = 85%
  - Maximum allowed operating pressure = 3,600 psi
- Minimum allowed annular fluid velocity = 120 ft/min
- Slope of line for friction pressure loss vs. flow rate: m = 1.5
- Mud: μ_p = 25 centipoise; τ_y = 10 lbs/100 ft^2
- Formation fracture gradient = 0.75psi/ft^2

(a) Using the maximum bit hydraulic horsepower criterion for optimum bottom-hole cleaning, determine the optimum flow rates and corresponding total nozzle areas at a depth of 8,000 ft to 11,500 ft for a depth interval of 1,000 ft.

(b) What optimum nozzle sizes would you recommend at 11,000 ft?

(c) If the friction pressure loss in the annulus while drilling at 9,000 ft is given by ΔP_{afdp} = 0.021 and ΔP_{afdc} = 0.126, will a kick occur?

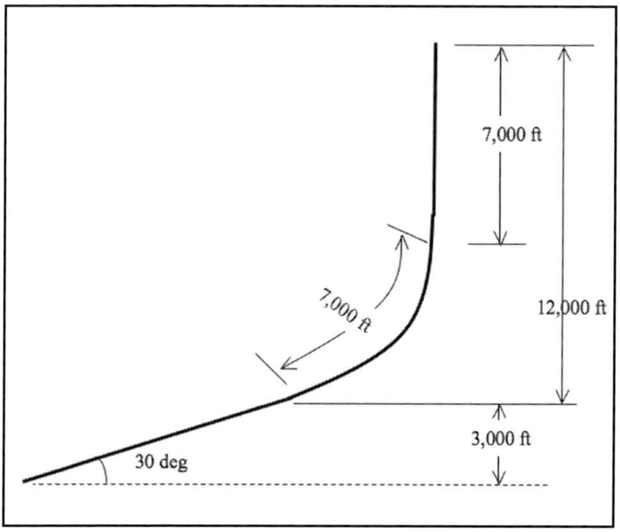

Fig. 4–7. Well conditions in problem 4–10

Problem 4–11. You are given the following well data:

- Drill pipe:
  - OD = 4.5 in.
  - ID = 3.64 in.
  - Weight = 20 lbs/ft
- Drill collar:
  - OD = 7 in.
  - ID = 2 in.
  - Weight = 120.3 lbs/ft
  - Length = 1,000 ft
- Mud:
  - Bingham
  - $\theta_{300} = 21$; $\theta_{600} = 29$ (Fann V-G meter)
  - Weight = 15.5 lbs/gal; to 12,000 ft TVD
- Pump: National duplex (double acting)
  - Maximum allowed operating pressure = 5,440 psi
  - Hydraulic horsepower = 1,600 hp
  - Volumetric efficiency = 80%
- Drilled cuttings:
  - Cuttings average settling velocity = 40 ft/min
  - Average minimum required cuttings rise velocity for annular hole cleaning = 120 ft/min
- Well geometry:
  - Shape as in problem 4–1 (see fig. 4–2);
  - Last intermediate casing is 9⅞ in. at 12,000 ft TVD
- Geological formation:
  - Formation pore and fracture pressure gradients are 0.80 psi/ft and 0.831 psi/ft, respectively.
  - Target vertical depth = 15,000 ft
- Drill bit:
  - 12⅞ in. tri-cone bit with 3-14 nozzles to 12,000 ft
  - Next bit to be used is 8⅞ in. tri-cone (assume hole washed out to 9⅞ in.)
- Field data at 12,000 ft, using the 12⅞ in. bit:
  - At $Q_1 = 300$ gpm, $P_{p1} = 2{,}800$ psi
  - At $Q_2 = 400$ gpm, $P_{p2} = 4{,}622$ psi

(a) Specify a mud weight to be used to drill from 12,000 ft to 15,000 ft TVD at a differential pressure of 200 psi. How much barite (lbs/bbl) must be added to original mud?

(b) Using the impact force criterion, determine the optimum hydraulics to be used to drill the next interval of depth, starting at 12,000 ft.

(c) Determine the maximum total hydraulic horsepower that will be utilized under the drilling conditions in part (b). Assume bit life of 100 h and drilling rate of 20 ft/h; friction pressure losses at the flow rate found in part (b) are as follows: P_{fdc} = 1.7 psi/ft; P_{fadp} = 0.0126 psi/ft; P_{fadc} = 0.0256 psi/ft.

(d) What is the maximum weight on bit that you can apply under the given mud weight and drill string conditions?

Problem 4–12. You are given the following well data:

- Drill pipe:
  - OD = 4.5 in.
  - ID = 4 in.
  - Weight = 12.75 lbs/ft
- Drill collar:
  - OD = 7.5 in.
  - ID = 4 in.
  - Weight = 107.3 lbs/ft
  - Length = 1,000 ft
- Intermediate casing:
  - OD = 10.75 in.
  - Wall thickness = 0.4 in.
- Current depth = 10,000 ft TVD
- TD = 15,000 ft
- Setting depth = 6,500 ft TVD
- Drill bit: 8 ½ in. tri-cone roller cone jet bit with 9-9-9 nozzles
- Mud:
  - Bingham plastic
  - γ = 10 lbs/gal

Problem 4–13. Using the concept of jet impact pressure, explain why at a given flow rate, the ROP increases for extended nozzles and blanked-off nozzles.

Nomenclature

A_a	annulus cross-sectional area
C	Constant
C_d	Coefficient of discharge
$d_{n\,opt}$	Optimum nozzle diameters
ECD	Equivalent circulating density of mud
H_h	Hydraulic power
H_{hb}	Drill bit hydraulic power
H_{hf}	Hydraulic power owing to friction pressure loss in circulating system

H_{hs}	Hydraulic pump power
P_{af}	Friction pressure loss in annulus
P_b	Bit pressure drop
P_f	Total frictional pressure drop
P_{frac}	Formation fracture pressure
P_h	Mud hydrostatic pressure
P_{fadc}	Annulus friction pressure loss around drill collar
P_{fadp}	Annulus friction pressure loss around drill pipe
P_{fdc}	Drill collar friction pressure loss
P_{fdp}	Drill pipe friction pressure loss
P_{fs}	Surface friction pressure loss
P_s	Maximum allowed surface pressure
Q	Flow rate
Q_{opt}	Optimum flow rate
$v_{a\,min}$	annular velocity
α	Flow exponent
η_v	Volumetric efficiency

References

1 Smalling, D. A., and Key, T. A. 1979. "Optimization of Jet-Bit Hydraulics Using Impact Pressure." Paper SPE 8440, presented at the 54th Annual Fall Technology Conference, Las Vegas, Nev.

5

Transport of Drilled Cuttings

Introduction

One of the primary functions of circulating drilling fluid in rotary drilling is to maintain good hole cleaning. This covers both cleaning—removal of drilled cuttings at the hole bottom and from the bit teeth (see chap. 4)—and efficient transportation of cuttings to the surface. The ability to transport such cuttings is generally referred to as the *carrying capacity* of a drilling fluid.

Inadequate hole cleaning can lead to costly drilling problems, such as

- pipe sticking
- premature bit wear
- slow drilling rate
- formation fracturing
- high torque and drag
- mud losses

For example, the rock fragments (drilled cuttings) beneath the bit, if not removed immediately, will cause wasteful bit wear and retard drilling. An excessive concentration of annular cuttings will increase the effective mud weight and therefore may cause excessive pressures that can fracture the formation and decrease penetration rate. During tripping operations, the settling of cuttings around the bottom-hole assembly can cause pipe sticking. These problems hinder the continuation of normal drilling operations. In recognition of these problems and their effect on overall well cost, a basic understanding of the cuttings transport phenomenon of drilling fluids is imperative.

Factors That Affect the Transport of Drilled Cuttings

The flow of cuttings in the annulus is a dynamic process and is subject to many forces in existence, such as those of gravity, buoyancy, drag, inertia, friction, and interparticle contact. The movement of a particle in the annulus is dictated by the dominating force. The forces acting on a single particle are shown in figure 5–1.

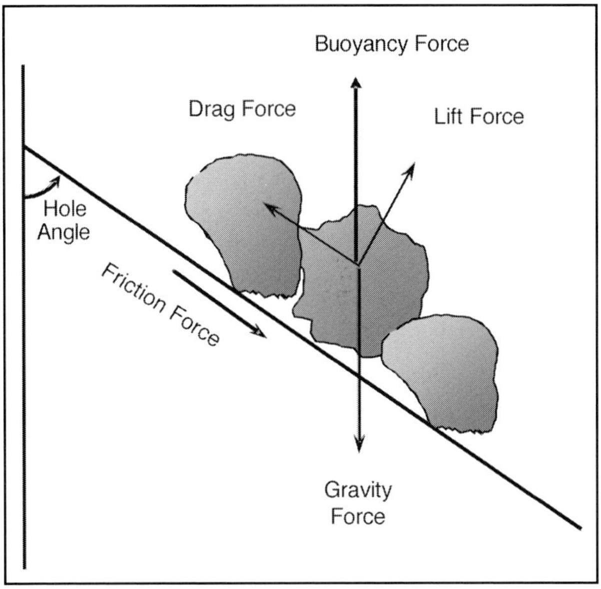

Fig. 5–1. Forces acting on a particle

Numerous factors affect the cuttings transport in the annulus. The factors that are believed to affect the ability of drilling fluids to transport drilled cuttings through the annular space are the following:

- cuttings slip velocity
- annular mud velocity
- flow regime of fluid and cuttings slippage
- annular velocity profile (affected by fluid rheology, annulus eccentricity, ratio of hole to pipe diameter)
- cuttings-bed formation (in inclined annuli only)
- drill pipe rotary speed
- drilling rate
- fluid rheological properties
- hole inclination

Other factors related to cuttings (size, density, shape, concentration, size distribution, etc.), fluid density, and annulus geometry are already accounted for by the aforementioned major factors. In the following sections, a brief description of each is given.

Cuttings slip velocity

Drilled cuttings have the tendency to fall down (slip) through the fluid medium at a velocity referred to as the *cuttings slip velocity.* Many correlations have been presented in the literature that concern the determination of particle slip velocity. These correlations will be discussed later.

For the fluid to lift the cuttings to the surface, the fluid annular average velocity, υ_a, must be in excess of the cuttings average slip velocity, υ_s. The relative velocity between υ_a and υ_s is termed the average cuttings transport (rise) velocity, υ_t; that is,

$$\bar{\upsilon}_t = \bar{\upsilon}_a - \bar{\upsilon}_s \tag{5.1}$$

$$\frac{\bar{\upsilon}_t}{\bar{\upsilon}_a} = 1 - \frac{\bar{\upsilon}_s}{\bar{\upsilon}_a} = R_t \tag{5.2}$$

where R_t is the cuttings transport ratio, as defined by Sifferman et al.[1]

In vertical well drilling, it is recommended that R_t be a minimum of 0.5–0.55. In the case of vertical annuli, there is only one axial component of slip velocity,

$$\bar{\upsilon}_s = \bar{\upsilon}_{sa} \tag{5.3}$$

in contrast, when an annulus is inclined at an angle θ from vertical, there will be two components for the slip velocity.

$$\bar{\upsilon}_{sa} = \bar{\upsilon}_s \cos\theta \tag{5.4a}$$

$$\bar{\upsilon}_{sr} = \bar{\upsilon}_s \sin\theta \tag{5.4b}$$

where υ_{sa}, υ_{sr} are the axial and radial components of the average slip velocity, respectively, as shown in figure 5–2.

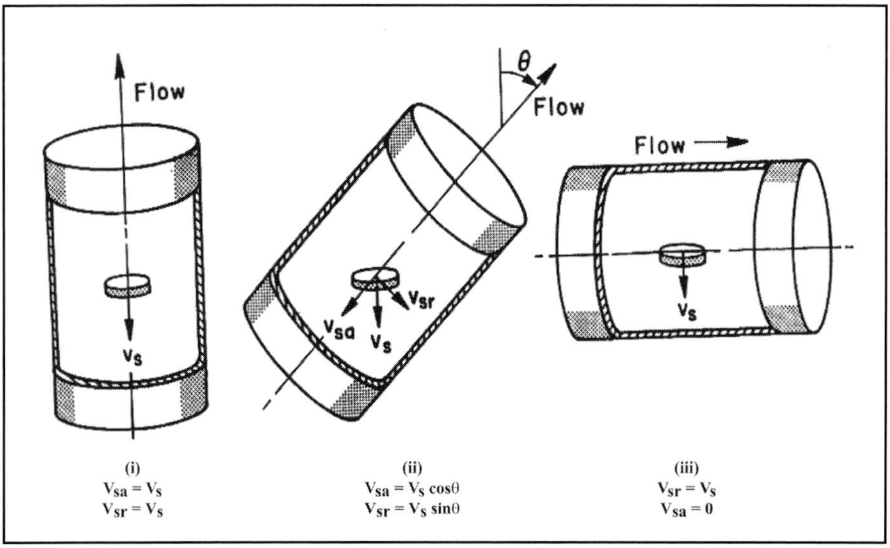

Fig. 5–2. Particle settling velocity in an inclined annulus

When the angle of inclination is increased, the axial component of the slip velocity decreases, reaching zero value at the horizontal position of the annulus. At the same time, the radial component reaches a maximum at the position mentioned. When these conditions are taken into account, all factors that may lead to improved cuttings transport by a reduction of the particle slip velocity will have a diminishing effect while the angle of inclination is increasing.

Annular fluid velocity

The annular fluid velocity in vertical drilling has to be sufficient to avoid cuttings settling and to transport these cuttings to the surface in reasonable time. As discussed earlier, in the case of an inclined annulus, the axial component of particle slip velocity plays a less important role. Thus, it might seem that for satisfactory transport, the annular mud velocity in this case could be lower than in the vertical annulus; this, however, would be a mistaken conclusion. The increasing radial component of particle slip velocity pushes the particle toward the lower wall of the annulus, causing a cuttings bed to form. Consequently, the annular fluid velocity has to be sufficient to avoid (or at least to limit) the bed formation. Studies have shown that for limitation of cuttings-bed formation, the annular fluid velocity in directional drilling has to be generally much higher than in vertical drilling.[2]

Flow regime and regime of particle slippage

When the phenomenon of cuttings transport is considered, the regime of the flowing fluid and vertical slippage should be considered simultaneously. A fluid in turbulent flow always induces a turbulent regime of particle slippage, independent of the cuttings shape and dimensions. Therefore, in this case, the only factor that determines the particle slip velocity is the momentum forces of the fluid; fluid viscosity has little or no influence.

If the mud flows in the laminar regime, then depending on the cuttings shape and size, either the turbulent or the laminar regime of slippage may be expected. The laminar regime of slippage will always provide a lower value of particle slip velocity, indicating that laminar flow usually will provide better transport than turbulent flow. However, recall that in the case of an inclined annulus, the significance of the axial component of particle slip velocity decreases; thus, the advantage of laminar flow will be nullified while the angle of inclination is increased, as confirmed by several investigators.[3]

Velocity distribution profile in laminar flow as a result of mud rheological properties

The power law index, n, in the power law model, has a bearing on the velocity profile distribution, as shown in figure 5–3. As n decreases, the velocity profile of the flowing mud becomes more flat. In vertical wells, this is advantageous to cuttings-transport efficiency. However, in inclined wells with high eccentric annuli, the decrease in n causes an increase in hydraulic requirements for effective hole cleaning. This is due to the increase in diversion of fluid flow from the narrow gap of the annulus to the wider gap as n decreases.

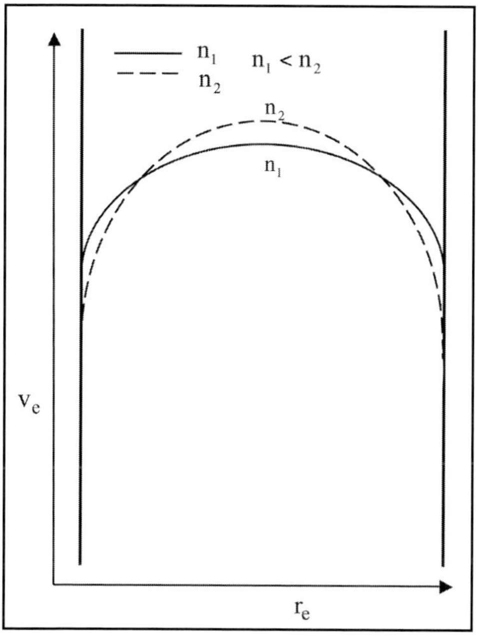

Fig. 5–3. Effect of parameter n *on annular velocity profile*

Velocity distribution profile as a result of eccentricity and ratio of outside to inside pipe diameter

With the slot differential approach for non-Newtonian power law fluids, Iyoho and Azar[4] have developed an equation for the velocity profile that occurs in laminar flow inside an inclined annulus:

$$v_a = \frac{n}{n+1}\left(\frac{h}{2}^{\frac{n+1}{n}} - |y|^{\frac{n+1}{n}}\right) \tag{5.5}$$

Eccentricity ratio is defined as the ratio of the eccentricity to the difference between the hole radius and the pipe radius:

$$E = \frac{e}{r_h - r_p} \tag{5.6}$$

Thus,

$$e = E(r_h - r_p) \tag{5.7}$$

The gap between the pipe and wellbore wall can be given as:

$$\delta = (1 - E)(r_h - r_p) \tag{5.8}$$

If there is a tool joint, the outer diameter (OD) of the tool joint should be used instead of the pipe OD. The parameters introduced in equations (5.6)–(5.8) are illustrated in figure 5–4.

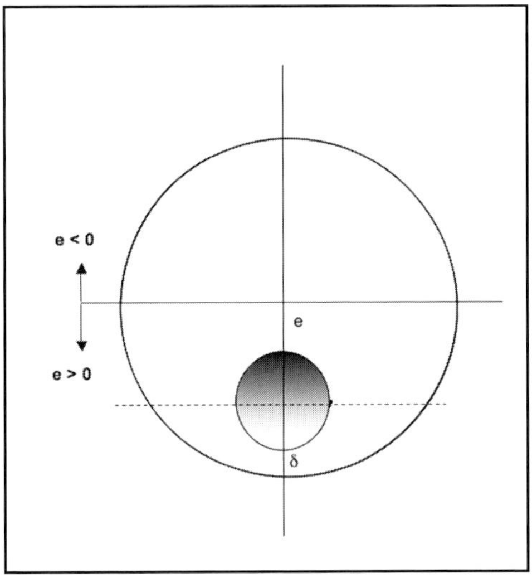

Fig. 5–4. Annular eccentricity

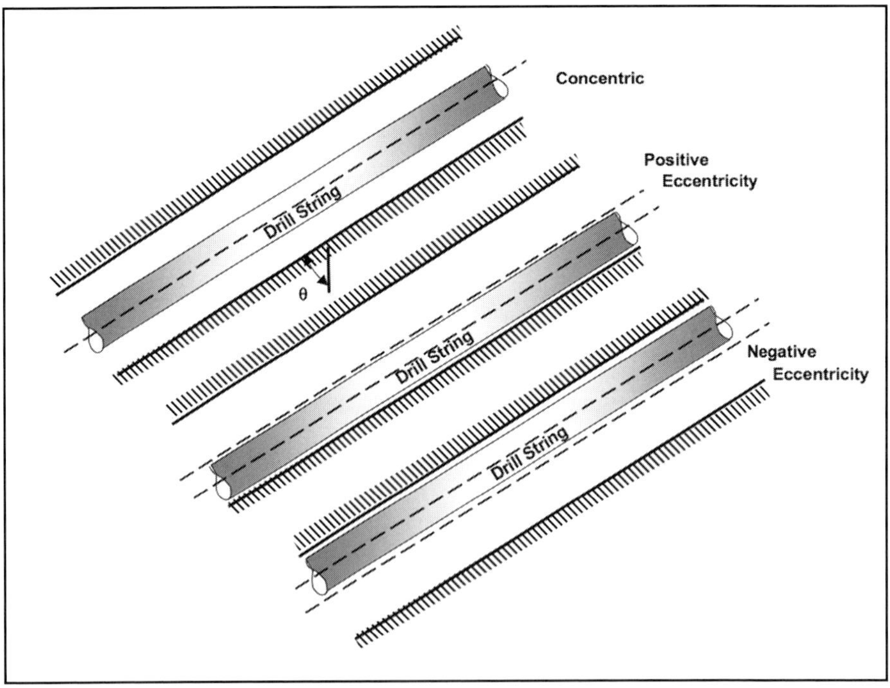

Fig. 5–5. Concentric annulus versus positive and negative eccentricity

The position of the inside pipe may vary, as shown in figure 5–5. Equation (5.5), developed to describe the velocity profile, indicates that displacement of the inside pipe toward the lower wall of the annulus (positive eccentricity) reduces the mud velocity in this area, as is shown schematically in figures 5–6 and 5–7. Therefore, for inclined wellbores with positive eccentricity, cuttings-transport problems become more acute.

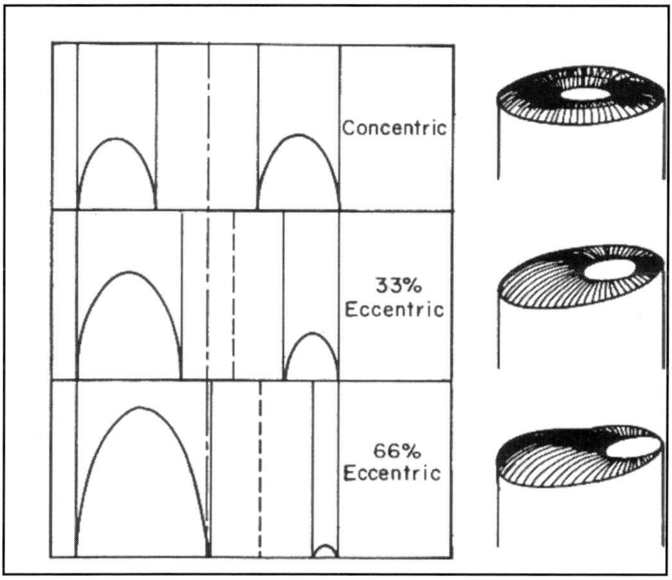

Fig. 5–6. Effect of eccentricity on annular velocity profile

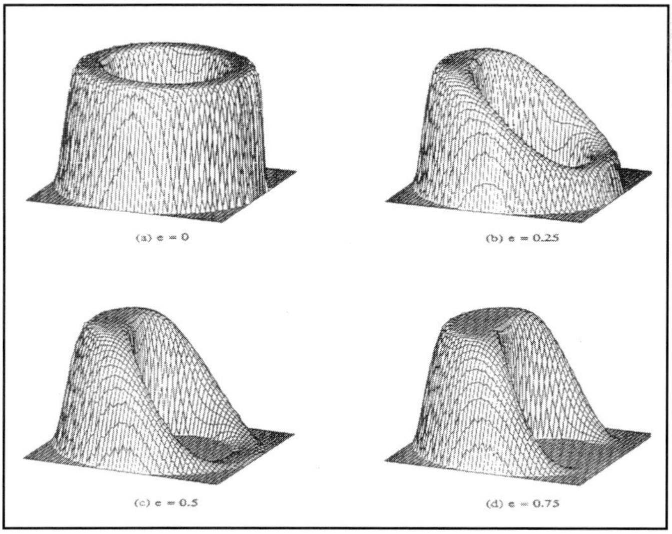

Fig. 5–7. Three-dimensional velocity profile for yield power law fluid versus eccentricity[5]

Solid/liquid flow in inclined annuli

In directional well drilling, the solid/liquid flow in the inclined section of the wellbore may be pseudo-homogeneous, heterogeneous, saltation, or a flow with stationary bed. (See fig. 5–8.)

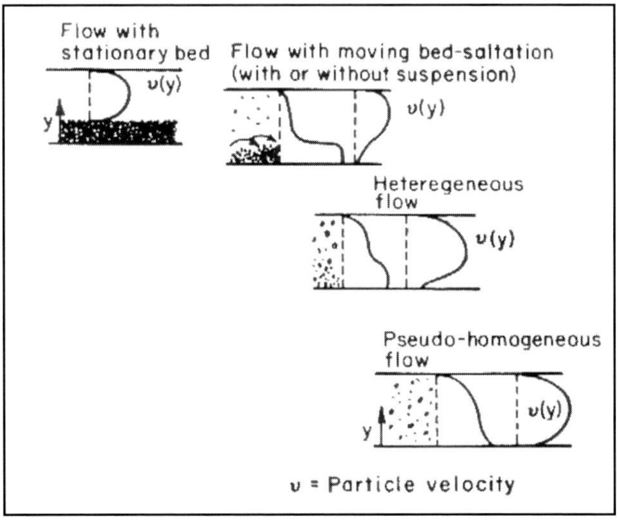

Fig. 5–8. Qualitative solid/liquid flow pattern

Drilling rate

The drilling rate has an important effect on the quantitative aspect of cuttings transport. As the drill rate increases, the cuttings concentration in the annulus also increases (for details, see the later section "Cuttings Transport in Directional Well Drilling"). For effective removal of drilled cuttings, as the drilling rate increases, the hydraulic requirement also increases.

Rotary speed

For vertical wells, increases in drill pipe rotary speed has a small to moderate effect on the enhancement of cuttings transport, depending on hole and pipe diameter sizes. For larger hole sizes, an increase in pipe rotary speed causes a moderate increase in mud carrying capacity.[6]

Rheological properties of flowing fluid

This aspect of the problem will be discussed in later sections.

Cuttings Transport in Vertical Well Drilling

Removal of drilled cuttings from annuli during vertical well drilling has been a subject of investigation for many years. Various methods have been proposed to describe annular hole cleaning, most of which are based on three variables:

1. Average cutting slip velocity (v_s)
2. Average cutting transport velocity (v_t)
3. Average annular volumetric cuttings concentration (C_a)

An equation for the volumetric cuttings concentration, C_a, can be derived on the basis of volume balance. When steady-state conditions prevail, the volume of cuttings, Q_{cg}, that are being generated by a drill bit per unit time must equal the volume of cuttings being transported out of the annulus, Q_{ce}. The volume of cuttings generated by the drill bit can be formulated as follows:

$$\frac{dQ}{dt} = \left(\frac{dh}{dt}\right)A \tag{5.9a}$$

$$\text{Rate of penetration (ROP)} = \frac{dh}{dt} \tag{5.9b}$$

$$\int dQ = RA \int dt \tag{5.9c}$$

$$Q_{cg} = R_d\left(\frac{\pi}{4}D_b^2\right) \tag{5.9d}$$

where R_d is the drilling rate and D_b is the bit diameter. The volume of cuttings exiting the annulus per unit time can be written as:

$$Q_{ce} = \bar{v}_t \bar{C}_a \frac{\pi}{4}(D_h^2 - D_{op}^2) \tag{5.10}$$

By exploitation of equation (5.1), equation (5.10) becomes

$$Q_{ce} = \left(\bar{v}_a - \bar{v}_s\right)\bar{C}_a \frac{\pi}{4}\left(D_h^2 - D_{op}^2\right) \tag{5.11}$$

Since $Q_{cg} = Q_{ce}$,

$$R_d \frac{\pi}{4} D_b^2 = (v_a - v_s) \frac{\pi}{4} \bar{C}_a (D_h^2 - D_{op}^2) \tag{5.12}$$

or

$$\bar{C}_a = \frac{R_d D_b^2}{(\bar{v}_a - \bar{v}_s)(D_h^2 - D_{op}^2)} = \frac{R_d D_b^2}{v_a \left[1 - (\bar{v}_s / \bar{v}_a)\right](D_h^2 - D_{op}^2)} \tag{5.13}$$

However,

$$R_t = 1 - \frac{\bar{v}_s}{\bar{v}_a} \tag{5.14}$$

and

$$\bar{v}_a = \frac{Q_m}{(\pi/4)(D_h^2 - D_{op}^2)} \tag{5.15}$$

Therefore, equation (5.13) becomes

$$\bar{C}_a = \frac{R_d D_b^2}{1.27 R_t Q_m} \tag{5.16}$$

where Q_m is the mud flow rate.

The cuttings slip velocity is generally taken as a constant average cutting slip velocity in a quiescent fluid of given density and rheology. Although several empirical correlations can be used to predict the slip velocity, these will give different answers. Nevertheless, some of these correlations are given in the following sections. As a guideline, it is recommended that the concentration of annular cuttings be kept below 5% by volume for trouble-free drilling.[7] This number has been determined on the basis of statistical analysis of field history experiences.

Chien correlation

$$\bar{v}_s = 0.458\beta \left[\sqrt{\left(\frac{36{,}800 d_s}{\beta^2}\right)\left(\frac{\rho_s - \rho_m}{\rho_m}\right) + 1} - 1 \right] \quad (\text{for } \beta < 10) \tag{5.17}$$

$$\bar{v}_s = 86.4\, d_s \sqrt{\left(\frac{\rho_s - \rho_m}{\rho_m}\right)} \tag{5.18}$$

where

$\beta = \mu_a / \gamma_m d_s$

$\mu_a = \mu_p + (300\, \tau_y d_s / \bar{v}_a)$

ρ_m is the mud density, in lbs/gal

ρ_s is the cuttings density, in lbs/gal

μ_a is the mud apparent viscosity, in centipoise

μ_p is the mud plastic viscosity (PV), in centipoise

τ_y is the mud yield value, in lbs/100 ft^2

d_s is the equivalent spherical diameter of a cutting, in in.

$\bar{v}_a$ is the average annular fluid velocity, in ft/min

Moore correlation[8]

An alternative slip velocity was proposed by Moore, by using the equation for static fluid and applying it to flowing fluid. This method is based on the calculation of apparent Newtonian viscosity by use of the method proposed by Dodge and Metzner.[9] This method also involves equating the annular frictional pressure loss based on a power law fluid to that based on a Newtonian fluid and then solving it for apparent viscosity. Apparent viscosity is used in place of the Newtonian viscosity for computing the particle Reynolds number.

$$\bar{v}_s = 9.24\sqrt{\frac{\gamma_s - \gamma_m}{\gamma_m}} \quad (N_R > 2{,}000) \tag{5.19}$$

$$\bar{v}_s = 4{,}972(\gamma_s - \gamma_m)\frac{d_s^2}{\mu_a} \quad (N_R \le 1.0) \tag{5.20}$$

$$\bar{v}_s = \frac{174 d_s (\gamma_s - \gamma_m)^{0.667}}{(\gamma_m \mu_a)^{0.333}} \quad (1.0 \le N_R < 2{,}000) \tag{5.21}$$

$$\mu_a = \frac{K}{144}\left[\frac{(D_h - D_{op})/\bar{v}_a}{60}\right]^{1-n}\left[\frac{2 + (1/n)}{0.0208}\right]^n \tag{5.22}$$

where

K is the mud consistency index

n is the mud power law index

D_h is the hole diameter, in in.

D_{op} is the outer pipe diameter, in in.

Walker and Mayes correlation

Walker and Mayes developed an alternative slip velocity equation in which cuttings are assumed to be disklike shapes and fall flat side down. This represents a condition for high settling velocity:

$$v_s = 131.4\sqrt{h_s \frac{\gamma_s - \gamma_m}{\gamma_m}} \tag{5.23}$$

$$\mu_a = 511\frac{\tau}{\gamma} \tag{5.24}$$

$$\tau = 7.9\sqrt{h_s\left(\gamma_s - \gamma_m\right)} \tag{5.25}$$

$$v_s = 1.22\tau\sqrt{\frac{D_e\gamma}{\sqrt{\gamma_m}}} \quad \left(N_R < 100\right) \tag{5.26}$$

where
h_s is the cutting thickness
γ is the shear rate corresponding to the shear stress τ

Effect of mud rheology on cuttings transport in vertical wells

On the basis of full-scale laboratory testing,[10] it has been shown that

1. Increasing the ratio of yield point (YP) to PV increases the carrying capacity of muds, in vertical wellbores.
2. Apparent viscosity, YP, and initial gel strengths have an effect on carrying capacity of muds at low and medium annular fluid velocities (60–90 ft/min). Their effects diminish at higher annular fluid velocities (in excess of 120 ft/min).

Cuttings Transport in Directional Well Drilling

It is widely recognized that cuttings transport in high-angle wells is more complex than the transport in vertical wells. The presence of cuttings beds, eccentric flow regimes, and the variable impact of gravity make the physics of transport far more complicated. The comments and observations on hole cleaning in high-angle wells that are given in the following section have been predominantly based on experiments conducted in flow loops.[11]

For vertical and near-vertical drilling (where hole angle, θ, falls in the range of 0–10°), the cuttings-transport problem appears to have an adequate and acceptable solution, as discussed previously. However, as the hole angle deviates from the vertical (θ > 10°), the inclined annulus (generally eccentric) poses problems in cuttings transport that are not encountered in vertical wells. For example, as the hole angle approaches the 20–30° range and annular fluid velocity is at some low value (generally less than 120 ft/min), a cuttings bed begins to form on the low side of the wellbore (fig. 5–9), which phenomenon does not occur in near-vertical drilling.

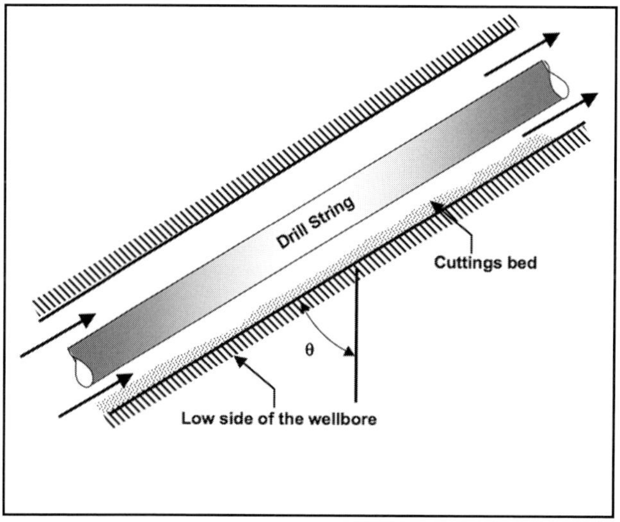

Fig. 5–9. Critical flow condition

To determine the proper hydraulics for effective hole cleaning in inclined wells, the concept of critical transport annular fluid velocity is used. This velocity is defined to be the minimum annular fluid velocity for adequate annular hole cleaning. For wells that are inclined at θ < 35°, the critical transport fluid velocity corresponds to the annular fluid velocity that will result in a concentration of annular cuttings of no more than 5% by volume. For wells that are inclined at θ > 40°, the critical transport fluid velocity corresponds to the minimum annular fluid velocity that will not allow a stationary cuttings bed to form.

The following sections describe, in brief, the factors affecting cuttings transport in directional well drilling.

Effect of mud rheology

On the basis of full-scale laboratory data,[12] it has been shown that mud rheology has slight to no effect on the transport of drilled cuttings at an annular fluid velocity greater than 120 ft/min, regardless of the wellbore angle of inclination. At high angles from vertical (θ > 40°), clear water drilling is slightly more effective in hole cleaning as compared to a mud having a ratio of YP to PV of 7/7 or greater. (See figs. 5–10 and 5–11.)

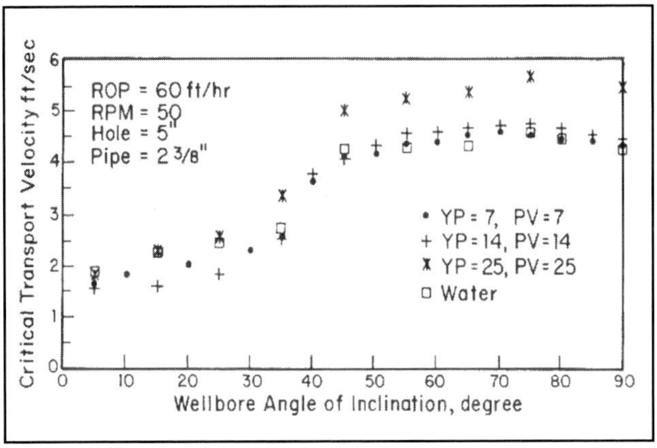

Fig. 5–10. Effect of mud rheology on critical fluid velocity in a five-inch hole

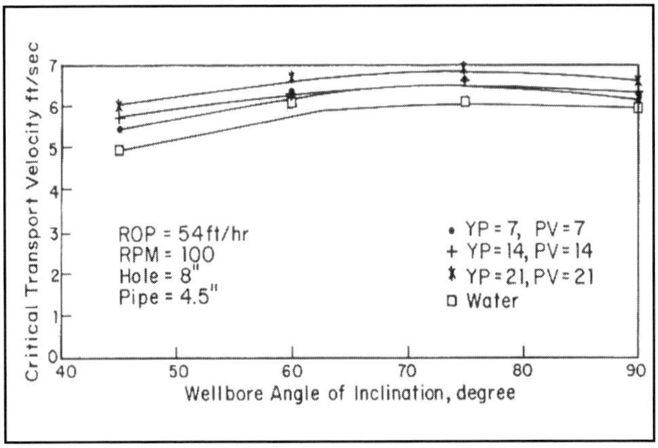

Fig. 5–11. Effect of mud rheology on critical fluid velocity in an eight-inch hole

Effect of eccentricity

For viscous muds, the position of the drill string in the wellbore becomes important in regard to hole cleaning. At high angles from vertical, as eccentricity goes from positive (pipe on the low side of the wellbore wall) to negative (pipe on the high side of the wellbore wall), the requirement of hydraulics for hole cleaning decreases. The opposite is true in clear water drilling. At low angles from vertical, eccentricity has minimal effect on hole cleaning.

Effect of mud weight

An increase in mud weight slightly enhances cuttings transport as long as there is not an accompanying increase in viscosity. The effect that mud weight has on hole cleaning is more pronounced for high angle-wells.

Effect of size of drilled cuttings

At high angles of inclination, drilled cuttings of smaller size are harder to transport. By contrast, at low angles of inclination, medium-sized cuttings are easier to transport than the smallest or the largest cuttings.

Effect of drilling rate

As the drilling rate increases, the hydraulic requirement for effective hole cleaning increases. The relationship between the drilling rate and the required critical transport fluid velocity has been shown to be linear.[13]

Effect of pipe rotation

In a vertical well, the impact that drill pipe rotation has on hole cleaning is relatively small. For inclined wells, the effects of drill pipe rotation have been shown to be more significant. Field evidence[14] has shown that rotating the drill pipe can dramatically increase the rate at which cuttings are removed from high-angle and horizontal wells. The mechanism for this increase is not certain but may be related to the redistribution of flow in an eccentric annulus, together with mechanical agitation of the cuttings bed created by the rotating drill pipe.

Empirical Correlations for Cuttings Transport in High-Angle Wells ($\theta > 50°$)

A number of mathematical models have been published that can be used to determine hole-cleaning requirements in high-angle wells.[15] These models have typically been derived from a combination of flow-loop experiments, physically based modeling, and field verification.

The following empirical equations for wells inclined at $\theta > 50°$ from the vertical have been derived using full-scale laboratory test data[16] (for 5 in. hole and 2⅜ in. drill pipe). Note that scaling up these correlations to larger hole sizes may introduce significant errors.

Critical transport fluid velocity, $\overline{V}_{ca}$, is given by

$$\overline{V}_{ca} = \overline{V}_{cr} + \overline{V}_{cs} \tag{5.27}$$

where $\overline{V}_{cr}$, $\overline{V}_{cs}$ are the cuttings rise velocity and the cuttings average slip velocity, respectively:

$$\overline{V}_{cr} = \frac{1}{\left[1-\left(D_{op}/D_{h}\right)^{2}\left[0.64+\left(18.16/\text{ROP}\right)\right]\right]} \tag{5.28}$$

$$\overline{V}_{cs} = \overline{V}_{cs} C_{ang} C_{size} C_{mwt} \tag{5.29}$$

In equation (5.29),

$$V_{cs} = \begin{matrix} 0.00516\mu_a + 3.006 & \mu_a \le 53 \\ 0.02554\mu_a + 3.280 & \mu_a > 53 \end{matrix} \tag{5.30}$$

where

$$\mu_a = \begin{matrix} PV + 1.12(YP)(D_h - D_{op}) & PV < 20\,\text{centipoise} & YP < 20\,\text{lbs}/100\,\text{ft}^2 \\ PV + 0.9(YP)(D_h - D_{op}) & PV > 20\,\text{centipoise} & YP > 20\,\text{lbs}/100\,\text{ft}^2 \end{matrix} \quad (5.31)$$

Also in equation (5.29), C_{ang}, C_{size}, C_{mwt} are the correction factors for change in hole angle, cuttings size, and mud weight, respectively:

$$C_{ang} = 0.0342\theta - 0.000233\theta^2 - 0.213 \quad (5.32)$$

$$C_{size} = -1.04D_{50} + 1.286 \quad (5.33)$$

$$C_{mwt} = 1 - 0.0333(\gamma_m - 8.7) \quad (5.34)$$

where θ is the hole inclination from vertical (in degrees) and γ_m is the mud weight (in lbs/gal).

For the subcritical flow, the concentration of nonmoving cuttings in the annulus is found using the following equations:

$$C_{corr} = C_{cal}C_{bed} \quad (5.35)$$

$$C_{cal} = \left(1 - \frac{\bar{V}_{opr}}{\bar{V}_{crit}}\right)(1 - \phi)100 \quad (5.36)$$

$$C_{cal} = \left(1 - \frac{Q_{opr}}{Q_{crit}}\right)(1 - \phi)100 \quad (5.37)$$

where

$\bar{v}_{opr}$, Q_{opr} are the operating annular fluid velocity and the operating flow rate, respectively
ϕ is the bed porosity (30–40%)
C_{bed} is a correction factor: $C_{bed} = 0.97 - 0.00231\mu_a$

Example 5–1[17]

You are given the following data:

- Mud
  - YP = 7
  - PV = 7
  - Weight = 8.57 lbs/gal
- Hole inclination from vertical: $\theta = 65°$
- Cuttings average size: Medium ($D_{50} = 0.175$ in.)

- $Q_{opr} = 100$ gpm
- Drilling rate = 54 ft/h
- Diameter of the hole = 5.0 in.
- Drill pipe: OD = 2 3/8 in.
- Cuttings-bed porosity = 36%
- ROP = 60 ft/h

Find the cuttings rise velocity, the critical transport fluid velocity, and the cuttings concentration at the given flow rate.

Solution: Using equations (5.27)–(5.29) and (5.32)–(5.34), since YP, PV < 20,

$$\overline{V}_{ca} = \overline{V}_{cr} + \overline{V}_{cs}$$

$$\overline{V}_{cr} = \frac{1}{\left[1-\left(D_{op}/D_{h}\right)^{2}\left[0.64+\left(18.16/\text{ROP}\right)\right]\right]}$$

$$\overline{V}_{cs} = \overline{V}_{cs} C_{ang} C_{size} C_{mwt}$$

$$C_{ang} = 0.0342\theta - 0.000233\theta^2 - 0.213$$

$$C_{size} = -1.04D_{50} + 1.286$$

$$C_{mwt} = 1 - 0.0333(\gamma_m - 8.7)$$

$$\begin{aligned}\mu_a &= \text{PV} + 1.12 \times \text{YP}(D_h - D_{op}) \\ &= \text{PV} + 1.12 \times 7(5 - 2.375) = 27.6 \text{ centipoise}\end{aligned}$$

$$V_{cs} = (0.00516 \times 27.6) + 3.006 = 3.15 \text{ ft/s}$$

$$V_{crit} = 1.32 + 3.57 = 4.89 \text{ ft/s}$$

Using equation (5.31), for the given well data,

$$\overline{V}_{opr} = \frac{Q_{opr}}{A_{ann}} = 2.32 \text{ ft/sec}$$

$$C_{cal} = \left(1 - \frac{2.32}{4.89}\right)\left(1 - 0.36\right)100\% = 33.7\%$$

$$\begin{aligned}C_{bed} &= 0.97 - 0.00231\mu_a \\ &= 0.97 - (0.00231 \times 27.6) = 0.91\end{aligned}$$

$$\begin{aligned}C_{corr} &= C_{cal} C_{bed} \\ &= 33.\% \times 0.91 = 30.5\%\end{aligned}$$

Supplementary Problems

Problem 5–1. You are given the following well data:

- Mud:
  - YP = 12
  - PV = 12
  - Density = 9 lbs/gal
- Hole:
  - Diameter = 9 in.
  - Inclination: $\theta = 70°$
- Cuttings average size = 0.174 in.
- Drill pipe: OD = 3.5 in.
- Bed porosity = 40%
- ROP = 60 ft/h

(a) Determine the critical flow rate.

(b) Determine the amount of cuttings (% by volume of annulus) if the operating flow rate is 70% of critical.

Problem 5–2. Repeat problem 5–1 for hole angles of 55° and 90°.

Problem 5–3. Repeat problem 5–1 for a hole angle of 0°.

Problem 5–4. You are given the following data:

- Mud
  - YP = 18
  - PV = 18
  - Density = 14 lbs/gal
- Hole size to target = 7⅞ in. diameter
- Kick-off depth = 4,000 ft
- Build rate angle = 8°/100 ft
- Final angle:
  - $\theta = 90°$ from vertical
  - Horizontal reach = 3,000 ft
- Cuttings average size = 0.25 in.V
- Drill pipe: OD = 4 in.
- Bed porosity = 45%
- ROP = 50 ft/h

(a) Determine the minimum flow rate for effective hole cleaning.

(b) Determine the cuttings concentration in various parts of annulus (i) at the critical flow rate found in part (a) and (ii) at 70% of that flow rate.

Nomenclature

$C_{ang}, C_{size}, C_{mwt}$	Correction factors for change in hole angle, cuttings size, and mud weight, respectively
C_{bed}	Correction factor
D_b	Bit diameter
D_h	Hole diameter, in in.
D_{op}	Pipe outer diameter (OD), in in.
d_s	Equivalent spherical diameter of cutting, in in.
h_s	Cutting thickness
K	Mud consistency index
n	Mud power law index
Q_m	Mud flow rate
R_d	Drilling rate
$\bar{v}_a$	Average annular fluid velocity, in ft/min
$\bar{V}_{ca}$	Critical transport average annular fluid velocity
$\bar{V}_{cr}$	Cuttings rise velocity
$\bar{V}_{cs}$	Cuttings average slip velocity
$\bar{V}_{opr}, Q_{opr}$	Operating annular fluid velocity and operating flow rate, respectively
v_{sa}, v_{sr}	Axial and radial components of the average slip velocity
γ	Shear rate corresponding to shear stress (τ)
γ_m	Mud weight, in lbs/gal
θ	Hole inclination from vertical, in degrees
μ_a	Mud apparent viscosity, in centipoise
μ_p	Mud plastic viscosity, in centipoise
ρ_m	Mud density, in lbs/gal
ρ_s	Cuttings density, in lbs/gal
τ_y	Mud yield value, in lbs/100 ft^2
ϕ	Bed porosity (30–40%)

References

1 Sifferenman, T. R., G. M. Myers, E. L. Haden, and H. A. Wahl. 1974. "Drill Cutting Transport in Full-Scale Vertical Annuli." *Journal of Petroleum Technology,* November, 1295–1303.

2 Okranji, S. S., and J. J. Azar. 1986. "The Effect of Mud Rheology on Annular Hole Cleaning in Directional Wells." *SPE Drilling Engineering,* August, 297–308. Iyoho, A. W., and J. J. Azar. 1981. "An Accurate Slot-Flow Model for Non-Newtonian Fluid through Eccentric Annuli." *SPE Journal,* October, 556–71.

3 Okranji and Azar, *SPE Drilling Engineering,* August 1986, 297–308. Iyoho and Azar, *SPE Journal,* October 1981, 556–71. Becker, T. E., J. J. Azar, and S. S. Okranji. 1991. "Correlations of Mud Rheological Properties with Cuttings Transport Performance in Directional Drilling." *SPE Drilling Engineering,* March. Larsen, Thor F. 1990. "A Study of the Critical Fluid Velocity in Cuttings Transport for Inclined Wellbores." Thesis, University of Tulsa.

4 Larsen, thesis.

5 Hopkin, E. A. 1967. "Factors Affecting Cuttings Removal during Rotary Drilling," *Journal of Petroleum Technology,* June, 807–14.

6 Azouz, Idir. 1991. "Numerical Simulation of Drilling Fluids in Annuli of Arbitrary Cross-Section." Advisory board meeting report, University of Tulsa Drilling Research Projects. November 12.

7 Ward, C., and E. Andreassen, "Pressure-while-Drilling Data Improve Reservoir Drilling Performance." *SPE Drilling and Completion,* 13 (1), 19–24.

8 Moore, P. L. 1974. *Drilling Practices Manual.* Tulsa: Petroleum Publishing Company.

9 Dodge, D. W., and A. B. Metzner. Journal 5, 189, 1959. "Turbulent Flow of Non-Newtonian Systems," *Journal of the American Institute of Chemical Engineers,* 5, 189.

10 Hussaini, S. M., and J. J. Azar. 1983. "Experimental Study of Drilled Cuttings Transport Using Common Drilling Muds." *SPE Journal,* February, 11–20. Tomren, P. H., A. W. Iyoha, and J. J. Azar. 1986. "An Experimental Study of Cuttings Transport in Directional Wells." *SPE Drilling Engineering,* February.

11 Okranji and Azar, *SPE Drilling Engineering,* August 1986, 297–308. Iyoho and Azar, *SPE Journal,* October 1981, 556–71. Becker, Azar, and Okranji. *SPE Drilling Engineering,* March 1991. Larsen, thesis. Hussaini and Azar, *SPE Journal,* February 1983, 11–20.

12 Okranji and Azar, *SPE Drilling Engineering,* August 1986, 297–308. Larsen, thesis. Hussaini and Azar, *SPE Journal,* February 1983, 11–20.

13 Iyoho and Azar, *SPE Journal,* October 1981, 556–71.

14 Luo, Y., P.A. Bern and B.D. Chambers. 1992. "Flow Rate Predictions for Cleaning Deviated Wells," New Orleans: Paper SPE 23884, presented at the IADC/SPE Drilling Conference.

15 Iyoho and Azar, *SPE Journal,* October 1981, 556–71.Rasi, M. 1994. "Hole Cleaning in Large, High-Angled Wellbores." Paper IADC/SPE 27464, presented at the IADC/SPE Drilling Conference. Clark, R. K., and K. L. Bickman. 1994. "A Mechanistic Model for Cuttings Transport." Paper SPE 28306, presented at the 69th SPE Annual Technical Conference and Exhibition, New Orleans.

16 Iyoho and Azar, *SPE Journal,* October 1981, 556–71.

17 Larsen, T.F., "A Study of the Critical Fluid Velocity in Cuttings Transport for Inclined Wellbore. Thesis, University of Tulda, 1990.

6

Prevention and Control Mechanics of Well Blowouts

Introduction

During drilling operations, an intrusion of formation fluids (water, oil, gas, or a combination of the three) into the wellbore is termed a *kick*, and the fluid is called *kick fluid*. When a kick is taken, there is the potential for blowout occurrence. If the well can be successfully shut in with equipment (blowout preventers [BOPs]) that will prevent any further flow of formation fluid into the wellbore, then the situation is referred to as a controlled kick; otherwise, it is an uncontrolled kick and therefore referred to as a blowout, which is potentially dangerous. If the formation fractures while the well is shut in and kick fluid goes into the fractured zone, then the blowout is called an *underground blowout*. If surface equipment fails to control the migration of kick fluid to the surface, then this is referred to as a *surface blowout*.

Blowouts can occur during drilling, tripping, casing, or workover operations. They are regarded as accidents, and their occurrence can endanger human lives, the monetary investment, and the environment. Therefore, their prevention is undoubtedly the most important task in any drilling venture. This chapter will address the causes, detection, control mechanics, and equipment that may prevent blowout occurrence.

Kick Causes

In general, a kick will occur when the pressure of formation fluid, P_{ff}, becomes greater than the wellbore pressure induced by the drilling fluid, P_{df}. In other words, a kick will occur when the differential pressure, ΔP (the pressure difference between drilling-fluid pressure and formation fluid pressure), becomes negative. Mathematically, the differential pressure is expressed as

$$\Delta P = P_{df} - P_{ff} \tag{6.1}$$

Therefore, depending on the algebraic value of ΔP, one of three conditions may obtain:

- $\Delta P > 0$, overbalanced condition—no kick
- $\Delta P = 0$, balanced condition—no kick
- $\Delta P < 0$, underbalanced condition—kick

It is under the third condition (i.e., $\Delta P < 0$) that formation fluids will flow into the wellbore. Figure 6–1 illustrates the differential pressure concept and the associated kick occurrences.

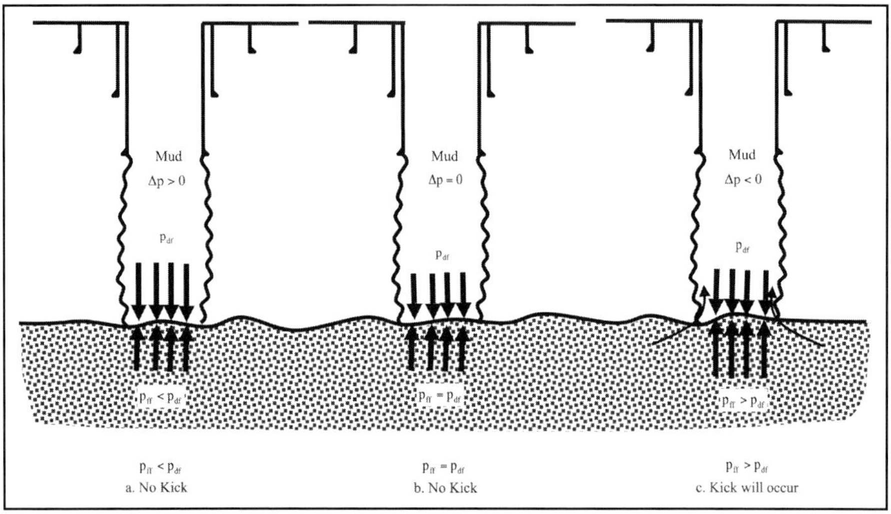

Fig. 6–1. Drilling conditions. a, *Overbalanced.* b, *Balanced.* c, *Underbalanced.*

The following conditions can cause the *differential* pressure to become negative (i.e., $\Delta P < 0$) and, therefore, kick occurrence:

- Encountering an unexpected high-pressure zone where the pore pressure becomes higher than the mud weight in use
- Wellbore fluid pressure (for reasons that will be discussed later) becomes less than formation fluid pressure

Formation pore pressure may be categorized as follows:

- *Normal* pressure (gradient = 0.433–0.465 psi/ft)
- *Subnormal* pressure (gradient < 0.433 psi/ft)
- *Abnormal* pressure (gradient > 0.433 psi/ft)

There is no exact method of determining formation pore pressures; therefore, their estimation is often subjected to uncertainties. Most estimations of formation pressure are based on seismic data, drilling data, and/or log data. There is inherent error in any data that are measured, interpreted, and/or calculated; hence, the exact value of these pressures is uncertain and may lead to improper mud weight selection. Abnormal formation pressures are known to be due to the following phenomena:

- Rapid sediment deposition
- Around salt domes and below salt beds
- Bottom aquifer below a gas column
- Biogenic gas generation
- Gas migration to higher zones through faults or improper cement job
- Tectonic activities in the area

The causes of a decrease in mud column hydrostatic pressure such that it may be less than that of the pore pressure are discussed in the following sections.

Lost circulation

Lost circulation is defined here as total loss of drilling mud into a formation, causing a decrease in the mud hydrostatic head, as shown in figure 6–2. Potential zones for lost circulation are characterized as

- Highly permeable
- Cavernous
- Inherently fractured
- Fractured due to improper drilling, casing, or tripping practices.

While the conditions in the first three zones are unavoidable, the conditions of the last can be prevented by proper drilling practices. For example, during tripping in, surge pressure (an increase in annular pressure due to pressure friction loss) has to be maintained at a value such that $\gamma_{frac} > \gamma_{df} + \gamma_{surge}$. Likewise, during drilling, the friction pressure loss gradient in the annulus, γ_{af}, should be maintained at a value such that $\gamma_{frac} > \gamma_{df} + \gamma_{af}$, where γ_{frac} is the formation fracture pressure gradient, γ_{df} is the drilling fluid pressure gradient, and γ_{surge} is the surge pressure gradient.

If the aforementioned conditions are not properly maintained, formation fracturing may occur, leading to loss of circulation. This in turn will lead to kick occurrence, as shown in figure 6–2.

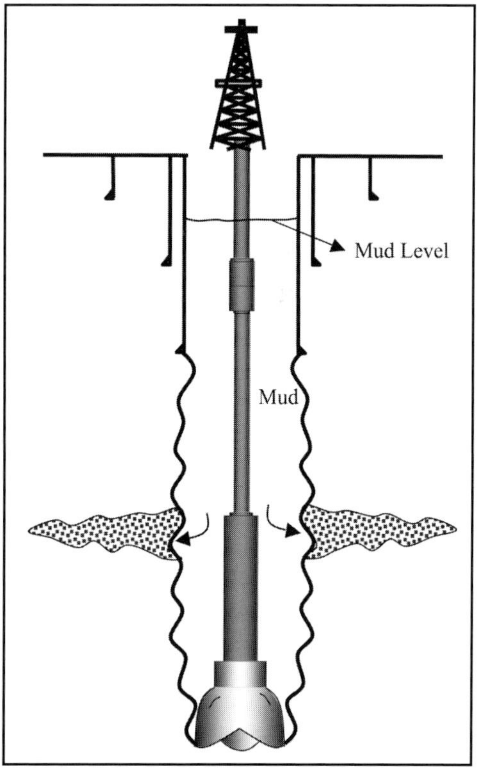

Fig. 6–2. Lost circulation zone

Improper tripping-out practices

Tripping out can result in the reduction of drilling-fluid pressure, P_{df}; hence, a kick will occur. This pressure reduction is caused either by wellbore swabbing or by not adequately refilling the hole with a mud volume equal to that of the pipe being pulled out. (See fig. 6–3 for illustration.) Proper tripping out is a situation dictated by the equation $P_{ff} < P_{df} - P_{swab}$, where P_{ff} is the formation fluid pressure.

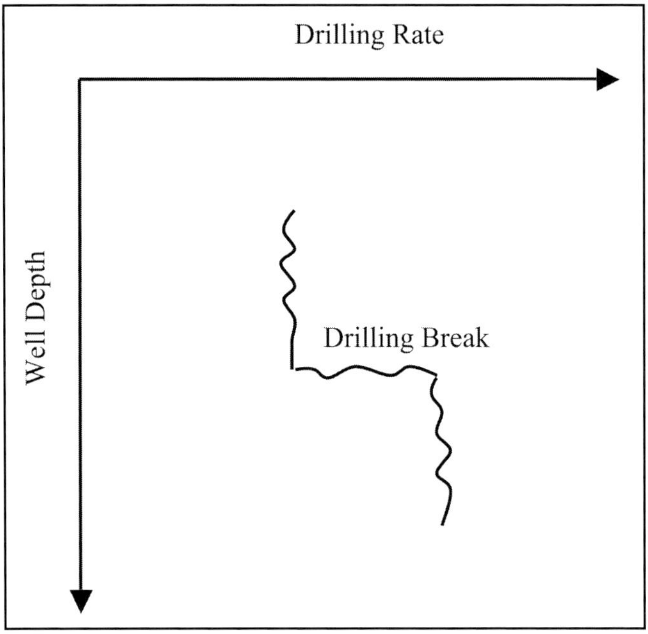

Fig. 6–3. Drilling break

Inadequate drilling-mud weight

Inadequate mud weight may mean that the mud pressure gradient is lower than the formation fluid pressure gradient; as a result, a kick will occur. Also, it may mean having a pressure gradient greater than the formation fracture gradient; consequently, the formation will break, leading to loss of circulation and, hence, kick occurrence. Therefore, it is important to know the formation pore and fracture pressures prior to selecting a mud weight.

Gas-cut mud

Gas-cut mud can be the result of any combination of the following situations:

- Drilling through high-pressure shale that contains gas
- Trip gas during tripping or making connections
- Drilling through low-permeability/high-pressure gas reservoirs

In general, gas-cut mud does not lead to significant mud weight reduction and therefore is not considered to be a major problem. Even if the mud surface density drops to half its original density, the reduction in hydrostatic pressure at the hole bottom is small, since most of the gas expansion occurs near the surface. However, when drilling a shallow formation at a high drilling rate, the presence of gas can reduce the hydrostatic pressure to a point that may cause a kick. The reduction in bottom-hole pressure owing to gas-cut mud may be approximated by the equation $\Delta P = 2.3(\gamma_{om} - \gamma_{rm})(\log P_{omh})/\gamma_{rm}$.

Kick Detection

Early detection of kicks is extremely important in well control operations and in the prevention of possible uncontrolled blowouts. The following are common indications that a kick may have been taken:

- A sudden increase in mud return to surface
- Sudden increase in drilling rate (*drilling break*)
- Sudden decrease in circulating drill pipe pressure
- Hole refilling is less than normal during tripping out

Other signs of possible kick occurrence include oil/gas traces in mud returning to the surface or a moderate change in mud density and rheological properties.

Sudden increase of mud returning to surface

When kick fluid enters into a wellbore, there will be a sudden increase of mud returning to the surface. This increase may be detected by monitoring with a flowmeter, which measures mud flow rate at the return line, or by using a pit volume *totalizer,* which measures the increase in mud level in the return tank.

Sudden increase in drilling rate

A sudden increase in drilling rate is commonly referred to in the field as a *drilling break* (fig. 6–3). It has been well proved that as the differential pressure decreases, the drilling rate increases. Therefore, when a drilling break suddenly occurs, a formation has been penetrated that either has a pore pressure gradient higher then before or is a softer formation than has been encountered previously. If the well is shut in and there is an indication of shut-in drill pipe pressure and casing pressure, then a kick has been taken place. Otherwise, a softer formation has been penetrated, and drilling can be continued without mud weight adjustment.

Sudden decrease in circulating standpipe pressure

A sudden decrease in circulating standpipe pressure can be due to either the occurrence of a kick or parted drill pipe. When a kick is taken, the differential pressure aids surface mud pumps in lifting drilling fluids in the annulus; as a result, standpipe pressure will drop. Also, if a break develops in the drill string, the portion of the circulating system below the break, including the drill bit, will no longer contribute to pressures required from surface pumps; therefore, a drop in standpipe pressure will again be observed. As in the previous case of a drilling break, the well must be shut in to determine whether a kick has been taken or whether a break in the drill string has occurred.

Less than normal hole refilling during tripping out

When tripping out, the mud volume that the hole should be taking must be approximately equal to the volume of pipe being pulled out of the hole. Otherwise, some formation fluid is invading the wellbore.

Kick Prevention

Kick prevention is extremely important because it eliminates the occurrence of potentially dangerous and very costly blowouts. The major factors in kick prevention are

- Accurate knowledge of formation pore pressure and fracture pressure gradients
- Proper tripping, drilling, and casing practices
- Proper mud maintenance
- Rig personnel

Proper mud selection and case setting depths are dependent on the availability and accuracy of data related to pore and fracture pressure gradients of formation tops and types that will be penetrated. Once these subsurface geological conditions are known accurately, the mud program and the casing program can be properly designed.

Good tripping practices (avoiding excessive surge and swab pressures) and drilling practices (avoiding excessive friction pressure losses in the annulus) can always minimize the potential of kick occurrence. Proper maintenance of mud weight will have a bearing on the pressure gradient it exerts against the formation. Also, proper maintenance of mud viscosity will have a bearing on the equivalent circulating mud weight during drilling, tripping, and casings operations.

Educated and well-trained rig personnel are a key factor in the prevention of any accident in drilling. Knowing the whats, the hows, and the whys are a must in education.

Fundamentals of Well Control

Under normal conditions, when a kick is taken, the well is shut in by blowout prevention equipment, as will be discussed later. After the well stabilizes, the shut-in drill pipe pressure, P_{sidp}, and the shut-in casing pressure, P_{sic}, are recorded. To bring the well under control, two fundamental operations are conducted:

- Circulating kick fluid out of the wellbore
- Displacing the original light mud with heavier mud

Both operations must be accomplished under the following conditions:

- No further kick fluid intrusion into the wellbore is to occur
- Formation fracture pressure gradient and maximum allowed operating pressures of equipment are not to be exceeded

Well control methods—an introduction

The following methods are available to control a kick:

- Driller's method
- Wait-and-weight method
- Concurrent method
- Volumetric method
- Dynamic-killing method

In the driller's method, the kick fluid is circulated out of the annulus, using the original mud weight, and then the heavy mud (kill mud) is circulated in, to displace the original mud that is inside the drill string and the annulus.

In the wait-and-weight method, the heavy mud (kill mud) is circulated in, to displace the original mud. This simultaneously circulates the kick fluid out of the annulus and to the surface.

In the concurrent method, the original mud is weighted incrementally; that is,

$$\gamma_{m1} = \gamma_{m0} + \Delta\gamma_m \tag{6.1a}$$

$$\gamma_{m2} = \gamma_{m1} + \Delta\gamma_m \tag{6.1b}$$

$$\gamma_{mn} = \gamma_{m(n-1)} + \Delta\gamma_m \tag{6.1c}$$

For the first incremental increase, the concurrent method is identical to the wait-and-weight method. From the second increment on, however, it is similar to the driller's method in the second circulation.

In the volumetric method, the kick fluid is allowed to migrate to the top by successively bleeding the original mud through a surface choke valve. Afterward, the gas on top is displaced by successively injecting heavy mud into the well while bleeding out the gas.

The dynamic-killing method uses the concept of the equivalent circulating density (ECD), which is a function of the friction pressure drop in the annulus. Circulation of light mud or even clear water at high enough flow rates can cause enough of an increase in mud weight such that a kicking well can be brought under control.

All these methods of kick control will be discussed in more detail later in this chapter.

Well control safety

The safety of well control operations is predicated on the following:

- Early detection of kick (kick volume effect)
- Knowledge of type of the kick fluid (kick-type effect)
- Immediate circulation of annular kick fluid and replacement of light fluid by the heavier kill mud (time-delay effect)
- Knowledge of annular pressure profile during well control, specifically pressures at the last casing shoe and at the surface, near the choke

Kick volume effect. Generally, the kick fluid is lighter than the mud being used; therefore, the greater the kick volume is, the higher will be the shut-in casing pressure. In particular, if it is a gas kick, the greater the initial gas volume is, the higher will be the pressure in the annulus as the kick fluid is being circulated out. Thus, it is more likely that the formation fracture gradient or the allowable operating surface pressures will be exceeded when there is a high initial kick volume. (See fig. 6–4.)

Kick type effect. If the kick fluid that is being circulated out is liquid, the casing pressure will remain constant. By contrast, in a gas kick, the casing pressure will increase significantly and becomes a maximum when the top of the gas column reaches the surface.

Time delay effect. In general, the sooner the kick fluid is circulated out of the annulus and heavy mud displaces the light mud, the safer will be the well control operation.

Annular pressure profile. Accurate knowledge of the annular pressure profile during well control will safeguard against the fracturing of open hole formations and/or the possibility that operating allowable pressures of equipment (casing, BOP, choke, etc.) will be exceeded.

Well control equipment

As mentioned previously, kicks can occur during drilling, during tripping, or after trip has been completed. If a kick is taken during any of the situations described previously, the kick fluid can percolate to the surface through the annular space and/or inside the drill string. Following is a brief description of what may happen and what may be needed to prevent the kick fluid from exiting to the atmosphere.

During a drilling operation, the only exit to the atmosphere that is available to the kick fluid is through the annular space. Therefore, the only preventer(s) needed will shut out the annular space. (See fig. 6–4.)

During tripping, two situations may exist:

- No subsurface check valve (internal preventer) inside drill string above drill bit
- Subsurface check valve inside drill string

During the first situation, a kick fluid can exit to the atmosphere through the drill string and the annular space around the portion of drill string left in the hole. Therefore, what is needed to prevent the kick fluid from exiting to the surface is a preventer that shuts in the wellbore completely. By contrast, the second situation is similar to conditions encountered during drilling.

When there is no pipe in the hole (i.e., when tripping is completed), kick fluid can exit through the casing. Hence, a preventer is needed to completely shut in the wellbore.

In addition, as mentioned before, the fundamentals of well control are such that the kick fluid may be circulated out without allowing any further fluid to enter the wellbore and such that the original mud is replaced with heavier mud. Therefore, to control a kick that has occurred under any of the aforementioned situations and to safely conduct well control operations, well control equipment is mandatory. The following section is devoted to brief discussion and illustration of this equipment.

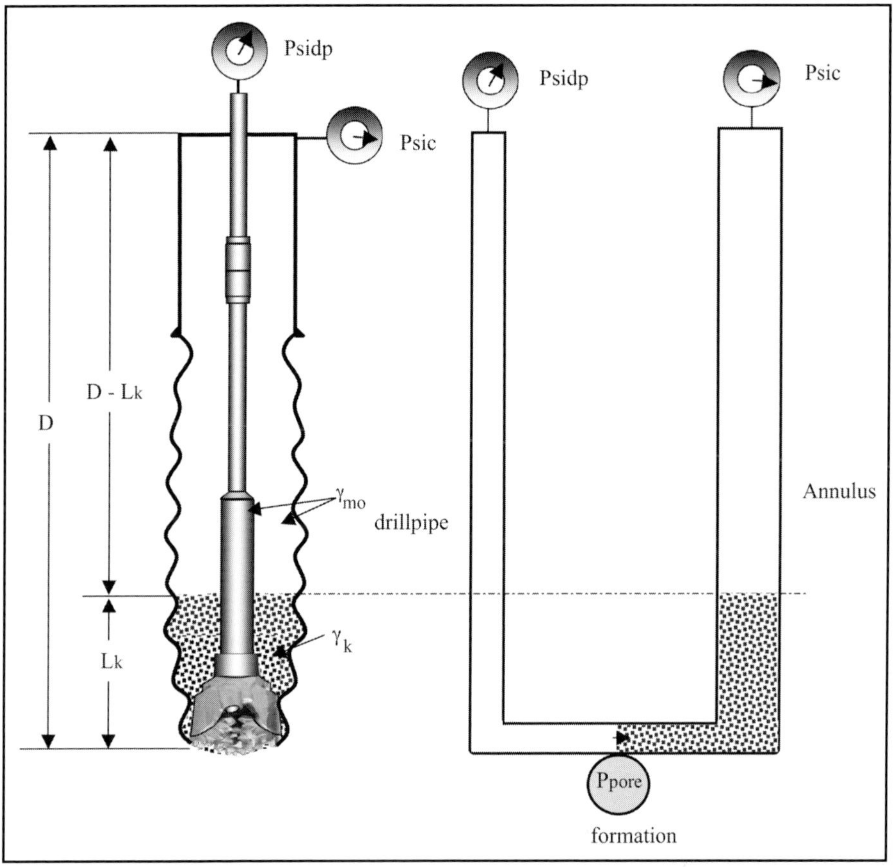

Fig. 6–4. Shut-in well during a kick

Well Control System

The well control system (fig. 6–4) prevents kicks from becoming imminent uncontrolled blowouts. The system requirements allow

- Shutting in the well at the surface
- Removal of kick fluid from the wellbore
- Replacement of original mud with heavier mud
- Stripping/snubbing the drill pipe into or out of the hole

The following are the basic components of the well control system:

- BOP stack (fig. 6–5)
  - Annular preventer
  - Ram preventers
  - Spools
  - Internal preventers
- Casing head
- Kill/choke lines and connections
- Facilities for mud and gas handling (separator)
- Accumulators
- Diverter system
- Choke manifold
- Buffer tank
- Degasser
- Monitoring equipment

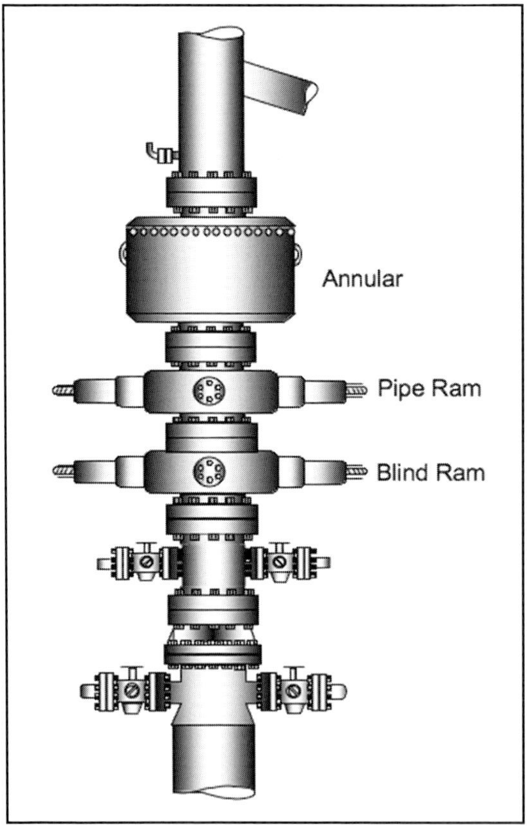

Fig. 6–5. BOP stack with annular preventer, pipe rams, and blind rams

Ram preventers

There are three types of *ram preventers:*

- Pipe ram or casing ram
- Blind ram
- Shear ram (fig. 6–6)

Pipe or casing ram preventers are mechanical devices that close the annular space around the drill pipe or casing, respectively, when they are hydraulically activated. Pipe or casing rams have semicircular openings that match the outer diameter of the drill pipe or the casing. Rams that are designed to close in the annulus when no pipe is present are called blind rams. These rams, if activated accidentally while the drill string is in the hole, will not shut in the annulus. Rams that are designed to shear off drill pipe while string is in hole are referred to as shear rams. These kinds of rams are activated only if all other preventers fail to shut in the well.

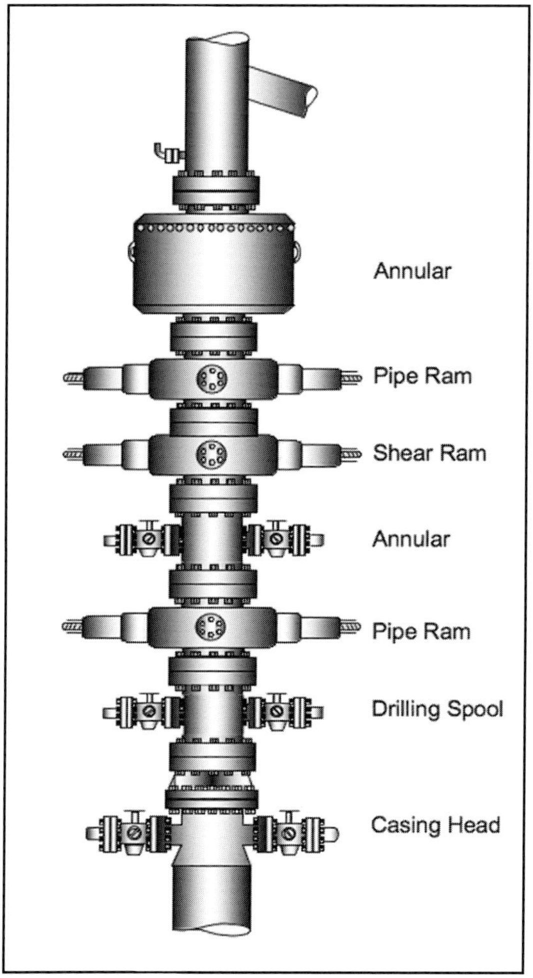

Fig. 6–6. BOP stack with annular preventer, pipe ram, shear rams, and drilling spool

Annular preventers

Annular preventers (fig. 6–6) are well control devices that employ a ring of reinforced synthetic rubber as a packing unit, surrounding the wellbore to cause the shut-off. In the full open position, the inner diameter of the packing unit is equal to the diameter of the preventer bore. These preventers will close and shut in the well regardless of the shape or the diameter of the conduit that might be in the hole.

Drilling spools

Drilling spools (fig. 6–6) are drill-through–type fittings that are placed in the BOP stack assembly to provide a space between two consecutive pipe rams, for temporary storage of tool joints during stripping operations, and to allow attachment of the kill and choke lines.

Casing head

A casing head (fig. 6–6) is part of the permanent well equipment. It provides a means of landing and sealing around the casing string. The casing head also supports the BOP stack.

Diverter system

A *diverter* system is designed to divert the kick fluid to flow through large-diameter vent lines and away from the rig floor. A diverter system is usually not designed to withstand high pressures. Rotating diverter control heads, as shown in figure 6–7, are mounted at the top of the BOP.

Fig. 6–7. Purpose-built diverter system (Courtesy Smith International Inc.)

Choke/kill lines (flow lines)

During well control operations, it is important that circulation can be carried out through some external conduits other than the normal circulating system used during drilling. These external flow lines are referred to as choke/kill lines and are shown in figures 6–8 and 6–9, respectively.

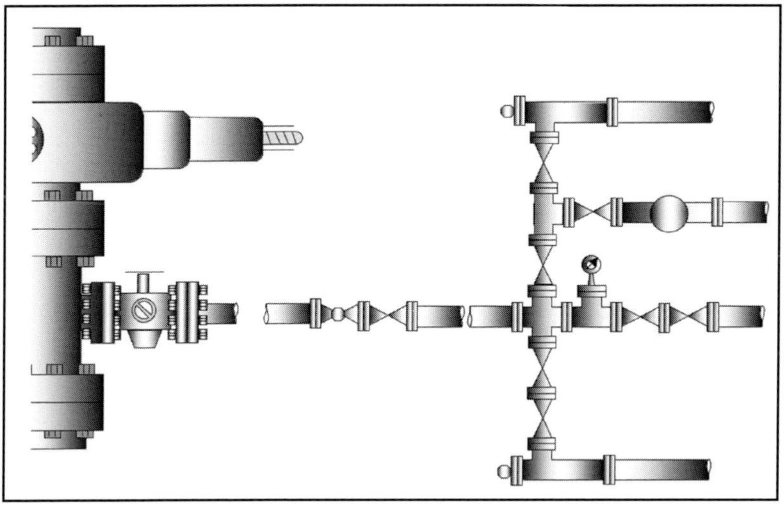

Fig. 6–8. Typical choke manifold

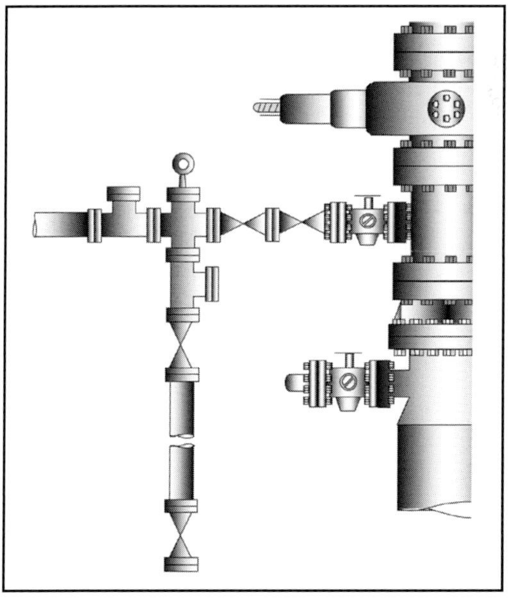

Fig. 6–9. Typical kill manifold

Choke manifold

A *choke manifold* is an arrangement of chokes, valves, and lines (fig. 6–8) that allows the controlled release of fluids from the wellbore. It is designed such that the flow can be diverted through any one of the available chokes. The operation of a choke manifold is either manual or automatic or both.

Buffer tanks

Buffer tanks are installed downstream of the choke manifold to allow gas expansion. If a buffer tank is installed, it should be connected in a manner that permits the chokes in the manifold to be isolated, if necessary, without disrupting the flow of fluids.

Degassers

Degassers are used to remove gas from the mud by creating a vacuum. It is necessary to remove gas from the mud before the mud is returned to the active pit, to prevent low-density gas-cut mud from being circulated down the drill string.

Pit-level systems

Numerous pit-level indicator systems are available to monitor mud pit levels. Many of these include audible alarms and/or chart recorders that provide a permanent record of the pit level.

Flow-line monitors

Flow-line monitors measure the flow rate at the flow line. Most flow-line monitors have an audible alarm that warns of an increase or a decrease in the flow rate. An increase in flow rate is an indication that a kick is occurring. A flow-line monitor may detect the increased flow rate owing to a kick before the pit level has increased enough to be detectable.

Trip tanks

A trip tank is used to measure the amount of fluid required in order to fill the hole when pipe is pulled out and to measure the amount of fluid displaced when pipe is run in. Although trip tanks vary in construction, they all have a small cross-sectional area, so that changes in fluid level may be detected with small changes in fluid volume. A change of one barrel should cause at least a one-inch change in fluid level. A float and marker indicator can be used to allow a visual inspection of the fluid level from the driller's console.

Fill lines

A fill line allows the hole to be filled with drilling fluid. It is installed in the bell nipple above the flow line.

Internal preventer

An inside BOP valve should be available to install on the drill string above the drill bit, to prevent backflow through the drill string if a kick is taken while tripping. However, inside BOP valves have certain disadvantages, including the following:

- They prohibit wireline work
- They may become plugged if lost circulation material is injected
- They hinder the measurement of drill pipe pressure during a kick
- The string must be filled from the surface while running in

However, it is recommended that a downhole backpressure valve be used when drilling from a floating vessel and approaching or drilling through a reservoir.

Kelly cock preventers

Kelly cock preventers are valves installed below and above the kelly to protect the swivel packing and gooseneck (see chap. 1 for illustration) from high pressures.

Accumulators

Accumulators are pressurelike vessels that allows the storage of high pressure in the hydraulic system required in order to open or close the annular or ram-type preventers. The major components of an accumulator unit are the fluid reservoir, high-pressure hydraulic pumps, the accumulator bottle, the piping, manifolds, valves, and pressure regulators.

Gas detectors

Gas detectors monitor the gas concentration in the mud when it returns to surface. An audible alarm will be activated if the gas concentration in the return mud exceeds a preset concentration. Some gas detectors provide a permanent record of the gas concentration by using strip chart recording.

BOP stack arrangement

The arrangement of the various components of the BOP stack must be such that the following well control operations can be conducted:

- Shut in the well with or without drill string in the hole
- Circulate kick fluid out of hole and replace mud with kill mud
- Have the capability of hanging drill string if its shearing is mandated

There are many common arrangements that are generally recommended. Some typical arrangements are shown in figures 6–10 and 6–11.

Well Control and Blowout Prevention: Important Elements in Kick Control

Although kicks may be taken at any time other than while drilling, it is extremely important that the bit be stripped or snubbed back to the bottom if possible. Most attempts to control a kick while the bit is far above the bottom are bound to fail. Therefore, if it is assumed that the bit is on the bottom, the necessities of well control consist of the following actions, which will be defined mathematically in the next section:

- Record shut-in drillpipe and casing pressures
- Calculate the encountered formation fluid pressure
- Calculate the required kill mud weight
- Calculate the required circulating drill pipe pressure or pump pressure schedule during well control
- Determine the induced wellbore pressure profile during well control—specifically, the pressures at the last casing shoe and at the surface choke
- Establish operating pressure limitations of the blowout equipment
- Determine or assume the type of kick or composition (gas, liquid, or gas and liquid)
- Determine the required bottom-hole pressure that is to be maintained during well control operations

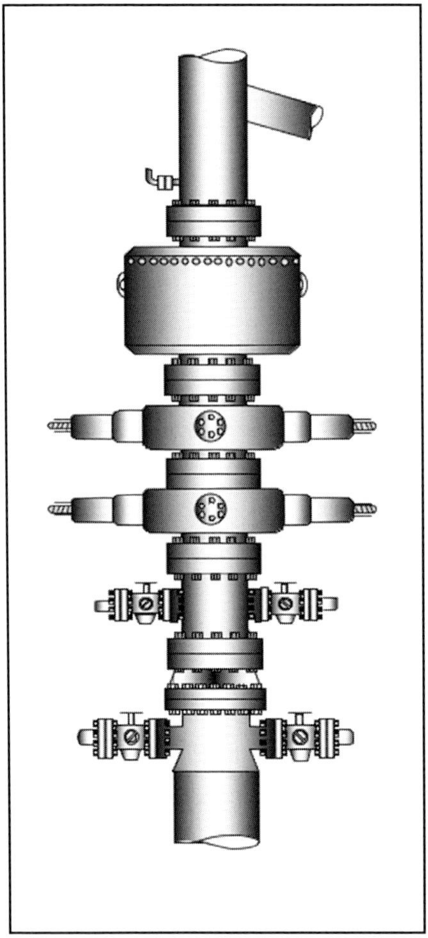

Fig. 6–10. Typical arrangement of a BOP stack (annular preventer, two rams and spool)

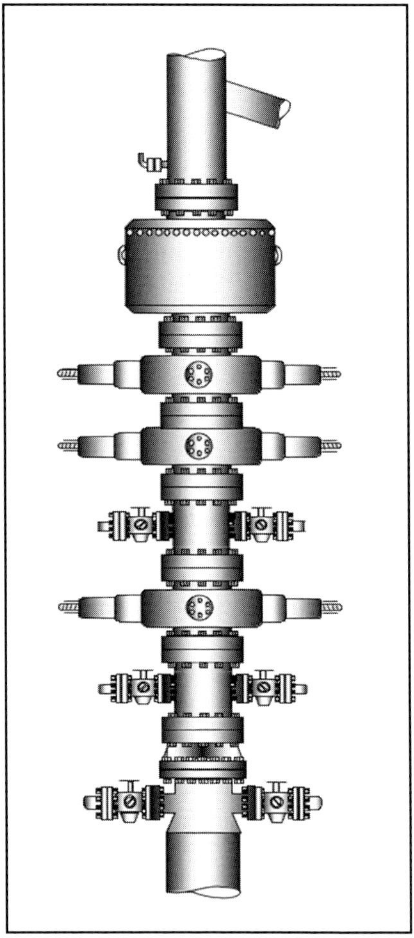

Fig. 6–11. Typical arrangement of a BOP stack (annular preventer, two ramps, spool, ram preventer, spool)

Principles of Well Control

Record shut-in drill pipe pressure, P_{sidp}, and shut-in casing pressure, P_{sic}

When a kick is taken, the well is shut in. If the kick is a gas (i.e., $P_{sidp} < P_{sic}$), after allowing some time (approximately 10–20 min) for the shut-in drill pipe to stabilize (as diagrammed in fig. 6–12), P_{sidp} and P_{sic} are recorded (as illustrated in fig. 6–13). Note that in the case of shallow gas kicks, it may not be advisable to shut in the well; instead, the well should be allowed to flow through the diverter system, to prevent possible formation fracturing that may lead to an underground blowout situation.

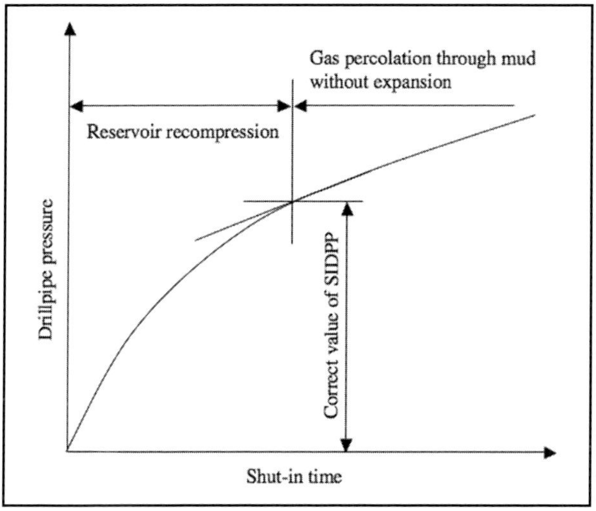

Fig. 6–12. Drill pipe pressure buildup with well shut in (gas kick assumed, for a high-permeability formation)

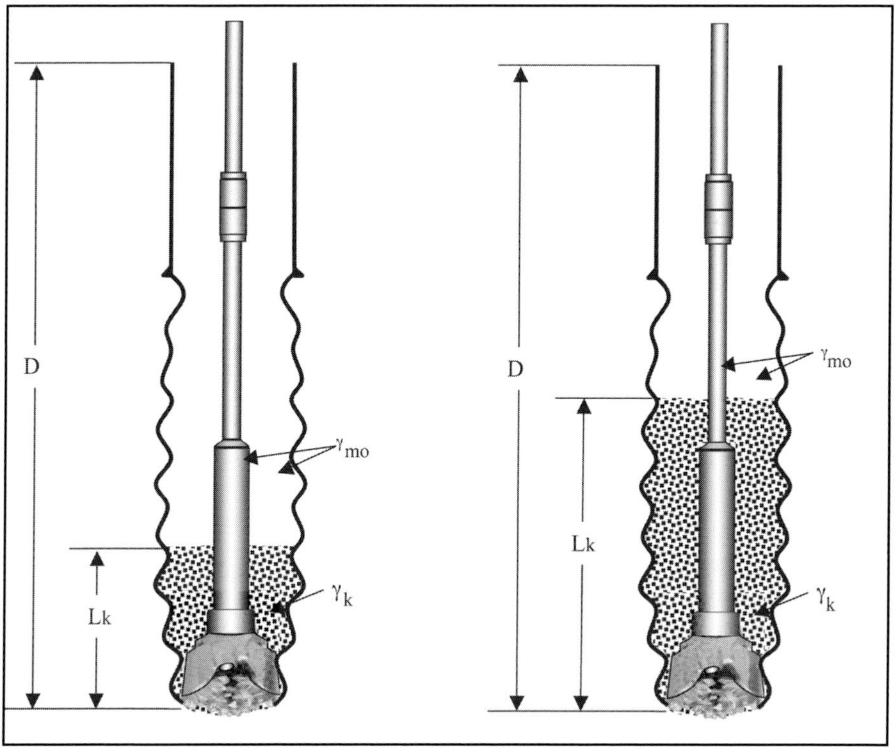

Fig. 6–13. Kick fluid intrusion into annulus

Calculation of encountered formation fluid pressure, P_{ff}

By exploitation of the pressure equilibrium concept, as shown in figure 6–4, the formation fluid pressure, P_{ff}, can be calculated using either of the following equations:

$$P_{ff} = P_{sidp} + k\gamma_{mo}D \tag{6.2a}$$

$$P_{ff} = P_{sic} + k\gamma_{mo}(D - L_k) + k\gamma_k L_k \tag{6.2b}$$

where

k is a unit conversion constant

L_k is the kick length

γ_k is the kick density

Since the kick composition and the kick length are never known exactly, equation (6.2b) is not used to determine P_{ff}. Instead, equation (6.2a) is used since the mud density inside the drill string and the depth, D, at which the kick was taken is known.

If the kick is taken while tripping or when the drill string is completely out of the hole, then accurate calculation of formation pore pressure is not possible. The reason for this is that the mud has been contaminated with kick fluid, whose density and amount of intrusion in the wellbore are not known. In this case, well control is best achieved by using the concurrent method, while the original mud weight is increased incrementally until the well is killed.

Determine the bottom-hole pressure, P_{bh}, that needs to be maintained constant during well control operations

As stated previously, during well control, no further kick fluid should be allowed to enter the annulus. Prevention of kick fluid entry can be accomplished by maintaining a wellbore pressure just in excess of encountered pore pressure—that is,

$$P_{bh} = P_{ff} + \Delta P_s \tag{6.3a}$$

where ΔP_s is some safety differential pressure. By use of equation (6.2a), equation (6.3a) can be rewritten as

$$P_{bh} = P_{sidp} + P_{moh} + \Delta P_s \tag{6.3b}$$

where P_{moh} is the mud hydrostatic pressure: $P_{moh} = k\gamma_{mo}D$

From the start of circulating the kick fluid out of the wellbore and replacing the light fluid with the kill fluid, the bottom-hole pressure in equation (6.3) must be held constant by continuously adjusting the choke, so as to maintain the required circulating drill pipe pressure within a given number of pump stroke increments. This will be illustrated later.

Calculation of mud weight required in order to bring the well under control

The mud weight required in order to resume overbalanced drilling is given by

$$\gamma_{mf} = \gamma_{mo} + \frac{P_{sidp}}{kD} + \Delta\gamma_s \tag{6.4}$$

where $\Delta\gamma_s$ is the safety margin.

Determination of kick fluid type

There is no exact theoretical method for predicting kick fluid composition; however, an equation can be derived to approximately identify kick fluid type. With reference to figure 6–24 and in accordance with the pressure balance concept, the following is obtained:

$$P_{sidp} + k\gamma_{mo}D - k\gamma_k L_k - k\gamma_{mo}(D - L_k) - P_{sic} = 0 \tag{6.5}$$

Solving for γ_k yields

$$\gamma_k = \gamma_{mo} - \frac{P_{sic} - P_{sidp}}{kL_k} \tag{6.6}$$

where

γ_k, L_k are the kick density and the length, respectively

P_{sidp}, P_{sic} are the shut-in drill pipe pressure and the pressure casing, respectively

D is the depth at which kick is taken

γ_{mo} is the original mud weight

Note that in the derivation of equation (6.5), it has been assumed that there is no mixing between kick fluid and the mud in the annulus.

If L_k is known, then the kick density, γ_k, can be calculated. To determine L_k, consider the well schematics in figure 6–13. Let

- C_{adp} be the annulus capacity behind drill pipe (in bbl/ft)
- C_{adc} be the annulus capacity behind drill collar (in bbl/ft)
- L_{dc} be the drill collar length (in ft)
- V_{adp} be the annulus volume behind drill pipe (in bbl)
- V_{adc} be the annulus volume behind drill collar (in bbl)
- V_k be the kick volume

Two cases need to be considered:

- Case 1:

$$L_k = \frac{V_k}{C_{adc}} \quad \text{for } V_k < V_{adc} \tag{6.7}$$

- Case 2:

$$L_k = L_{dc} + \frac{V_k - V_{adc}}{C_{adp}} \quad \text{for } V_k > V_{adc} \tag{6.8}$$

Substitution of equations (6.7) and (6.8), respectively, into equation (6.5) yields

for $V_k < V_{adc}$

$$\gamma_k = \gamma_{mo} - \frac{(P_{sic} - P_{sidp})}{k(V_k / C_{adc})} \tag{6.9}$$

and for $V_k > V_{adc}$

$$\gamma_k = \gamma_{mo} - \frac{P_{sic} - P_{sidp}}{k\left\{L_{dc} + \left[\left(V_k - V_{adc}\right)/C_{adp}\right]\right\}} \tag{6.10}$$

Note that the kick volume, V_k, is recorded on the surface by a pit volume totalizer (PVT) or some other means, before the well is shut in.

By use of equation (6.9) or (6.10), if the calculated kick density is less than approximately 4 lbs/gal, then the kick fluid is assumed to be gas. If it is greater than approximately 6 lbs/gal, then kick fluid is assumed to be liquid.

Importantly, by observation of the comparative difference between P_{sidp} and P_{sic}, one can estimate whether kick fluid is mainly gas or liquid. For example, if P_{sic} is much greater than P_{sidp}, then the kick fluid most likely is gas. If $P_{sic}$and P_{sidp} are fairly close in magnitude and depending on the mud weight in use, then the kick fluid most likely is liquid. For instance, if the mud in use is seawater and P_{sic} and P_{sidp} are almost equal when the well is shut in, then the kick is a formation water kick.

Determination of circulating drill pipe pressure (pump pressure schedule) during kick control

To prevent any further intrusion of formation fluid into the wellbore during well control, the bottom-hole pressure must be maintained constant at a value given by the following equation:

$$P_{bh} = P_{ff} + \text{safety} = k\gamma_{mo}D + P_{sidp} + P_s \tag{6.11}$$

This bottom-hole pressure, P_{bh}, is kept constant during the entire well control operation by continuously adjusting a surface choke until the well is killed. Under the assumption that the kick fluid is being circulated out of the annulus by use of the original mud, the initial and final required circulating drill pipe pressures are given by

$$P_{cdpi} = P_{cdpf} = P_{pr} + P_{sidp} + P_s \tag{6.12}$$

whereP_{cdpi}, P_{cdpf} are the initial and final circulating drill pipe pressures, respectively, and P_{pr} is the pump pressure at a reduced flow rate Q_r, which is equivalent to that used during well control. P_{pr} is the pump pressure required in order to circulate the fluid at a flow rate of Q_r; that is,

$$P_{pr} = P_{fr} + P_{br} \tag{6.13}$$

where P_{fr} is the total friction pressure loss in the circulating system at the reduced flow rate Q_r and P_{br} is the pressure drop across the drill bit, also at Q_r.

It is common field practice to measure P_{pr} prior to making a trip. Otherwise, this pump pressure has to be determined by calculating the friction pressure losses and the dynamic pressure change in the well control circulating system by using available engineering equations. In onshore drilling and in offshore drilling from bottom-supported units, the drilling and well control circulating system is practically the same. However, in floating units, it is not the same, as will be illustrated in a later section.

If the original mud weight, γ_{mo}, is being displaced by the kill mud weight, γ_{mf}, the initial circulating drill pipe pressure must be reduced, depending on the location of heavy mud inside the drill string. When the kill mud reaches the drill bit, the reduction in circulating drill pressure can be approximated as follows:

$$\Delta P_{cdp} = k(\gamma_{mf} - \gamma_{mo})D - P_{pr}\left(\frac{\gamma_{mf} - \gamma_{mo}}{\gamma_{mo}}\right) \tag{6.14}$$

Therefore, the final circulating drill pipe pressure at the time the kill mud reaches the bottom of the hole is

$$P_{cdpf} = P_{cdpi} - \Delta P_{cdp} = P_{pr} + P_{sidp} + P_s + P_{pr}\left(\frac{\gamma_{mf} - \gamma_{mo}}{\gamma_{mo}}\right) - k(\gamma_{mf} - \gamma_{mo})D \tag{6.15a}$$

or

$$P_{cdpf} \approx P_{pr}\frac{\gamma_{mf}}{\gamma_{mo}} + P_s \tag{6.15b}$$

When the initial and final circulating drill pipe pressure requirements are known, linear interpolation between the two values can yield the pump pressure schedule required for well control. The equation can be derived easily from the straight-line equation:

$$P_{cdpn} = \left(\frac{P_{cdpf} - P_{cdpi}}{\overline{\text{TNPS}}}\right)\overline{\text{NPS}} + P_{cdpi} \tag{6.16}$$

In equation (6.16), $\overline{\text{TNPS}}$ denotes the total number of pump strokes to pump mud inside the drill string until it reaches the drill bit:

$$\overline{\text{TNPS}} = \frac{L_{dp}C_{dp} + L_{dc}C_{dc}}{F_p} \tag{6.17}$$

where

$\overline{\text{TNPS}}$ is any number of pump strokes

F_p is a pump factor, in bbl/stroke

L_{dp}, L_{dc} are the respective drill pipe and drill collar lengths, in ft

C_{dp}, C_{dc} are the respective drill pipe and drill collar capacities, in bbl/ft

Importantly, the mechanics of well control is the same for offshore and onshore drilling. However, because of equipment changes, the methods of implementation may be quite different in offshore well control. For example, when drilling from a floating rig, the circulating system while drilling is not the same as is used while conducting well control. The friction pressure losses in the choke lines and manifold, which are ignored in onshore well control, cannot be ignored in offshore well control.

Calculation of annular pressure during well control

As stated previously, exceeding the open hole formation fracture pressure gradient and/or the working pressures of the well control equipment during well control operations can result in very costly accidents. Therefore, it is essential that the induced pressure profile in the annulus be known during well control operations. In particular, the pressures at the casing-shoe depth, D_{cs}, and the surface choke depth, D_{sc}, must be determined. These pressures are maximum when the top of the gas reaches these respective depths. Two cases will be considered:

1. Kick fluid is circulated out with original mud (driller's method)
2. Kick fluid is circulated out with weighted mud (wait-and-weight method).

Both cases will be considered under the following assumptions:

- Kick is predominantly gas
- Kick does not mix with the mud
- Ideal gas law applies
- The annulus cross section is a constant
- Gas density is neglected
- Bottom-hole pressure, P_{bh}, is held constant
- Friction pressure losses in annulus are neglected

Driller's method. The assumption is made that the top of the gas column has reached a depth of interest, D_i, measured from the top of the well, as shown in figure 6–14. A pressure balance yields the following: $P_i + k\gamma_k L_{ki} + k\gamma_{mo}(D - D_i - L_k) = P_{ff}$, where P_{ff} is the formation fluid pressure. Since γ_k is very small compared to γ_{mo}, the term $k\gamma_h L_{ki}$ can be neglected. Therefore, solving for the pressure at the depth of interest, D_i, yields

$$P_i = P_{ff} - k\gamma_{mo}(D - D_i - L_{ki}) \tag{6.18}$$

where P_i is the pressure at the point of interest and L_{ki} is the kick length after the gas-bubble top reaches a certain point, at depth D_i. The length, L_k, can be expressed as $L_k = V_{ke}/C_a$, where V_{ke} is the expanded gas kick volume when the top of the gas reaches a desired point and C_a is the average annular capacity (volume per unit length).

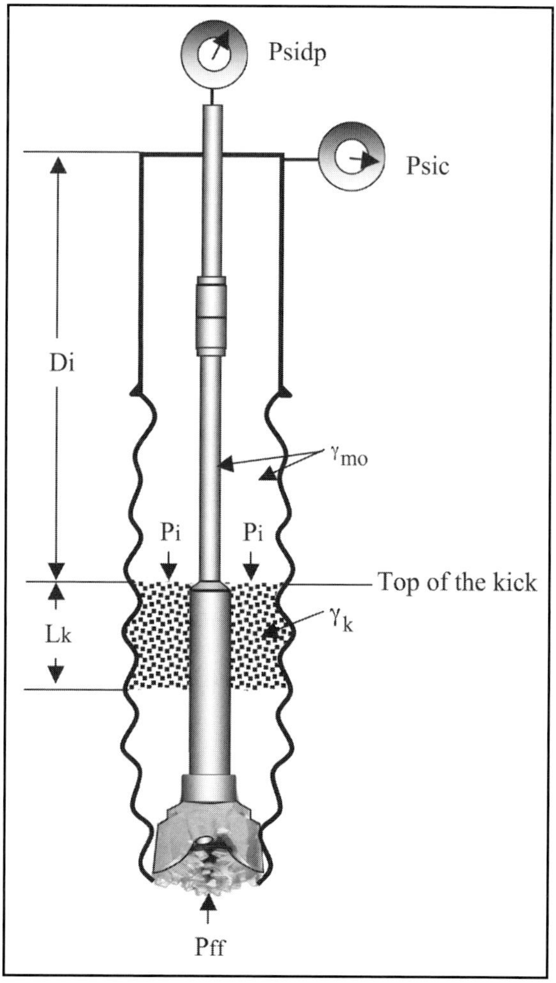

Fig. 6–14. Schematic of a kick (driller's method)

The expanded kick volume, V_{ke}, may be found by considering the ideal gas law—that is, PV/ZT = constant. Thus, taking the bottom hole as one reference point, where the conditions of pressure, P_{bh}, and temperature, T_{bh}, and the gas deviation factor, Z_{bh}, are known and taking as another point the depth D_i, representing the location where top of gas bubble has reached and under the conditions P_i, T_i, and Z_i, the following can be written:

$$\frac{P_{bh}V_{ko}}{Z_{bh}T_{bh}} = \frac{P_iV_{ke}}{Z_iT_i} \tag{6.19}$$

Solving for V_{ke} yields

$$V_{ke} = \eta\frac{P_{bh}}{P_i}V_{ko} = \frac{\eta(P_{ff} + P_{safety})}{P_i}V_{ko} \tag{6.20}$$

where $\eta = Z_iT_i/Z_{bh}T_{bh}$.

The safety pressure that is used during well control is generally very small compared to the formation fluid pressure and therefore can be neglected. Hence, the kick length at any point in the annulus as the gas bubble rises is given by

$$L_k = \frac{\eta P_{ff}V_{ko}}{P_iC_a} \tag{6.21}$$

Substituting equation (6.21) into equation (6.18) yields

$$P_iP_{ff} - k\gamma_{mo}\left(D - D_i - \frac{\eta P_{ff}V_{ko}}{P_iC_a}\right) \tag{6.22}$$

Expressing equation (6.22) in a quadratic form,

$$P_i^2 - \left[P_{ff} - k\gamma_{mo}\left(D - D_i\right)\right]P_i - \frac{k\gamma_{mo}\eta P_{ff}V_{ko}}{C_a} = 0 \tag{6.23}$$

Solving for P_i in equation (6.23),

$$P_i = \frac{1}{2}\left[P_{ff} - k\gamma_{mo}\left(D - D_i\right)\right] + \frac{1}{2}\sqrt{\left[P_{ff} - k\gamma_{mo}\left(D - D_i\right)\right]^2 + \frac{4\eta k\gamma_{mo}P_{ff}V_{ko}}{C_a}} \tag{6.24}$$

Thus, for any desired depth D_i, the pressure, P_i, can be calculated. Of special interest are the pressures at the last casing-shoe depth and at the surface choke when the gas-bubble top reaches these points (i.e., $D_i = D_{cs}$, and $D_i = D_{sc} = 0$, respectively, where D_{cs} is the casing-shoe depth). At the casing shoe, pressure calculation may be performed by assuming that the temperature and gas deviation factors are relatively constant. Thus, $\eta = 1$ in equation (6.24).

Wait-and-weight method

With reference to the driller's method and figure 6–15, the following equation can be obtained:

$$P_i = \frac{1}{2}\left[P_{ff} - k\left(D - D_i - L_{ok}\right)\gamma_{mf} - k\gamma_{mo}L_{ok}\right] + \frac{1}{2}\sqrt{\left[P_{ff} - k\left(D - D_i - L_{ok}\right)\gamma_{mf} - k\gamma_{mo}L_{ok}\right]^2 + \frac{4\eta k P_{ff} V_{kO}\gamma_{mf}}{C_a}} \tag{6.25}$$

where L_{ok} is the length of original mud between the bottom of the gas column and the top of the final mud.

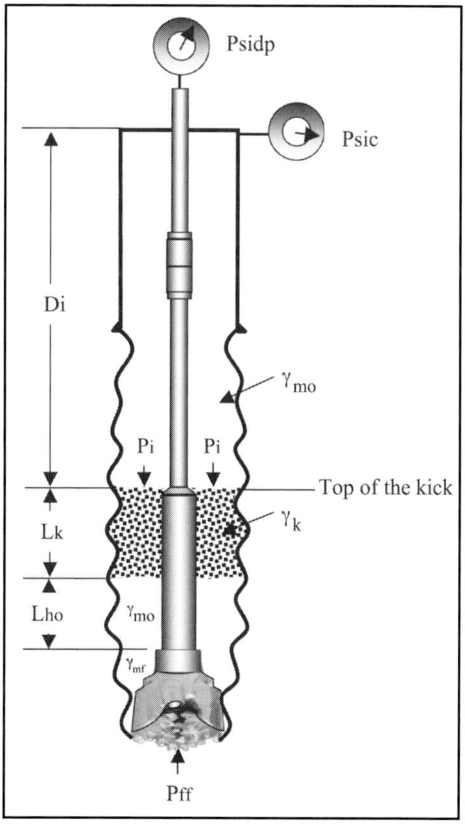

Fig. 6–15. Schematic of a kick (wait-and-weight method)

Well Control Procedures: Kick Control with Drill Bit on Bottom of Hole

When a kick is taken, the kick fluid must be circulated out of the well, and the original light mud must be totally displaced with heavier mud that has a hydrostatic pressure gradient, ΔP_{df}, slightly higher than that of the encountered formation pressure gradient, ΔP_{ff} (i.e., $\Delta P_{df} > \Delta P_{ff}$).

Regardless of the kick control method, the principles are generally the same as discussed previously. These comprise the following actions:

- Record the shut-in drill pipe pressure, P_{sidp}, and the shut-in casing pressure, P_{sic}
- Calculate the encountered formation fluid pressure, P_{ff}
- Calculate the mud density required in order to kill the well
- Calculate the bottom-hole pressure, P_{bh}, to be maintained constant during kick control operations
- Calculate the initial circulating drill pipe pressure, P_{cdpi}, and the final circulating drill pipe pressure, P_{cdpf}, when the kill mud reaches the hole bottom
- Determine the pump pressure schedule as heavy (kill) mud is being circulated to displace light (original) mud
- Determine the pressure profile while circulating the gas kick out of the well—specifically, the pressures when the top of the gas kick reaches the last casing shoe and the surface choke

Commonly Used Well Control Methods

Although there are several ways to circulate a kick fluid out of a well, the two most commonly used in the drilling industry are the driller's method and the wait-and-weight method. Both use the concept of maintaining a constant bottom-hole pressure during the entire well control operation. In addition, the friction pressure losses in the annulus are assumed to be negligible. However, in offshore drilling while using mobile rigs, friction pressure losses in the choke line are significant and cannot be ignored.

Driller's method

The driller's method consists of two separate circulations. In the first circulation, the kick fluid is circulated out of the well while using the original mud. In the second circulation, after the well is free from kick fluid, the light mud is totally displaced by the heavier kill mud. During the first circulation, the pump pressure (initial circulating drill pipe pressure) stays constant and is equal to $P_{cdpi} = P_{pr} + P_{sidp}$. However, during the second circulation, the pump pressure starts at P_{cdpi} and is reduced to a final circulating pressure P_{cdpf}, as indicated in figures 6–16 and 6–17.

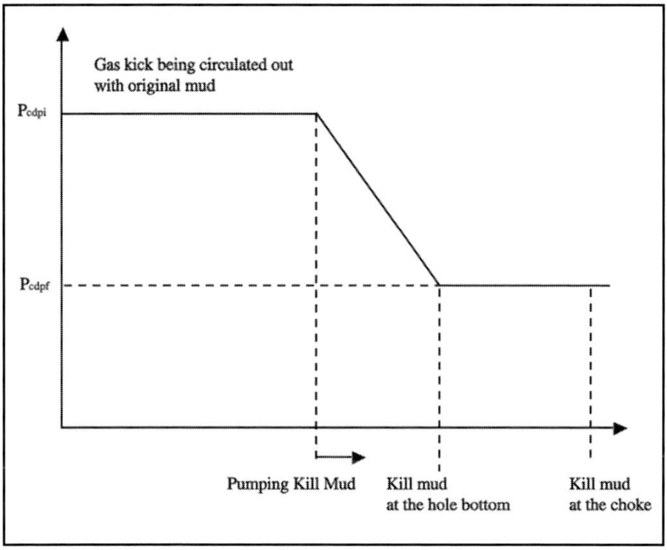

Fig. 6–16. Driller's method

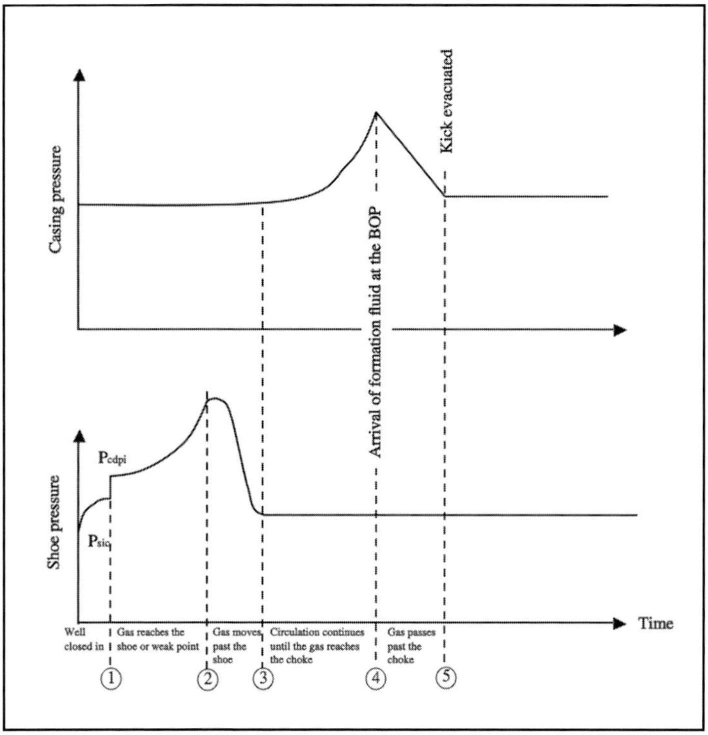

Fig. 6–17. Pressures at casing shoe and choke

Operational procedure. Under the assumption of a gas kick, during the first circulation, the kick fluid is circulated out of the annular space by using the original light mud and without allowing any further kick fluid intrusion; this is done by keeping the circulation drill pipe pressure at $P_{cdpi} = P_r + P_{sidp}$. Afterward, the choke is closed and P_{sidp} and P_{sic} are read. They should be equal. If not, continue circulating until all gas is removed from the well.

During the second circulation, a volume of the heavy kill mud that is equal to the total drill string volume capacity is circulated while keeping the choke pressure at P_{sidp}. The circulating drill pipe pressure should drop from P_{cdpi} at the beginning of circulation to P_{cdpf} when the new mud reaches the bit. The circulation of heavy mud is continued until all the original mud in annulus is displaced. Afterward, the choke is closed and P_{sidp} and P_{sic} are read. They should be equal to 0 psi. If so, open the well and resume the drilling operations; if not, continue the well control operation. $V_m = V_{dp} + V_{dc} + V_{adp} + V_{adc}$, or in terms of required number of strokes, $\overline{\text{TNPS}} = V_m/F_p$.

Example 6–1

A gas kick was taken (see fig. 6–18) while drilling at 10,000 ft TVD with a mud weight of 10 lbs/gal. The pit gain was recorded at 10 bbl. The recorded shut-in drill pipe and casing pressures are 500 psi and 590 psi, respectively. The safety margin is 0.3 lbs/gal. In addition, the following data are given:

- $C_{adp} = 0.049$ bbl/ft
- $C_{adc} = 0.0292$ bbl/ft
- $C_{dp} = 0.0178$ bbl/ft
- $C_{dc} = 0.0061$ bbl/ft
- Pump factor: $F_p = 2.6$ gal/stroke
- Pump volumetric efficiency = 95%
- Pump pressure at reduced flow rate: $P_{pr} = 1{,}500$ psi at 60 strokes/min
- Casing-shoe depth = 6,000 ft
- Drill collar length: $L_{dc} = 1{,}000$ ft
- $k = 1$

(a) Determine the pressures at casing shoe and surface choke during first circulation with the original mud.

(b) Determine the pump pressure schedule during second circulation with the kill mud to replace the original mud.

Solution: Formation fluid pressure is determined as follows:

$$\begin{aligned} P_{ff} &= k\gamma_{mo}D + P_{sidp} \\ &= 0.052(10)1{,}000 + 500 \\ &= 5{,}700 \text{ psi} \end{aligned}$$

Required kill mud weight is

$$\gamma_{mo} + \frac{P_{sidp}}{kD} + \text{safety}$$
$$= 10 + \frac{400}{0.052(10{,}000)} + \text{safety}$$
$$= 10.96 + 0.3 = 11.3 \text{ ppg}$$

The pressure, P_i, at any depth D_i in the annulus is

$$P_i = \frac{b + \sqrt{b^2 + 4c}}{2}$$

where

$b = P_{ff} - 0.052\gamma_{mo}(D - D_i)$
$c = 0.052\gamma_{mo}kP_{ff}(V_k/C_{adp})$
$D_i = D_{cs} = 6{,}000 \rightarrow P_{cs}$ (casing shoe)
$D_i = 0 \rightarrow P_{sc}$ (surface choke)

At the casing shoe,

$$b = \begin{array}{l} 5{,}700 - 0.052(10)(10{,}000 - 6{,}000) \\ 3{,}620 \end{array}$$
$$c = \begin{array}{l} 0.052(10)(1)(5{,}600)\dfrac{10}{0.046} \\ 644{,}348 \end{array}$$

At the surface choke,

$$b = \frac{5700 - 0.052\,(10)\,(10{,}000)}{500}$$
$$c = 644{,}348$$

Therefore,

$$P_{cs} = \frac{3{,}620 + \sqrt{3{,}620^2 + (4 \times 644{,}348)}}{2}$$
$$= 3{,}790 \text{ psi}$$
$$= 2{,}000 - \left(0.0923 \times \overline{\text{NPS}}\right)$$

where $\overline{\text{NPS}} = 0.2710$. This can be plotted as shown in figure 6–19; for sequences of the driller's and wait-and-weight methods, see also figures 6–20 and 6–21.

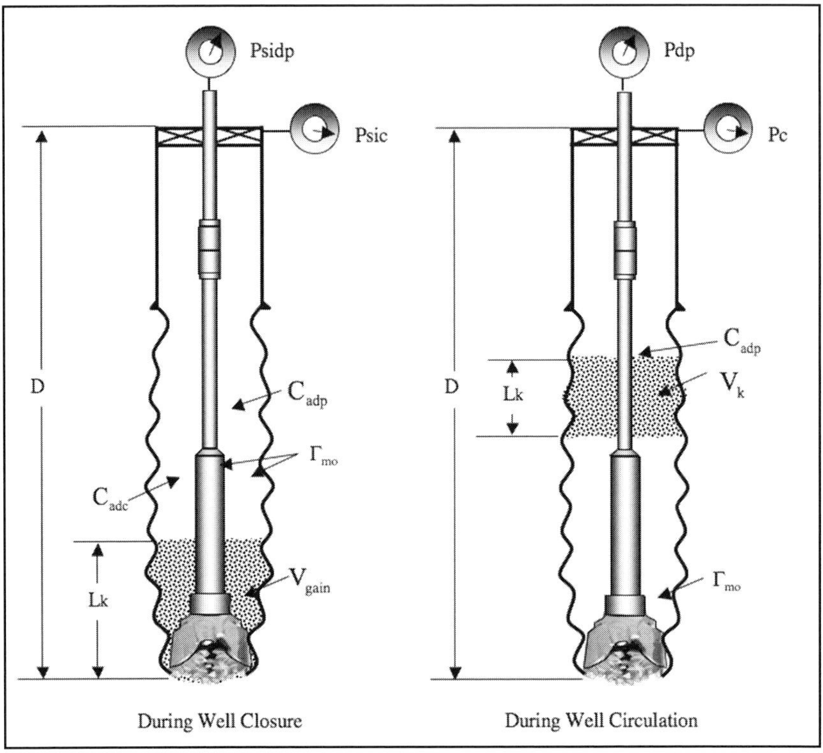

Fig. 6–18. Gas kick taken in example 6–1

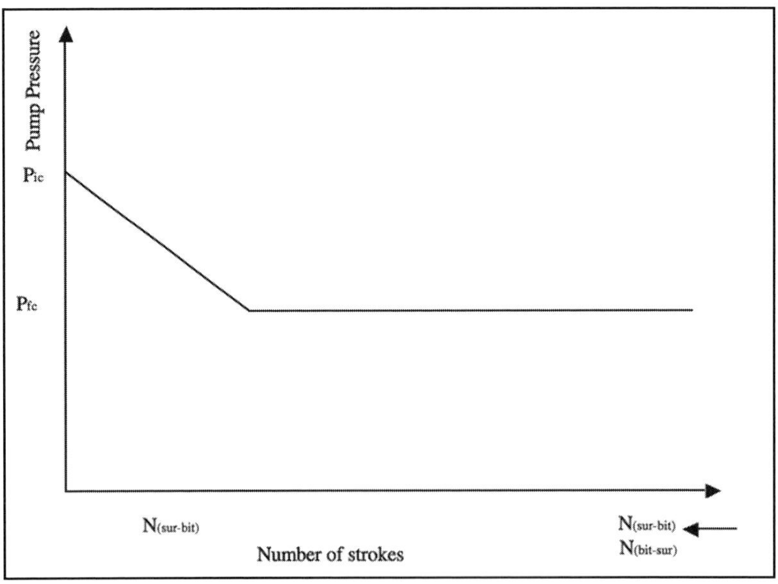

Fig. 6–19. Plot of pump pressure versus number of strokes

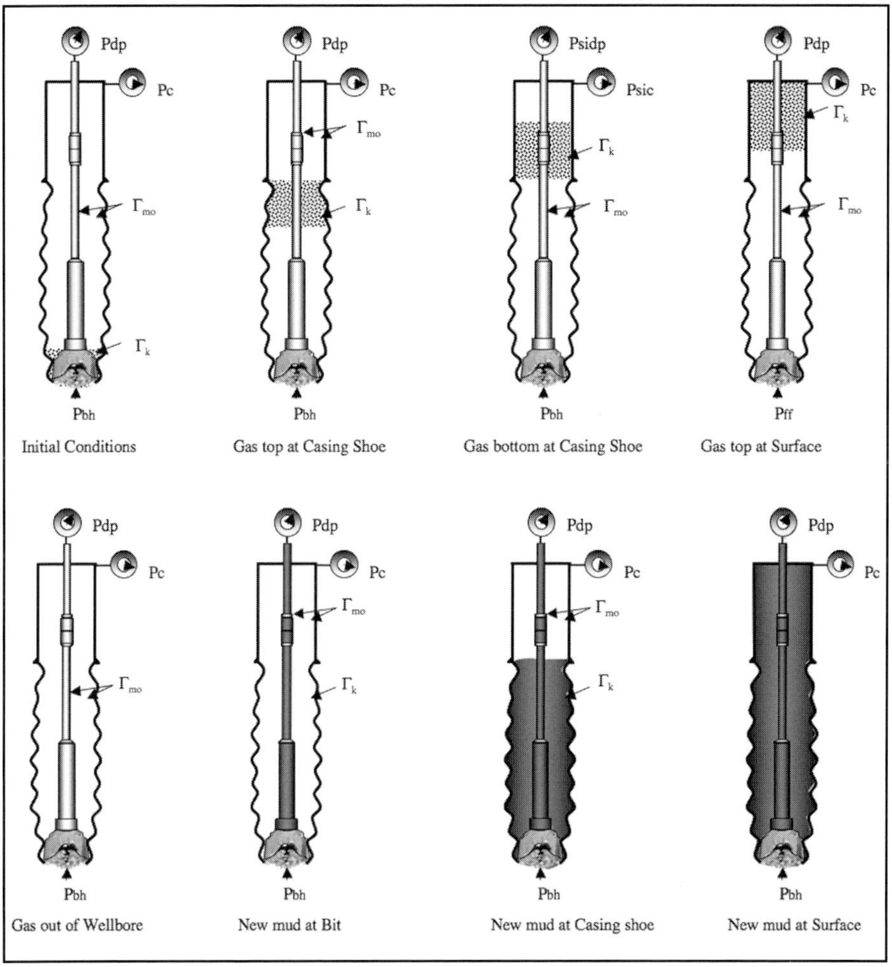

Fig. 6–20. Driller's method sequence

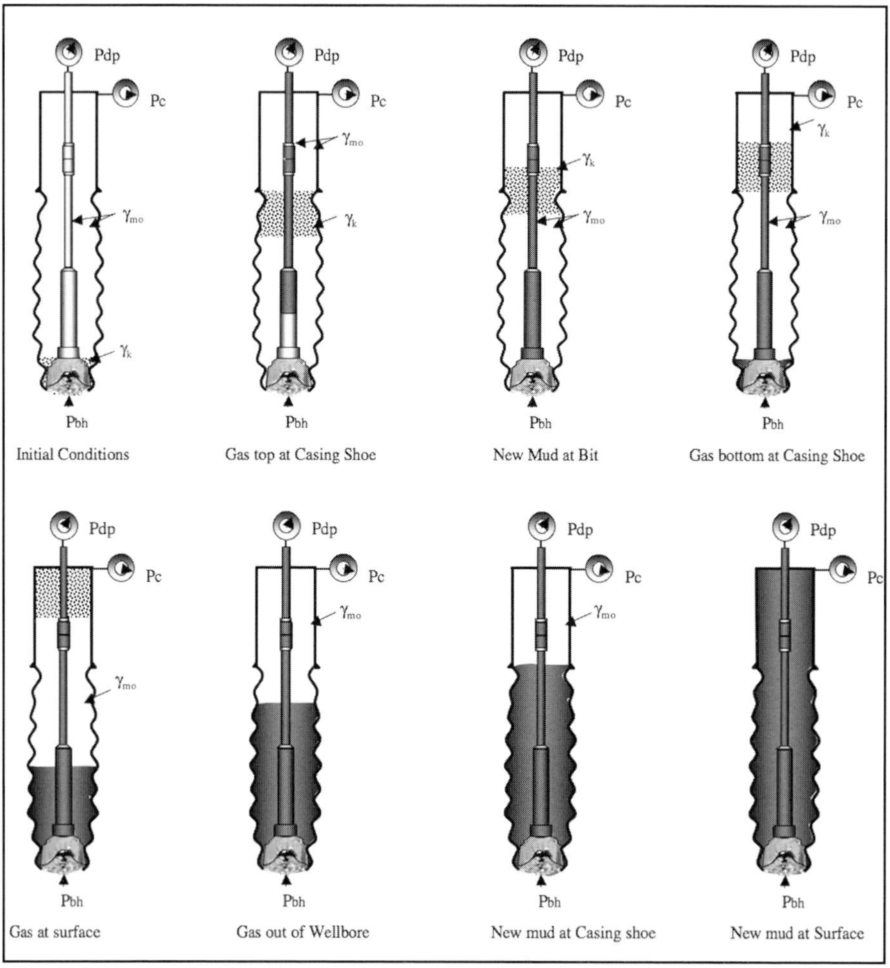

Fig. 6–21. Wait-and-weight method sequence

Wait-and-weight method

The wait-and-weight method, which is also referred to as the once-circulation method, consists of circulating the kick fluid out while the original mud is being displaced simultaneously with the kill mud. During the circulation of kill mud from the surface to the drill bit, the circulating drill pipe pressure depends only on the average mud density inside the drill string. The bottom-hole pressure is held constant by incrementally reducing the circulating drill pipe pressure, P_{cdp}, from the initial pressure, $P_{cdpi} = P_{pr} + P_{sidp}$, to the final pressure, $P_{cdpf} = P_{pr}(\gamma_{mf}/\gamma_{mo})$, when the kill mud reaches the bit. This final pressure is held until all original mud has been displaced with the heavy mud. (See figs. 6–22 and 6–23.)

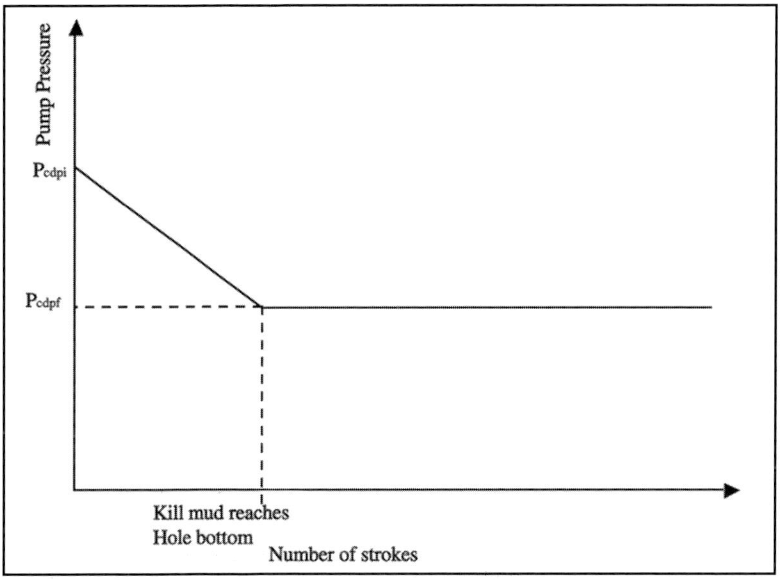

Fig. 6–22. Wait-and-weight method

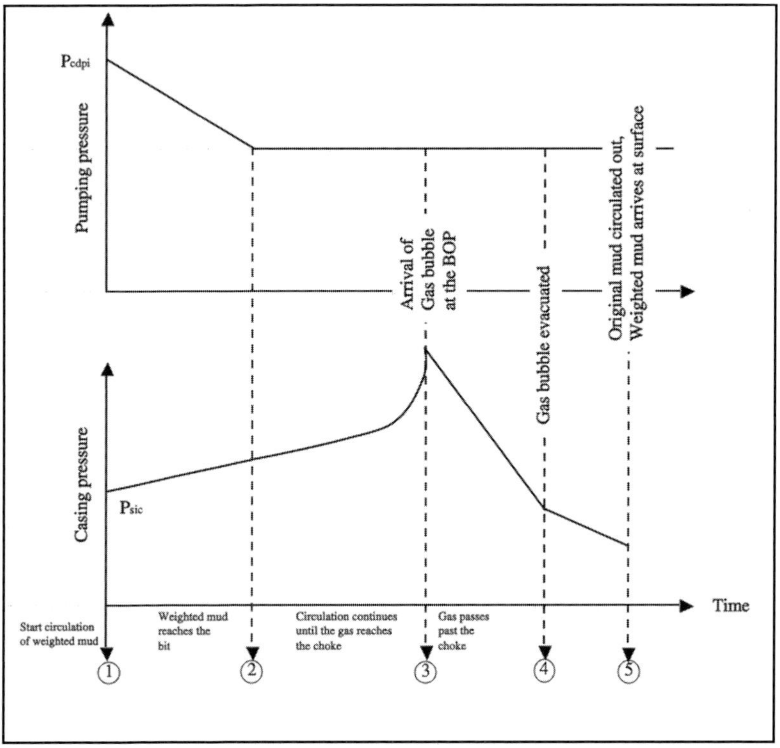

Fig. 6–23. Well control sequence of one cycle of mud

Volumetric method

The volumetric method is applied when kick circulation is not possible (because of plugged drill bit, drill string out of the hole, mechanical problems with the circulation system, etc.). The method consists of two phases. In the first phase, the gas is allowed to migrate up to the surface with successive mud-bleeding periods, to maintain the bottom-hole pressure nearly constant. In the second phase, also called top kill, the gas is replaced by a weighted mud through successive injection of mud and bleeding of gas. Changes in annular capacities and difficulty in measuring surface mud volumes are the most important limitations of the method.

Implementation. The first phase encompasses four steps:

1. Allow the casing pressure to increase by 100 psi (safety margin) owing to the gas migration.
2. Allow an additional increase in casing pressure Δ_p (normally 50 or 100 psi).
3. Bleed through the choke, at constant pressure, a volume of mud (V_m) that corresponds to the additional increase from step 2. If this increment is 50 psi, use the equation $V_m = 962C/\gamma$, where V_m is in bbl, C is in bbl/ft, and γ is in lbs/gal.
4. Repeat steps 2 and 3 until the gas reaches the surface.

The second phase consists of three further steps:

1. Calculate the mud density to kill the well (γ_{mf}). Inject through the kill line a volume of this mud (ΔV_m) until the casing pressure is increased by 100 psi. Measure this volume.
2. Convert this volume to pressure in psi (Δ_p) by using the equation $\Delta_p = 0.052\Delta V_m / C\gamma_{mf}$.
3. Allow some time for this kill mud to fall down. Afterward, bleed the gas through the choke until the casing pressure drops by $100 + \Delta_p$.

Concurrent method

The concurrent method, also referred to as the multi-circulation method, consists of increasing the mud weight incrementally from its original weight, γ_{mo}, to a final weight, γ_{mf}, that will kill the well. This is expressed by the equation

$$\gamma_{mi} = \gamma_{m(i-1)} + \Delta\gamma \quad i = 1, n \tag{6.26a}$$

that is,

$$\begin{aligned} \gamma_{mi} &= \gamma_{m0} + \Delta\gamma \\ \gamma_{m2} &= \gamma_{m1} + \Delta\gamma + \gamma_{m0} + 2\Delta\gamma \\ \gamma_{mn} &= \gamma_{m(i-1)} + \Delta\gamma = \gamma_{m0} + i\Delta\gamma \end{aligned} \tag{6.26b}$$

The first circulation is similar to the wait-and-weight method, except that the first increase in mud weight is a small increment of what may be needed to kill the well. From then on, the method is similar to the second circulation of the driller's method.

Other methods of well control

Other methods of well control that will not be discussed here include:

- Dynamic killing
- Low choke pressure
- Partition
- Bull-heading

Determining the best method for well control is not an easy approach and requires several considerations, some of the more important of which are

- Time requirement for implementation
- Surface pressure limitation
- Open hole fracture gradient limitation
- Procedural complexity
- Personnel

Nonconventional Kick Situations

Kicks can occur whether the drilling operations are for vertical wells, directional wells, or horizontal wells in onshore or offshore areas. They can occur while the bit is on bottom, off bottom, or totally out of the hole. The equipment and the fundamentals of kick control, as discussed in the previous sections, are generally the same. However, there are some basic differences in the engineering aspects for handling the kicks and bringing the well under control for different kick occurrence situations.

Conventional kick situations are defined here as those that occur while the bit is on bottom or can be stripped back to bottom and the drilling is onshore or offshore on bottom-supported units. Nonconventional kick situations include drilling from floating units offshore, horizontal drilling, shallow gas, hazardous gas (e.g., H_2S), drill bit off bottom, drill bit out of hole, kick volume too large, fractured zone (and therefore possible underground blowout), uncontrolled surface kicks, extremely high-pressure wells, underbalanced drilling, and slim hole drilling. This section is devoted in its entirety to the discussion of some of these nonconventional situations. The reader is encouraged always to seek the latest technological advances to the topics that are presented here.

Kick occurrence in horizontal wells

Kick volume. One of the many reasons for drilling horizontally into the reservoir is to increase production. Thus, if kick occurs while drilling the horizontal interval, the kick volume will be several times greater than that encountered while drilling vertically into the reservoir. When such a large kick volume reaches the vertical section of the well during kick circulation, the induced pressures on the casing shoe and the surface equipment could easily exceed the allowed pressures (formation fracture pressure or surface choke working pressure).

Kick detection time. Unlike kicks in vertical wells, kick detection in horizontal wells can be delayed significantly owing to possible enlarged or wavy sections that allow the accumulations of gas pockets, as shown in figure 6–24. Thus, the sudden increase in mud volume on the surface is not a true indication of the total gas influx into the annular space.

Shut-in pressure. If a gas kick is taken in vertical wells while drilling, the shut-in casing pressure is always greater than the shut-in drill pipe pressure owing to the gas influx that displaces part of the mud in the annular space. However, when a kick is taken during drilling into the horizontal section and the gas is contained within that section, the shut-in casing and drill pipe pressures will be the same. (See fig. 6–25.)

Drilling through vertical fractures. One of the applications of horizontal drilling is to intersect as many naturally fractured zones in the reservoir as possible. However, the pressures in these natural fractures may not all be of the same magnitude. Thus, taking a kick in the horizontal section that is connecting many of these fractures can cause underground blowouts in the lower-pressure zones. (See fig. 6–26.)

Bottom-hole assembly (BHA). The BHA in horizontal drilling is quite different than that in vertical sections. The portion right above the drill bit is the lightest and can be the smallest in terms of outside diameter. This is followed by heavier drill pipe in the build section (heavy-weight drill pipe [HWDP]) and then by drill collars, which are the heaviest and largest in outside diameter in the vertical section.

During kick control operations, when the kick fluid (assumed to be a gas) reaches the HWDP and the drill collars, the gas bubbles will elongate significantly owing to the small annular clearances and their increased velocities. This will lead to a decrease in hydrostatic pressure that can cause more kick fluid intrusion into the wellbore. In addition, as gas velocity increases, it will expand rapidly, causing a sudden increase in flow across the choke, and consequently, the casing pressure will increase. Therefore, proper adjustment of the choke has to be made quickly, to avoid excessive pressure and to maintain the proper bottom-hole pressure. (See fig. 6–27.)

Equivalent circulating density (ECD). The ECD of the mud in the horizontal portion of the well increases as the length of the horizontal section increases. However, in contrast to vertical wells, the formation fracture pressure and the pore pressure remain the same. This may cause possible formation fracturing during drilling (annular friction pressure loss) or tripping in (surge pressure) and kicks while tripping out (swab pressure). (See fig. 6–28.)

Kick control operations. The fundamentals of kick control in horizontal well drilling are generally the same as in vertical wells—that is, to circulate the kick fluid out of the hole and to replace the light mud with the heavier mud. Both of these operations must be completed without allowing any further formation fluid intrusion into the wellbore and without exceeding open hole fracture pressure gradients or surface equipment pressure tolerances. However, the basic difference from vertical well drilling is that the assumption of single-phase flow (gas as a slug) can lead to calculation errors that may not be significant in a vertical well but will be very significant in a horizontal well.

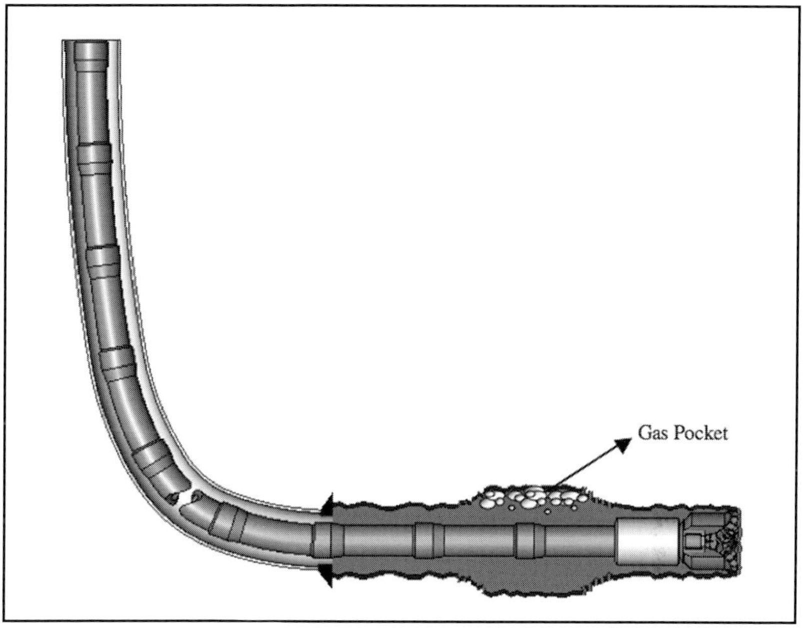

Fig. 6–24. Horizontal well section, showing gas accumulation on top

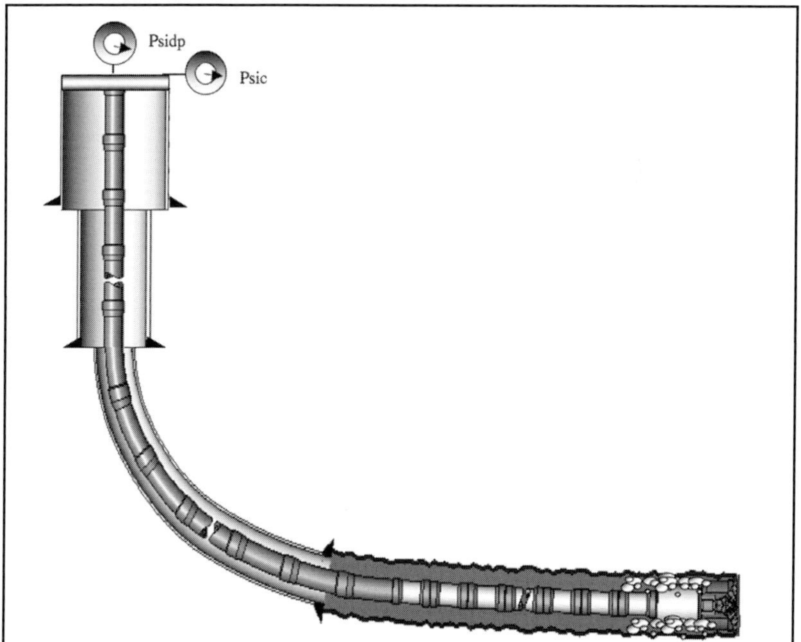

Fig. 6–25. Gas kick contained in the horizontal section

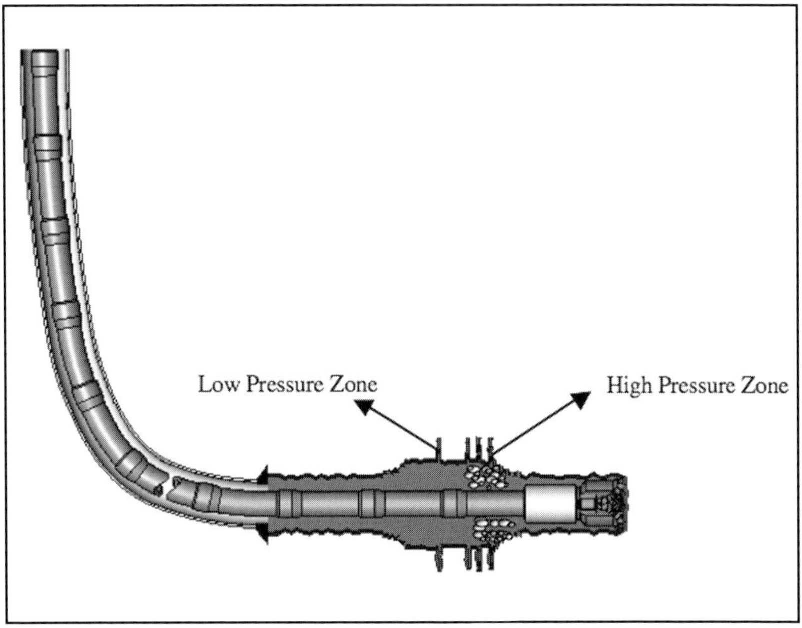

Fig. 6–26. Naturally fractured reservoir

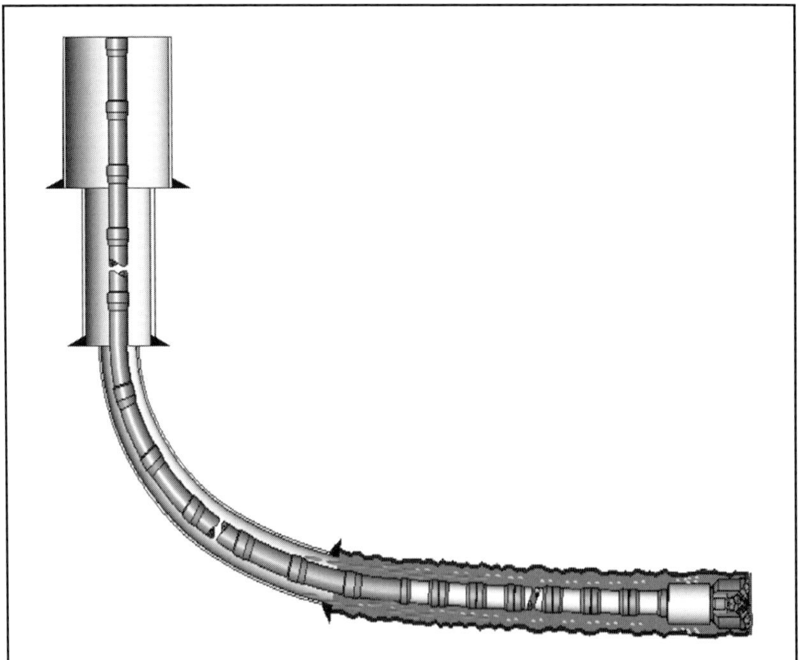
Fig. 6–27. Elongation of the gas bubble in the narrow section of the annulus

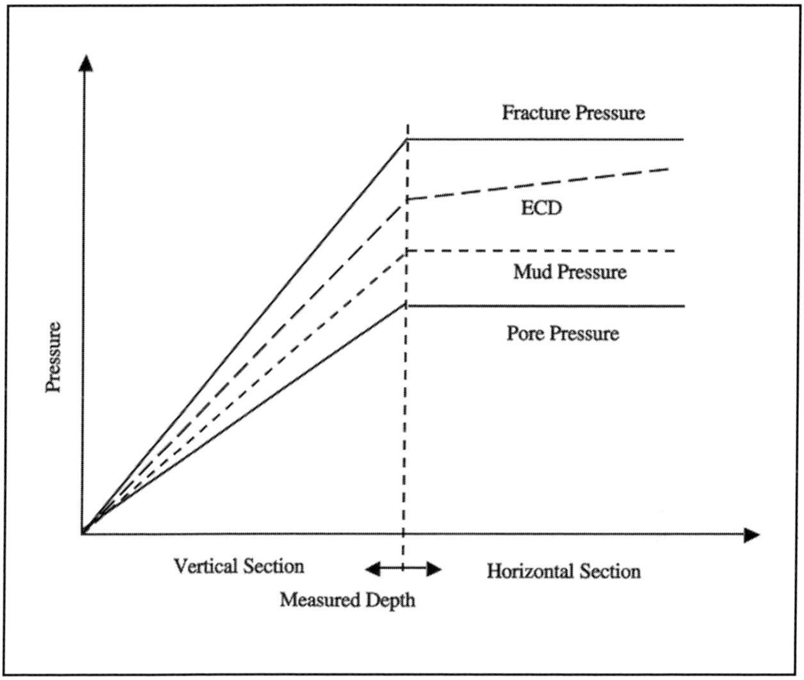

Fig. 6–28. ECD in vertical and horizontal section of well

Offshore drilling using mobile rig units

The most critical issue in kick control when drilling from mobile units in deepwater are the friction pressure losses in kill and choke lines. In onshore drilling and offshore drilling from fixed platforms, the kill and choke lines are in the vicinity of the BOP stack; therefore, their friction pressures losses can be ignored. However, in deepwater drilling, these lines can be several thousands of feet in length, depending on water depth. Thus, for a given water depth, line inside diameters, mud viscosity, and flow rate, the friction pressure losses in these line can be significantly large. A major concern is the friction pressure loss inside the choke line, which can lead to excessive ECD and potential fracturing of open hole sections. Some of the other important issues in offshore well control operations are

- Low fracture pressure gradient of formations
- Formation of hydrates in the choke line
- Gas in the riser system
- Quick change in choke pressure when gas enters and exit the choke line

It is highly recommended that the reader stay abreast of the current technologies as they appear in various publications from the Society of Petroleum Engineers (SPE) and in other sources in the literature.

Supplementary Problems

Problem 6–1. A kick was taken at 10,000 ft while drilling with a 14 lbs/gal mud. Shut-in drilling pressure (SIDP) = 600 psi; shut-in casing pressure (SICP) = 800 psi.

(a) Determine amount of barite to be added to 800 bbl of original mud to contain formation pressure such that a differential pressure of 300 psi will be achieved.

(b) Determine the pressure at the casing shoe (3,000 ft) if the kick is (i) saltwater and (ii) gas.

(c) If the formation fracture gradient is 0.86 psi/ft, will formation fracturing occur in either case in part (b)?

Problem 6–2. How many pounds of barite and how many pounds of bentonite must be used to make up 800 bbl of 11.5 lbs/gal mud? The material is mixed in the ratio of four sacks of barite to one sack of bentonite, with freshwater as the liquid phase. The weight density of the additives are as follows:

- $\gamma_W = 8.33$ lbs/gal (water)
- $\gamma_B = 36$ lbs/gal (barite)
- $\gamma_c = 20.8$ lbs/gal (clay)

Problem 6–3. You are given the following well data and the plan in figure 6–29:

- TD = 10,000 ft TVD
- Last intermediate casing:
  - Set depth = 5,000 ft TVD
  - OD = 9⅞ in.
  - Wall thickness = 0.3125 in.
- Next hole size: 8½ in.; tri-cone roller cone jet bit (hole washed out to 9¼ in.)
- Minimum velocity required for annular hole cleaning: 120 ft/min
- Drill pipe:
  - OD = 4½ in.
  - ID = 3¾ in.
  - Air weight = 18.10 lbs/ft
- Mud:
  - Bingham plastic
  - $\theta_{600} = 37$; $\theta_{300} = 35$; $\tau_y = 13$; lbs/100 ft^2; 10 s/10 min gel = 7/16 lbs/100 ft^2
- Steel weight = 66.3 lbs
- Coefficient of friction between formation and pipe = 0.2
- Formation fracture pressure gradient at casing shoe = 17 lbs/gal

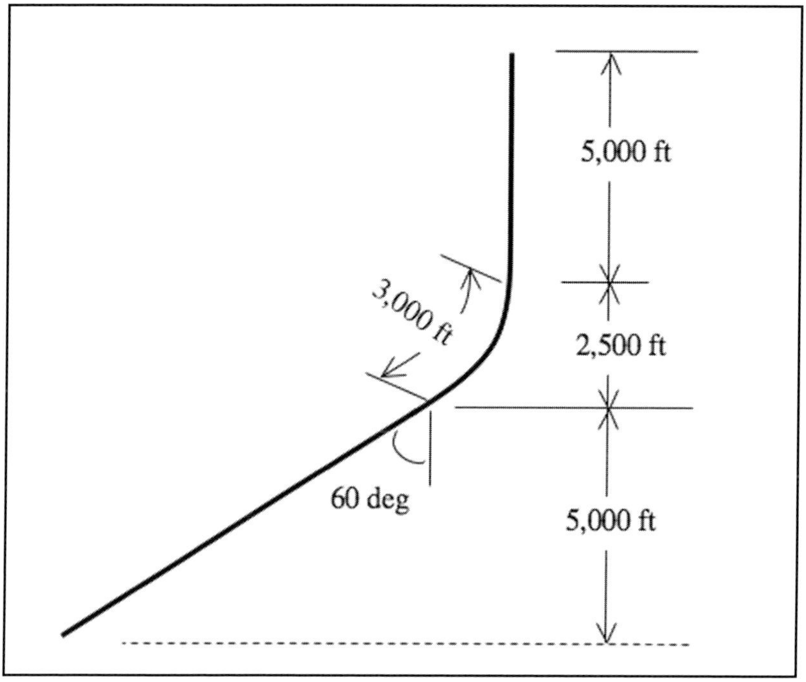

Fig. 6–29. Drilling plan in problem 6–3

(a) A kick was taken at a TMD of 9,500 ft while drilling 12 lbs/gal mud. The well SICP was recorded. Determine the amount of iron oxide (specific gravity = 5.3) to add to a barrel of original mud to resume drilling at a differential pressure of 500 psi.

(b) Determine the pump pressure schedule for well control, using (i) driller's and (ii) wait-and-weight methods.

Problem 6–4. You are given the following well data (see also fig. 6–30):

- Casing-setting depths: 1,000, 8,000, 18,000, 20,000 ft TMD
- Last hole size = 7⅞ in. at 18,000 ft TMD
- Drill pipe:
  - OD = 4½ in.;
  - ID = 3½ in.
  - Air weight = 26.88 lbs/ft
- Drill collar:
  - OD = 6½ in.
  - ID = 3½ in.
  - Air weight = 100.8 lbs/ft
  - Length = 1,500 ft

- Mud:
  - Bingham plastic
  - Specific gravity: 1.5 between 8,000 and 18,000 ft TMD; 2.0 to TD
  - $\theta_{600} = 55$; $\theta_{300} = 35$; 10 s/10 min gel = 18/30 lbs/100 ft^2 (Fann V-G meter)
- Formation below 18,000 ft TMD:
  - Fracture gradient = 17.2 lbs/gal
  - Pore pressure gradient = 0.83 psi/ft
- Surface pressure limitation = 4,000 psi

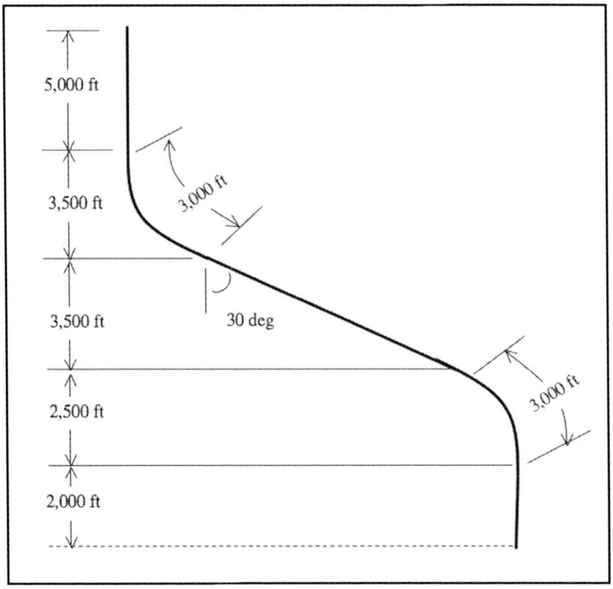

Fig. 6–30. Well conditions in problem 6–4

A kick was taken while drilling at 20,000 ft TMD. The initial pit gain was recorded at 32 bbl. The recorded SIDP and SICP were 600 psi and 800 psi, respectively.

(a) Determine the total cost of barite needed to raise mud weight to have a differential pressure of 200 psi. (Total original mud volume = 2,000 bbl; barite cost = \$8/100 lb sack; γ_{barite} = 36 lbs/gal.)

(b) Determine the kick type.

(c) Determine the pump pressure schedule for the driller's method.

(d) Determine whether any problem will arise during well control, using (i) driller's and (ii) wait-and-weight methods. (The reduced pump rate for well control is 100 gal/min.)

Problem 6–5. You are given the following well data (see also fig. 6–31):

- Last casing setting = 14,000 ft TMD
- TD = 13,500 ft TVD
- Hole size to TD = 7⅞ in.
- Drill pipe:
  - OD = 4¼ in.
  - ID = 3½ in.
  - Air weight = 26.88 lbs/ft
- Drill collar:
  - OD = 6 in.
  - ID = 3½ in.
  - Air weight = 100.8 lbs/ft
  - Length = 1,500 ft
- Pumps: 2,000 hp; E_i = 85%; P_{max} = 3800 psi
- Minimum annular fluid velocity for hole cleaning = 120 ft/min
- Coefficients of friction at walls: 0.1, cased sections; 0.2, open hole section
- Mud:
  - Bingham plastic
  - Specific gravity = 2.0 for next depth interval
  - θ_{600} = 55; θ_{300} = 35; 10 s/10 min gel = 18/30 lbs/100 ft^2 (Fann V-G meter)
- Formation below 14,000 ft TMD:
  - Fracture gradient = 18.0 lbs/gal
  - Pore pressure gradient = 0.83 psi/ft

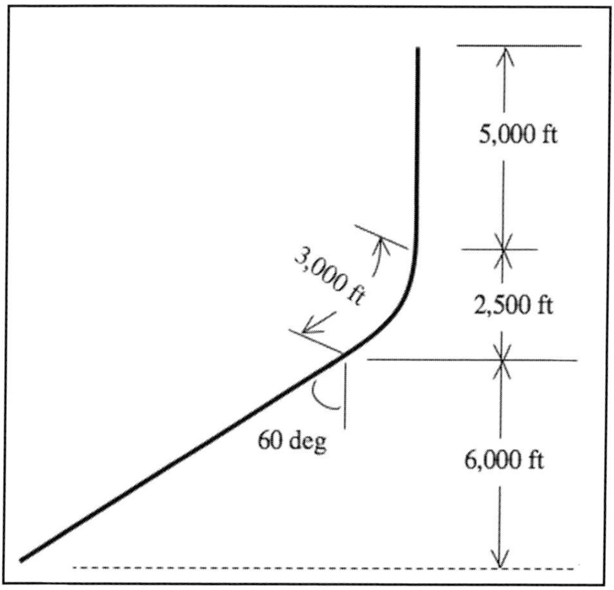

Fig. 6–31. Well conditions in problem 6–5

(a) A kick was taken while drilling at 16,000 ft TMD. The recorded SIDP and SICP were 400 psi and 800 psi, respectively. Determine the pump pressure requirements to bring the well under control, using the wait-and-weight method. (Measured pump pressure at reduced flow rate for well control is 1,867; pump factor = 4 gal/stroke.)

(b) If in part (a), the kick was a gas kick and the measured sudden increase in mud volume on the surface was 20 bbl, determine whether fracture of formation will occur during controlling the well.

Problem 6–6. You are given the following well data:

- 8 in. casing-shoe depth = 10,000 ft
- Formation fracture gradient at shoe = 0.9 psi/ft
- Hole size below shoe = 7 in.
- Drill pipe: ID = 3½ in. OD = 4 in.
- Gas kick was taken at 11,000 ft
- Pore pressure gradient = 0.8 psi/ft

(a) Determine the maximum kick volume that can be tolerated without exceeding the maximum allowed surface pressure of 8,000 psi.

(b) What mud weight would be required if the kick volume was 4,200 gal.

Nomenclature

C_{adp} Annulus capacity behind drill pipe, in bbl/ft
C_{adc} Annulus capacity behind drill collar, in bbl/ft
D Depth at which kick is taken
F_p Pump factor, in bbl/stroke
k Unit conversion constant
L_{dc} Drill collar length, in ft
L_{dp} Drill pipe length, in ft
L_k Kick length
$\overline{NPS}$ Number of pump strokes
ΔP Differential pressure
P_{df} Wellbore fluid pressure
P_{ff} Formation fluid pressure
P_{moh} Mud hydrostatic pressure
ΔP_s Safety differential pressure
P_{sic} Shut-in casing pressure (SICP)
P_{sidp} Shut-in drill pipe pressure (SIDP)
V_{adc} Annulus volume behind drill collar, in bbl
V_{adp} Annulus volume behind drill pipe, in bbl
V_k Kick volume

γ_{df} Drilling-fluid pressure gradient
γ_{frac} Formation fracture pressure gradient
γ_k Kick density
γ_{mo} Original mud weight
γ_{surge} Surge pressure gradient

References

1 *Blowout Prevention and Well Control.* Tr. by Paul W. Ellis. 1981. Editions Technip, 27 Rue Ginoux 75737, Paris Cedex 15.

2 Goins, W.C., Riley Sheffielf, *Blowout Prevention, 2nd ed.*, Gulf Publishing Co., Book Division. Houston, London, Paris.

3 *IADC Drilling Manual*, International Association of Drilling Contractors. Houston.

4 Mitchell, Bill, *Oil Well Drilling Engineering, 8th ed.* 1974: Golden, CO.

7

Directional and Horizontal Well Drilling

Directional Well Drilling

Since the advent of rotary drilling, a major concern for the petroleum industry has been—and still is—the ability to drill a hole economically and maintain its direction along a predetermined trajectory, thousands of feet below the surface. In the early days, the main concern was to maintain a vertical course down to a target area located directly underneath the rig floor. However, in recent years, the concern has been extended to include the ability to drill a hole down to a pay zone target that may be located thousands of feet of horizontal departure away from the surface location under the rig floor. This is what is commonly referred to as directional well drilling, which consists of drilling a vertical hole section beneath the rig floor to a certain preselected *kick-off depth* (KOD) and then intentionally deviating the wellbore along a preselected trajectory to reach the geological target zone (see fig. 7–1).

Applications of directional well drilling

Directional well drilling has many applications, the most common of which are as follow:

- Drilling multiple wells from a single offshore location
- Reaching inaccessible target areas
- Drilling relief wells to control blowing wells
- Drilling to sidetrack
- Drilling to bypass potential problem zones
- Drilling from an offset location to achieve minimum-cost drilling
- Horizontal well drilling into a reservoir

Drilling multiple wells from a single offshore location is the most common and economical application of directional drilling. Its application to the drilling of horizontal wells has gained great prominence in recent years. It is expected that a future trend will be for whole offshore or onshore fields to be developed from a single site through the technology advancements made in the drilling of horizontal and multilateral wells.

Reaching inaccessible producing zones (e.g., under cities or mountains) requires a remote rig location and directional drilling in order to reach the pay zones.

Often, in uncontrolled blowouts, when conventional methods of putting out well fires fail, the last resort is to offset and directionally drill a relief well that will hydraulically communicate with the producing formation.

Drilling problems (e.g., unsuccessful retrieval of lost tools in the wellbore) may be encountered that dictate plugging the bottom interval of the hole and sidetracking using directional drilling to reach the target zones. In addition, sidetracking is done in drilling reentry into new wells and in old depleted wells.

The bypassing of potential problem zones is an important application of directional drilling. Examples of problem zones include salt domes and faults.

Offsetting a well to take advantage of natural bit walk in a dipped formation, rather than continuously correcting for hole deviation, can be a cost-saving approach using directional drilling. Finally, the recent interest in drilling horizontal and multilateral wells into certain reservoirs has brought directional drilling technology and its application to a climax.

Requirements for directional well drilling

With the exception of some special addition requirements, the equipment and the engineering that are used to economically and successfully drill a vertical well are generally also used to drill a directional one. Some of those unique additional requirements are

- Use of downhole steerable systems and *measuring-while-drilling* (MWD) tools
- Use of rubber protectors on drill string to protect pipe/casing wear in highly deviated wells
- Use of special lubricating additives in drilling muds to reduce torque and drag requirements
- Use of higher hydraulics for effective annular hole cleaning
- Use of articulated drill pipe in short-radius drilling
- Engineering calculations of directional plans for the preselected wellbore trajectory
- Unique makeup of bottom-hole assemblies (BHAs) for maintaining on-course trajectory and effective transfer of weight to the drill bit
- Special steering tools to direct drill bit along preselected paths
- Deflection tools
- More centralization of casing string

Directional well profiles

Planning is an important factor in minimizing and controlling the overall cost, for directional wells in particular. Planning begins with the selection of one of the typical wellbore patterns (profiles) shown in figure 7–1. These profiles are generally identified as

- Build and hold
- Build, hold, and drop (*S* shape)
- Continuous build

Once the profile type has been selected, detailed calculations are made to define the desired trajectory of the well to be drilled. Design of the directional plans involves the utilization of mathematical equations that require the following information:

- TVD and horizontal departure (HD) to target
- KOD
- Turn-off depth (TOD)
- Build rate angle (BRA)
- Drop rate angle (DRA)
- Lead angle (LA)

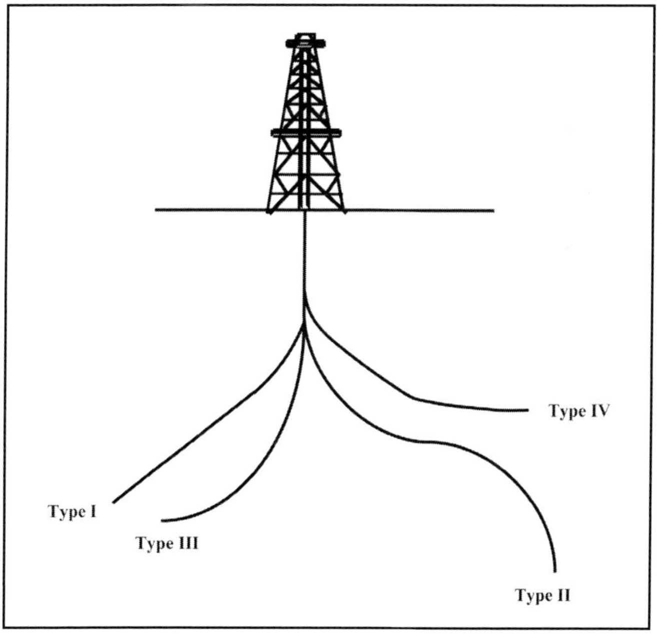

Fig. 7–1. Wellbore patterns

Target location. The desired target location is specified in terms of its TVD, its departure in the horizontal plane, and the desired radius target, as shown in figure 7–2.

KOD. The KOD is a point below the surface at which the incremental building of the hole angle of inclination begins. The hole-angle building is generally specified in degrees per 100 feet of drilled section. (See figure 7–3.)

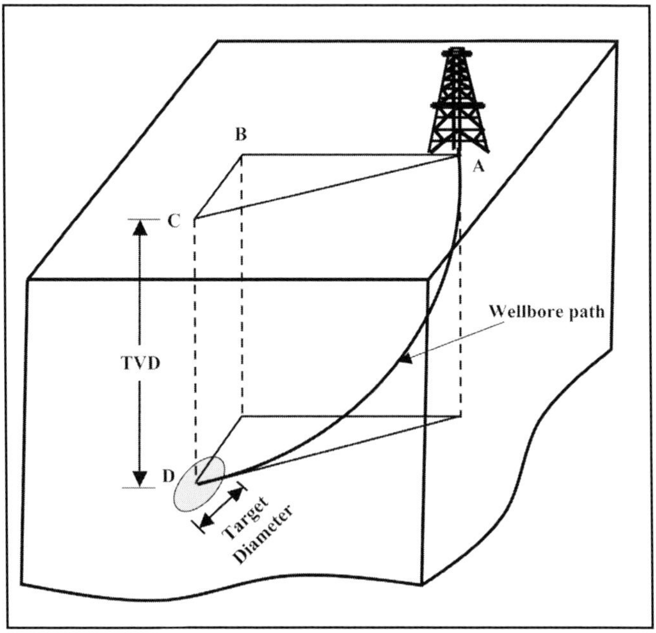

Fig. 7–2. Target location. (The line AC shows HD.)

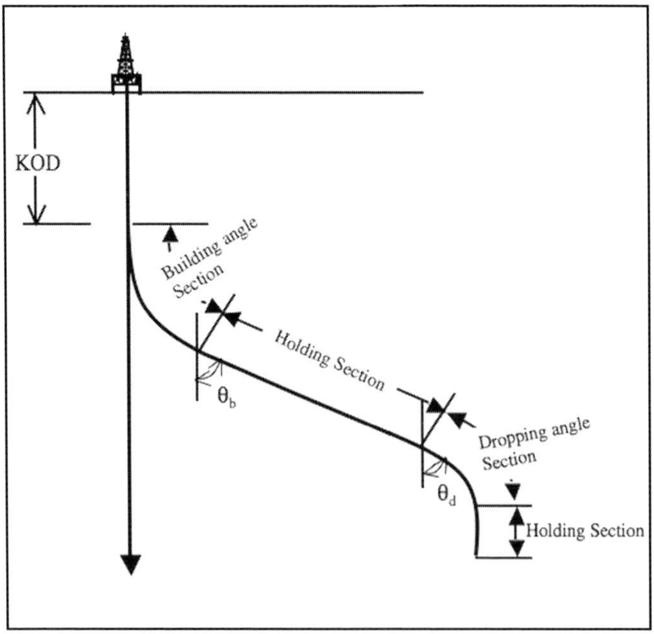

Fig. 7–3. Building, holding, and dropping configuration

TOD. The TOD is a point below the surface at which the direction of the hole is changed. For example, a target located N20°E, N20°W, S20°E, or S20°W means the hole direction is 20° in the first, second, third, or fourth quadrant, respectively, as shown in figure 7–4. Also, hole direction can be described by an azimuth angle that is measured clockwise from north. Note that when magnetic survey instruments are used to measure direction in the horizontal plane, the reference is to magnetic north, which is different from true north. Corrections have to be made to the reference azimuth angle to reflect true north. The difference between magnetic and true north is an angle called the magnetic declination. The declination is normally known for a given region where drilling is taking place. It is specified as east declination or west declination, as measured from north. If declination is east, the declination angle is added to the measured azimuth; if the declination is west, the declination angle is subtracted from the measured azimuth.

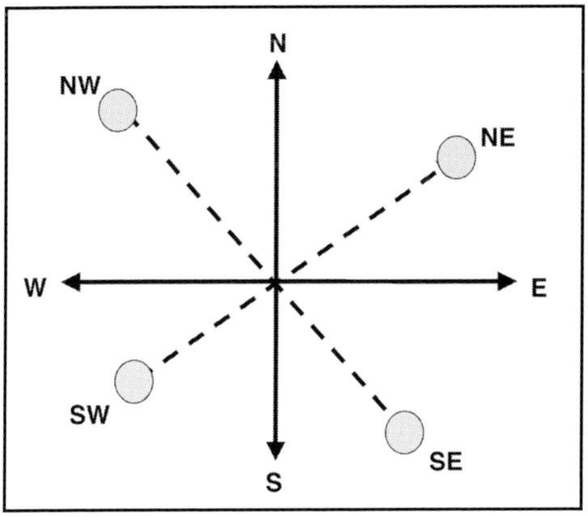

Fig. 7–4. Hole direction configuration

BRA. The BRA is an incremental increase of the hole inclination angle, as measured from vertical. This rate is specified in degrees per 100 feet of hole section. For example, in figure 7–5, in the building section, the hole is being drilled with an incremental increase of angle until the final inclination of θ_b is reached. The range of BRAs can be subdivided into three categories:

- 1–6°/100 ft of drilled section (long-radius drilling)
- 6–35°/100 ft of drilled section (medium-radius drilling)
- 1.5–3° per foot of drilled section (short-radius drilling)

DRA. The drop rate angle is an incremental decrease of hole inclination. (See fig 7–3.)

TRA. The turn rate angle is an incremental change in direction of hole (azimuth). It is specified in degrees per 100 ft of hole section.

LA. Because of drill string clockwise rotation, some drill bits, for example, have the tendency to walk to the right in the horizontal plane; others may walk to the left. To account for this natural bit walk, it is common practice to initiate the turn of the well path with a specified LA to the left or the right of the target area, depending on whether the bit walks to the right or left, respectively. The magnitude of the angle is generally based on experience documented in past records of wells drilled in the area. (See fig. 7–5.)

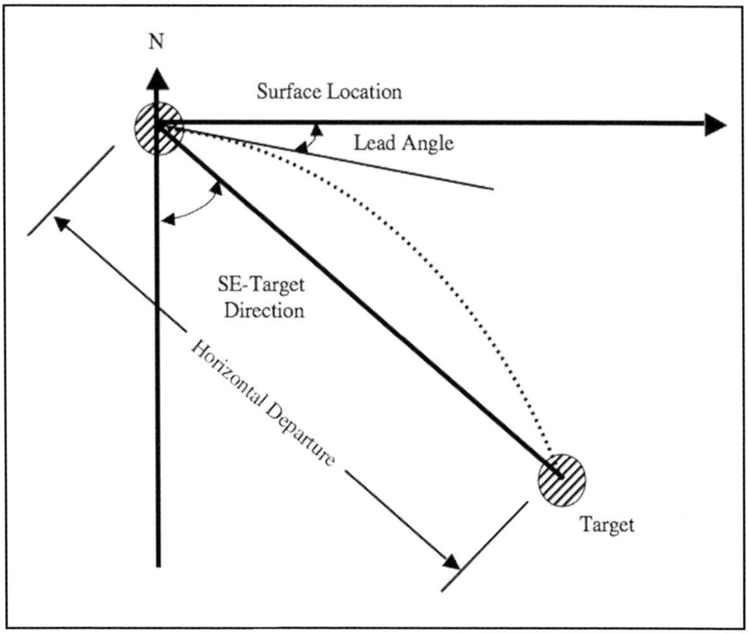

Fig. 7–5. Well path in the horizontal plane

Wellbore trajectory design

The following sections outline the basic trigonometric equations that allow the drilling engineer to carry out his or her well plan design in the vertical plane.

Circular arc

Arc length: $$L_c = \frac{\theta_j - \theta_i}{\text{BRA}} \tag{7.1}$$

Vertical distance: $$V = R_b(\sin\theta_j - \sin\theta_i) \tag{7.2}$$

Horizontal distance: $$H = R_b(\cos\theta_i - \cos\theta_j) \tag{7.3}$$

where $R_b = 190/(\pi \times \text{BRA})$.

Tangent section

Vertical distance: $$V = L_t \cos\theta \tag{7.4}$$

Horizontal distance: $$H = L_t \sin\theta \tag{7.5}$$

where L_t is the tangent length.

Build-and-hold pattern. The following data, as defined in figure 7–6, are required: KOD, target HD in the *y* plane, target TVD, and BRA in degrees per 100 ft.

$$\text{TVD} = \text{KOD} + R_b \sin\theta_b + L_t \cos\theta_b \qquad (7.6)$$

$$\text{HD} = R_b(1 - \cos\theta_b) + L_t \sin\theta_b \qquad (7.7)$$

$$\text{TMD} = \text{KOD} + \frac{\theta_b}{\text{BRA}} + L_t \qquad (7.8)$$

In equations (7.6) and (7.7), respectively, the TVD and the HD to geological target are always known. Two of the four parameters (BRA, L_t, KOD, and θ_b) can be determined if the other two are assumed.

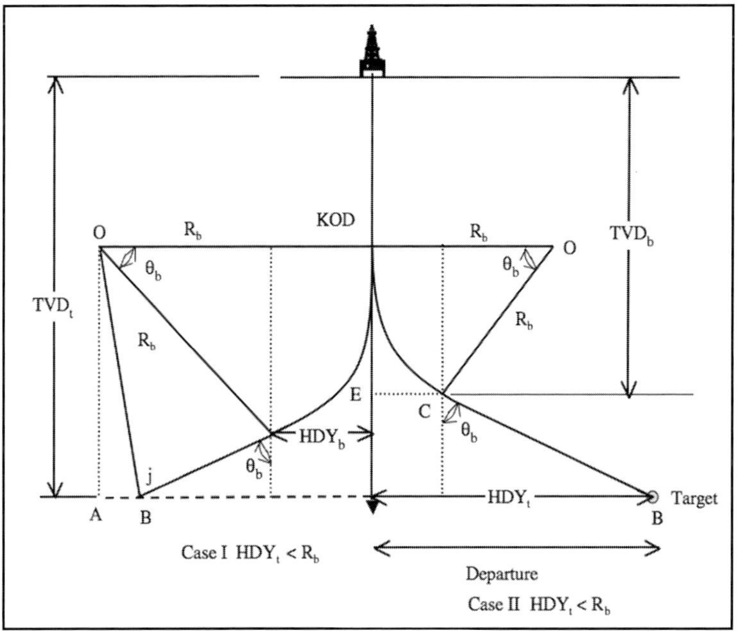

Fig. 7–6. Build-and-hold design

Continuous build. The data required are similar to the build-and-hold case. With reference to figure 7–7, the HD in the vertical plane and the TVD to the geological target are given by the following equations:

$$\text{TVD} = R_b \sin\theta_b + \text{KOD} \qquad (7.9)$$

$$\text{HD} = R_b(1 - \cos\theta_b) \qquad (7.10)$$

$$\text{TMD} = \text{KOD} + \frac{\theta_b}{\text{BRA}} \qquad (7.11)$$

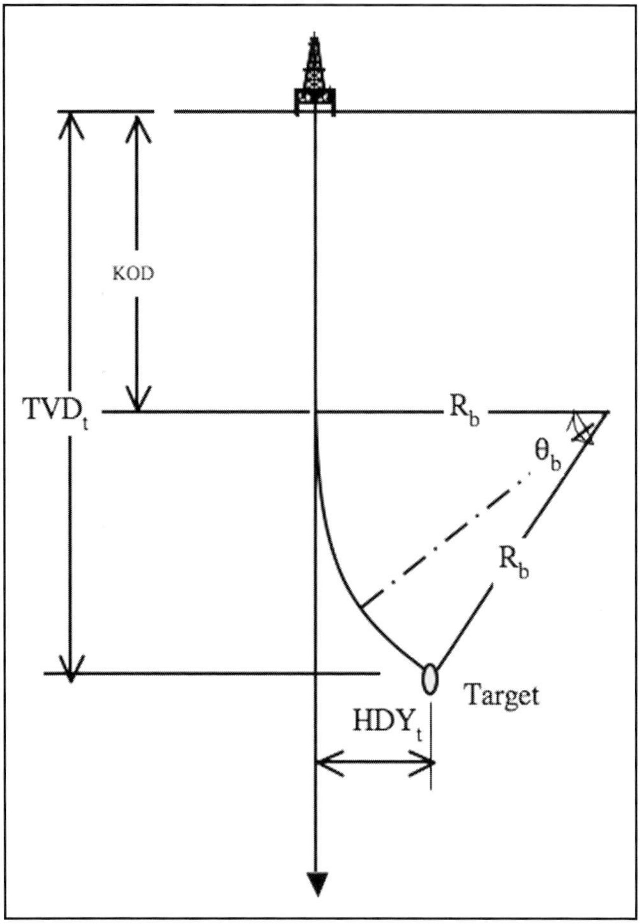

Fig. 7–7. Continuous build

Build, hold, and drop. The required data are similar to previous well patterns. (See fig. 7–8.)

$$TVD = KOD + R_b \sin\theta_b + L_t \cos\theta_b + R_d(\sin\theta_d - \sin\theta_b) \tag{7.12}$$

$$HD = R_b(1 - \cos\theta_b) + L_t \sin\theta_b + R_d(\cos\theta_b - \cos\theta_d) \tag{7.13}$$

$$TMD = KOD + \frac{\theta_b}{BRA} + L_t + \frac{(\theta_d - \theta_b)}{BRA} \tag{7.14}$$

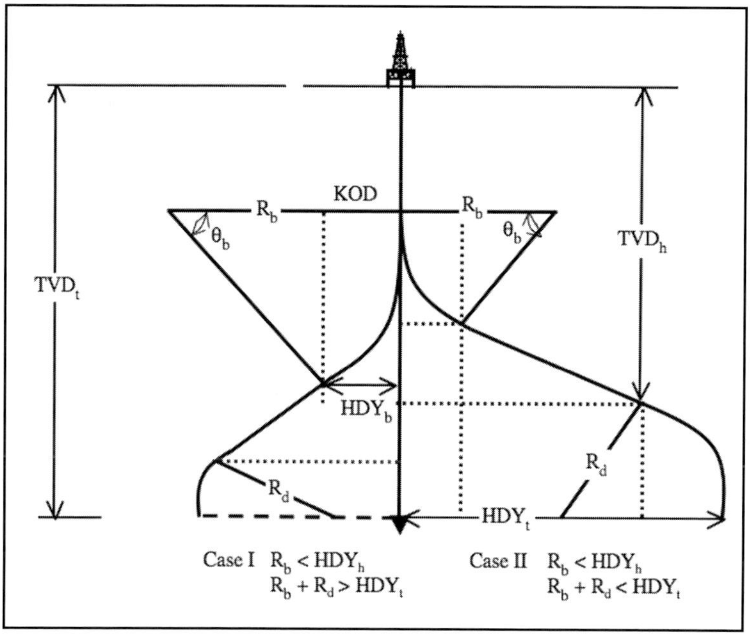

Fig. 7–8. Build, hold, and drop

As in the build-and-hold pattern, the TVD and the HD to the geological target are always known. Therefore, equations (7.12) and (7.13) can be solved simultaneously for two unknowns, assuming that the others are known.

The calculations that will define the well path in the horizontal plane can be performed by using the following equations:

$$(\mathrm{NS})_i = \Delta d \left\{ \left[\frac{\sin(\theta_i + \theta_{i-1})}{2} \right] \left[\frac{\cos(\alpha_i + \alpha_{i-1})}{2} \right] \right\} \tag{7.15}$$

$$(\mathrm{EW})_i = \Delta d \left\{ \left[\frac{\sin(\theta_i + \theta_{i-1})}{2} \right] \left[\frac{\sin(\alpha_i + \alpha_{i-1})}{2} \right] \right\} \tag{7.16}$$

and the incremental true vertical depth can be determined from

$$(\mathrm{TVD})_i = \Delta d \left(\frac{\cos(\theta_i + \theta_{i-1})}{2} \right) \tag{7.17}$$

where

$(NS)_i$ is the incremental distance measured along the north-south direction
$(EW)_i$ is the incremental distance measured along the east-west direction
Δd is an assumed incremental distance of hole section generally equal to 100 ft
θ_{i-1}, α_{i-1} are the respective inclination and azimuth angles at point $(i-1)$
θ_i, α_i are the respective inclination and azimuth angles at point (i)

The total distances can be found by summing the total number if increments, n—that is,

$$(\text{TNS}) = \sum_{i=1}^{n} (\text{NS})_i \qquad (7.18)$$

$$(\text{TEW}) = \sum_{i=1}^{n} (\text{EW})_i \qquad (7.19)$$

$$(\text{TVD}) = \text{KOD} + \sum_{i=1}^{n} (\text{TVD})_i \qquad (7.20)$$

Deflection tools in directional well drilling

Deflection tools are devices that are used to cause a drill bit to deviate and follow a preselected trajectory of a well. The tools that are currently available and are in use include

- Directional wedges
- Jet bits with oriented nozzles
- Downhole steerable systems (motors and bent subs, rotary steerable system, etc.)
- Conventional BHA hookups

Directional wedges. Figure 7–9 provides a schematic of the wedge technique, which intentionally causes the desired wellbore deviation. The wedge is rigidly attached to the BHA by means of a shear pin. The assembly is then lowered to the bottom and oriented in the proper direction; proper weight is applied to set the wedge and shear the pin; 10–15 ft of undergauge hole is drilled; and a trip is made to change the tool with a full-gauge hole opener. Afterward, the the hole is surveyed to ensure proper direction, and the process is continued until the build section of the wellbore is complete. This method is time consuming and often has limited applications.

Jet bits. The technique of using jet bits (shown in fig. 7–10) to intentionally cause the desired hole deviation is applicable in soft formations. The bit is lowered to bottom, the jet is oriented in the direction desired, and mud flow is initiated to erode approximately few feet pilot hole; then, conventional rotary drilling is initiated to open the hole, and the process is repeated. Hole surveys are made after drilling 10–15 ft of build section. The advantage of this method over the wedge method is that tripping is not needed to open the pilot hole to a full gauge.

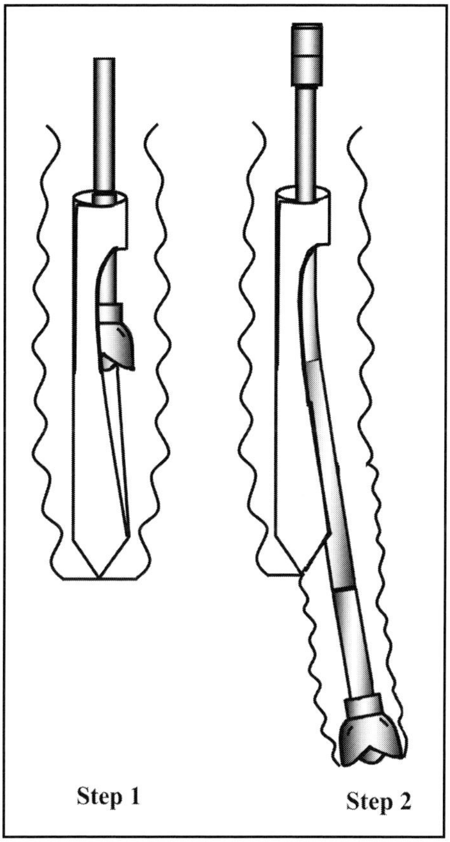

Fig. 7–9. Wedge technique

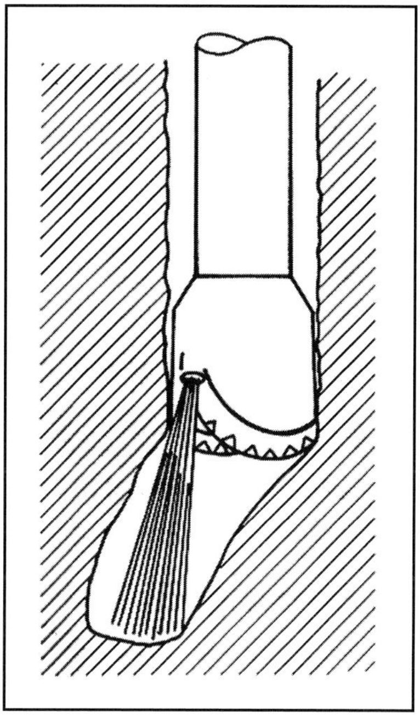
Fig. 7–10. Jet bit technique

Downhole steerable systems. Downhole steerable systems can be either downhole mud motors or rotary steerable systems.

Mud motors are hydraulically driven by circulating the mud, as is done in normal drilling. There are two types:

- Positive displacement mud motors (PDMs)
- Turbine mud motors

The main advantage of PDMs over turbines is that they have bent housing. This allows the motor to be used as a deviating assembly when in the oriented mode or as a holding assembly when in the rotary mode. The makeup of the BHA is similar in both types of mud motors, consisting of a full-gauge drill bit, a downhole motor, a bent sub, nonmagnetic drill collars, and stabilizers. (See fig. 7–11.)

Rotary steearable systems represent a recent advance in directional (including horizontal) well drilling. They are designed with steel ribs that can be activated from the surface, causing a deviating force against the side of the wellbore to push the drill bit inward along the preselected well path. Their main advantage over downhole mud motors is that drilling is always in rotary mode, whether building or holding angle; by contrast, in the case of mud motors, the system must be in sliding mode while building angle and in rotary mode while holding angle. The advantage of drill string rotation is its significant aid in hole cleaning while drilling directional wells.

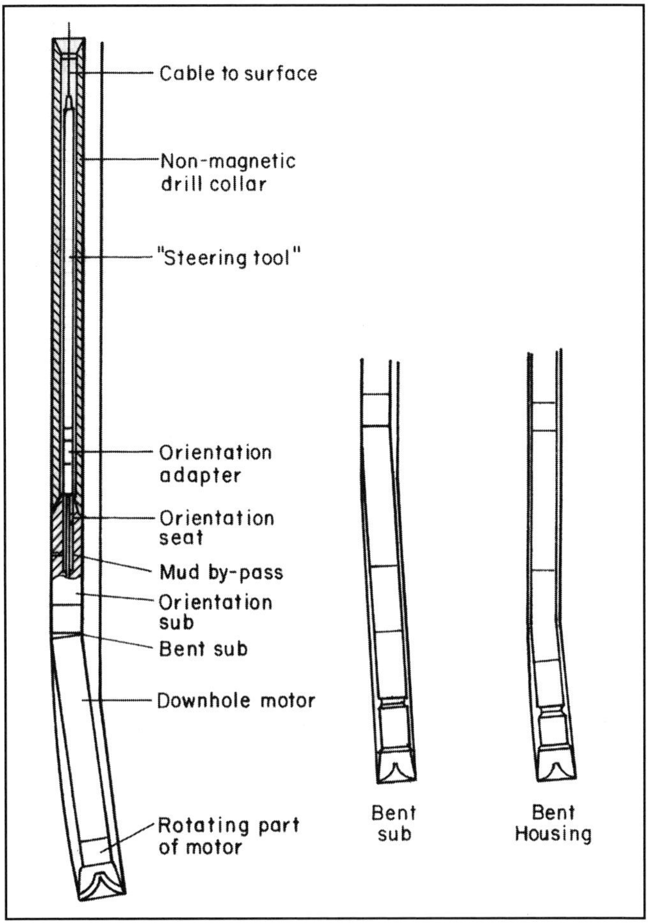

Fig. 7–11. Mud motor assembly

BHA hookups. In development well drilling, where geology, dipping, and strength of formations may be well known, proper design of BHA allows the drilling of a directional well from the surface to the target without use of mechanical or hydraulic techniques. A typical BHA consists of drill collars, a full-gauge drill bit, a reamer, and stabilizers. (E.g., see fig. 7–12.) There are several commercial software packages based on finite element analysis that can be used in BHA design.

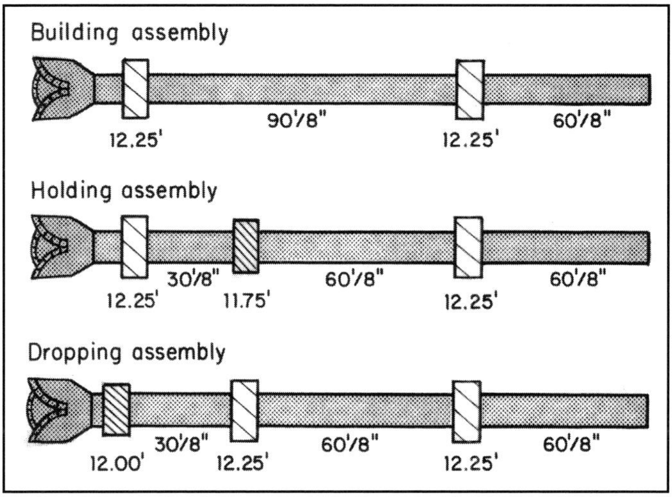

Fig. 7–12. Typical BHA hookups

Survey tools in directional well drilling

To ensure that the correct well trajectory is being drilled, directional surveys must be conducted and survey calculations must be performed. Permanent records of hole locations can have a significant impact on future drilling and completion operations. Acurate knowledge of the well path can be very important in many technical and geology-related decisions. The measuring instruments available to determine the wellbore course (inclination and azimuth) may be categorized among two groups:

- Magnetic survey instruments
- Gyroscopic survey instruments

Both groups have single-shot and multi-shot survey capability. They can be dropped, lowered on a wireline, or used as part of MWD packages. The basic components of both systems consist of time device and compass or electronic sensor device. Single-shot instruments record only one point at a given well depth during the measuring process, whereas multi-shot instruments record several points along the well depth.

Magnetic survey. *Magnetic survey* instruments, whether electronic or mechanical, utilize the magnetic field to determine the hole path. The earth's magnetic field is described by the following components, which vary considerably with the geographic location and thus must known for the area:

- Magnetic north
- Vertical and horizontal components of the local magnetic field
- Total field strength of local magnetic field
- Variation between true north and magnetic north
- Dip angle of local magnetic field in reference to the measured horizontal field

Mechanical survey instruments are based on the compass principle and therefore uses only the earth's local horizontal magnetic component to reference magnetic north. Mechanical survey instruments cannot be used in conjunction with MWD tools.

Electronic magnetic survey instruments use magnetometers to measure the earth's magnetic field and accelerometers to measure the earth's gravitational field. These electronic survey tools can be used to measure hole inclination and azimuth and tool face. They are used in MWD tool packages. The output data is transmitted to the surface through pressure mud pulse telemetry systems and is then decoded as actual data.

Gyroscopic survey. *Gyro survey* tools are generally used when the accuracy of the magnetic survey system could be affected by the presence of magnetic objects (e.g., casings), owing to the inability to place the instrument in the proper location in the BHA, or because of geographic location. There are three basic types of gyro systems:

- *Free gyro.* The free gyro system is the first survey instrument used, dating back to the late 1920s. It consists of a motor-driven spinning mass, referred to as the rotor, that is mounted in a set of gimbals.
- *Rate gyro.* The rate gyro system, also referred to as inertial grade gyro, has a very accurate drift rate of 0.01°/h. Therefore, it can be used to detect the earth's rotation, thereby allowing the calculation of the earth' spin axis and, hence, the geographic north.
- *Inertial navigation system.* The inertial navigation survey system differs from other traditional survey instruments in that it measures movements of hole inclination and azimuth through the three-dimensional Cartesian coordinates, *x, y,* and *z,* to yield a plotted well path trajectory. This system uses a group of gyros to orient the system to the north and accelerometers to detect movements in the *x, y,* and *z* planes.

Although the inertial navigation system is thought to be the most accurate, the accuracy of all of these gyro survey systems is affected by the following:

- Calculation methods
- Well depth
- Tool axial misalignment
- Magnetic interferences, in the case of magnetic survey instruments

Wellbore trajectory calculations from survey data

During directional well drilling, survey data (hole inclination, θ_1, and hole azimuth, α_1) are taken at various stations, *i,* at increments of measured drilled distance, Δd_1, along the wellbore. (See fig. 7–13.) To plot the actual trajectory, coordinates (x_i, y_i, z_i) need to be calculated for each station, *i*.

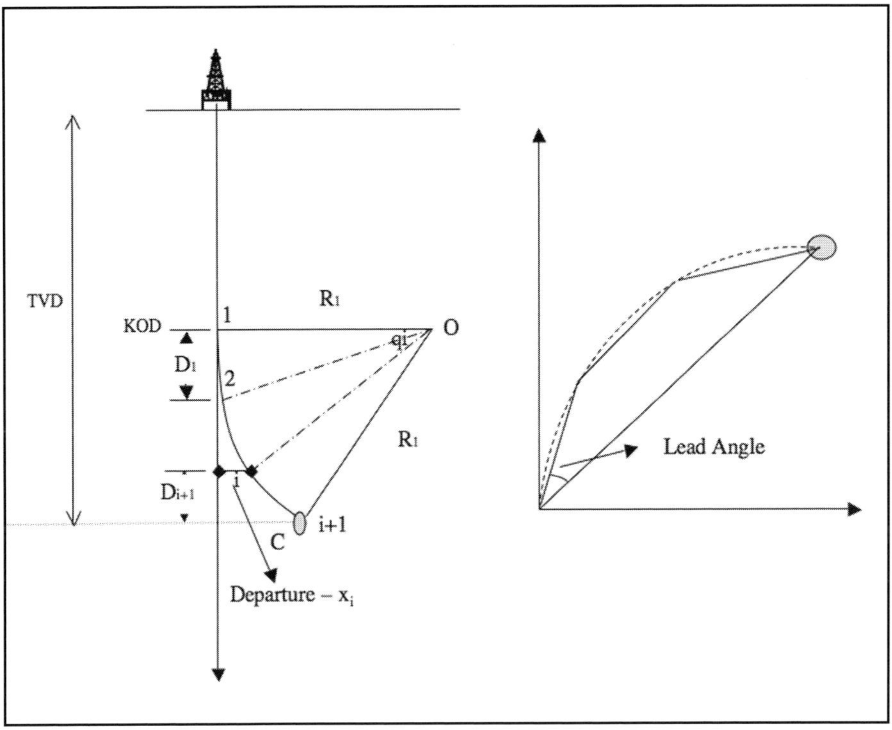

Fig. 7–13. Schematic for wellbore trajectory calculations

There are five available methods that may be used to accomplish this task:

- Minimum-curvature method
- Radius-of-curvature method
- Angle-averaging method
- Tangential method
- Balanced tangential method

The least accurate of these is the tangential method, and most accurate are the radius-of-curvature and minimum-curvature methods. A summary of equations for the first three of these methods will be given in the following sections.

Minimum-curvature method. The minimum-curvature method use the angles measured at two consecutive stations, $i - 1$ and i, to describe a smooth circular curve that represents the wellbore path. This method uses a dogleg severity ratio factor, R_{ei} corresponding with each section of the curve.

The coordinates x_i, y_i, and z_i, which represent the west-to-east, south-to-north, and vertical (downward) depths, respectively, are given by

$$x_i = \frac{\Delta d_i}{2}\left(\sin\theta_{i-1}\sin\alpha_{i-1} + \sin\theta_i \sin\alpha_i\right)R_{e_i} \tag{7.21}$$

$$y_i = \frac{\Delta d_i}{2}\left(\cos\theta_{i-1}\sin\alpha_{i-1}\sin\theta_i + \cos\alpha_i\right)R_{e_i} \tag{7.22}$$

$$z_i = \frac{\Delta d_i}{2}\left(\cos\theta_{i-1} + \cos\theta_i\right)R_{e_i} \tag{7.23}$$

In equations (7.21)–(7.23), $R_{e_i} = (2/e_i)(\tan)e_i/2)$, where e_i is the angle change of drill string between station i and $i-1$: $e_i = \arccos(\sin\theta_i \sin\theta_{i-1} \cos(\alpha_i - \alpha_{i-1}) + \cos\theta_i \cos\theta i{-}1)R_{e_i}$

The x, y, and z coordinates of the nth point along the well path can be determined by algebraically summing over the total number of survey points, n—that is,

$$x_n = \sum_{i=1}^{n} x_i \tag{7.24}$$

$$y_n = \sum_{i=1}^{n} y_i \tag{7.25}$$

$$z_n = \text{KOD} + \sum_{i=1}^{n} z_i \tag{7.26}$$

Radius-of-curvature method. As in the minimum-curvature method, it is assumed that the wellbore has the shape of a smooth curve represented by circular or spherical segments. On the basis of survey data at two consecutive stations $i-1$ and i, the coordinates x_i, y_i, and z_i can be determined using the following equations:

$$x_i = \frac{180\Delta d_i}{\pi\left(\theta_i - \theta_{i-1}\right)\left(\alpha_i - \alpha_{i-1}\right)}\left(\cos\theta_{i-1} - \cos\theta_i\right)\left(\cos\alpha_{i-1} - \cos\alpha_i\right) \tag{7.27}$$

$$y_i = \frac{180\Delta d_i}{\pi\left(\theta_i - \theta_{i-1}\right)\left(\alpha_i - \alpha_{i-1}\right)}\left(\cos\theta_{i-1} - \cos\theta_i\right)\left(\sin\alpha_i - \sin\alpha_{i-1}\right) \tag{7.28}$$

$$z_i = \frac{180\Delta d_i}{\pi\left(\theta_i - \theta_{i-1}\right)}\left(\sin\theta_i - \sin\theta_{i-1}\right) \tag{7.29}$$

Note that $\theta_i \neq \theta_{i-1}$ and $\alpha_i \neq \alpha_{i-1}$. Equations (7.24)–(7.26) can be used to calculate the final coordinates.

Angle-averaging method. This method takes the mean of two measured sets of values for the inclination and azimuth, θ_i, α_i and θ_{i-1}, α_{i-1}. It is assumed that the wellbore follows a tangential path, the coordinates of which at some station *i* are given by

$$x_i = \Delta d_i \sin\frac{\theta_i + \theta_{i-1}}{2}\cos\frac{\alpha_i + \alpha_{i-1}}{2} \tag{7.30}$$

$$y_i = \Delta d_i \sin\frac{\theta_i + \theta_{i-1}}{2}\sin\frac{\alpha_i + \alpha_{i-1}}{2} \tag{7.31}$$

$$z_i = \Delta d_i \cos\frac{\theta_i + \theta_{i-1}}{2} \tag{7.32}$$

Equations (7.24)–(7.26) can be used to calculate the final coordinates.

Example 7–1

You are given the following data for a build-and-hold pattern–type well:

- KOD = 2,500 ft
- TVD = 12,500 ft
- HD to target (THDY) = 9,200 ft
- BRA = 3°/100 ft

Maximum hole angle inclination = 45°. Determine the total measured depth.

Solution: From equation (7.6),

$$R_b = \frac{1}{\text{BRA}}\left(\frac{180}{\pi}\right) = \frac{100}{3}\left(\frac{180}{\pi}\right) = 1{,}910 \text{ ft ft}$$

TMD, from equation (7.8), is

$$\text{TMD} = \text{KOD} + \frac{\theta_b}{\text{BRA}} + \frac{\text{TVD}_t - \text{KOD} - R_b \sin\theta_b}{\cos\theta_b}$$

$$= 2{,}500 + \frac{45}{3/100} + \frac{12{,}500 - 2{,}500 - 1{,}910 \sin 45°}{\cos 45°}$$

$$= 2{,}500 + 1{,}500 + 12{,}233 = 16{,}233 \text{ ft}$$

Horizontal Well Drilling

Introduction to horizontal drilling

Horizontal drilling is the process of steering a drill bit to follow a horizontal path oriented approximately 90° from vertical through the reservoir rock. The interest in drilling horizontal wells can be attributed to the following major reasons:

- Enhancement in primary production
- Enhancement in secondary production
- Enhancement in ultimate recovery of hydrocarbons in place
- Significant reduction in number of wells to develop an entire field
- Significant increase in production

Horizontal well drilling, coupled with multilateral and extended-reach drilling, has unveiled many opportunities for the oil and gas industry to economically recover hydrocarbon reserves from fields that otherwise would not have been feasible. In addition to the economic gains, reducing footsteps in the area has a very significant positive impact on the environment. The success in drilling and completing horizontal wells has been attributed to significant technological breakthroughs, innovation in design approaches, effective team work, utilization of the learning process, effective planning, proper program implementation, real-time field monitoring of drilling data, and updating of program designs.

Whether drilling vertical or directional (including horizontal) wells to exploit underground hydrocarbons, the same elements are needed to drill the well successfully and economically; they differ only in terms of requirements. These elements (as discussed in chap. 1) include

- Downward force on drill bit
- Drill bit rotation
- Fluid circulation

Force. In vertical well drilling, WOB (force) is provided by the heavy drill collars located directly above the bit, with practically no weight loss due to sliding friction. However, in directional well drilling, there is inherent contact between the drill string and the walls of the wellbore; consequently, considerable friction forces (drag) can be encountered, reducing the amount of weight needed to be transferred to the bit. This means that tubulars placed above the bit should be of a weight variation such that their contribution to drag forces will be minimized and their contribution to WOB will be maximized. This will be discussed in more detail in chapter 9.

Rotation. Drill bit rotation may be induced at the surface, through the conventional rotary table or top-drive motor, or at the bottom, through the use of downhole mud motors. In vertical well drilling, the major source of torque is the drill bit, with an almost negligible torque contribution from drill string frictional forces. However, in directional well drilling, rotation is induced at surface, and the portion of the drill string that is in contact with the walls of the wellbore will cause friction torque, in addition to bit torque, that can be five to ten times the friction torque encountered when drilling vertical wells. As hole angle increases from vertical to horizontal, the drag and torque due to friction forces will likewise increase. In the horizontal section of the wellbore, the total weight of that part of the drill string will cause friction torque and drag, which is an undesirable situation. Excessive torque may limit a rig's available rotary power; by contrast, when there is excessive drag, advancement of the bit may become the limiting factor in reaching the desired target.

Circulation. Flow rate requirements in directional (including horizontal) drilling can be two to four times higher than the flow rate requirements in vertical well drilling. This is due to the proven need of higher annular fluid velocities for the effective removal of drilled cuttings from the annulus to the surface. Higher flow rates cause high friction pressure losses and, therefore, higher rig hydraulic horsepower requirements.

Applications of horizontal well drilling

The reservoirs that may be considered as possible candidates for horizontal drilling are

- Reservoirs that may have potential water/gas-coning problems
- Tight reservoirs (permeability << 1 millidarcies [md])
- Natural vertically fractured reservoirs
- Economically inaccessible reservoirs
- Heavy oil reservoirs
- Channel and reef sand reservoirs
- Coal bed methane reservoirs
- Thin reservoirs
- Layered reservoirs with high dip angle
- Partially depleted reservoirs

Although several reservoir types may be candidates for horizontal drilling applications, the ultimate deciding factor is economics—that is, the best return on investment. Overall well cost comparison between drilling a vertical or drilling a horizontal well should be based not on dollars per foot ($/ft) cost but rather on dollars per potential produced barrel of oil ($/bbl).

Methods of horizontal drilling

Horizontal drilling starts at a surface location with a vertical section, followed by a deviated section at some preselected KOD, starting at 0° from vertical and ending approximately at 90° at the entry point into the reservoir. The following methods are used to drill the deviated build sections to reach the entry point into the reservoir:

- Long-radius (LR) method
- Medium-radius (MR) method
- Short-radius (SR) method
- Ultrashort-radius (USR) method

Tables 7–1 and 7–2 compare the basic characteristics of the first three methods. A brief description of each of these methods is undertaken in the following sections.

Table 7–1. Comparison of horizontal drilling methods—drilling process (Courtesy SPE/IADC)

	Long Radius	Medium Radius	Short Radius
Build Rate	Up to 6° per 100'	6° to 35° per 100'	1.5° to 3° per 1'
Build Radius	1000' to 3000'	955' to 286'	20' to 40'
Hole Size	No Limitation	4¾", 6⅛", 8½", 9⅞"	4¾" 6½"
Drilling Method	Rotary or steerable motor systems for curved and horizontal sections	Special design motors for angle—build section; rotary or steerable motor system for horizontal sections	Special design deflection tool or articulated motor for angle—build section. Rotary tools and special drillpipe for horizontal sections
Tubular Type	Conventional	Heavy weight drillpipe up to 15° per 100'. Compressive service drillpipe above 15° per 100'	Special articulated tubulars. Special drillpipe with short articulated motors
Drill Bit	Unlimited	Unlimited	Rotary = Unlimited Motor = Diamond or Synthetic Diamond
Drilling Fluids	Unlimited	Unlimited	Unlimited
Survey Type	Unlimited	MWD limited in hole sizes smaller than 6⅛"	Special
Coring	Conventional Unlimited	Conventional Unlimited	3' core barrel, 1" core

Table 7–2. Comparison of horizontal drilling methods—completion (Courtesy World Petroleum Congress)

	Selective Completion of Portions of the Drain Hole	Multiple Pay Zones	Artificial Lift	Workover
Short Radius	No	Yes	Rod pumps in vertical portion	Yes
Medium Radius	Yes	Yes	All Types	Yes
Long Radius	Yes	No	All Types	Yes

LR method. The long turn radius is defined as a drilling method that will result in a wellbore with at least one section that has been deviated at a build rate of 1–6°/100 ft. The LR method has applications in offshore locations, inaccessible locations, and reservoirs that require geological isolation.

The LR method utilizes either of two systems in the process of drilling horizontal wells. These are the conventional rotary BHA system and the steerable system. (See fig. 7–14.) The use of conventional rotary assemblies to drill long-radius wells is quite common. BHA software is available that allows for the optimization and prediction of BHA operating parameters. Use of the software can be quite successful, assuming that there is accurate information on formation anomalies; otherwise, repeated required corrections for the well path can lead to a high cost of drilling. For this reason, the steerable systems (downhole motors or rotary systems) with MWD tools may be more cost effective than the conventional BHA system. The decision whether to use conventional or steerable systems is based on economics.

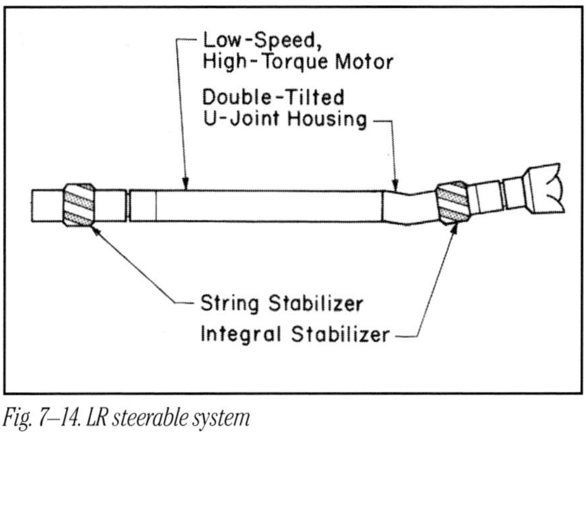

Fig. 7–14. LR steerable system

MR method. The medium turn radius is defined as a drilling method that will result in a wellbore with at least one section that has been deviated at a build rate of 6–35°/100 ft. The MR method has applications in channel/reef reservoirs, fractured reservoirs, and potential gas- or water-coning reservoirs.

The MR method utilizes mainly steerable systems. Figure 7–15 shows a typical building assembly that allows for continuous, smooth curvature, built up to a desired angle of inclination. The angle-hold assembly may be of the conventional rotary type or may contain a downhole motor. (See fig. 7–15.)

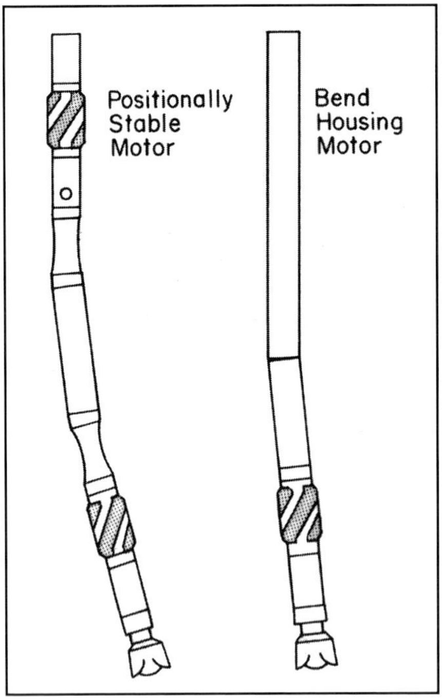

Fig. 7–15. MR angle-build assembly

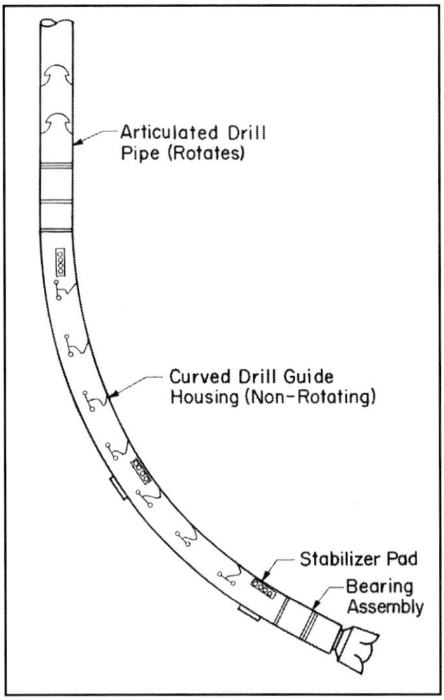

Fig. 7–16. SR lateral-drilling angle-build assembly with curved drill guide

SR and USR methods. These two methods utilize unique tools. Whereas the USR method utilizes specialized hydraulic jetting techniques, the SR method utilizes specialized "wiggly" drill collars and drill pipes, known as articulated pipes and/or knuckle joints (see figs. 7–16 and 7–17). The use of these two methods is very limited.

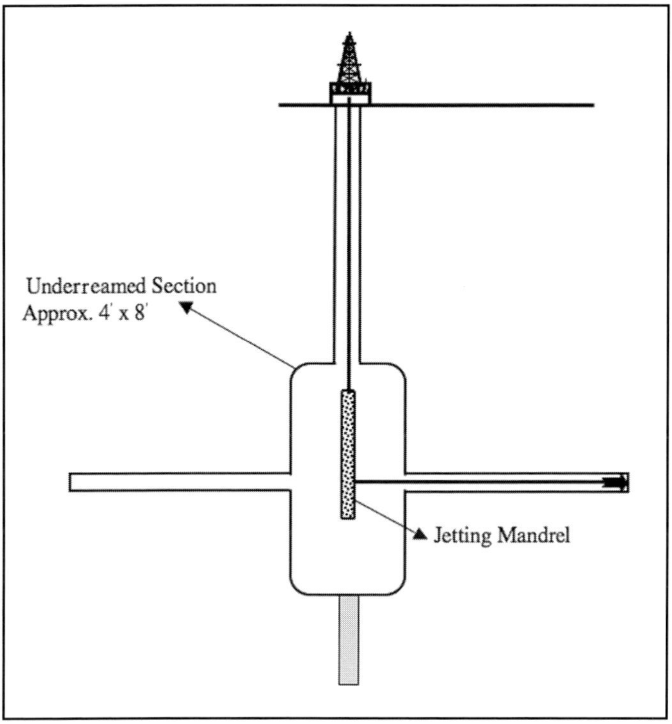

Fig. 7–17. USR assembly

Considerations in selection of drilling method

The following considerations enter into the process of selecting the method to be used in drilling horizontal wells:

- Cost
- Well spacing/lease restrictions
- Conditions of reentry holes
- Reservoir rock characteristics
- Production methods
- Company objectives
- Problem-causing lithologies above the desired zone
- Amount of reach
- Completion methods
- Availability of survey, coring, and other downhole tools
- KOD constraint
- HD constraint

Planning the drilling operation of a horizontal well

Planning for horizontal well drilling generally consists of specifying an entry depth (ED) into the reservoir and the minimum horizontal length (MHL) of reach inside the reservoir. Therefore, the objectives of well planning are twofold:

- Economically reach the target
- Economically drill the horizontal section

To reach the target, planning consists of three phases:

- Vertical phase
- Directional phase
- Horizontal phase

These phases of horizontal well drilling are illustrated in figures 7–18 and 7–19.

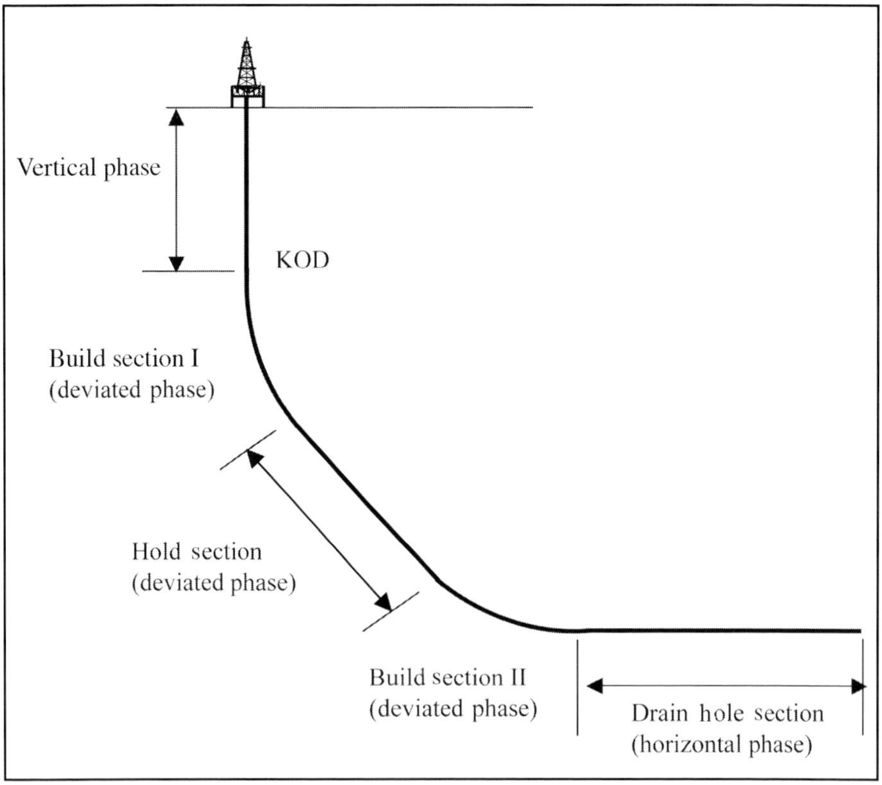

Fig. 7–18. Phases of horizontal well drilling—two builds

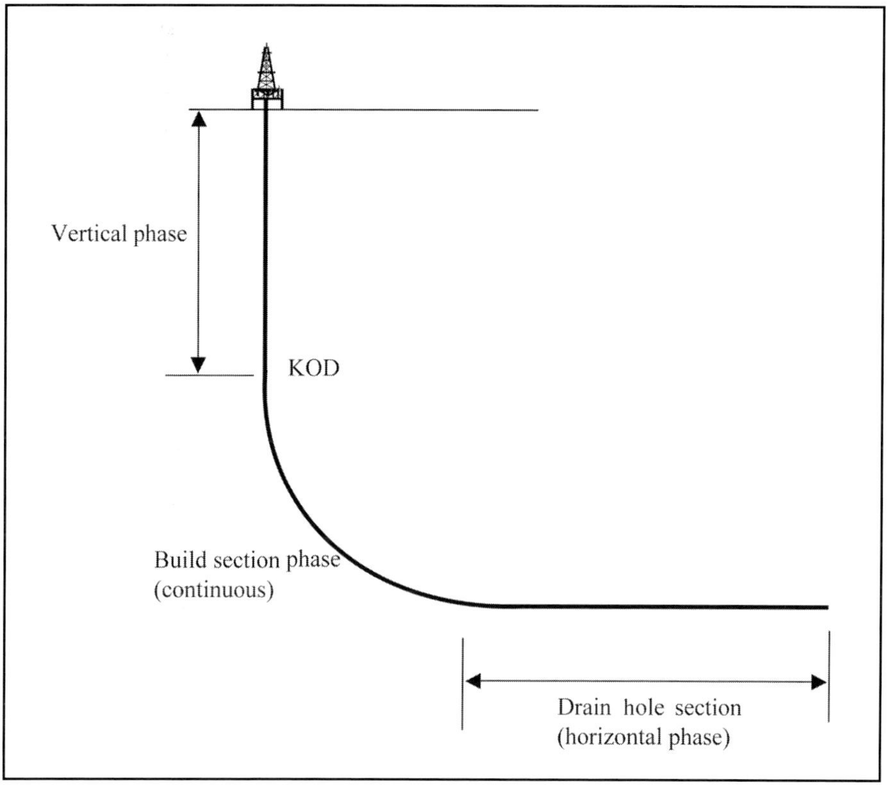

Fig. 7–19. Phases of horizontal well drilling—one build

Vertical phase. All horizontal wells start with the drilling of a vertical section down to a specific KOD below the surface.

Directional phase. The directional phase of horizontal well drilling consists of one of the following:

- One continuous-build drilling
- Build-and-hold and build drilling

At the KOD, the plans have already been made either to continuously build until the target entry point into the reservoir or to build and hold and then build until the entry point into the reservoir. The choice of wellbore trajectory is generally governed by minimizing the following well-drilling problems.

- Annular hole cleaning
- Drag and torque
- Wellbore instability

Also of importance in the selection of well profile is the allowance for easy completion.

The two major problems that may occur during this phase of drilling are

- Inability to build angle
- Inability to correct the azimuth

Generally, troublesome formations and improper BHA design are the causes of these problems.

Build curve types

In horizontal well drilling, the following build curve types are used to reach the entry point into the reservoir at approximately 90° from vertical:

- Single build curve (fig. 7–20)
- Simple tangent build curve (fig. 7–21)
- Complex tangent build curve (fig. 7–22)
- Ideal build curve (fig. 7–23)

The mathematical equations given in the previous section can be used in the design of the well directional plans.

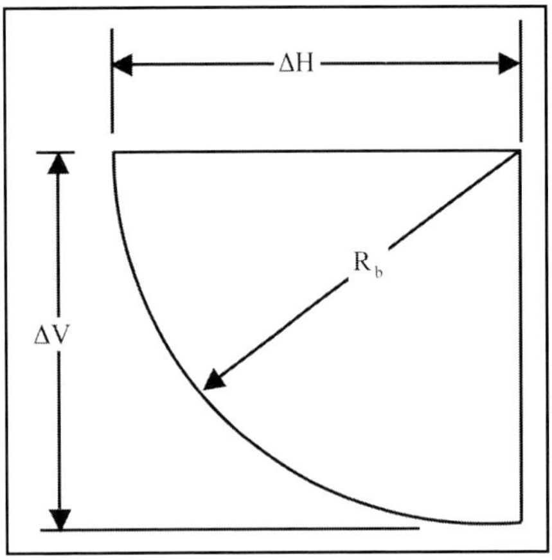

Fig. 7–20. Single build curve

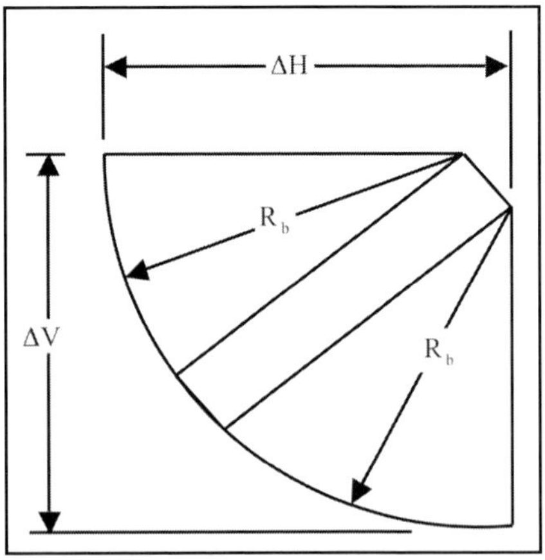

Fig. 7–21. Simple tangent build curve

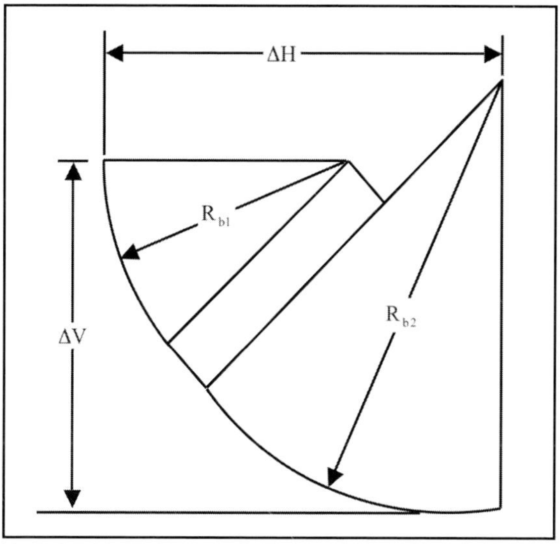

Fig. 7–22. Complex tangent build curve

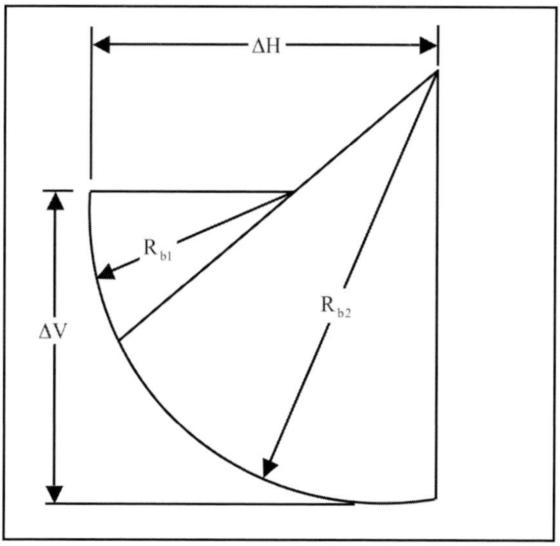

Fig. 7–23. Ideal build curve

Planning for well trajectory changes

During the drilling process of a directional well, changes in the trajectory may become necessary after a survey—either to correct or to simply change the course of the wellbore for some other reason. Millheim et al.[1] derived equations to calculate where to properly reposition the drilling tool (tool face angle) to take on a different course. Their derivations will not be repeated here.

The equations that Millheim et al. derived determine the relationships between the following parameters:

- Original hole inclination, θ_o
- New hole inclination, θ_n
- Original hole azimuth, α_o
- New hole azimuth, α_n
- Overall hole angle change, β
- Required tool face rotation from the original course direction, λ

These parameters are defined by equations (7.33)–(7.35). For directional change,

$$\Delta\alpha = \alpha_n - \alpha_o = \arctan\left(\frac{\tan\beta\sin\lambda}{\sin\theta_o + \tan\beta\cos\theta_o\cos\lambda}\right) \tag{7.33}$$

For new inclination angle,

$$\theta_n = \arccos\left(\cos\beta\cos\theta_0 - \sin\beta\sin\theta_0\cos\lambda\right) \tag{7.34}$$

For tool face angle,

$$\lambda = \arccos\left(\frac{\cos\theta_o\cos\beta - \cos\theta_n}{\sin\theta_n\sin\beta}\right) \tag{7.35a}$$

Alternatively,

$$\lambda = \arcsin\left(\frac{\sin\theta_n\sin\Delta\alpha}{\sin\beta}\right) \tag{7.35b}$$

Further, the dogleg severity, ψ, is defined as

$$\psi = \frac{\beta L}{\Delta d} \tag{7.36}$$

where

$L = 100$ ft if dogleg severity is expressed in degrees per 100 ft

Δd is the length of the drilled interval between two surveys

Finally, the overall angle change is defined as

$$\beta = \arccos\left(\cos\theta_0\cos\theta_n + \sin\theta_0\sin\theta_n\cos\Delta\alpha\right) \tag{7.37}$$

Example 7–3

Calculate the dogleg severity (DLS) and angle of tool face rotation given the following data:

Measured Depth	Angle	Direction
5,000 ft	5.5°	150°
5,120 ft	7.5°	148°

S*olution:* Overall angle change is given as

$$\begin{aligned}\beta &= \text{arc cos } (\cos 2 \sin 7.5 \sin 5.5 + \cos 5.5 \cos 7.5)\\ &= 2.01^\circ\end{aligned}$$

Thus,

$$\text{DLS} = \frac{2.01}{120} \times 100 = 1.67^\circ/100\,\text{ft}$$

With θ, β, θ_n all known, the tool face rotation angle is

$$\begin{aligned}\lambda &= \text{arc cos}\left(\frac{\cos\theta_o \cos\beta - \cos\theta_n}{\sin\theta_n \sin\beta}\right)\\ &= \text{arc cos}\left(\frac{\cos 2.01 \cos 5.5 - \cos 7.5}{\sin 7.5 \sin 2.01}\right) = 43.27^\circ\end{aligned}$$

Tortuosity

Introduction to tortuosity

Tortuosity is a critical factor to consider for complex directional well trajectories, complicated build rates, precise steering in thin reservoirs, and extended reach wells. When planning a well, well path modeling commonly generates smooth curves; however, an actual well contains severe doglegs and other irregularities. Determining the tortuosity factors to apply in the well path is always a challenge during the planning phase.

At present, no quantification method is available; moreover, field tests have shown that the present calculations gravely underestimate the drill string torque and drag values for highly deviated (tortuous) deep wells. Problems occur mainly when the drill string lies in a strongly curved well path. In the long transitional region that covers most of the operational range of field applications, the conventional equations used for torque and drag calculation result in underprediction. For increased extended and steered wells, the torque and drag calculations need to correct for the tortuous effects of the wellbore.

Hole spiraling and wellbore oscillation

The occurrence of borehole spiraling, or oscillation of wellbore path, has been known for many years in the drilling industry. Characterization of this behavior and its effects are increasingly discussed and analyzed with improved borehole imaging techniques and logging tools. Increased evidence of hole spiraling has been reported.[2] Hole spiraling, or threading, has a deep impact on the well construction and completion process.

During the planning phase, the hole spiraling or oscillation of the wellbore can effectively be described by applying the desired tortuosity factor to the wellbore. The difference between the planned smooth well profile, based on well path modeling during planning, and the actual well drilled, with doglegs and other irregularities, can have a major impact on the torque and drag losses for the well. Models apply different rippling or roughness techniques to a planned well path to simulate the variations found in actual well path surveys. This option renders planned (smooth) well profiles in a form that more realistically predicts loads. Note that tortuosity is usually applied only when the surveys represent an unrealistically smooth path. Currently, the industry uses different generations of rotary steerable systems and adjustable downhole tools to reduce wellbore tortuosity.

The industry has no standard for quantifying tortuosity. Tortuosity is usually expressed in degrees/100 ft, similar to the expression of dogleg severity. The calculation of *running tortuosity* is the station-to-station summation of the total curvature normalized to a standard wellbore course length between survey stations. In general, tortuosity is defined as the ratio of the summation of the total curvature, including build and walk, to the survey stations' length. Rippling or undulation can be applied using the different methods that modify the inclination and azimuth of the survey points. Furthermore, *micro-tortuosity*[3] caused by hole spiraling results in a spiraled hole axis, instead of a straight line. This can be taken into account during the planning phase by applying tortuosity over and above the well path that has been already tortured with the desired magnitude and amplitude or by superimposing a helix with the desired pitch and magnitude.

Tortuosity methods

Different methods are used to apply roughness to a planned well path, to simulate the variations found in actual well path surveys. These options render planned (smooth) well profiles in a form that gives rise to a more realistic prediction for the analysis. It is expressed as the difference between the actual and planned curvatures divided by the respective distance between the survey stations and can be written as

$$T = \frac{\sum_{j=1}^{n} DL_{\mathrm{a}} - \sum_{i=1}^{m} DL_{\mathrm{p}}}{\mathrm{MD}_j - \mathrm{MD}_i} \tag{7.38}$$

and is expressed in degrees per 100 ft.

Rippling or undulation is applied using the following methods:

- Sine wave method
- Helical method
- Random inclination and azimuth method
- Random inclination–dependent azimuth method

Sine wave method. The *sine wave method* modifies the inclination and azimuth of the survey point, based on the concept of a sine wave–shaped ripple running along the wellbore and using the magnitude (amplitude) and period (wavelength) specified. The angle change is modified using the following relationship

$$\Delta\alpha = M\sin\left(\frac{\mathrm{MD}}{P}2\pi\right) \quad (7.39)$$

where MD is the measured depth (in ft), *P* is the period, and *M* is the magnitude.

The magnitude, *M*, is the maximum variation of angle that will be applied to the inclination and azimuth of the native (untortured) well path. The determination of magnitudes should be based on the evaluation of historical data from offset or similar wells. The values of magnitude can vary depending on many factors, including hole type (cased or open hole), hole size, drilling performance, and hole curvature, among other parameters. Further, the inclination angle is modified so that it does not become less than zero, since negative inclination angles are not allowed. The new angle and azimuth are given as follows:

$$\alpha_n = \alpha + \Delta\alpha \quad (7.40a)$$

$$\varepsilon_n = \varepsilon + \Delta\alpha + \psi_{xvc} \quad (7.40b)$$

where α_n is the new angle, ε_n is the new azimuth, and ψ_{xvc} is the cross vertical correction.

Also, while applying tortuosity, ensure that the measured depth, MD, of the survey point is not an exact integer multiple of the period as

$$\Delta\alpha = \sin\left(\frac{\Delta\mathrm{MD}}{P}2\pi\right) = 0 \quad (7.41)$$

In this situation, tortuosity will not be applied to the undithered data set. Caution should be exercised to avoid this situation.

Example 7–4

You are given the following information:

- MD = 3,725 ft
- Angle of inclination = 3.25°
- Azimuth = 165°

Apply the sine wave method to find the angle change and the new angle and azimuth, as well as the cross vertical correction.

Solution:

$$\Delta\alpha = \sin\left(\frac{3{,}725}{1{,}000}2\pi\right)\times 1 = -0.99$$

$$\alpha_n = 3.25 - 0.99 = 2.26^\circ$$

$$\varepsilon_n = 165 - 0.99 + 0 = 164.01$$

If $\alpha_n < 0$, then $\psi_{xvc} = 180°$ (i.e., $\alpha_n = |\alpha_n|$); if $\alpha_n \geq 0$ then $\psi_{xvc} = 0°$.

Note: If the measured depth of the survey point is an exact integer multiple of the period, then equation (7.39) equals zero.

Helical method. The *helical method* modifies the inclination and azimuth of the survey points by superimposing a helix along the wellbore path and using the magnitude (radius of the cylinder in the parametric equation) and period (pitch) specified. This method uses the circular helix defined as:

$$f(\alpha) = a\cos(\alpha) + a\sin(\alpha) + b\alpha \tag{7.42}$$

The generalized parametric set of equations for the helix used to superimpose the wellbore path is given by

$$x(\alpha) = M\cos(\alpha) \tag{7.43a}$$

$$y(\alpha) = M\sin(\alpha) \tag{7.43b}$$

$$z(\alpha) = \frac{P}{2\pi}\alpha \tag{7.43c}$$

Random inclination and azimuth method. The *random inclination and azimuth method* is similar to the random inclination-dependent azimuth variation method described in the next section except that the azimuth variation is independent of inclination. The new angle and azimuth are given as follows:

$$\alpha_n = \alpha + \Delta\alpha \tag{7.44a}$$

where

$$\Delta\alpha = \frac{\Delta MD \times M}{P}$$

$$\varepsilon_n = \varepsilon + \Delta\alpha + \psi_{xvc} \tag{7.44b}$$

where ψ_{xvc} is the cross vertical correction.

Example 7–5

You are given the following information:

- $MD_1 = 3{,}725$ ft
  - Angle of inclination = 3.25°
  - Azimuth = 165
- $MD_2 = 3{,}900$
  - Angle of inclination = 5.15°
  - Azimuth = 166
- Random number: $\zeta = 0.375$

Apply the random inclination and azimuth method method to find the angle change and the new angle and azimuth, as well as the cross vertical correction.

Solution:

$$\Delta\alpha = \frac{0.375 \times (3{,}900 - 3{,}725)}{100} \times 1 = 0.66$$

$$\alpha_n = 5.15 + 0.66 = 5.81^\circ$$

Since $\alpha_n > 0$, $\psi_{xvc} = 0^\circ$. Thus,

$$\varepsilon_n = 166 + 0.66 + 0 = 166.66^\circ$$

Random inclination-dependent azimuth variation method. The *random inclination-dependent azimuth variation method* applies a random variation to the survey inclination and azimuth within the magnitudes specified. The random numbers used may be between –1.0 and +1.0. In this method, the azimuth variation is inversely proportional to the inclination, resulting in higher inclination with lower azimuth variation and lower inclination with higher azimuth variation. The change in angle, the new angle, and the new azimuth are given by equations (7.45)–(7.47):

$$\Delta\alpha = \zeta \times \delta \tag{7.45}$$

where $\delta = M(\Delta\text{MD})/\text{P}$ and ζ is a random number; further,

$$\alpha_n = \alpha + \Delta\alpha \tag{7.46}$$

$$\varepsilon_n = \varepsilon + \frac{\Delta\alpha}{2\sin\alpha_n} + \psi_{xvc} \tag{7.47}$$

where ψ_{xvc} is the cross vertical correction.

Example 7–6

You are given the following information:

- MD_1 = 3,725 ft
  - Angle of inclination = 3.25°
  - Azimuth = 165°
- MD_2 = 3,900 ft
  - Angle of inclination = 5.15°
  - Azimuth = 166°
- Random number: $\zeta = 0.375$

Find the angle change and the new angle and azimuth, using the random inclination-dependent azimuth variation method.

Solution:

$$\Delta\alpha = \frac{0.375 \times (3{,}900 - 3{,}725)}{100} \times 1 = 0.66^\circ$$

$$\alpha_n = 5.15 + 0.66 = 5.81^\circ$$

$$\varepsilon_n = 166 + \frac{0.66}{2\sin 5.81} + 0 = 169.26^\circ \qquad = 169.26$$

Calibration of tortuosity factors

Qualitative assessment and quantitative characterization of the torque and drag are critically important in many phases of well construction. They allow not only subsequent adjustments to the well plan when combined with pay zone geological steering tools but also successful completion of extended-reach and complex wells. Increasingly, more difficult wells are being drilled; thus, accurate estimation of torque and drag calculations is required. Operating outside the safety limits may lead to costly complications.

The calibration methodology uses actual survey data and known surface hook loads for different operating conditions—such as tripping in, tripping out, and rotating off bottom—to arrive at appropriate friction factor values, which can subsequently be used in torque and drag calculations. Once the friction factors for both cased hole and open hole are determined, sensitivity analyses can be done on the planned survey. Appropriate tortuosity values are derived by comparing predicted results with the field driller's data. The following section describes the steps for calculating tortuosity factors that better represent the actual curved wellbore.

Methodology. The following is a simple procedure to estimate the tortuosity factors for a particular hole size.

Step 1. Collect the relevant drilling data. These include the original well plan (casing program, directional plan, BHA designs, etc.) and actual drilling data (well configuration, hole section BHAs, definitive survey data, mud logging data, any downhole torque/WOB measurements, etc.).

Step 2. For each hole section, calibrate the friction factor through casing, based on the actual drilling data while drilling out the shoe and the definitive survey data without any tortuosity correction.

Step 3. Repeat step 2 for the open hole in each of the hole sections. Note that the friction factor through the casing should be that from step 2.

Step 4. Once the appropriate friction factors have been established for both cased and open holes, these can be used in conjunction with the original planned well profile to define the most appropriate tortuosity factor for the well by matching the torque and drag predictions with the actual drilling data. Note that the tortuosity factor obtained will be applicable to the BHA used in the hole section.

Supplementary Problems

Problem 7–1. The geological TVD and the HD (east-west) of a well target are 10,000 ft and 2,500 ft, respectively. KOD = 2,000 ft. Find the maximum angle of inclination for a build-and-hold profile at the following build rates:

(a) 1°/100 ft

(b) 3°/100 ft

(c) 5°/100 ft

Also find the TMD in each case.

Problem 7–2. In problem 7–1, if the maximum angles of continuous build to target are to be 60°, 70° and 80°, what are the respective build rates (BRA), the corresponding required KOD, and the TMD?

Problem 7–3. Determine the KOD, the total horizontal departure THDY and the TMD at a TVD of 7,000 ft for the following build rates:

(a) 1°/100 ft

(b) 15°/100 ft

(c) 2°/ft

(d) 90°/ft

Problem 7–4. You are given the following information: HDY = 2,000 ft; TVD = 12,000 ft; BRA = 2°/100 ft; maximum inclination angle (θ_{max}) = 60°. Determine the corresponding KOD values.

Problem 7–5. In problem 7–4, find the maximum hook load during tripping out of the hole and the maximum torque while drilling. Assume that the average coefficient of friction is 0.25 and the average effective weight of drill string is 15 lbs/ft.

Problem 7–6. Construct a plan for a build-and-hold–type well, given the following data:

- Build rate = 3°/100 ft
- KOD = 2,000 ft
- TVD to target = 9,000 ft
- Target coordinates: 5,500 ft N35°E of rig

Problem 7–7. A well is to be drilled to a target that is 7,000 ft away from the rig at a vertical depth of 15,000 ft. If a build rate of 2°/100 ft is to be used, what is the maximum angle that will be obtained for the following KOD values?

(a) 1,500 ft

(b) 2,000 ft

(c) 2,500 ft

Problem 7–8. Construct a plan for a build-and-hold–type well. Assume that the maximum angle of inclination is to be between 40° and 45°. The maximum BRA is 3.5°/100 ft. The KOD is to be between 1,500 ft and 2,500 ft. You are also given the following information:

- Rig coordinates: 2,000 ft N800W
- Target coordinates: 4,500 ft N500E
- Target TVD = 12,000 ft

Problem 7–9. Determine the KOD required, given the following data for a build-and-hold type well:

- Target TVD = 9,600 ft
- Total HD = 2,000 ft
- BRA = 2°/100 ft
- Maximum angle at total depth = 40°

Problem 7–10. Determine the KOD required, given the following data for a continuous-build–type well:

- Target TVD = 15,000 ft
- Total HD = 4,000 ft
- BRA = 3°/100 ft
- Maximum angle at target = 20°

Problem 7–11. Derive equation (7.30–7.32).

Problem 7–12. An azimuth angle, α, that can have values between 0° and 360° is used to specify the direction of a target. Express β, a quadrant angle (quadrature system), in terms of α.

Problem 7–13. Plan the vertical trajectory of a horizontal well assuming (a) MR drilling and (b) LR drilling. For each part, consider the following curve design options: (i) single build curve; (ii) ideal build curve; (iii) simple tangent curve; (iv) complex tangent curve.

The following well data are given:

- Total TVD to the reservoir entry point = 10,000 ft.
- The reservoir has zero dip
- Desired horizontal reach into the reservoir = 2,000 ft
- Reservoir average thickness = 100 ft
- Surface area runs 20 miles east and 12 miles north from the lease lines of west and south

Problem 7–14. A rig is located at south y ft and west x ft. A proposed target is at L ft $S\theta E$of the rig location. Determine the rectangular coordinates of the target with respect to the rig.

Problem 7–15. You are given an east or west declination angle, η. For a given azimuth angle, α, determine the corrected azimuth angle, α_n, in terms of α and η for east and west declinations.

Problem 7–16. List the following:

(a) Applications of directional drilling

(b) Reservoir candidates for horizontal drilling

(c) Well profiles in directional drilling

Problem 7–17. For a build-and-hold wellbore trajectory, derive expressions for TTVD, THD, and TMD to the target. Assume: kick-off—depth = KOD; build rate angle BR; maximum inclination angle = θ_{max}; length of slant section L_s.

Problem 7–18. A survey shows 10° inclination with direction far off course. A deflection assembly for a 2.4° dogleg is run and oriented for maximum turn. Determine the magnitude of tool rotation (γ) required in order to obtain maximum turn, the hole direction change ($\Delta\varepsilon$), and the new inclination (θ_n)

Problem 7–19. A survey shows 21.3° inclination and N40E direction. It is desired to turn 5° to the left and build angle. Determine the new hole inclination, the tool rotation from original course direction, and the tool direction if a deflection tool with a 2.5° dogleg is used.

Problem 7–20. A directional survey shows that the original hole direction and inclination are S10W and 5°, respectively. To obtain a new hole direction of S20E, a deflection tool with the deflection angle 3° will be used. Determine the required deflection tool direction if the anticipated walk to the right is 15°.

Problem 7–21. The original hole direction and hole inclination are N20E and 10°, respectively. It is desired to deviate the borehole so that the new inclination angle and direction will be N30E and 13°, respectively. What tool orientation and tool deflection angle (dogleg) are necessary to achieve this turn and build?

Nomenclature

BRA	Build rate angle
Δd	An assumed incremental distance of hole section generally equal to 100 ft
DRA	Drop rate angle
e_i	Angle change of drill string between stations i and $i - 1$
$(EW)_i$	Incremental distance measured along the east-west direction
H	Horizontal distance
HD	Horizontal departure to target
KOD	Kick-off depth
L_t	Length of tangent
LA	Lead angle
LR	Long radius
M	Magnitude
MD	Measured depth, in ft
MR	Medium radius
n	Number of increments
$(NS)_i$	Incremental distance measured along the north-south direction
P	Period
R_b	Radius of curvature
R_{ei}	Dogleg severity (DLS) ratio factor
SR	Short radius
TOD	Turn-off depth
TVD	True vertical depth to target
USR	Ultrashort radius
V	Vertical distance
x_i, y_i, z_i	Coordinates representing west to east, south to north, and vertical depth, respectively
α_i	Azimuth angle at point (i)
α_{i-1}	Azimuth angle at point ($i - 1$)
α_n	New hole azimuth
α_o	Original hole azimuth
β	Overall hole angle change
ζ	Random number
θ_i	Inclination angle at point (i)
θ_{i-1}	Inclination angle at point ($i - 1$)
θ_n	New hole inclination
θ_o	Original hole inclination
λ	Tool face rotation from original course direction
ψ_{xvc}	Cross vertical correction

References

1 Millheim, K. K., F.H. Eubler, and H.B. Zaremba. 1979. "Evaluation and Planning Directional Wells Utilizing Post Analysis Techniques and 3-Dimensional Bottomhole Assembly Program." Paper SPE 8339, presented at SPE ATCE, Las Vegas. Bourgoyne, A. T., M. E. Chenevert, K. K. Millheim, and F. S. Young. 1991. *Applied Drilling Engineering.* SPE Textbook Series, Vol. 2.

2 Stroud, D., S. Peach, and I. Johnston. 2004. "Optimization of Rotary Steerable System Bottomhole Assemblies Minimizes Wellbore Tortuosity and Increases Directional Drilling Efficiency." Paper SPE 90396, presented at the SPE Annual Technical Conference, Houston. Gaynor, T., D. Hamer, D. Chen, and D. Stuart. 2002. "Quantifying Tortuosities by Friction Factors in Torque and Drag Model." Paper SPE 77617, presented at the SPE Annual Technical Conference, San Antonio, Tex. Pastusek, P., and V. Brackin. 2003. "A Model for Borehole Oscillations." Paper SPE 77617, presented at the SPE Annual Technical Conference, Denver. Luo, Y., K. Bharucha, G. R. Samuel, and F. Bajwa. 2003. "Simple Practical Approach Provides a Technique for Calibrating Tortuosity Factors." *Oil and Gas Journal,* 15. Samuel, G. R., K. Bharucha, and Y. Luo. "Tortuosity Factors for Highly Tortuous Wells: A Practical Approach." Paper SPE 92565, presented at the SPE/IADC Drilling Conference, Amsterdam. 2005.

3 Gaynor, T., D. Hamer, D. Chen, D. Stuart, and B. Comeaux. 2001. "Tortuosity versus Micro-Tortuosity—Why Little Things Mean a Lot." Paper SPE 67818, presented at the SPE/IADC Drilling Conference, Amsterdam.

4 Mitchell, Bill. *Oil Well Drilling Engineering*, 8th ed. Golden, Colo., 1992.

5 Schuh, Frank. Class notes.

8

Drill Bit Mechanics

Introduction

One of the most fundamental tasks fulfilled by the drilling engineer is selecting, operating, and evaluating the performance of the drill bit. There are a wide variety of bit types available. Their differences range from subtle manufacturing details to completely different mechanics of cutting rock. The purpose of this chapter is to provide a basic understanding of the factors that affect bit selection, operation, and ultimately, performance. The drilling engineer's responsibilities consist of the following:

- Collect all relevant offset bit data and other data that affect bit selection and performance
- Identify the two best (lowest drilling cost) bits for every hole section
- Evaluate bit wear and causes thereof, to improve on next bit runs
- Investigate all controllable drilling variables that affect bit performance
- Determine the best operating parameters (WOB, rotary speed, and hydraulics) to ensure optimized drilling
- Conduct post-analysis, comparing actual performance versus expected, and make recommendations
- Understand the design, application, and drilling techniques suited for the bit selected

Bit Selection

Although there is no exact scientific theoretical approach to the proper selection of drill bits, the following useful methods may provide a close estimate to the best bit for the given formation interval to be drilled:

- Thorough evaluation and comparison of offset well bit records
- Bit run cost equation
- Drill-off test
- Specific energy equation

Among all of these methods, the best is that based on experience and the comparison of offset bit records in the area. The bit run cost equation is generally what provides the drilling engineer with a quick estimate of the offset bit run cost and, thus, the ability to compare bits.

Bit run cost equation

The drilling cost, C_d, or hole making cost, needs to be synthesized on the basis of individual bit run performances. Bit run cost may be expressed as follows:

$$C_{\mathrm{d}i} = \frac{C_{\mathrm{b}i} + C_\mathrm{r}\left(T_{\mathrm{d}i} + T_{\mathrm{t}i} + T_{\mathrm{c}i}\right)}{\Delta D_i} \tag{8.1}$$

where

$C_{\mathrm{d}i}$ is the drilling cost, in \$/ft, for bit run i
$C_{\mathrm{b}i}$ is the cost of bit run i, in \$
C_r is the rig cost, in \$/h
$T_{\mathrm{d}i}$ is the drilling time, in h, for bit i
$T_{\mathrm{t}i}$ is the trip time, in h, for bit i
$T_{\mathrm{c}i}$ is the connection time, in h, for bit i
ΔD_i is the formation interval drilled, in ft, by bit i

The bit cost and rental cost are reasonably known for a given well, and trip time can be closely estimated. The trip time can be closely estimated. This means that predicting operating cost for a particular selection and operation is determined by one's knowledge of the average penetration rate and bit life. Both the bit life and the ROP are governed by a combination of factors that are controlled by the driller, such as operating conditions, and factors that are beyond the driller's control, such as formation characteristics. Even though it is mathematically possible to optimize a drilling operation, most often, the inexactness of the parameters that are incorporated in the process gives inadequate overall accuracy. Consequently, bit selection and operation are usually based on offset well records. However, an understanding of the fundamental drilling process will aid the engineer in making better choices in the selection of drill bits.

Specific energy equation

The *specific energy* is defined as the amount of energy required in order to remove a unit volume of rock. Thus, for bit performance evaluation, the lower the specific energy is, the higher will be the bit performance. The equation used is

$$\text{Specific Energy} = \frac{4 \times \mathrm{WOB}}{\pi D_\mathrm{b}^2} + \frac{480NT}{\mathrm{ROP} \times D_\mathrm{b}^2} \tag{8.2}$$

where

WOB is the weight on bit, in lbs
D_b is the bit diameter, in in.
N is the rotary speed, in rpm
ROP is the rate of penetration, in ft/h
T is the rotary torque, in ft-lbs

Drill-off test

A common method used in the field to determine the optimum WOB and rotary speed is referred to as the drill-off test. It is generally conducted at the start of a new bit run, or when a new formation is encountered or a decrease in ROP under current bit operating conditions and mud density is observed. The test assumes that the drill string is a linearly elastic rod whose length varies with the amount of applied tension. With the hydraulics and rotary speed maintained constant, the general procedure for conducting the test is as follows:

- Apply the maximum allowed WOB
- Lock the break on the drum
- Maintain a constant flow rate and rpm
- Begin drilling
- Record the time to drill off an increment of WOB, normally 2,000–5,000 lbs
- Record the footage drilled between successive increments
- Conduct an additional test to make sure there was no change in lithology
- Plot ROP versus WOB on Cartesian coordinates
- Select the WOB that resulted in the highest ROP and repeat the test, varying rpm
- Plot drilling rate versus rpm and select the rpm that gave the highest ROP
- Use that weight and rpm for that bit

These data can be used to determine the WOB and rotary speed exponents that appear in the drilling-rate equation, which will be discussed later.

Types of Drill Bits

Drill bits may be categorized into two major groups:

- Roller cone bits
- Drag bits

Roller cone drill bits

Roller cone bits are classified as either milled tooth or tungsten carbide insert tooth. (See fig. 8–1.) Their design features consist of the following:

- Conventional jet and extended jet bits
- Roller bearings, sealed bearings, and journal bearings
- Application is in soft to very hard formations
- Gauge protection
- Small to large diameters

Fig. 8–1 Roller cone milled-tooth and insert bits (Courtesy Halliburton DBS)

Milled-tooth bits. Milled tooth bits are available in three series, selection among which is based on formation hardness:

- IADC series 1. Application in soft to medium formations
- IADC series 2. Application in medium to hard formations
- IADC series 3. Application in hard to very hard formations

The rock-cutting mechanisms of milled-tooth bits are by gauging (series 1), scraping (series 2), and chiseling (series 3). The failure mechanisms of milled-tooth bits are tooth wear, bearing failure, or both.

Insert drill bits. Roller insert bits are available in five series, numbered from 4 to 8. These bits are designed to drill into very hard formations. Their rock-cutting mechanism is by chipping/crushing. Their only failure mechanism is bearing failure.

Drag drill bits

Drag bits are classified as follows:

- Polycrystalline diamond compact (PDC) bits
- Thermally stable polycrystalline (TSP) bits
- Diamond matrix bits

Their design features consist of the following:

- Shape and metallurgy of head
- Size and shape of compacts, for PDC and TSP bits, and karat size, for diamond matrix bits
- Fluid passages
- Side and back rake angles of compacts

The failure mechanisms of PDC and TSP bits are cutters breakage and wear. For diamond matrix bits, the failure mechanism is loss of the diamond matrix.

Classification of Roller Cone Drill Bits

The International Association of Drilling Contractors (IADC) has developed a standard system for the classification of roller cone drill bits. The IADC system classifies roller cone bits as milled or insert tooth, according to degrees of formation hardness, common mechanical features, and manufacturer.

The series numbers designate a class of bits to drill a formation with the described characteristics. Numbers 1 through 3 refer to milled-tooth bits. Numbers 4 through 8 refer to insert bits.

The column marked "type" is a designation of formation hardness within the series described. Type 1 bits are to be used in the softest formation of the series. Type 4 is a bit for the hardest formation of the series.

Up to nine design features of modern milled-tooth and insert bits may be listed in the column marked "features":

1: Standard roller bearings
2: Roller bearings—air cooled
3: Roller bearings—gauge protected
4: Sealed roller bearings
5: Sealed roller bearings—gauge protected
6: Sealed friction bearings
7: Sealed friction bearings—gauge protected
8: [Reserved for future use]
9: [Reserved for future use]

The resulting three-digit numeric system designates series, type, and feature. In 1987, the IADC revised its classification system by adding a fourth character, to describe additional design features. Explanation of the IADC code for fourth character is as follows:

A: Air application (journal bearing bits with air-blast nozzles)
C: Center jet
D: Deviation control
E: Extended jets (bits with full extension jet nozzles [welded tubes with nozzles])
G: Extra gauge protection
J: Jet deflection
M: Metal face seal
R: Reinforced (for geothermal, motor, or percussion applications)
S: Standard model (standard-type steel-tooth bit with no additional features)
X: Chisel tooth
Y: Conical tooth
Z: Other tooth shape

The fourth character in the IADC code indicates special or additional design features of a particular bit type. For example, a 114S (standard model) would be designated as a 114E if extended nozzles were added. A 517X (chisel tooth) and a 517Y (conical tooth) are distinguished by the fourth character, indicating a difference in tooth shape. A deviation control bit would be distinguished by the fourth character "D"—for example, as a 517D.

When necessary, two or more letters can be used to indicate the special feature. For example, 517EY would indicate a bit with extended nozzles. The major bit feature is always indicated first.

Drill Bit Operating Parameters

The important parameters that affect the optimum performance of drill bits are the WOB, the bit rotary speed, and the hydraulics. To induce rock failure, as will be discussed later, the force applied by the bit to the formation must be higher than the strength of the rock. Most rocks fail in direct proportion to the force applied. It is agreed that an increase in WOB and rotary speed will result in an increase in ROP, assuming that the bit hydraulics level is proper. In fact, if the level of hydraulics is not high enough, a decrease in ROP can result in increased rotary speed and WOB. In addition, too high WOB and rotary speed can decrease the bit life and result in higher dollars-per-foot cost. Extensive theoretical work and field tests have helped determine the best combinations of WOB and rotary speed with the proper hydraulics that can lead to the lowest drilling cost. In general, it has been found that moderate weights and rotary speeds are best for fast drilling of nonabrasive formations, whereas heavier weights and lower rotary speeds are better for the slow drilling of hard formations.

Grading of Dull Drill Bits

The grading and evaluation of dull bits is extremely important in the overall drilling process. Grading is important so that any unusual wear is noted and appropriate precautions are taken on the next bit program design.

The key reasons for grading dull bits are

- To improve bit type selection
- To identify those operating conditions (weight, rpm, hydraulics, stabilization, mud, etc.) that can be changed to optimize performance
- To determine the break-even cost to get the maximum use of the drill bit

The IADC has devised grading systems for roller cone and drag bits. The IADC system for grading roller cone bits is described in the following section.

Roller cone drill bit grading system

- Tooth dullness (see fig. 8–2)
  - T1: Tooth height ⅛ worn out
  - T2: Tooth height ¼ worn out
  - T7: Tooth height ⅞ worn out
  - T8: Tooth height all worn out
- Expanded bearing life
  - B1: Bearing life used is ⅛
  - B2: Bearing life used is ¼
  - B7: Bearing life used is ⅞
  - B8: All life gone (locked or lost)
- Gauge condition
  - I: Bit is in gauge
  - O: Bit is out of gauge, followed by amount of out of gauge in fractions of an inch

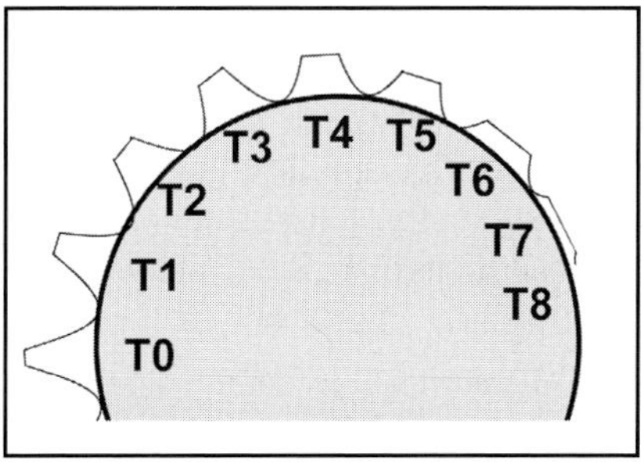

Fig. 8–2. Schematic of tooth wear

Inserts that are broken flat at the root or below the surface of the land usually indicate that the torque limit has been exceeded. Inserts that are chipped or broken at an angle longitudinal to the axis of the insert or are shaped in an erratic manner usually indicate that the impact load limit has been exceeded.

Classification of Drag Drill Bits

The IADC-adopted classification system for drag bits (diamonds, PDC, TSP, etc.) is similar to that for roller bits. The earlier three-digit code for roller cone bits was replaced with a four-character format. The new identification system describes the bit design features, rather than the drilling application.

The first character code describes the primary cutter type and body material:

D: Natural diamond—matrix body

M: PDC—matrix body

S: PDC—steel body

T: TSP—matrix body

O: Other

The profile (numbered 1 through 9) in the second character code refers to the cross-sectional profile of the fixed-cutter bits. The number 0 is used for a bit profile that cannot be described by the 3×3 matrix.

The numbers 1 through 9 in the third character code refer to the hydraulic design of the bit. The hydraulic design is described by two components: the type of fluid outlet and the flow distribution. The 3×3 matrix is of the orifice type, and flow distribution defines the nine numeric hydraulic design codes.

The numbers 1 through 9 and 0 in the 4th character code of the fixed-cutter classification system refer to the cutter size and placement density on the bit.

Definitions

Cutters. The term cutters describes the diamond materials affixed to the face of the bit. Either synthetic or natural diamond cutters are used to drill the formation.

Apex. The apex is the geometric center of the bit.

Profile. The profile refers to the distinctive shape of the bit when viewed from the side. The bit head is referred to as the crown. The basic objective of a profile is to optimize bit stabilization and achieve wellbore trajectory objectives, such as dropping, building, or holding angle. Profiles have five basic components: shoulder, flank, nose, cone, and gauge.

Cone. The recessed central area of the bit is called the cone. The recessed cone makes more surface area available to be set with diamond cutters. It also provides additional stabilization for the bit to help prevent it from moving sideways. A 90° angle is considered relatively deep, and a 150° angle is considered relatively flat or shallow. Deeper cone angles provide more area for placing cutters but also increase the distance cuttings must travel from the apex to the rock surface.

Nose. The nose is the leading edge of the profile and is rounded. Two aspects of the nose are important: the nose radius and the nose location. The radius determines the degree of sharpness of the nose. Nose location describes the distance of the tangent point on the nose radius from the bit centerline.

Flank. Where the outside diameter of the bit begins is called the flank or taper. Some flanks are flat, resulting in a relatively sharp corner at the gauge and nose. Some fixed-cutter bits use a parabolic flank to provide roundness.

Shoulder. The shoulder provides a smooth transition from the vertical gauge into the flank. The shoulder is often rounded to provide increased profile strength, resulting in greater cutter density and greater wear resistance.

Gauge. The fundamental purpose of the gauge is to maintain hole size. Standard fixed-cutter bit models have predetermined lengths, for optimum efficiency. However, the length of the gauge can be varied depending on the needs of the application. For example, extending the gauge length provides more wall contact for improved deviation control. In contrast, shorter gauge lengths are more advantageous for directional drilling and mitigation of the bit-whirling effect.

Profile length. The total profile length of a bit results from its shape. Longer profile lengths will produce greater facial surface areas for the placement of cutters and hydraulic layouts or nozzles. They also increase the area to be cleaned. Shorter lengths are generally less stable but reduce the distance drilled cuttings must travel to reach the annulus.

Profile shape. The profile shape is the side view configuration of the bit. It is one of the most important characteristics of every fixed-cutter bit. The profile shape has a direct bearing on typical cutter densities, cutter placement patterns, hydraulic layouts, and bit stability during drilling. Directional characteristics, WOB considerations, and bit durability are additional factors related to the shape of the profile.

Upper section. The upper section of a fixed-cutter bit and a roller cone bit are similar. This section of the bit includes the shank, the breaker slots, and the pin connection. For fixed-cutter bits, the shorter the shank is, the less whirling motion of the bit will result.

Fluid courses. Fluid courses provide a patterned outlet for the drilling fluids to exit bit nozzles and flow across the face of the bit. They are designed to make optimum use of the available hydraulic energy, for maximum bit cleaning and cooling.

Void areas. Void areas are designed into fixed-cutter bits to enable the drilled cuttings and drilling fluids to move freely across the bit face to the gauge and up the annulus.

PDC drill bits

The PDC bit represent a new generation of drill bit that drill into rock by means of a shearing action. The concept was developed by General Electric in the early 1970s, and engineering design development and testing was conducted by the University of Tulsa Drilling Research Projects, as subcontractors for the United States Department of Energy. The diamond compact is a polycrystalline layer is about $\frac{1}{32}$ in. thick, mounted on a tungsten carbide slug that is inserted into a predrilled hole in the bit body. (E.g., see fig. 8–3.)

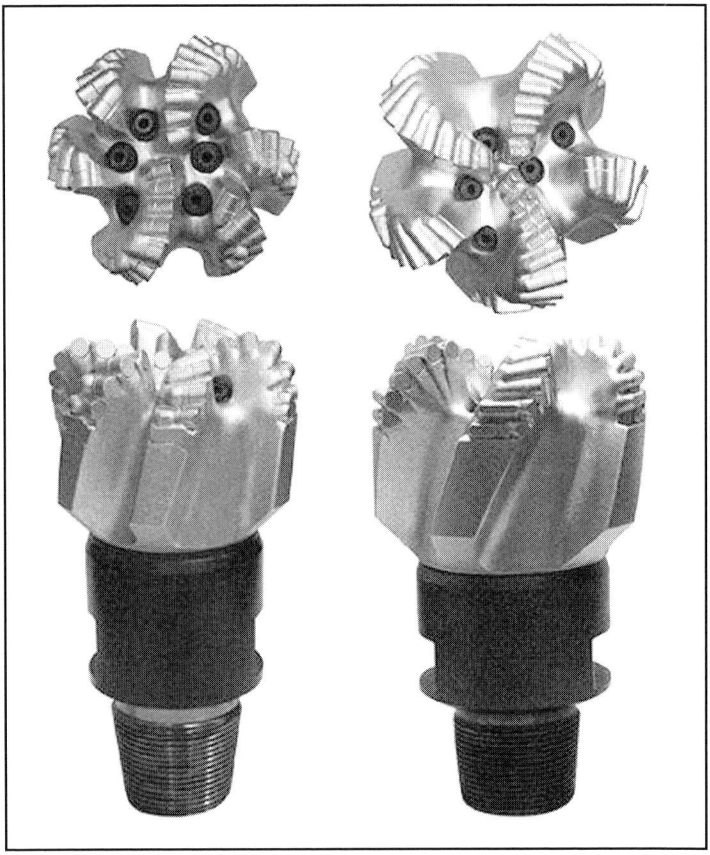

Fig. 8–3. Typical PDC bits (Courtesy Halliburton DBS)

Cutter shapes and sizes. PDC cutters commonly come in round shapes. Different vendors promote variations of this round shape for specific applications. The larger size cutter (greater than one inch) provides larger formation cuttings for improved cleaning in soft formations where bit balling is anticipated. The smaller cutter promotes longer bit life in medium-soft and medium-hard formations.

Cutter density. Since PDC cutters are much larger than natural diamonds, far fewer of the former are needed on a bit of comparable size. The terms light, medium, and heavy are used to describe the various densities available for PDC bits. With PDC bits, density is a function of profile shape and cutter size.

Bit profile. Bit profile affects cleaning and stability of the hole and gauge protection. Five profile types are commonly associated with PDC bits:

- *Ballistic.* This is a long, parabolic profile with a shallow cone, a short nose located close in, and a long curved flank blending into a long shoulder. The ballistic profile is applicable in soft homogeneous sticky rocks. This profile generally requires higher rpm for optimum application.

- *Shallow cone.* This is a shorter profile with a shallow cone. The nose is broad, resulting in a short shoulder and no flank. The shallow cone profile provides the greatest drilling versatility. It is applicable in transitional, interbedded, or homogeneous intervals of soft, medium-soft, and medium-hard formations.
- *Deep cone.* This profile is characterized by a deep cone, a wide nose location, and a short shoulder. The deep cone is a special-application profile designed to enhance stabilization necessary for deviation control in dipping formations. It is applicable to transitional or interbedded medium-soft formations.
- *Long taper.* This profile is characterized by a medium-depth cone, a shorter nose radius and closer nose location, and a long, straight flank. The long taper is applicable in homogeneous or interbedded intervals of medium-hard formations.
- *Near flat.* This profile has a shallower cone angle, a very wide nose location, and almost no flank or shoulder. It is a special-application profile used for sidetracking operations.

Gauge protection. Gauge protection is provided by carbide inserts near the edge of the bit and/or diamonds, depending on the application.

Compact cutters. Cutters are produced in the following shapes: standard cylindrical, chisel or parabolic, and convex.

Cutter placement. Cutters are purposely arranged so as to achieve optimum cleaning and cooling. The following are typical cutter placements:

- *Bladed.* In a bladed placement, cutters are positioned in a straight line from the apex to the gauge. The number of blades will vary with model and size. The advantages of a bladed cutter placement include better hydraulics, good shear strength, and effectiveness in sticky formations.
- *Ribbed.* This is a combination-style pattern with several blades interspersed with shorter placements that begin out from the apex. Ribbed design provides good hydraulics for cleaning of the bit face and removal of drilled cuttings. This pattern also provides for greater cutter densities than other placements.
- *Synchronous.* In a synchronous cutter placement, each cutter pocket stands independently of all the others. This placement allows cuttings to swirl off each cutter face without riding down a rib or a blade. This approach is relatively resistant to abrasion but is a poor alternative for sticky applications.
- *Back rake angle.* Each PDC cutter is positioned on the bit so that the top of the cutter leans backward at some optimum rake angle. A back rake angle helps reduce the impact force on the cutter and increases wear resistance.
- *Side rake angle.* When ribbed or synchronous cutter placements are used, the PDC cutters are also side raked. When side raked, the cutters are placed on the bit in a way that mechanically induces the cuttings to move outward. Side rake is most necessary in medium-soft and medium formations where balling tendencies are anticipated.

Drag drill bit wear modes

Whether synthetic or natural, all diamond materials will eventually wear. The amount of wear is directly related to the amount of frictional heat at the cutters. As formation hardness increases, frictional heat and diamond wear increases. Having a better understanding of how and where the various diamonds will wear gives the engineer the ability to anticipate and modify which cutter should be chosen for the specific application. Fixed cutters on drag bits wear differently from the teeth of roller cone bits, and under the right circumstances, they can last significantly longer.

The following wear modes on PDC cutters are caused by heat:

- *Heat checking.* A checkerboard pattern of cracks in the tungsten carbide substrate is a form of heat checking.
- *Spalling.* Spalling, caused by heat and/or cutter impact, is the separation of a partial thickness of the total diamond layer from the face of the substrate. It is visually identified by the presence of a rough area, usually along the top portion of the diamond layer.
- *Delamination.* Overheating can cause entire layers of diamond to come off the cutter. In the worst case, gross delamination, the whole diamond layer separates from the substrate. This condition can be visually identified.
- *Worn cutter.* Despite a high abrasive resistance, PDC cutters will eventually wear away. Individual or small groups of crystals will be worn off. This is sometimes referred to as micro-chipping. It is also possible that a strong impact will break a PDC cutter.

TSP drill bits

TSP bits are man-made polycrystalline structures consisting of a multitude of synthetic diamond crystals sintered together at very high temperature and pressure. They are resistant to thermally induced diamond lattice breakout.

Benefits associated with TSP bits include the following:

- The TSP material drills many of the same formations as natural diamonds.
- The crystalline structure is radial and random. As a result, microchipping of the material continually exposes fresh diamond as it drills, providing a self-sharpening effect.
- There are no cleavage planes to fail by gross fracture.
- Since the material is synthetic, it can be engineered in a number of exact geometric shapes and sizes.
- TSP cutters have a greater thickness of synthetic diamond for longer life in tougher drilling.

Shapes and sizes. The most commonly used shapes are cube, disk, and cylinder. For TSP material, as with natural diamonds, size is determined by weight.

Cutter density. When referring to TSP bits, the terms medium and heavy are used to describe cutter density.

Cutter placement. TSP cutters are generally set into a ribbed cutter placement. The ribbed cutter placement provides balanced bit cleaning and cooling to provide optimum drilling efficiency in medium, medium-hard, and hard formations.

Wear mode. Because TSP cutters are also polycrystalline, the wear mode is a gradual wearing of the cutter (micro-chipping), leaving fresh diamonds for cutting. TSP cutter fracture does occur in some cases, but their high thermal stability precludes delamination failure.

Diamond matrix drill bits

Diamond matrix bits have been in use for many decades, since well before the new generation of PDC and TSP bits were introduced into the drilling industry. They consist of a solid steel head impregnated with a large number of small single diamonds as cutters, as shown in figure 8–4. The application of matrix diamond bits begins when the economics of drilling with insert bits in very hard formations ceases. The following drilling conditions justify the use of diamond matrix bits over roller cone insert bits:

- Very short roller cone insert bit life
- Very low ROP (less than 5 ft/h)
- Deep and small-diameter hole
- Limited WOB
- Downhole motor application possible

Fig. 8–4. Typical natural diamond bit (Courtesy Baker Hughes)

Rock Mechanics

To understand the performance of a drill bit, knowledge of the material being drilled is necessary. Rocks have a number of properties that affect the performance of a bit, such as porosity, permeability, matrix composition, strength, and ductility. One of the most significant characteristics of rock is that its properties are strongly dependent on *in situ* stresses. Field experience has demonstrated that the ROP (i.e., drilling rate, in ft/h) decreases as depth increases. The stress environment that exists at the bottom of a well must be defined, and how these stresses affect the rock drillability must be examined, to provide an understanding of the process. A brief discussion on rock failure mechanics is presented in the following section.

Rock failure

The strength of rocks is generally determined by testing cylindrical specimens in a compressive loading frame (fig. 8–5), where an axial compressive load may be applied at a controlled rate, along with confining pressures. The compressive load may be increased until the rock fails, while keeping confining pressures constant. Under this type of loading, the induced stresses on the rock are principal maximum, intermediate, and minimum stresses and are denoted by σ_1, σ_2, and σ_3, respectively. As the load on the rock increases, the rock will deform (strain) before it fails. Figure 8–6 shows a schematic of the three types of stress-strain curves that are obtained when testing rocks in compression.[1] Curve 1 is an example of a typical brittle failure. The strain varies almost linearly with applied stress until the rock breaks. Typically, the failure under these conditions will be a along a single plane inclined about 30° to the maximum applied stress. This type of failure occurs when the minimum principal stress σ_3 is low.

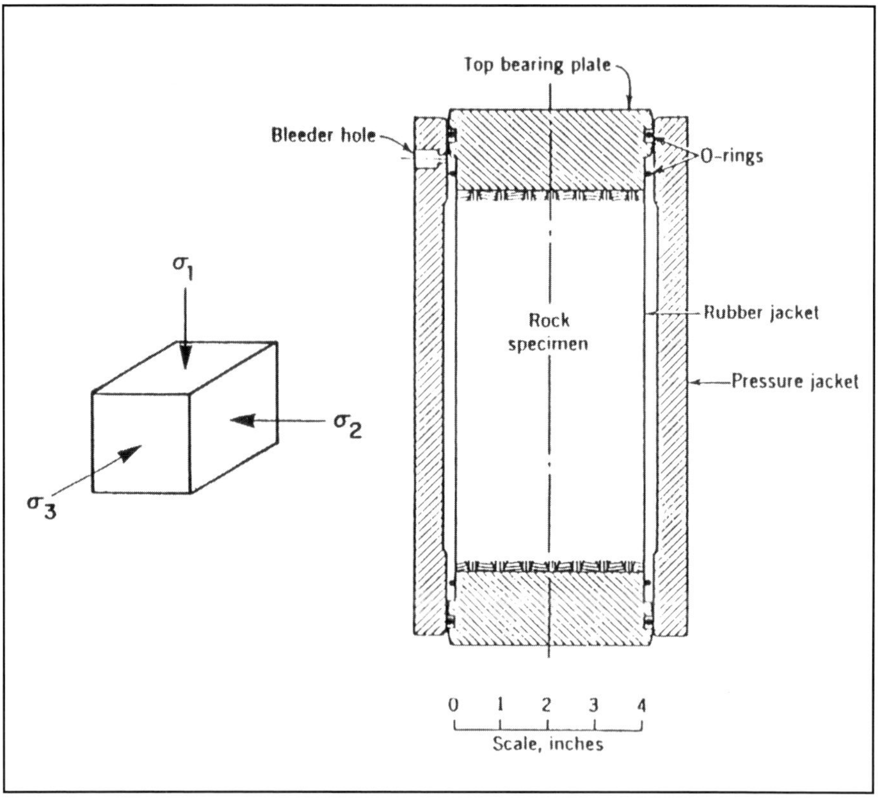

Fig. 8–5. Schematic of compressive strength determination test cell[2]

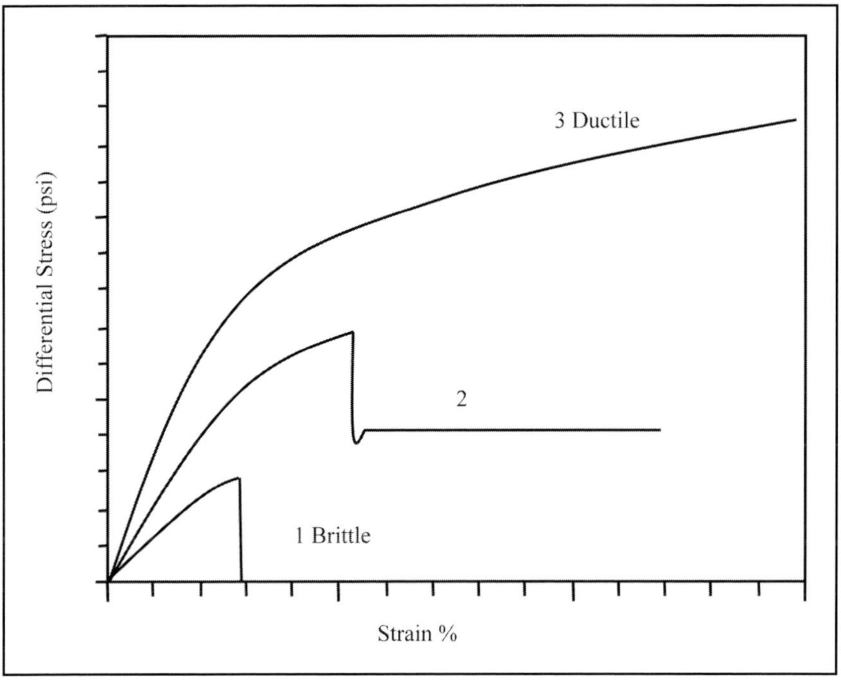

Fig. 8–6. Typical stress-strain curves for rock failure

As the minimum principal stress is increased by increasing the confining pressure, failure occurs along a zone of rock inclined about 30° to the maximum principal stress. The failure zone will appear to encompass a series of parallel failures, rather than a single plane. Curve 2 is typical of this type of failure. Both the stress and strain at the failure point for curve 2 are greater than those for curve 1.

As the minimum principal stress is increased even further, there will no longer be any distinct failure point. In this case, the rock can strain a considerable amount, and there will be little resistance to the applied stress. This response is referred to as plastic behavior. As the effects of stress and strain are discussed, it should be kept in mind that the increased stresses are caused by the teeth penetrating the rock sufficiently to cause chipping. On the basis of figure 8–7, an analogy can be drawn between the respective stresses caused by the compressive strength test and by tooth penetration.

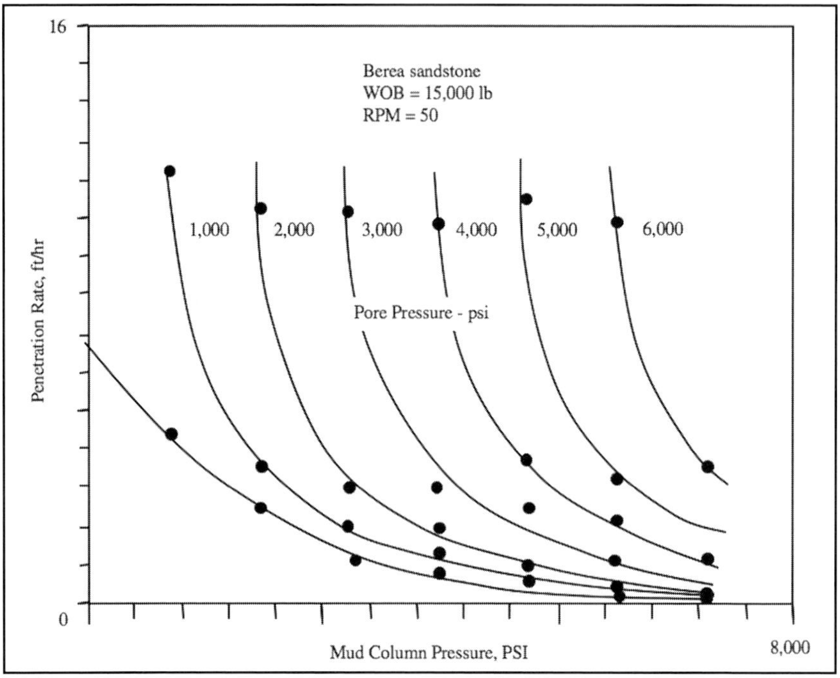

Fig. 8–7. Stresses in compressive strength test and tooth penetration

Failure envelope

If the compressive stress applied on a rock at failure is plotted versus the confining pressure, the resulting curve is referred to as the *failure envelope*. On the basis of this information, rock strength at a given depth can be predicted. A model that is commonly used to this end is the Coulomb-Mohr model of failure. The model is referred to as Mohr's envelope and describes a failure that occurs along a single plane of shear. In many cases, it seems to predict adequately the failure mechanism for rocks. A basic assumption of this model is that the intermediate principal stress has no effect on the failure strength. Principal stresses are plotted on the abscissa, and shear stresses are plotted on the ordinate. The failure envelope is defined as the line tangent to circles drawn through the minimum and maximum principal stresses at failure, for various confining pressures. The failure envelope can be generally be described by

$$\tau = c_0 + \sigma_{min} \tan\mu \quad (8.3)$$

where

c_0 is the cohesive strength
μ is the angle of internal friction
σ_3 is the minimum principal stress
τ is the shear strength

Pore pressure effects

The pore fluid pressure in rocks has a significant effect on the failure strength of rock because it supports part of the load that would normally be carried by the grain in contact if there was no pore pressure. This effect is generally accounted for by utilizing the principle of effective stress, which is defined as the principal stress minus the pore pressure.

Rock mechanics tests on porous rock can be classified as either *drained* or *undrained* tests. Drained tests are conducted by maintaining the pore fluid at a constant pressure throughout the test. Undrained tests are more representative of an impermeable rock, because the pore fluid is locked into the pore spaces. When porous rocks are stresses at levels below the failure point, the pore volume changes. Gertsma[3] has described this change with the following equation:

$$\frac{\Delta V_{\mathrm{p}}}{V_{\mathrm{p}}}=\frac{1}{\phi}\left[C_{\mathrm{c}}-C_{\mathrm{mc}}\right]\Delta\sigma+\frac{1}{\phi}\left[C_{\mathrm{bc}}-(1+\phi)C_{\mathrm{mc}}\right]\Delta p \qquad (8.4)$$

where

ϕ is the porosity
C_{bc} is the bulk rock compressibility
C_{mc} is the rock matrix compressibility
Δp is the pore pressure change
$\Delta\sigma$ is the mean stress: $\Delta\sigma = (\sigma_1 + \sigma_2 + \sigma_3)/3$

The change in pore volume given by equation (8.4) will cause a change in pore pressure for impermeable rocks. The magnitude of the pore pressure change for a given pore volume change depends on the compressibility of the pore fluid. The mathematical definition of compressibility can be combined with equation (8.4) to give the following equation for the change in pore pressure:

$$\Delta p=\frac{(1/\phi)(C_{\mathrm{bc}}-C_{\mathrm{mc}})\Delta\sigma}{C_{\mathrm{wc}}+(1/\phi)\left[C_{\mathrm{bc}}-1(1+\phi)C_{\mathrm{mc}}\right]} \qquad (8.5)$$

where C_{wc} is the pore water compressibility.

Bottom-hole stress evaluation

When drilling into a rock, the stresses at the hole bottom are altered because the borehole fluid pressure is less than the overburden stress that originally acted on the rock. In general, both the minimum principal effective stress and the mean effective stress are reduced. This results in a change in the strength of the rock that is exposed to the bit. To predict the strength of a particular rock, it is necessary to determine the local stresses and the pore pressures that are immediately adjacent to the drill bit teeth.

The maximum principal stress at any depth below the earth's surface is usually assumed to be the weight of the overburden and is approximately equal to 1 psi/ft. A value of 0.7 psi/ft is a reasonable value for the minimum principal stress. In areas of strong thrust faulting and for formations with high Poisson's ratio, the minimum principal stress will be greater than 0.7 psi/ft.

In areas with normal pore pressure, a pore pressure gradient of 0.465 psi/ft is usually used. This value can range up to approximately 1 psi/ft in cases of extremely abnormal pressure.

Figure 8–8 shows contours of the mean effective stress around a borehole, representing a typical permeable sandstone at 10,000 ft when the borehole pressure equals the *in situ* pore pressure. The local pore pressure is assumed to remain equal to the in *situ pore* pressure to simulate a permeable formation. The constant pore pressure assumption is valid because any change in local pore volume will be offset by the large volume of fluid in the interconnected pores. The mean effective stress for this case is reduced from 3,300 psi, away from the borehole, to 0 psi, near the surface of the bottom of the hole. The altered stress zone is generally limited to a distance less than one borehole radius from the borehole surface. The three principal stresses 0.1 in. below the borehole surface are shown in figure 8–9.

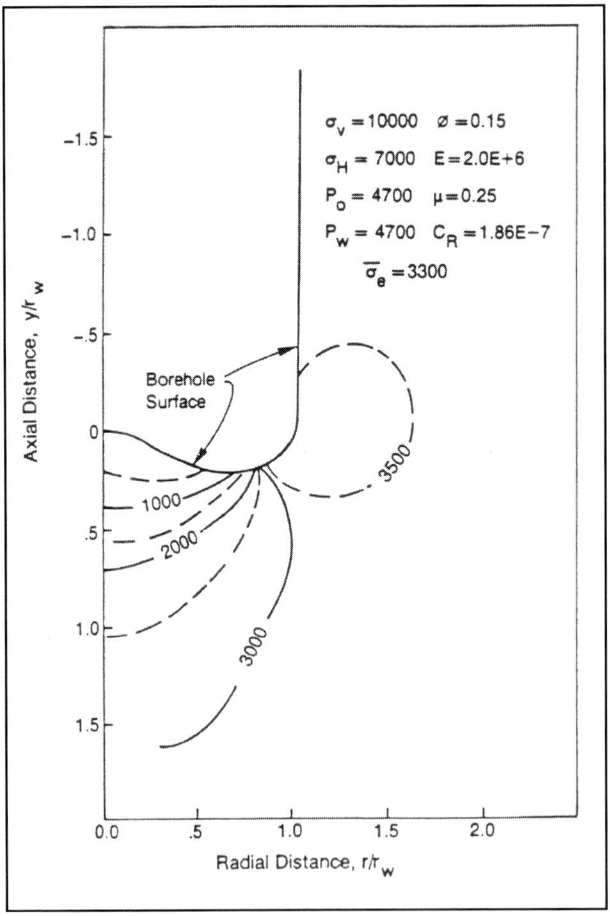

Fig. 8–8. Mean effective stress around a borehole in permeable sandstone[4]

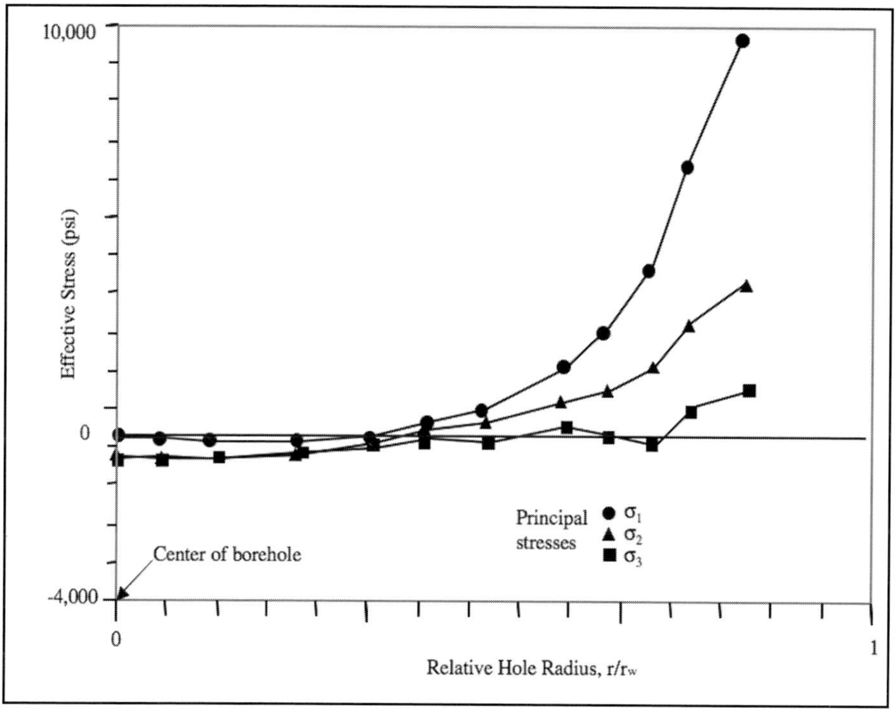

Fig. 8–9. Principal effective stress across the bottom of a 10,000 ft borehole in permeable sandstone[5]

Effect of local stress on drilling rate

In general, the minimum principal stress for most practical drilling situations is perpendicular to the borehole bottom. It is equal to the differential pressure, which is the difference between the borehole pressure (mud pressure) and the local pressure. The other two principal stresses (confining pressure) are parallel to the borehole surface. As previously stated, the strength of a rock is largely controlled by the minimum principal stress. Thus, for a permeable rock, the increase in strength owing to depth is caused primarily by the difference between wellbore pressure and *in situ* pore pressure.

Single-tooth impact tests have been reported by Maurer[6] for which the crater volume was measured at various stress conditions. As the borehole pressure increases above the pore pressure, the crater volume is significantly reduced. When the borehole pressure and the pore pressure are held constant and the horizontal stress parallel to the rock surface is increased, crater volume remains constant.

Effect of local stress—roller cone drill bits. A roller cone bit is composed of cones that have numerous teeth that impact the rock as the bit rotates. Each of these teeth forms a small crater and removes some rock. To begin to understand the bit's performance, look at what happens when an individual tooth impacts the rock. Single-tooth penetration tests and theory indicate that the crater volume from a single-tooth impact is proportional to the square of the applied force and inversely proportional to the square of the rock strength. If there is no interference between the teeth and each tooth impact is independent, then

$$R \approx \frac{N^b W^2}{S^2} \tag{8.6}$$

where
- R is the ROP, in ft/h
- N is the rotary speed, in rpm
- W is the WOB, in thousands of pounds (kips)
- S is the relative rock strength
- b is a dimensionless exponent for the rotary speed effect on ROP

From dimensional analysis, the bit diameter effect is included, and the proportionality constant is added, to give

$$R = \frac{N^b W^2}{aS^2 D^3} \tag{8.7}$$

where
- D is the bit diameter, in in.
- a is the dimensionless bit constant

Equation (8.7) is not a realistic expression for the penetration, because it implies that there is no upper limit to ROP as WOB is increased. Limitations are imposed on the ROP because of considerations of both mechanical design and cuttings removal. As the teeth penetrate deeper into the formation, the load is distributed over more teeth. Assuming that the fraction of weight that is supported by adjacent teeth is proportional to the total depth of penetration and inversely proportional to the tooth length gives the following expression for ROP:

$$R = \frac{N^b \left[W - \left(cR / ND\right)W\right]^2}{aS^2 D^3} \tag{8.8}$$

where c is a dimensionless bit constant. Equation (8.8) can be simplified to

$$R = \left(\frac{aS^2 D^3}{N^b W^2} + \frac{2c}{ND} - \frac{c^2 R}{N^2 D^2}\right)^{-1} \tag{8.9}$$

For reasonable values of R, N, and D, the last term in equation (8.9) is negligible and the ROP becomes

$$R = \frac{aS^2 D^3}{N^b W^2} + \frac{c'}{ND} \tag{8.10}$$

where $c' = 2c$.

The first term in equation (8.10) describes the volume per bit revolution of the crater created by wedge-shaped teeth. It is supported by numerous single-tooth impact tests. The second term accounts for the increase in tooth area in contact with the rock as the tooth penetration increases and provides the maximum limit to tooth penetration. Figure 8–10 shows a simplified cross section of a bit cone. As a tooth penetrates deeper into rock, adjacent teeth begin to support the applied weight. The adjacent teeth may be on other rows of teeth, as well as on the same row of teeth (as shown in fig. 8–10). The general model given in equation (8.10) describes the drilling performance for both steel tooth and inserts bits when the appropriate bit constants are used.

Fig. 8–10. Variation of tooth contact with increasing depth of tooth penetration

Effect of hydraulics and mud on drilling rate

Under steady-state drilling conditions, the rate of cuttings removal under the bit must equal the rate at which new chips are formed. This implies that the ROP can be controlled by the process of cuttings generation or removal or by combinations of the two processes. The previous sections discussed the conditions for which the ROP is controlled completely by the process of cuttings generation; this section discusses conditions for which the ROP is partially controlled by the process of cuttings removal.

Parameters such as jet impact force, hydraulic horsepower, jet velocity, and jet Reynolds number have been used in attempts to quantify the effect that bit hydraulics has on ROP. All of these parameters refer to properties of the fluid at the time it exits the jet nozzle. It is assumed that these parameters also reflect the hydraulics energy at the bottom of the hole. There is no consensus on which parameter most directly governs bit hydraulic effects on ROP or on what the exact mechanism is by which hydraulics affects ROP.

The two parameters most commonly used to quantify the effect of hydraulics on penetration rate are jet impact force and hydraulic horsepower per square inch of bit area (HSI). Given a maximum allowable surface pump horsepower, standpipe pressure, and drill string geometry, there is a unique set of nozzles that maximize either hydraulic horsepower or jet impact force at the bit

It has been shown[7] that the peak impact pressure represents the hydraulic parameter that controls the effect that cuttings removal has on ROP. The peak impact pressure is proportional to the true impact force on the hole bottom; however, the true impact force (F_{jt}) on the bottom of the hole is not calculated by the normal impact force equation used in most hydraulics models, because of the energy loss between the nozzle and the bottom of the hole. An empirical technique was used to estimate the true impact force for various jets and flow conditions. The area available for fluid return from under a tri-cone bit is approximately equal to 15% of the total bit area. If α is the ratio of the jet velocity to the fluid return velocity, then α is given by

$$\alpha = \frac{V_j}{V_r} = \frac{0.15D^2}{d^2} \tag{8.11}$$

where
D is the diameter of the bit, in in.
d is the diameter of the jet, in in.

Since the bottomhole impact force is proportional to the peak impact pressure, an expression for true impact force can be found:

$$F_{jt} = f(\alpha)F_j \tag{8.12}$$

This expression was numerically evaluated for the data on the three jets to yield the following expression:

$$F_{jt} = (1 - {}_{a}{}^{-0.122})F_j \tag{8.13}$$

where F_j is the jet impact force, as given by equation (4.24) in the "Drill Bit Hydraulics" chapter.

Dimensional analysis was used to isolate a group of variables, consisting of the hydraulic parameter and mud properties, for incorporation into the penetration model. These factors were compared to experimental data until a suitable model form was obtained. This resulted in the following expression of the ROP:

$$R = \left(\frac{aS^2D^3}{N^bW^2} + \frac{b}{ND} + \frac{cD\rho_s\mu_p}{I_t} \right)^{-1} \tag{8.14}$$

The mud effect incorporated in model (8.14) includes only the effect that the drilling fluid has on bit and bottom-hole cleaning. This effect seems to be adequately represented by the plastic viscosity and the mud specific gravity.

Effect of efficiency of cuttings removal on drilling rate

The volume of new rock that is fractured by a tooth depends on the tooth geometry, rock properties, and the depth of tooth penetration into virgin rock. The force applied to the tooth for a particular situation determines the depth of tooth penetration. A bit tooth will always penetrate into the rock until the resisting force offered by the rock equals the force applied to the tooth. Since the bit is composed of multiple teeth with a rather complex geometry, the force on a particular tooth may change as the tooth penetrates deeper into the rock. If the differential pressure inhibits bottom-hole cleaning, the force applied to a particular tooth to penetrate virgin rock may be diminished by the buildup of a bed of cuttings. Examination of the total borehole pressure effect will be facilitated by dividing this into a local crater effect and a global cleaning effect.

Figure 8–11 shows a single tooth that has penetrated a rock until the resisting force equaled the force applied to the tooth. In this case, it is assumed that the original rock surface was clean and free of debris. When the pressure acting on the rock surface is low, the broken chips will not hinder further tooth penetration. When the differential pressure (minimum principal stress) is increased, the strength of the rock increases, as can be calculated by equation (8.3).

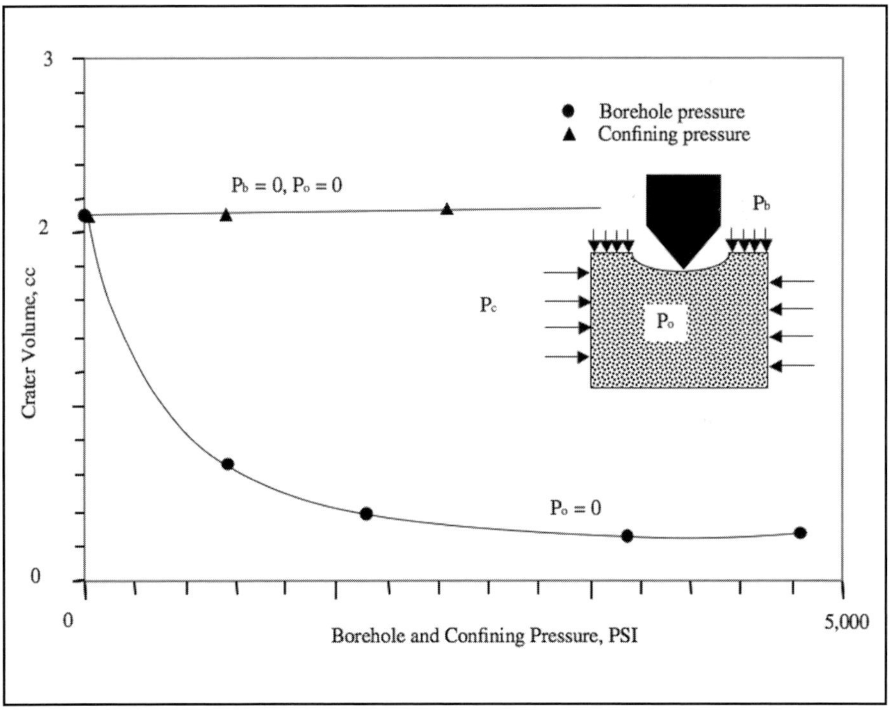

Fig. 8–11. Effect of normal and tangential stress on single-tooth crater volume[8]

The total force acting on the tooth can be estimated in a number of ways, but because of the complexity of the problem, an exact calculation of the force is not available. Methods based on the theory of plasticity[9] and brittle failure analysis[10] have been presented. Neither of these methods includes the effect that broken chips have on the tooth force. In fact, Maurer[11] has described plastic failure of rock as a series of brittle fractures with friction along the boundary.

Vertical tooth force[12] could be given as

$$F_{ubr} = c_0 d \frac{4\sin\theta \cos\mu}{1 - \sin(\theta + \mu)} \tag{8.15}$$

where

F_{ubr} is the force on a tooth
d is the depth of tooth penetration
c_0 is the cohesive strength
θ is the tooth wedge half-angle
μ is the angle of internal friction

Equation (8.15) is based on the assumptions that the shear stress all along the incipient crack exceeds the failure strength and that once the chip is formed, it no longer exerts a force on the tooth. When a differential pressure acts across the chips, an additional force resists the tooth penetration. The magnitude of the chip hold-down force can be calculated by assuming that a force equal to the product of the differential pressure and the chip area drives the chip against the unbroken rock.

Finally, the force resisting tooth penetration is given by

$$F_{br} = d' f \Delta p \frac{2\tan\mu \sin\theta}{\cos(\mu + \theta)} \tag{8.16}$$

The total tooth force, including the effect of the chips, can be expressed as

$$F_t = F_{ubr} + F_{br} \tag{8.17}$$

where F_t is the total tooth force; alternatively,

$$F_t = d[a(c_0 + \mu\Delta P_{ubr}) + bf\Delta P_{br}] \tag{8.18}$$

where

ΔP_{ubr} is the pressure differential across unbroken rock
ΔP_{br} is the differential pressure across chips
f is the coefficient of sliding friction
a, b are functions of tooth geometry and material properties

Since the crater volume is proportional to the square of penetration depth, the total crater volume can be expressed as

$$\text{Vol} \approx \frac{F_t^2}{\left[a\left(c_0 + \mu\Delta P_{ubr}\right) + bf\Delta P_{br}\right]^2} \tag{8.19}$$

The derivation of this expression is not rigorous and includes quite a few assumptions; further, it has utility in that it shows the functional form to be included in the ROP model to account for the effects that chip forces have on the penetration rate.

Global-cleaning effects on drilling rate

Figure 8–12 schematically shows the buildup of this rock debris into a considerable layer on the bottom of the borehole. The bit teeth tend to compress and squeeze water out of the crushed rock powder under each tooth. Once a tooth moves out of the crater, the crushed rock is still held in place by a pressure differential. The magnitude of the pressure differential is a time-dependent function of the compressibility and permeability of the compacted rock debris and fluid characteristics. The rock continues to build up until it reaches equilibrium, so that the rate of removal of the debris is exactly equal to the rate at which new chips are formed. This buildup of a bed of materials between the bit and virgin rock is referred to as a global-cleaning process. The turbulent energy of the hydraulics has a significant influence on the thickness of the bed at the equilibrium state. The thickness of the bed controls the ration of the total WOB that is applied to a particular tooth to form new crater volume.

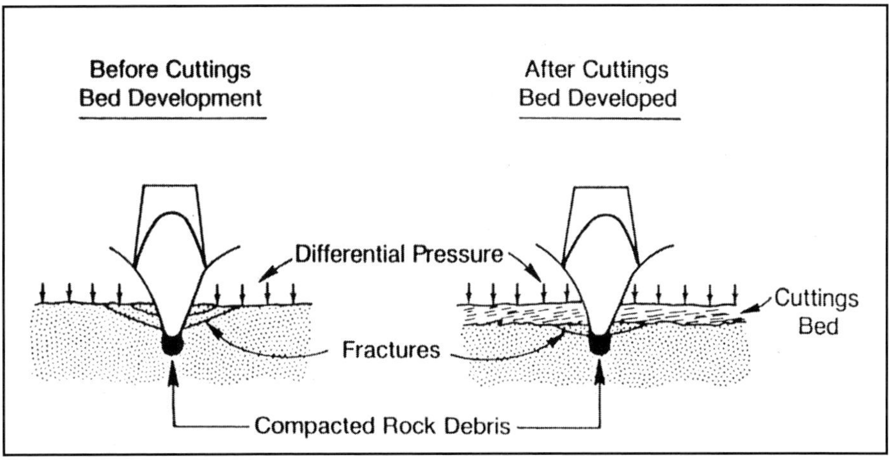

Fig. 8–12. Schematic of cuttings buildup under bit[13]

Performance Mechanics: Torque Estimation for Tri-cone Roller Cone Drill Bits

The torque required in order to rotate a roller cone bit is of interest for several reasons. First, it may give information about the formation being drilled and the condition of the bit. Second, bit torque exerts a significant influence on the bit walk experienced in directional wells. Finally, a prediction of bit torque is useful in matching a bit and a mud motor, for optimum performance.

A theoretical expression for the mean torque required in order to rotate a roller cone bit can be developed from a force balance of the bit. The torque applied to the bit shown in figure 8–13 is resisted largely by the bit face rolling resistance. A minor additional torque also results from bearing friction inside the cone.

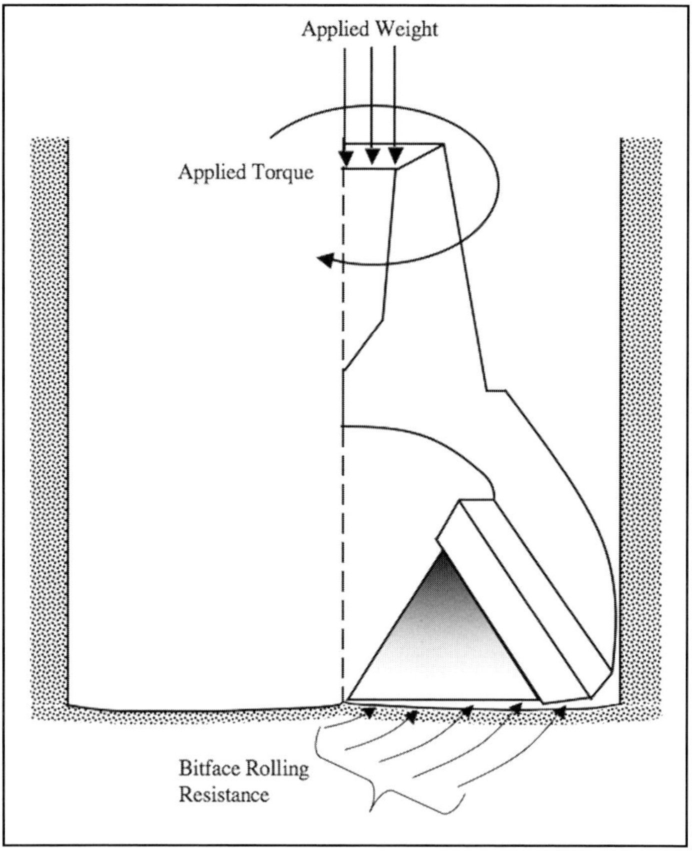

Fig. 8–13. Bit face rolling resistance to torque

Figure 8–14 is a diagram of the classical rolling resistance of a two-dimensional wheel.[14] When a vertical load F_V is applied through the center of the wheel, a force F_H is required in order to cause the wheel to roll at constant speed. The resistance to rolling is caused by a deformation of the wheel and/or the surface on which the wheel rolls. The contact reaction, *P*, in figure 8–13, is related to the surface deformation at the bearing and is initially assumed to be zero; thus, there is no moment about the wheel, and *P* acts through the center of the wheel.

The force F_H required in order to roll the cutter can be related to the applied vertical load, the tooth geometry, and the depth of tooth penetration by using the geometry shown in figure 8–13:

$$F_H = F_V\left(\frac{b}{c}\right) = F_V\frac{\sqrt{2ar - a^2}}{r - a} \tag{8.20}$$

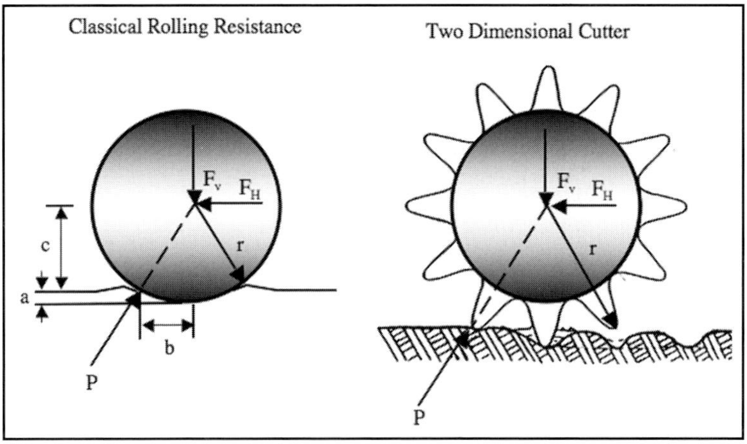

Fig. 8–14. Analogy between rolling resistance of wheel and two-dimensional cutter[15]

The teeth on a roller cone bit are arranged so that each row of teeth except the outermost row and cuts separate annular areas on the hole bottom. The resistance to rolling occurs at the teeth, because there is normally no contact between the cone shell and the rock. Three cones with discrete rows of teeth can be combined into a composite cone with continuous teeth; this simplifies the analysis. The total torque required in order to rotate the bit can be calculated by integrating the force acting on a circular cross section of the cone. If it is assumed that the maximum radius of the cone is r_{max} and d is the bit diameter, as shown in figure 8–15, then the radius of the cone as a function of the distance x along the borehole radius is

$$r = \frac{2r_{max}}{D} \quad 0 \le x \le \frac{D}{2} \tag{8.21}$$

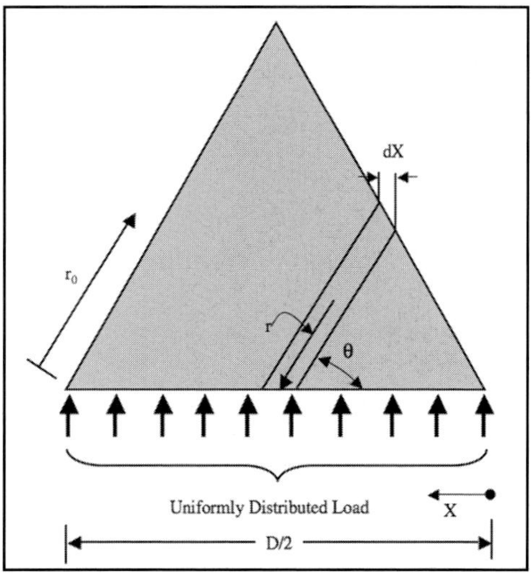

Fig. 8–15. Schematic of a uniformly loaded cone[16]

Under the assumption that the WOB is uniformly distributed along the single cone, the force dF_V acting on the element owing to the applied weight is

$$dF_V = \sin\theta \frac{2W}{D} dx \tag{8.22}$$

where W is the total applied WOB and $\sin\theta_{dx}$ is the differential element thickness. Combining equations (8.20) and (8.22) gives the force required in order to roll the element:

$$dF_H = \sin\theta \frac{2W}{D} \left[\sqrt{\frac{2a}{r} - \left(\frac{a^2}{r}\right)} \Big/ \left(1 - \frac{a}{r}\right) \right] dx \tag{8.23}$$

For most practical situations, the distance a will be much less than the cone radius, and equation (8.23) can thus be simplified as

$$dF_H = \sin\theta \frac{2W}{D} \sqrt{\frac{2a}{r}} dx \tag{8.24}$$

Tests[17] have been conducted with a single cutter wheel similar to the one shown in figure 8–14. A six-inch-diameter cutter with 36 teeth was rolled across Indiana limestone, and the horizontal and vertical forces, F_H and F_V, were measured for different depths of cut. The measurements were taken after several passes had been made, to establish an equilibrium surface condition.

For a vertical roller, test data[18] have showed that the horizontal force is proportional to the vertical force multiplied by the square root of the depth of cut. When the cutter was skewed from the direction of cut, the force relationship was offset by a constant. For a constant-diameter cutter, the horizontal force can be expressed as

$$F_H = F_V C_1 + C2 \sqrt{\delta} \tag{8.25}$$

The functional relationship expressed by equation (8.25) confirms the form given by equation (8.24) for a true rolling cutter. There is some sliding of the bit teeth on the formation because of the cone offset (0°–4°) and because the bit cone is not a true geometrical cone. Alternate sliding of various teeth has been discussed in greater detail by Fuji.[19] It is assumed that the effects of the bit tooth slippage, owing to both the cone offset and the varying cone angle, can be incorporated into the torque relationship by empirical constants used in equation (8.25). The total torque required in order to roll the cone about its apex is given by

$$T = \sin\theta \frac{2W}{D} \int_0^{\frac{D}{2}} \left(C_1 + C_2 \sqrt{\frac{2a}{r}} \right) dx \tag{8.26}$$

For a roller cone bit, the penetration per revolution, *R/N*, is equivalent to the depth of cut[20] and equivalent to *a* in equation (8.26). The cone radius, *r*, is assumed to be proportional to the bit diameter. Substituting these relationships into equation (8.26), integrating, and consolidating terms results in the following relationship describing bit torque:

$$T = \left(C_3 + C_4 \sqrt{\frac{R}{ND}} \right) WD \tag{8.27}$$

Performance Mechanics of Drag Drill Bits

The previous sections have discussed roller cone bits, for which the primary rock failure mechanism is crushing by teeth intermittently impacting the rock, more or less normal to the rock surface. Drag bits operate with the cutters continuously in contact with the rock and with the cutters moving essentially parallel to the rock surface. This provides a more direct shearing action for removing the rock.

Figure 8–16 shows two PDC bits. The cutters can be installed in either a tungsten carbide matrix body or a steel body. The cutters may also be installed on blades.

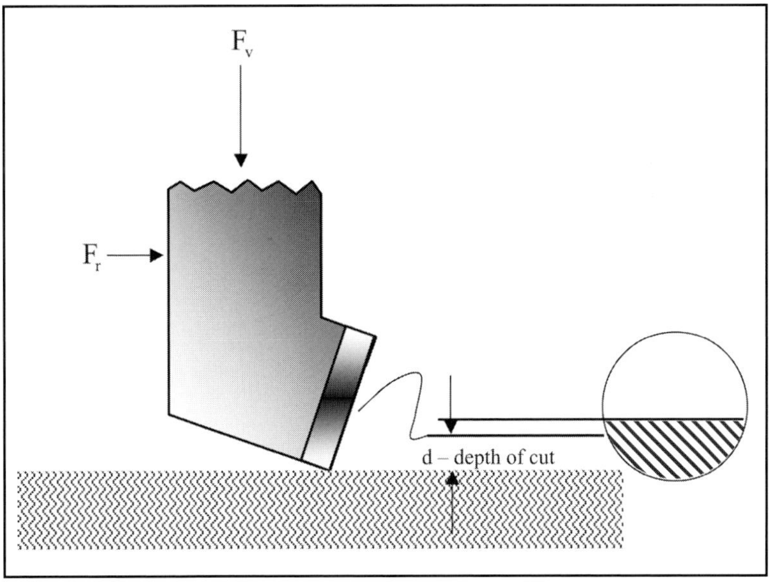

Fig. 8–16. PDC cutter engaged in rock

PDC bit design

The cutters on a PDC bit are designed to be continuously in contact with the rock and are expected to move in a simple spiral with a pitch equal to the penetration per revolution of the bit. Because of the relatively simple geometry, it is easy to relate the performance of single cutters to the performance of a bit composed of many cutters.

Figure 8–16 shows a single cutter that is taking a cut of depth *d* across a rock, corresponding to a cutting area *A*. A normal force of F_N is required in order to hold the cutter at depth *d*, and a cutting force F_c is required in order to move the cutter forward. Both the cutting force and the normal force depend on the area of cut, rock properties, the cutter orientation, and the geometry of the cutting edge. In general, the cutting force is about 50% of the normal force. The shape of the cutter has little effect on the forces as long as the area of the cut is constant, but the shape of the cutting edge has a large effect on the cutting force. Some manufacturers bevel the diamond edge to reduce the tendency for the diamond to chip. A beveled cutter requires about two times as much force for the same cutting area as a sharp-edged cutter.

A PDC cutter can be tilted backward with respect to its direction of motion (back rake) or canted sideways (side rake), as shown in figure 8–17. The cutters are generally installed on a bit with 15–30° back rake and little side rake. As the back rake angle increases, the cutter forces increase for a given area of cut, as demonstrated by Melaugh.[23]

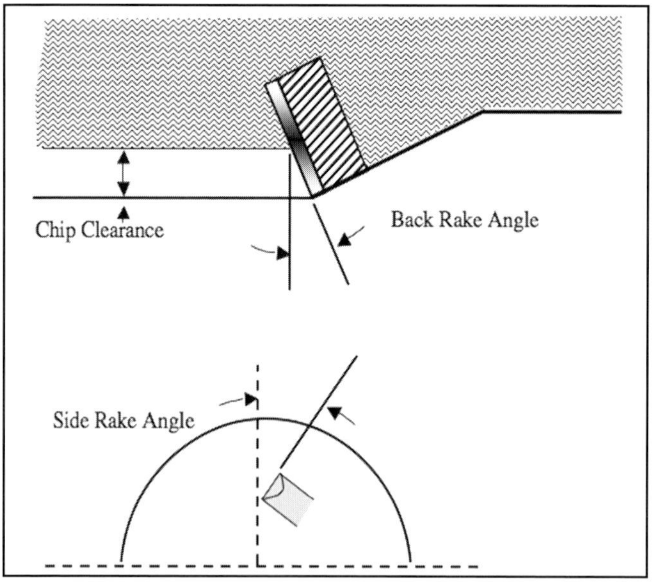

Fig. 8–17. Schematic of back rake and side rake

The design of a PDC bit can be graphically shown by use of two design drawings, referred to as the face view and the profile view. The face view shows the position of the cutters as if one were looking into the face of the bit, as shown in figure 8–18. The profile view shows the position of each cutter when it passes through a half-plane that includes the bit axis (e.g., see fig. 8–19).

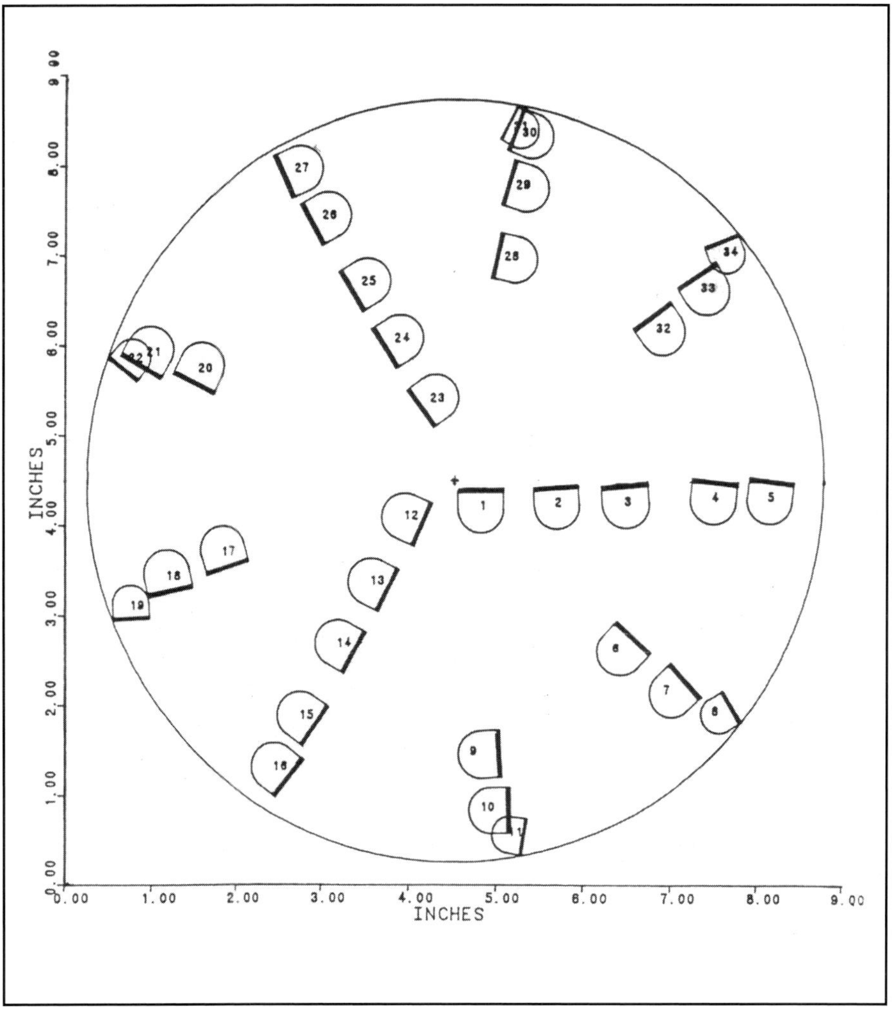

Fig. 8–18. Typical face view of PDC bit

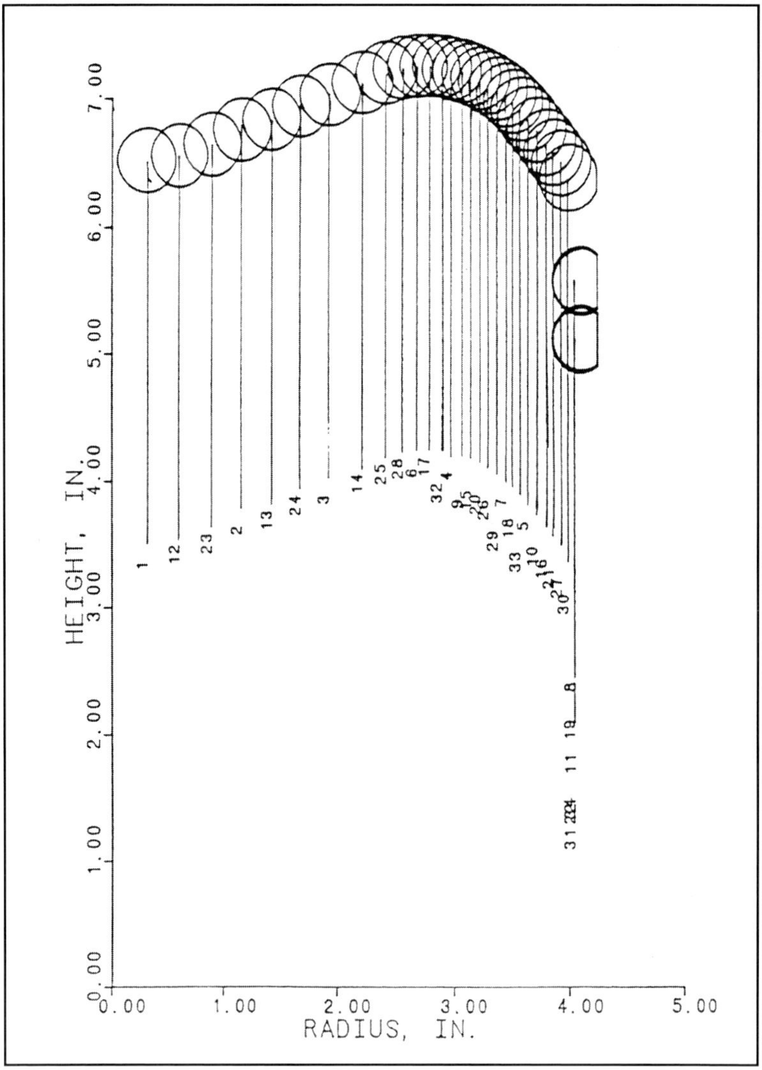

Fig. 8–19. Typical profile view of PDC bit

As shown in figure 8–20, the cutters do not generally cut on a horizontal plane. The normal force that is required in order to push the cutter into the rock at some angle from vertical (β) can be broken into a vertical component and a radial component. The cutting force is the circumferential component, and it is unaltered by the orientation. The sum of the vertical components for all cutters on a bit is the WOB, and the sum of the circumferential force times the radial position of the cutters is the bit torque.

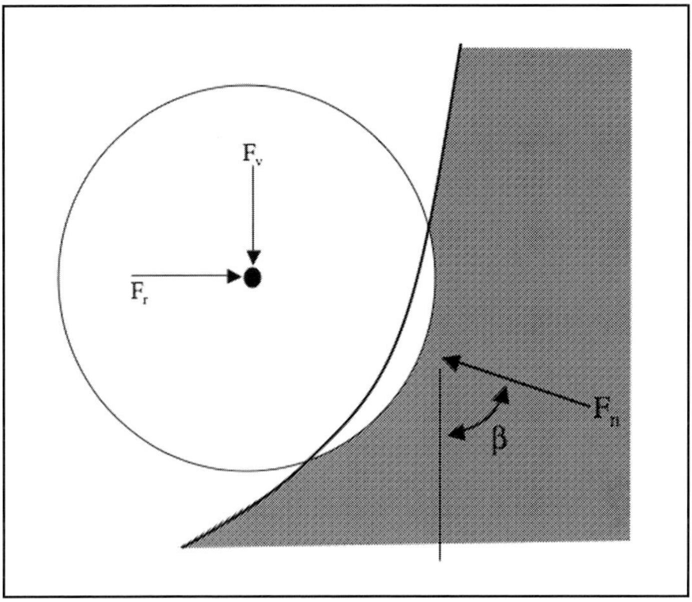

Fig. 8–20. Cutter forces for PDC cutter on inclined surface

In the design of a bit, the cutters must be positioned such that none are loaded high enough to break, so that they wear uniformly and so that the bit can be manufactured. In attempts to attain uniform wear, the criteria used have ranged from positioning of the cutters so as to achieve equal volume of rock removal for equal work to empirical positioning of the cutters, on the basis of observed wear.

In general, for sharp bits, there is little difference in the ROP achieved using various bit designs unless the bit balls up and vibrates. This has been demonstrated by removing cutters from a bit and observing that there is little change in the performance while up to 75% of the cutters are removed.[22] These observations apply only to a bit when it is sharp and do not apply when the bit is dulled, as discussed later.

Bit balling

Figure 8–21 shows the steps in chip generation for a PDC cutter while drilling a rock such as shale. When the first chip is formed, the pressure action on the top of the chip causes it to slide along the bottom of the hole, rather than be washed away. When the next chip forms, the first chip is stuck to the top of it and moves up the face of the cutter. This process continues until the stack of chips become unstable enough to be washed away. The tendency to form such stacks of chips is increased for fine-grained rock, low-strength rocks, high borehole pressure, and low-fluid-loss water-based muds.

If a bit is used to drill shale that forms chips, as described previously, and there is not sufficient clearance around the cutters, then the stacks of chips will be packed around the face of the bit until the ROP is severely reduced. This process of packing rock material around the cutters, referred to as bit balling, continues until the hydraulic energy is confined in such a small space that a critical velocity is reached and the balled material is washed away as fast as it is created. Further, the bit-balling process will progress to an equilibrium state. If the bit is cleaned and rerun under the same conditions, it will ball up in the same places, which is a function of the hydraulic design. For a given bit design, balling is a function of the flow rate.

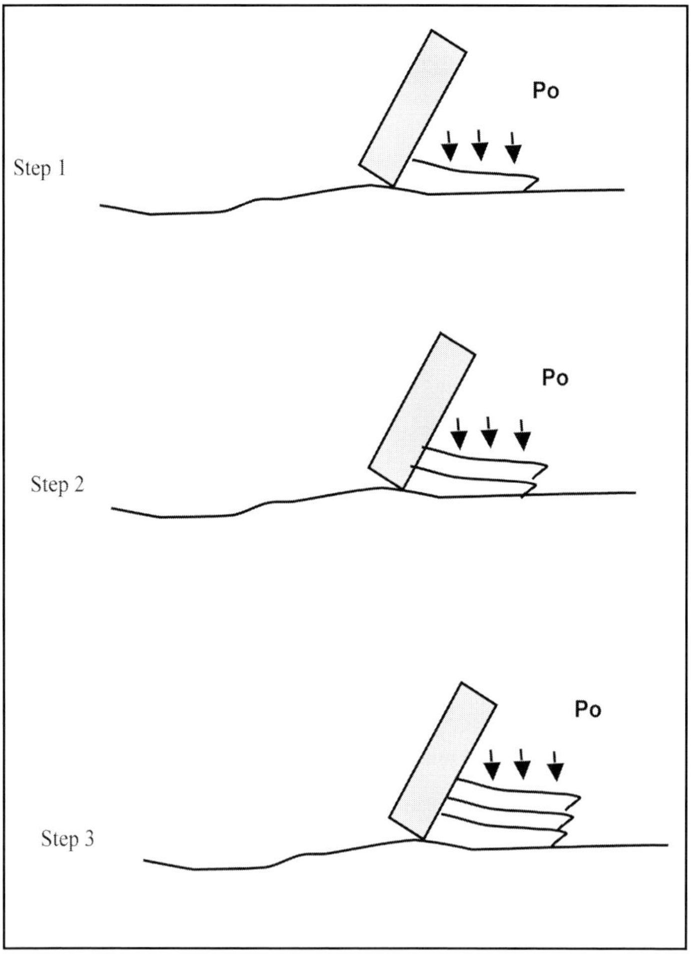

Fig. 8–21. Mechanism of cuttings generation

When drilling into a material that tends to ball PDC bits, the bit selection must be made appropriately to overcome balling. Bit designs that have large clearances above the cutters are much less susceptible to balling.

Bit wear

The diamond on PDC cutters is many times harder than the hardest rock that is drilled in normal oil and gas well drilling. When the bits were first developed, even though it seemed almost impossible, the cutters could sometimes be worn out after only a few feet of drilling. This accelerated wear is often attributed to the cutters becoming overly hot.

The PDC material generally contains small amounts of various metals located in the voids between diamond grains. Because of the differential thermal expansion of the diamond grains and the binder metals, heating of the cutters causes high stresses that makes individual diamond crystals easier to break away from the cutter. Additionally, at higher temperature, the diamond partially converts back into graphite, especially in an oxidizing atmosphere.

Much work has been undertaken to understand the thermal heating of PDC cutters. This work has resulted in an equation that determines cutter temperature

$$T_{wc} = T_f + \frac{K_f F_N \upsilon_c f}{A_w}\left(1 + \frac{3\pi}{4} f K_{hf} \sqrt{\frac{\upsilon_c}{\alpha_f L_w}}\right)^{-1} \tag{8.28}$$

where

T_{wc}, T_f are the wear-flat cutter and fluid temperatures, respectively
K_f is the friction coefficient between the rock and the cutter
υ_c is the cutter speed
L_w is the wear-flat length
F_N is the normal force on the cutter
F is a thermal response function
K_{hf}, α_f are the rock thermal conductivity and diffusivity, respectively

Equation (8.28) shows that the wear-flat temperature is a function of the normal force, the cutter velocity, the wear-flat area, hydraulic parameters, and the ambient temperatures around the cutter. This implies that the cutter temperatures will be higher for a high WOE and for higher rotary speeds, higher for cutters near the gauge of the bit, and higher for a given operating condition when the cutters are sharp. Often, when dulled PDC bits are pulled, thermal cracks in the carbide behind the diamond layer can be seen, indicating that the cutter did indeed become very hot.

Thermally accelerated wear is only one of the processes that are important in wearing away PDC cutters. PDC cutters are extremely strong when they are loaded with the diamond in compression but will chip and break at quite low loads if the diamond is loaded in tension. If a PDC bit rotates about its centerline, then the cutters will be loaded with the diamond in compression. Often, the cutters chip and break as if they are loaded in some other manner. Cutter breakage is a problem that has existed from the time when the very first prototypes of PDC bits were manufactured and is generally addressed by making the cutters, studs, and bonding processes stronger.

Drill bit vibrations, referred to as bit whirl, have been used to explain how cutters can move sideways and backward to cause the damage that is often seen with PDC bits. If the bit is rotated in a slightly overgauge hole and there is friction at the gauge of the bit, it will tend to rotate counterclockwise around the hole. When this occurs, the cutters may move in a path as shown in figure 8–22.

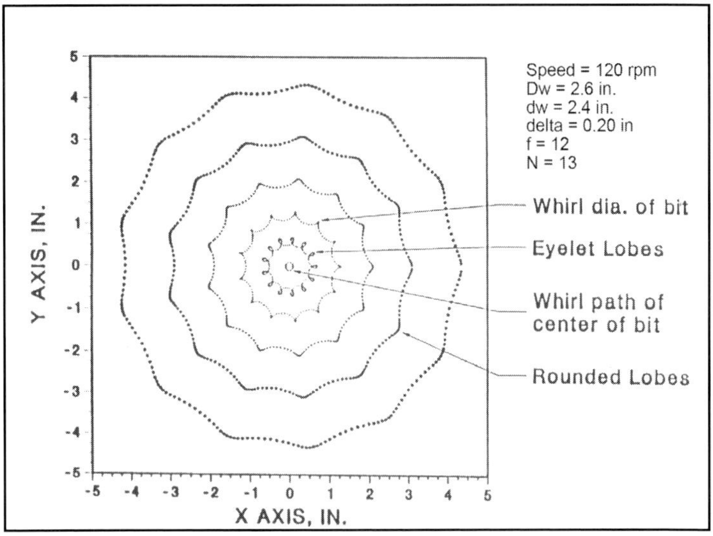

Fig. 8–22. Typical cutter paths on bit whirl

Using the geometry shown in figure 8–23, equations for the cutter velocity and the direction of motion relative to a line perpendicular to the cutter face can be derived. For the cutter velocity,

$$v_c = \omega\sqrt{r_c^{\,2} + R_w^2 - 2r_c R_w \cos\left(\frac{2R_w - r_c}{R_w - r_c}\right)\omega t} \tag{8.29}$$

For the direction of cutter motion,

$$\varsigma = \tan^{-1}\frac{R_w \sin\left[\left(2R_w - r_c\right)/\left(R_w - r_c\right)\right]\omega t}{r_c + R_w \cos\left[\left(2R_w - r_c\right)/\left(R_w - r_c\right)\right]\omega t} \tag{8.30}$$

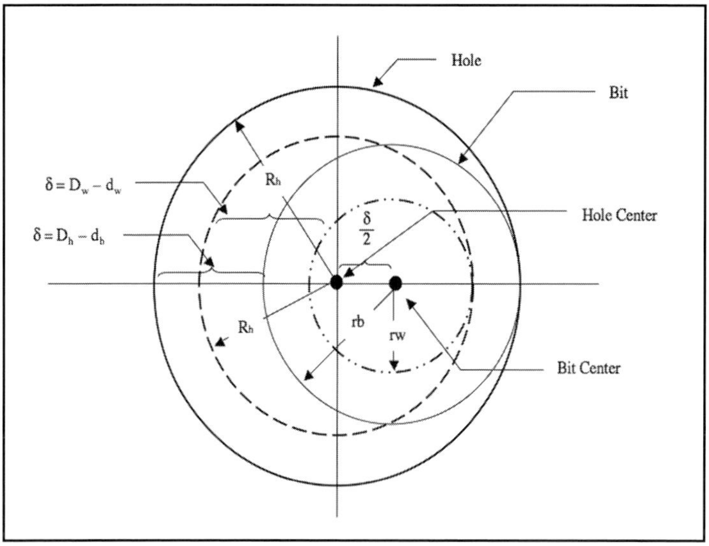

Fig. 8–23. Generic bit whirl geometry

Effect of wear on bit performance

As the cutters on a PDC bit develop wear flats, the ROP is reduced. The degree to which the ROP is reduced is a function of the geometry of the cutting edge and the rock that is being drilled. Weaker rocks are less affected by wear than stronger rocks.

In an earlier section, it was shown that the ROP for a bit was nearly independent of the number of cutters if the cutters are sharp. This is not the case for cutters with wear flats. When cutters are removed from a dulled bit, the ROP increases. A larger increase is obtained in harder rock.

The combination of the wear rate's dependence on cutter loading and the ROP of a dulled bit's dependence on the number of cutters makes the overall performance of a PDC bit more dependent on the number of cutters than the sharp-cutter data indicate. For a light-set bit, the cutter loading may be high; thus, the wear rate will be accelerated. However, because a light-set bit has fewer cutters, the ROP will not be affected as much by dulled cutters. For a heavy-set bit, the wear rate may not be as great, but the ROP will be more affected by a given degree of wear.

The effect of a particular size of wear flat is also dependent on the geometry of the cutting edge. The diamond layer is much more abrasion resistant than is the carbide support layer. Thus, the carbide will wear faster and can leave a diamond lip that protrudes above the carbide, as shown in figure 8–24. When this lip exists, the ROP for a worn cutter will be much higher than for the same wear-flat area without a lip.

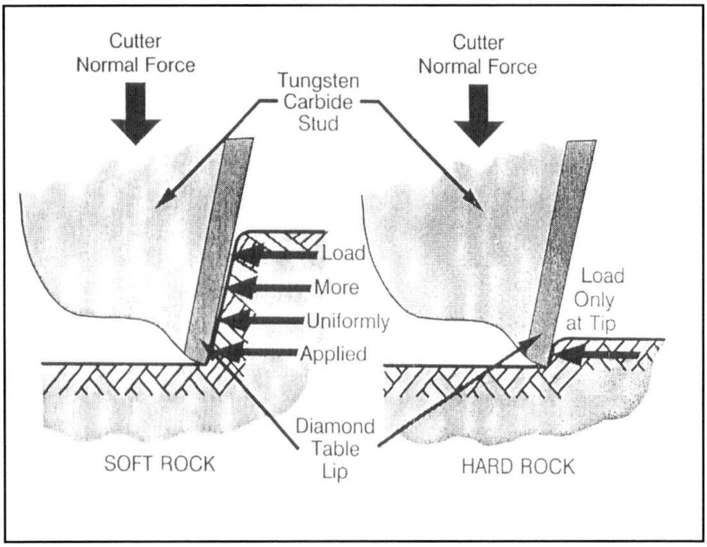

Fig. 8–24. Schematic of a wear-flat lip, in hard and soft rocks

In general, the ability of the cutter to wear in such a way as to have lips depends on both the rock that is being drilled and whether the bit runs without vibrations. In soft rock, for a given cutter loading, the cutter may take a relatively deep depth of cut as shown in the left-hand side of figure 8–24. This keeps the diamond in compression. If a hard rock is encountered with the same cutter loading, the penetration depth will be significantly reduced, and only the lip may be loaded, as shown in the right-hand side of figure 8–24. This will create high shear stresses and may create a bending moment about the lip that will put the diamond in tension, causing it to chip.

Bit whirl is also very detrimental to the maintenance of diamond lips. It is very easy for the diamond to be loaded in tension if the cutter moves sideways or backward. This often causes the diamond to spall down the face of the cutter. For most bit designs, the ability to sustain lips in hard rock is so low that the presence of lips has been used to describe a soft rock wear mode, whereas the absence of lips characterizes a hard rock wear mode.

A technique for reducing the tendency to whirl is to design a bit with a cutter-devoid area on one side, as shown in Figure 8–25. This will cause the bit to slide along a low-friction pad, providing much more uniform loading of the cutters. Lips on the PDC cutters may be maintained, even in hard rock, and much better overall bit performance may be attained.

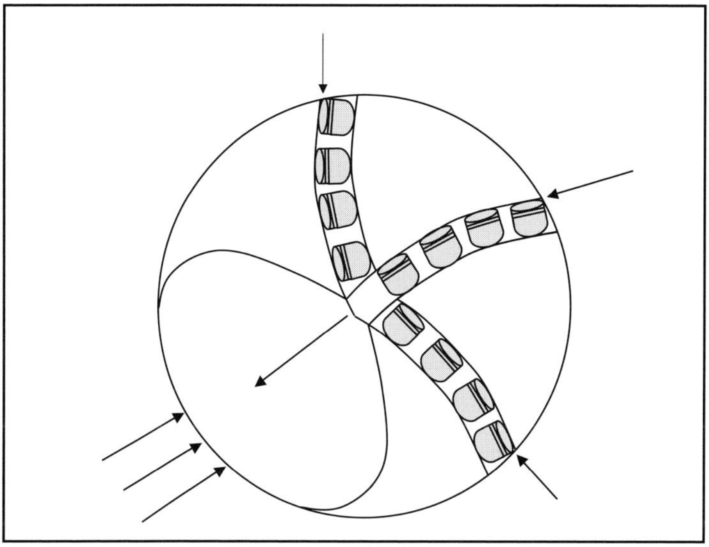

Fig. 8–25. Schematic of a low-friction gauge bit

Supplementary Problems

Problem 8–1. Using the bit model given (in equation 8.9), evaluate schematically the effect that each of the following variables has on drill bit performance:

(a) WOB

(b) Rotary speed

(c) Bit diameter

(d) Bit type

(e) Tooth wear

Problem 8–2. Using the bit model given in equation (8.14), evaluate schematically the effect that bit hydraulics has on drill bit performance.

Problem 8–3. When porous rocks are stressed at levels below the failure point, the pore volume and the pore pressure change. The strength and the ductility of the rock—and, thus, the ROP of a given drill bit—are governed largely by the vertical (overburden) stress. Explain the change and how significant is the ROP as depth increases for (a) permeable rock and (b) impermeable rock.

Problem 8–4. It has been demonstrated that for PDC drill bits, the ROP is independent of the number of sharp cutters being removed from a new bit; by contrast, for a dull bit, the ROP actually increases as more of the dulled cutters are removed. Explain why.

Nomenclature

a, b, c	Dimensionless bit constants
c_0	Cohesive strength
C_{bc}	Bulk rock compressibility
C_{bi}	Cost of bit i, in $
C_{di}	rilling cost, in $/ft, for bit run i
C_{mc}	Rock matrix compressibility
C_r	Rig cost, in $/h
C_{wc}	Pore water compressibility
d	Diameter of jet, in in.; also, depth of tooth penetration, in in.
D	Bit diameter, in in.
ΔD_i	Formation interval drilled, in ft, by bit i
f	Coefficient of sliding friction
F	Thermal response function
F_N	Normal force on cutter
F_t	Total tooth force
F_{ubr}	Force on tooth
K_f	Friction coefficient between rock and cutter
K_{hf}	Rock thermal conductivity
L_w	Wear-flat length
N	Rotary speed, in rpm
Δp	Pore pressure change
ΔP_{br}	Differential pressure across chips
ΔP_{ur}	Differential pressure across unbroken rock
R	Rate of penetration (ROP), in ft/h
S	Relative rock strength
SE	Specific energy, in lbs/in.^{3}
T	Torque, in ft-lbs
T_{ci}	Connection time, in h, for bit i
T_{di}	Drilling time, in h, for bit i
T_{ti}	Trip time, in h, for bit i
T_{wc}, T_f	Wear-flat cutter and fluid temperatures, respectively
v_c	Cutter speed
W	Weight on bit (WOB), in thousands of pounds
α_f	Rock diffusivity
θ	Tooth wedge half-angle
μ	Angle of internal friction
σ_3	Minimum principal stress
τ	Shear strength
ϕ	Porosity

References

[1] Peng, Y. H. 1972. "Experimental Study of Temperature and Strain Rate Effects on Deformation Behavior of Rocks Subjected to Triaxial Compression." Ph.D. dissertation, University of Texas.

[2] Obert, L., and W. Duvall. 1967. *Rock Mechanics and the Design of Structures in Rock*. New York: John Wiley & Sons.

[3] Geertsma, J. 1958. "The Effect of Fluid Pressure Decline on Volumetric Changes in Porous Rocks." *Journal of Petroleum Technology,* March, 210, 331–39.

[4] Warren, T. M., and M. B. Smith. 1985. "Bottomhole Stress Factors Affecting Drilling Rate at Depth." *Journal of Petroleum Technology,* August, 1523–33.

[5] Ibid.

[6] Maurer, W. C. 1965. "Bit Tooth Penetration under Simulated Borehole Conditions" *Journal of Petroleum Technology,* December, 1433–42.

[7] Warren, T. M. 1987. "Penetration Rate of Performance of Roller Cone Bits." *SPE Drilling Engineering,* March, 9–18.

[8] Maurer, *Journal of Petroleum Technology,* December, 1433–42.

[9] Cheatham, J. B., Jr. 1958. "An Analytical Study of Rock Penetration by a Single Bit Tooth." Paper presented at the 8th Annual Drilling and Blasting Symposium, University of Minnesota, Minneapolis.

[10] Sikarskie, D. L., and B. Paul. 1965. "Preliminary Theory of Static Penetration by a Rigid Wedge into a Brittle Material." *Transactions of the Society of Mining Engineers,* December, 372–83

[11] Maurer, W. C. 1965. "Shear Failure of Rock under Compression." *SPE Journal,* June, 167–76.

[12] Warren, *SPE Drilling Engineering,* March, 9–18.

[13] Ibid.

[14] Higdon, A., and W. B. Stiles. 1962. *Engineering Mechanics,* vol. 1, *Statics.* Englewood Cliffs, N.J.: Prentice Hall.

[15] Sikarskie and Paul, *Transactions of the Society of Mining Engineers,* December, 372–83.

[16] Warren, *SPE Drilling Engineering,* March, 9–18.

[17] Peterson, C. R. 1970. "Roller Cutter Forces." *SPE Journal*, March, 57–65,

[18] Ibid.

[19] Fuji, K. 1967. "A Study of Rotary Drilling by Means of Measuring the Torque on Cone Bits." *7th World Petroleum Congress Proceedings,* 3, 137–47.

[20] Peterson, SPE 2393.

[21] Sinor, L. A., and T. M. Warren. 1989. "Drag Bit Wear Model." *SPE Drilling Engineering,* June, 128–36

[22] Warren, T. M., and W. K. Armagost. 1988. "Laboratory Drilling Performance of PDC Bits." *SPE Drilling Engineering,* June, 125–35.

23 Melaugh, John, "Experimental Study of PDC Cutters," unpublished, University of Tulsa Drilling Research Projects, 1976–1978.

9

Drill String Design

Definition and Components

In general, drill strings consist of the following two major sections:

- Standard drill pipe and heavyweight drill pipe (HWDP)
- Bottom-hole-assembly (BHA)

The BHA consists of drill collars and stabilizers and a drill bit. For directional drilling, the BHA also includes a steerable mud motor or rotary steerable tool to steer the drill bit along the preselected well path. In many drilling operations, the BHA may also consist of MWD tools, reamers, crossover subs, and drilling jars, among other features. The drill collars, which are heavier than standard drill pipes or HWDP, provide total weight on bit (WOB), in vertical hole drilling, or additional weight relative to that of the HWDP, in highly deviated hole sections.

The following are the major functions of drill string in conventional rotary drilling operations:

- Transmits rotary motion from the surface to a drill bit
- Conveys drilling fluid to hole bottom
- Produces WOB for effective drilling action
- Provides control of borehole direction

Drill pipe

The major portion of the drill string is composed of drill pipes (fig. 9–1). Drill pipes are commonly made out of steel and butt-welded tool joints at each end. Tool joints provide a means for fastening the individual lengths of pipe together. To reinforce the ends of the pipe, the pipe is upset at both ends.

For design purposes, drill pipes are classified according to the following four characteristics:

- Outside diameter (OD): 2⅜ in., 2⅞ in., 3½ in., 4 in., 4½ in., 5 in., 5½ in., 5⅞ in., and 6⅝ in.
- Nominal unit weight (e.g., 19.5 lbs/ft)
- Steel grades: E, X, G, S, Z, and V (the most commonly used of which are X, G, and S, with minimum yield strengths of 95 kips per square inch [ksi], 105 ksi, and 135 ksi, respectively)
- Class: I (new, premium-class drill string; based on ≥ 80% wall thickness), II (based on ≥ 70% wall thickness), and III (based on < 70% wall thickness)

Fig. 9–1. Standard drill pipe (on the right) and HWDP (on the left)

Table 9–1 shows the drill pipe properties for selected sizes and grades. For example, the five-inch, 19.5 lbs/ft G-grade drill pipe has a minimum torsional strength of 39,166 ft-lbs and minimum tensile yield strength of 378,605 lbs. These values should be higher than the loads acting on the drill strings during drilling. Tool joints are made from high-alloy steel with a large thread (four to five threads per inch) on the pin and box. In general, tool joints are designed to have a lower strength than drill pipe, so that if a failure occurs, it will occur in the tool joint, not the drill pipe body.

Table 9–1. Properties of drill pipe (after API-RP-7G)

Size OD	New wt. nom. w/ threads & couplings	Torsional yield strength based on uniform wear, ft-lb				Tensile data based on uniform wear load at minimum yield strength, lb.			
in.	lb/ft	E	X	G	S	E	X	G	S
3½	9.50	9612	12,176	13,457	17,302	132,793	168,204	18,910	239,027
	13.30	12,365	15,663	17,312	22,258	183,398	232,304	256,757	330,116
	15.50	13,828	17,515	19,359	24,890	215,967	273,558	302,354	388,741
4	11.85	13,281	16,823	18,594	23,907	158,132	200,301	221,385	284,638
	14.00	15,738	19,935	22,034	28,329	194,363	246,193	272,108	349,852
	15.70	17,315	21,932	24,241	31,166	219,738	278,335	307,633	395,528
4½	13.75	17,715	22,439	24,801	31,887	185,389	234,827	259,545	333,701
	16.60	20,908	26,483	29,271	37,637	225,771	285,977	316,080	406,388
	20.00	24,747	31,346	34,645	44,544	279,502	354,035	391,302	503,103
	22.82	27,161	34,404	38,026	48,890	317,497	402,163	444,496	571,495
5	16.25	23,974	30,368	33,564	43,154	225,316	285,400	315,442	405,568
	19.50	27,976	35,436	39,166	50,356	270,432	342,548	378,605	486,778
	25.60	34,947	44,267	48,926	62,905	358,731	454,392	502,223	645,715

HWDP. HWDP (shown in fig. 9–1) is heavier than the standard drill pipe and has longer tool joints and additional upset(s). The longer upsets are designed to increase the stiffness of the pipes to better withstand the loads acting on the drill string. Placing HWDP between standard drill pipe and drill collar also lowers the bending stiffness ratio, which has shown to improve drill string fatigue life. The common sizes of HWDP are 3½ in., 4 in., 4½ in., 5 in., 5½ in., and 6⅝ in.. The inner diameter (ID) and weights vary among manufacturers because HWDP has not been standardized by API.

Drill collar

Drill collars, in most cases, provide the major part of the WOB. They are manufactured in various sizes and shapes. Conventional drill collars have a round cross-sectional area; however, square and spiral drill collars are also used. Drill collars with the square cross-sectional outside profile are used for increasing the stiffness of BHA. The spiral type are recommended while drilling in the areas where differential pressure sticking is a problem; the spiral grooves on the outside surface of the drill collars reduce the contact area between the borehole wall and the drill collars, which in turn decreases the sticking force. Corresponding data for non-API drill collars are available from manufacturers on request. Conventional drill collars are usually made with uniform OD; however, drill collars with slip and elevator recesses are also available. The string of drill collars is formed by connecting individual collars (usually with the length of about 30 ft) by means of rotary-shouldered connections, as depicted in figure 1–2. The connections must be made up with a proper make-up torque so that they will not separate under the downhole conditions. The recommended make-up torques for different sizes of drill collars are given in API-RP-7G ("Recommended Practice for Drill Stem Design and Operating Units").

Fig. 9–2. Drill collar

Length of drill collars. Calculation of the required length of drill collars in the drill string is important to provide the desired WOB. Required length of the drill collar to create bit loading is given by

$$L_{dc} = \frac{\mathrm{WOB} \times \mathrm{DF}}{w_{dc} \times \mathrm{BF} \times \cos\theta} \tag{9.1}$$

where

WOB is in lbs

DF is a design factor

w_{dc} is the buoyed weight of drill collar, in lbs/ft

BF is a buoyancy factor

θ is the inclination angle, in degrees

Note that drill collars are not placed in highly inclined sections of hole owing to the excessive torque and drag that they may cause.

Design Criteria

A drill string must be designed in accordance with the following criteria for the entire drilling operation:

- Static loads, such as tension, torsion, and pressure loads
- Torque and drag
- Buckling
- Fatigue

Static loads

Static loads acting on the drill strings are tension, compression, torsion, collapse, or burst pressures. The static loads change with the location of the drill string, as well as in different drilling operations. Seven load cases should be considered during drilling operations:

- Rotating drilling mode
- Slide drilling mode
- Tripping into the hole
- Tripping out of the hole
- Rotating off bottom
- Reaming into the hole
- Back-reaming out of the hole

For example, in tripping out of the hole, the tension load and the overpull at the top of drill string should be used for the design load. According to API-RP-7G, the tensile capacity of the top section of drill pipes should be greater than the total drill string weight plus at least 100,000 lbs overpull force. If five-inch 19.5 lb/ft G-grade drill pipes (minimum tensile yield strength of 378,605 lbs) are used, then the total drill string weight should not exceed 378,605 – 100,000 = 278,605 lbs. If the total weight is higher than 278,605 lbs, then a higher grade or a larger size of drill pipes should be used.

API specifications also provide alternative criteria during tripping out of the hole. API states that the drill string can be modeled with 50,000 lbs tension at the bit, to simulate the load required in order to activate the jars and to allow 50,000 lbs margin of overpull at the surface, instead of 100,000 lbs. However, one may need to use a computer program to calculate the maximum tensile loading in this case.

Indeed, using a computer program, such as the torque-and-drag model, is inevitable in designing the drill string. It is not practical and is nearly impossible to calculate static loads at various locations and load cases without the use of advanced computer analysis. In particular, the design (maximum) loads are usually from the combination of the various static loads (axial and bending, axial and pressure, etc.). The torque-and-drag program used to calculate the torque and drag on the drill string will be discussed in a later section of this chapter.

Required length of drill pipe

As part of the drill string, the maximum length of drill pipe that is allowed under a given set of drilling conditions can be calculated using the following relationship:[1]

$$L_{dp1} = \frac{0.9T_t - MOP}{w_{dp} \times BF} - \frac{w_{dc}L_{dc}}{w_{dp}} \tag{9.2}$$

where

T_t is the theoretical tension load given by the manufacturer
w_{dp} is the unit weight of the drill pipe, in lbs/ft
MOP is the margin of overpull, in lbs
BF is a buoyancy factor
w_{dc} is the unit weight of the drill collar, in lbs/ft
L_{dc} is the length of the drill collar, in ft

Margin of overpull

Margin of overpull (MOP) is defined as the additional allowed tension to be applied to free stuck pipe. It is the difference between the maximum allowable tension and the calculated hook load:

$$MOP = T_a - T_h \tag{9.3}$$

where

T_a is the allowable axial tension, in lbs
T_h is the calculated hook load, in lbs

The ratio of these tensions gives the safety factor (SF):

$$SF = \frac{T_a}{T_h} \tag{9.4}$$

Example 9–1 shows the drill string design based on the MOP criterion.

Collapse check

Sometimes drill string is subjected to lower internal pressure than the outside pressure that may result in collapse failure of the pipe. Thus, it is essential to check against the internal pressure resistance of the pipe by considering the worst-case scenario wherein the pipe is empty inside and there is drilling fluid outside. This collapse pressure can be calculated as follows:

$$P_c = \frac{L_c \rho_m}{19.251} \tag{9.5}$$

where

P_c is the collapse pressure, in psi
L_c is the depth at which collapse pressure is to be checked, in ft
ρ_m is the mud density, in lbs/gal

Alternatively, the collapse pressure can be calculated when the string is partially empty, with different fluid density inside than outside, using the following equation:[2]

$$P_c = \frac{L_c \rho_m - (L_c - x)\rho_{mi}}{19.251} \tag{9.6}$$

where

ρ_{mi} is the inside fluid density, in lbs/gal
x is the inside fluid depth, in ft

Example 9–1

Design a drill string for a hole size of 8½ in., to 15,000 ft, given the following additional data:

- MOP desired = 25,000 lbs
- Drill collar size: 6 in. × 3 in.; 72 lbs/ft; 720 ft
- Use class II drill pipe of E, G, or higher grade
- Mud weight = 12 lbs/gal
- Percentage of yield = 90 %

Solution: Select the lowest-grade pipe (i.e., E75 grade). The length of drill pipe that can be used with the desired MOP is then found.[3] From the drill pipe table for class II pipe T_t = 270,432 lbs, and BF = 1 – (12/65.4) = 0.817. Thus, using equation (9.2),

$$L_{dp1} = \frac{(0.9 \times 270,432) - 40,000}{20.85 \times 0.817} - \frac{72 \times 720}{20.85} = 9,453.5 \text{ ft}$$

To make the length equal to the complete stands, the length of the E75 drill pipe can be run only up to 9,450 ft.

The length of the next drill pipe required is 15,000 – 720 – 9,450 = 4,830 ft

Further, the total tension with the drill pipe can be calculated. The weights of small components (stabilizers, subs, mud motors, etc.) can be included in the drill collar weight. The total tension is

$$T_a = (L_{dp}w_{dp} + L_{dc} \times w_{dc}) \times \text{BF}$$
$$= [9,450 \times 20.85) + (720 \times 72)] \times 0.817 = 203,329 \text{ lbs}$$

The length of the second drill pipe, a G105 (next higher grade) drill pipe, is

$$L_{dp1} = \frac{(0.9 \times 378,605) - 40,000}{21.93 \times 0.817} - \frac{203,329}{21.93} = 7,513 \text{ ft}$$

which is more than the required length of 4,380 ft.

Thus, the final drill string design is as follows:

Component	Description	Length (ft)	Depth (ft)	Weight (lbs)
Drill collar	6 in. × 3 in.; 72 lbs/ft	720	15,000–14,280	42,353
First drill pipe	5 in.; 19.5 lbs/ft; E75	9,450	14,280–4,830	160,975
Second drill pipe	5 in.; 19.5 lbs/ft; G105	4,380	4,380 to surface	86,538

Collapse check: Collapse pressure, P_c, at the bottommost pipe depth, 14,280 ft, is calculated as a check:

$$P_c = \frac{14{,}280 \times 12}{19.251} = 8{,}901 \text{ psi}$$

From the reference table, the collapse pressure of E75 grade pipe is 5,514 psi, which is less than the calculated value. Thus, proper precautions should be taken to avoid this situation.

To calculate the maximum depth, the pipe can be run empty with a safety factor of 1.125 for collapse:

$$L_{max} = \frac{5{,}514 \times 19.251}{12 \times 1.125} = 7{,}863 \text{ ft}$$

Also, the depth at which the fluid can drop inside the drill string if the inside fluid density is 13 lbs/gal is calculated:

$$\frac{5{,}514}{1.125} = \frac{(14{,}280 \times 12) - (14{,}280 - x)\,12}{19.251}$$

where $x = 8{,}356$ ft.

Torque and drag

Torque and drag may be the critical factors in determining whether the desired well path can actually be drilled and cased. Typically, torque and drag of the drill string are calculated using computer models. These models consider well trajectory, drill string configuration, doglegs, friction factors, and casing depth. Accurate analysis for torque and drag is important for several reasons, including

- Optimizing the well path to minimize torque and drag
- Fine-tuning the well path to minimize local effects, such as excessive normal loads
- Providing normal force loads for inputs into other programs, such as casing-wear models
- Identifying the depth or reach capabilities or limitations, both for drilling and for running casing/tubing
- Matching the strength of drill string components to the loads (axial, torsional, or lateral) in the wellbore
- Identifying the hoisting and torque requirements of the drilling rig

The most commonly used torque-and-drag models are based on the soft-string model developed by Johancsik et al.[4] The drill string is modeled as a cable that can carry axial loads but not bending moments. The normal force at each calculation node has two components:

- Effective weight of the pipe in drilling fluid
- Lateral reaction force resulting from drill string tension through curved sections of the wellbore.

A simplified drill string element is shown in figure 9–3.

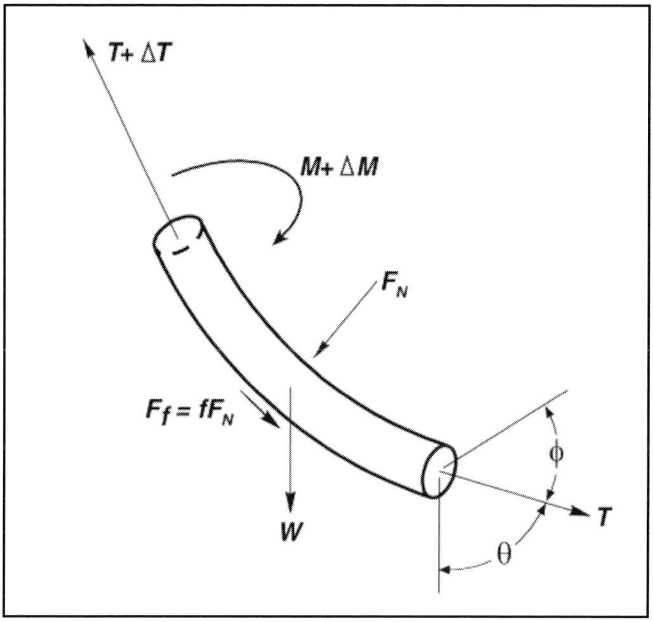

Fig. 9–3. Drill string element for soft-string torque-and-drag model[5]

The procedure for calculation of torque and drag consists of dividing the drill string into several elements. The calculation starts at the bottom of the drill string and proceeds stepwise upward to the surface. Each short element of the drill string contributes increments of axial and torsional load as the calculations proceed toward the top of the string. The normal force, F_N, is determined first by equation (9.7):

$$F_N = \sqrt{\left(T_{axial}\Delta\phi\sin\theta_{avg}\right)^2 + \left(T\Delta\phi + w\sin\theta_{avg}\right)^2} \tag{9.7}$$

where

F_N is the net normal force

T_{axial} is the axial tension at the lower end of the element

w is the buoyed weight of the element

θ is the inclination angle at lower end of the element

ϕ is the azimuth angle at lower end of the element

The calculations of tension and torsion increments are then made by using equations (9.8) and (9.9), respectively:

$$\Delta T = w \cos\theta_{avg} \pm fF_N \quad (9.8)$$

$$\Delta M = fF_N R \quad (9.9)$$

where

R is the characteristic radius of the element
M is the torsion at the lower end of the element
f is the coefficient of friction

In equations (9.7)–(9.9), $\Delta(T, M, \phi, \theta)T$ is the change in those values over the length of the element. Further,

$$F_F = fF_N \quad (9.10)$$

where F_F is the sliding friction force acting on the element. Note that in equation (9.10), the friction force, F_F, is considered positive when the drill string is moving or running into the hole and considered negative when the string is moving up or tripping out of the hole.

When accurate friction factors are derived from existing field data, the soft-string model yields reasonably accurate results. However, since the soft-string model does not consider the stiffness of the drill string, its accuracy will degrade as drill string diameter and/or hole curvature increases. Increases in pipe stiffness and hole curvature result in high normal forces and, therefore, an increase in torque and drag. The effect of stiffness is particularly important when analyzing casing/liner. Even though the use of a stiff-string model is recommended in this case, it has been shown that results are underconservative. A study by BP has showed that a higher friction factor is needed in casing/liner analysis, to compensate for the incorrect results in the common stiff-string models.

Neither type of model (soft- or stiff-string model) can accommodate localized hole and drill string/BHA mechanical interactions (e.g., a stabilizer hanging up on a ledge or a dogleg). Such a model would require much more information about actual hole geometry than is available from the survey data.

Hole spiraling (fig. 9–4) is a common cause of *tight hole,* which results in excessive torque and drag on the drill string. Unfortunately, spiraling is difficult to detect from conventional MWD survey.

Since local or micro-level of well geometry is often difficult to obtain with conventional MWD survey data, empirically derived macro-level friction factors are often used for predictive analysis, since these factors are calculated from wells with similar localized geometry. Friction factors should be derived from analogous case histories. The properties of the drilling fluids used in the baseline wells and that to be used in the planned well should be similar. If prior experience is unavailable, the ranges in table 9–2 can be used as starting points.[6]

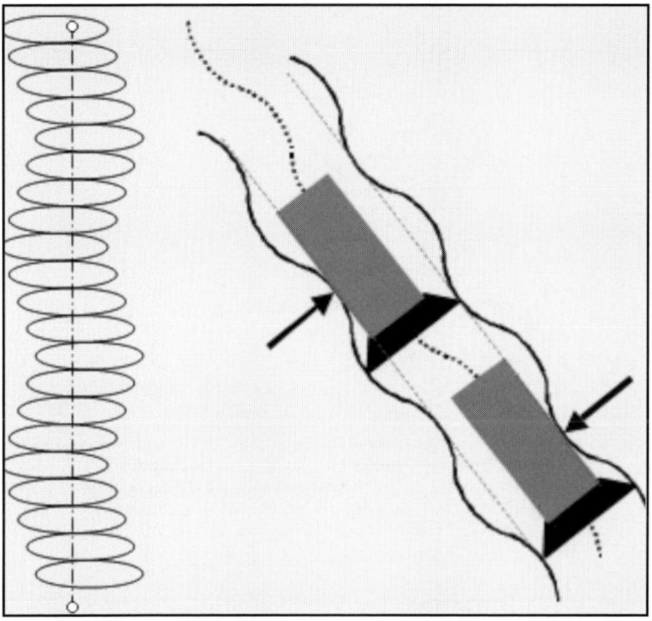

Fig. 9–4. Hole spiraling

Field experience has shown that axial drill string drag is reduced when the drill string is rotated. Torque-and-drag models account for this mathematically by the use of velocity vectors (fig. 9–5).[8] The resultant velocity V_R of a contact point on the drill string is the vector sum of two components: circumferential velocity, V_C (caused by rotation); and axial velocity, V_A (affected by drilling rate or tripping speed).

Table 9–2. Friction factors in the casing and the formation[7]

Drilling Fluid	*f* in casing	*f* in formation
Oil-based	0.16 to 0.20	0.17 to 0.25
Water-based	0.25 to 0.35	0.25 to 0.40
Brine	0.30 to 0.40	0.30 to 0.40

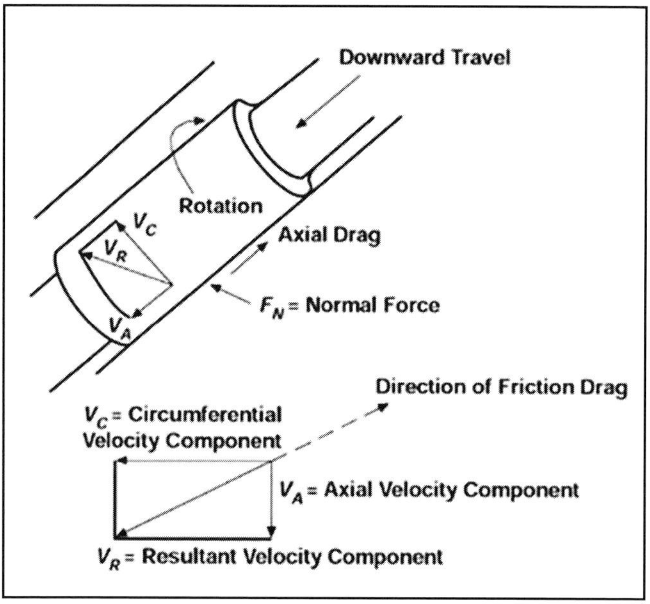

Fig. 9–5. Effect of drill string rotation on axial friction

The *direction* of the resultant frictional force is assumed to act in the direction opposite to that of the resultant velocity V_R; therefore, its vector components will be in proportion to those of resultant velocity. The *magnitude* of the resultant frictional force is simply the product of the normal force F and the friction coefficient f, and it does not vary with velocity. Since the magnitude of the vector sum of these components is a fixed quantity, the axial component must decrease as the circumferential component increases. Logically, as drill string rotation speed increases, it increases the circumferential component, which decreases axial friction.

Well planning should include torque-and-drag modeling with worst-case friction factors, to ensure that the drill string can be advanced, rotated, slid (if oriented drilling is necessary), and pulled out of the hole. Similar modeling should be used to ensure that friction will not prevent the casing from being run and that the casing can be pulled if necessary. In addition, the torque and tension/compression loads at any location in the drill string must be compared with the torsional, tension, and buckling capabilities of the drill string and tool joints.

The results from torque-and-drag analysis are usually expressed graphically, with torque and/or drill string tension on one axis and measured depth on the other (see figs. 9–6 and 9–7). The subject well has a build-hold-drop profile with a 5,000 ft tangent section at a 60° angle. A friction factor of 0.2 was used for cased holes, and a factor of 0.3 was used for open holes.

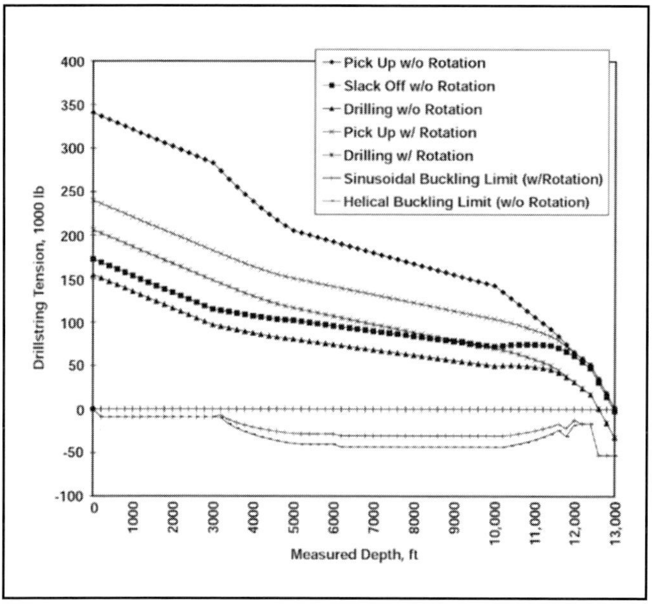

Fig. 9–6. Depth versus tension

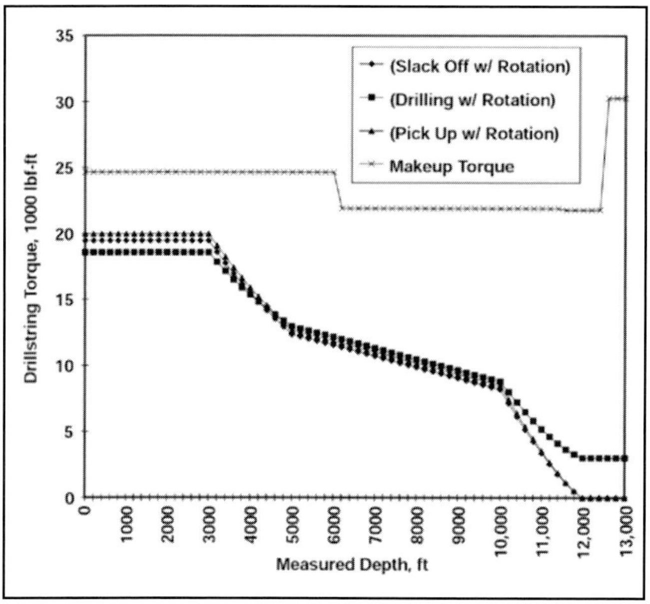

Fig. 9–7. Depth versus torque

The torque-and-drag analysis qualifies this well as drillable with the operating parameters used and allows the selection of drill string components with a reasonable safety margin as compared to loads. Axial load values never drop into the buckling region, even during sliding. Because of the relatively high hole angle, torque is nearly 20,000 ft-lbs, dictating the use of five-inch, 19.5 lb, high-strength (S135) drill pipe in the upper 6,000 ft of the well. Maximum tension in the drill string is about 340,000 lbs. According to API, adding another 100,000 lbs overpull force still leaves a reasonable safety margin below the yield strength of 486,700 lbs. On the basis of torque and drag, this well is drillable with standard drill string components and practices as long as the friction factors used in modeling are accurate and unusual hole problems do not occur.

During drilling, a log of predicted and actual torque and/or drag versus depth should be maintained (fig. 9–8). Such a log enables friction factors to be updated and verified. If actual friction factors are significantly different than planned, problems can be predicted and prevented instead of dealt with after the fact. The log will also show the effect of changing bit types or altering operational parameters. Note the increase in overall torque when PDC bits are run, rather than roller cone rock bits. The log can also reveal deteriorating hole conditions, such as the buildup of cuttings, and local hole features, such as doglegs. Remedial actions, whether preventive or after the damage has occurred, could include the following:

- enhancing hole cleaning through higher flow rates, rotation, or modified drilling-fluid rheology
- making short trips to condition the hole
- reaming out ledges, key seats, or doglegs
- changing the mud type
- as an extreme measure, altering the well profile or changing the casing or hole program

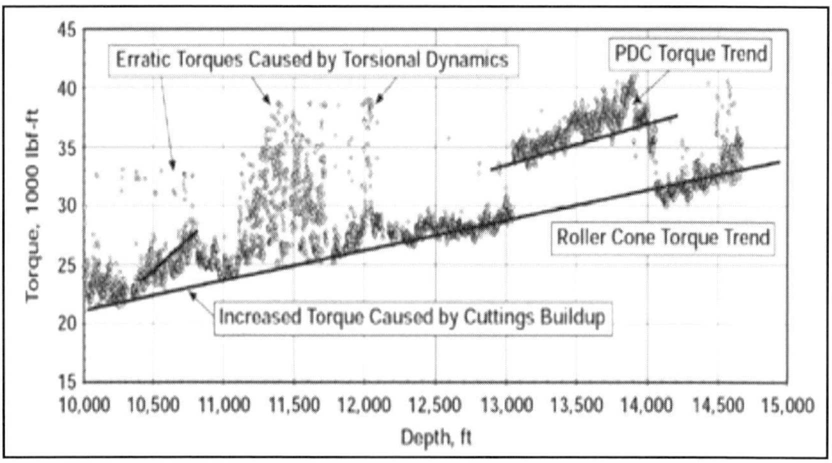

Fig. 9–8. Log of actual versus predicted drill string torque

Buckling of the Drill String

When a drill string is subjected to increasing compressive loads, it will undergo several stages in configurations. The first stage is referred to as *sinusoidal buckling.* The drill string assumes a two-dimensional waveform, resembling a sine wave, winding back and forth along the wellbore. The sinusoidal buckling load can be calculated using the Dawson and Pasley equation:[9]

$$F_{sin} = 2\sqrt{\frac{EIW \sin\theta}{r}} \tag{9.11}$$

where

F_{sin} is the sinusoidal buckling load
E is Young's modulus of elasticity
I is the cross-sectional area of moment of inertia
W is the buoyed unit weight of the drill string
θ is the inclination angle
r is the radial clearance between the drill string and wellbore

When compressive loads are increased further, the second stage, called *helical buckling* occurs. This causes the drill string to ride up the sides of the wellbore in the shape of a helix. The increase in the wall contact area increases drag, requiring more axial load in order to maintain the same bit weight. The added axial load causes higher well contact force, which further increases drag. Therefore, helical post-buckling should be avoided by all means. Unlike sinusoidal buckling, there are several equations to predict the helical-buckling load, among which the Chen and Cheatham equation is regarded as the most accurate:

$$F_{hel} = 2\sqrt{\frac{2EIW \sin\theta}{r}} = \sqrt{2}F \sin\theta \tag{9.12}$$

where F_{hel} is the helical-buckling load. Thus, the helical buckling load is about 1.41 times the sinusoidal buckling limit.

When the compressive loads are increased further, the increased wall contact force will eventually generate so much drag that no amount of slacking off will move the drill string. In the sliding mode, this third and final stage is commonly referred to as *lockup*. At this point, a change in the drill string design is required for drilling to continue.

Drill string buckling analysis is often integrated within the torque-and-drag software. The program should not only predict the onset of the buckling (either sinusoidal or helical) but also calculate the post-buckle normal force. The post-buckle force is an important feature in predicting the lockup.

When the drill string is rotated, most of the axial drag present while sliding is converted to rotational drag. This increases torque and decreases axial drag. The decrease in drag allows the drill string to move down the wellbore more freely. Thus, rotary drilling is usually possible beyond the point where a BHA would lock up in the sliding mode. The critical load required in order to helically buckle the drill string remains unchanged, but a greater bit weight is necessary in the rotary mode to reach that critical load.

The main difference between rotating and sliding, from a buckling standpoint, is that substantial fatigue damage will occur when the string is rotated while buckled. During rotation, drill string buckling—even sinusoidal buckling—greatly increases the risk of fatigue failure. By contrast, in the sliding mode, the drill string will experience very little or no damage, even if buckling occurs (as long as the drill string is not rotated). Therefore, from an operational perspective, it is important to pick up off bottom before starting to rotate if the drill string is buckled.

Drill String Fatigue

When the drill string components undergo cyclic stress over a period of time, fatigue damage may occur. In principle, fatigue damage occurs only when the drill sting is rotated while it is axially curved. As the drill string is rotated, the curved portions experience one stress cycle per revolution. The amplitude of the stress is directly proportional to the degree of curvature where the drill string is located. Typically, cyclic stress, which can be calculated using equation (9.13), is associated with the bending stress induced from the curvature.

$$\sigma_b = \frac{ED}{2R_c} = \sqrt{2}F_{sin} \tag{9.13}$$

where

σ_b is the the bending stress
E is Young's modulus of elasticity
D is the drill pipe OD
R_c is the radius of curvature

As equation (9.13) demonstrates, the larger the OD is, the higher will be the bending stress. Thus, a large drill pipe will have a higher bending stress than will a smaller drill pipe in the same hole curvature. This should be kept in mind when designing the drill string for a well with a high dogleg severity (DLS). From the standpoint of fatigue life, a smaller drill string size may be desirable.

For example, rotating a five-inch drill pipe in a 10°/100 ft DLS well, the bending stress can be calculated as follows:

$$\sigma_b = \frac{29E\left(6 \times 5\right)}{2 \times (5{,}730/10) \times 12} = 10{,}544 \tag{9.14}$$

The fatigue life for any particular bending stress can be found from the published *S-N* curve for the particular drill string. Ideally, the bending stress induced from the curvature should be less than the endurance stress limits from the fatigue (*S-N*) curve (see fig. 9–9), so that the drill string can have an unlimited number of rotations. Note that because drill pipes are also subjected to a high tension load near the surface, the effect of the axial load should be considered in the fatigue analysis.

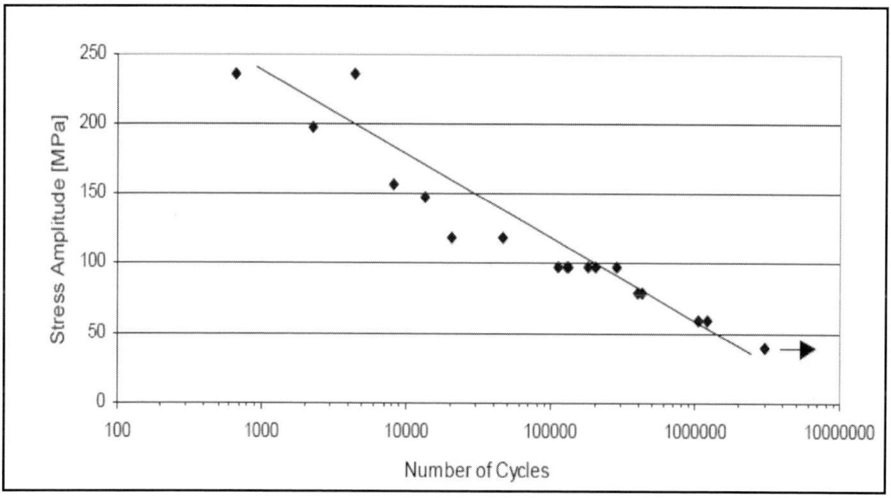

Fig. 9–9. Fatigue (S-N) curve for drill pipe with an endurance stress of 40 million Pascals (MPa)

As mentioned previously, rotating a buckled drill string should be avoided, to prevent fatigue damage. Buckling usually occurs near the bottom of the drill pipe but may occur at other locations under certain circumstances. In addition to buckling, vibration can create fatigue damage by inducing repeated cyclic bending stress on the drill string. The magnitude of bending stress from vibration is much higher than that induced by curvature. Vibration creates high impact loads to the drill strings, leading to low-cycle fatigue failure (e.g., drill pipe twist-off and washout and downhole component failures). Unfortunately, drilling vibration is a very complex issue. Not only is the dynamic stress hard to calculate, but the conventional fatigue analysis using the *S-N* curve is also not applicable. Thus, the best way to avoid vibration-related failures is to mitigate drilling vibration in the first place. A more detailed discussion on drilling vibration is given in the next section.

Drill String Vibration

Drilling vibration[10] has been shown to be a leading a leading cause of drill string component failures, such as MWD equipment failures,[11] drill string twist-off and washout, and tool joint and stabilizer wear. Drilling vibration can also result in premature bit damage and lower ROP, among other problems. Thus, it is important to understand drilling vibration and mitigate it through proper BHA/bit design, optimization of drilling parameters, and incorporation of best practices.

Vibration mechanisms

Although drill bits and drill strings can exhibit complex motions, such as backward or forward whirl, these motions can be simplified into three basic vibration modes (as illustrated in fig. 9–10):

- Lateral (transverse): side-to-side motion
- Axial (longitudinal): motion along drill string axis
- Torsional (stick-slip): motion causing twist-off and torque

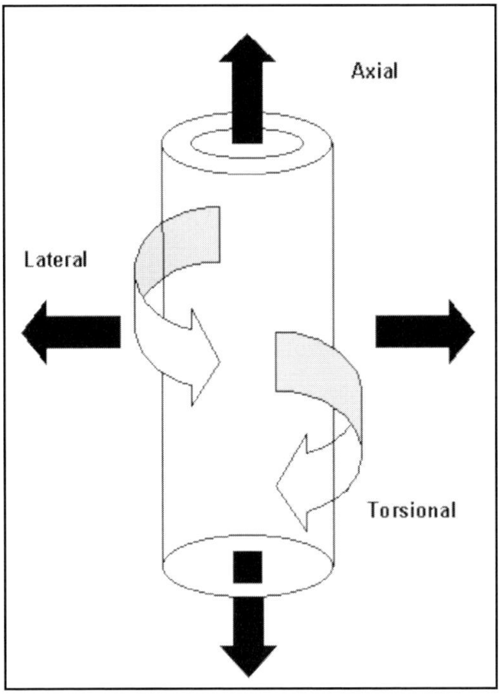

Fig. 9–10. Modes of vibration

Lateral vibration. Lateral vibration is the most important vibration mode, because it is responsible for most of the downhole tool and drill string failures. Lateral vibration can occur at the BHA, the bit, and drill pipes. Impacts of the BHA against the wellbore wall generate high shock loads and high cyclic bending stresses, resulting in MWD equipment, motor, and BHA component failures. Impacts on the bit lead to premature bit damage. Impacts on the drill pipes cause wear on drill pipes, twist-off (fig. 9–11), and washout.

Fig. 9–11. Twisted-off MWD tool and drill collar, owing to lateral vibration

Software Solution. One of the main causes of lateral vibration is resonance, which leads to self-excited high-magnitude vibration. Resonance occurs when the rotary speed is too close to one of the natural frequencies of the BHA. Thus, natural frequencies of the BHA, sometimes called critical rotary speeds, should be calculated before each bit run. Figure 9–12 illustrates how the BHA is modeled as a continuous beam, with pin supports at the bit and stabilizer. The BHA is terminated at some distance from the uppermost stabilizer when the collars are in contact with the wellbore wall. This distance is called the *tangential point* and is usually estimated from the static BHA model. The critical speeds can then be predicted with the dedicated software, which is usually finite-element based.

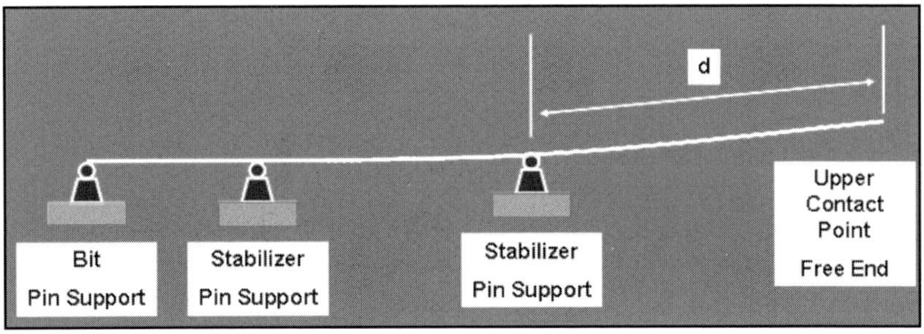

Fig. 9–12. Critical speed model

Importantly, resonance is not the only cause of drilling vibration. Thus, avoiding critical speeds should be considered only one portion of a vibration mitigation process.

Hardware solution. BHA eccentricity and buckling are the two most common hardware issues that cause high lateral vibrations. Downhole tools, such as bicentered bits and roller reamers, have to be carefully chosen to minimize the eccentricity forces.

Operating parameter solution. Lateral vibrations such as BHA and bit whirl sometimes can be mitigated by decreasing the rotary speed.

Axial vibration. Axial vibration causes the bit to repeatedly lift off and impact at hole bottom, resulting in large WOB fluctuations. Axial vibration is usually created by roller cone bits in hard rock. (PDC bits have not been shown to be associated with axial vibration.) The impact loads from the axial vibration can damage the drill bit, drill strings, or the surface hoisting equipment.

During a drilling operation, a decrease in WOB and/or a decrease in rpm will mitigate axial vibration. However, changing the hardware, such as the use of a less aggressive roller cone bit and/or a shock sub, has been shown to be more effective than changing the drilling parameters.

Torsional vibration (stick-slip). Torsional vibration is defined as a nonuniform bit rotation in which the bit stops rotating momentarily at regular intervals. As a result, it usually causes the string to periodically torque up and then spin free (i.e., stick-slip). The typical environment for stick-slip is in high-angle wells with aggressive PDC bits and high WOB, when the downhole frictional torque exceeds the rotary torque. Common issues resulting from stick-slip are PDC bit damage, lower ROP, connection overtorque, back off, and drill string twist-off.

Software solution. Stick-slip is initiated by inadequate torque. Thus, there is no critical rpm to calculate and avoid.

Changing the drilling parameters on the fly to mitigate stick-slip has been shown to be quite effective. As a rule of thumb, when stick-slip occurs, one should increase rpm or decrease WOB by about 15%. If stick-slip persists, stop the rotary and restart drilling under a higher rpm and/or lower WOB.

Hardware solution. Other solutions to torsional vibration, or stick-slip, include

- Reducing bit torque, through a less aggressive bit design
- Reducing BHA torque, through better wellbore quality, and reducing the number of stabilizers and gauge sizes
- Reducing drill string torque, through better hole cleaning and mud lubricant

Vibration detection

Detection of downhole vibration can be done by

- Downhole vibration sensors
- Surface sensors
- Post-run inspection

Downhole vibration sensors. Downhole vibration sensors, usually located in the MWD tool, are the most effective way to detect vibrations. The reason for their effectiveness is that lateral vibrations predominately occur in the BHA and do not propagate to the surface, making direct detection at the surface impossible. The disadvantages of downhole measurement are availability and low MWD transmission bandwidth, which may limit amount of information transmitted to the surface.

Downhole vibrations are usually quantified by the acceleration data. Typically, vibration sensors consist of triaxial accelerometers, as shown in figure 9–13. Some include additional magnetometers for monitoring downhole rpm, as cyclic rpm is a result of torsion (stick-slip) vibration. Lateral vibration, such as BHA whirl, is best monitored by downhole acceleration data, as shown in figure 9–14. The plots show same magnitude but out-of-phase lateral (X and Y) accelerations measured during BHA whirl. Applying the Fourier transform to the same acceleration data, the frequency data indicate a whirling frequency of 24 Hz.

In practice, the accelerometers are mounted on the MWD board such that they are offset from the center of the tool. X and Y are used to measure the lateral and torsional vibrations, and Z is used to measure the axial vibration.

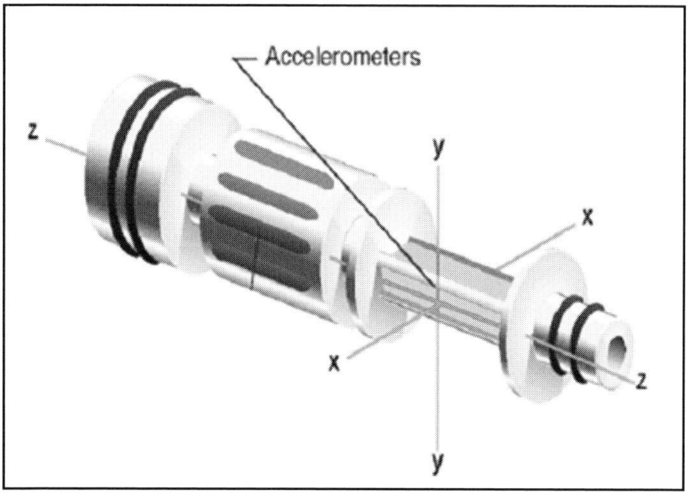

Fig. 9–13. Typical three-axis accelerometer and MWD vibration sensor

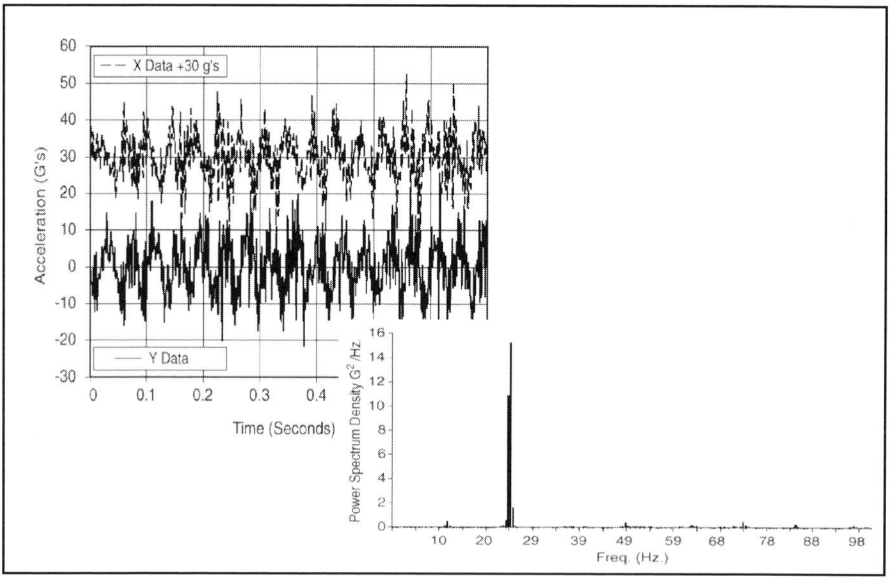

Fig. 9–14. Downhole acceleration data

Modern MWD vibration-monitoring systems usually include the surface display, as shown in figure 9–15. The commonly used traffic light display shows the vibration mode, as well as the vibration magnitude; green, yellow, and red lights represent low, medium, and high levels of vibration, respectively. The traffic light display is much more intuitive to rig personnel than simply displaying the vibration data. When a red light is on, some displays even suggest how the WOB and rpm should be adjusted to reduce vibration.

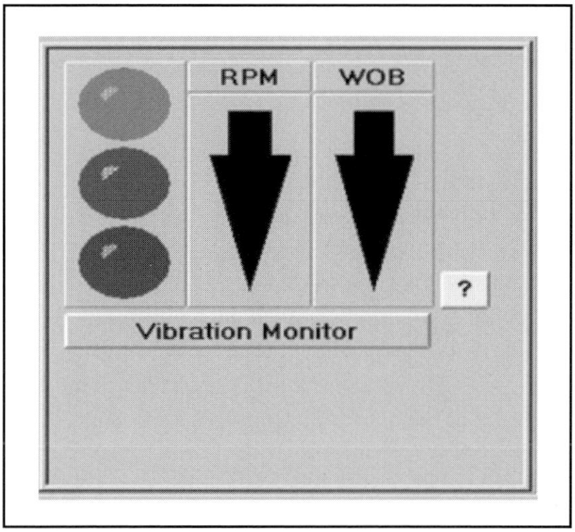

Fig. 9–15. Vibration-monitoring surface display, using traffic light display to show vibration mode and magnitude

Surface data. Rig site surface data (e.g., rpm, torque, and ROP) provide limited information about downhole vibration. They are valuable when downhole vibration sensors are not available or when the MWD data quality is poor. However, as mentioned earlier, surface data are limited to axial and torsional vibrations only, as lateral vibration will not propagate to the surface. The following surface data can be used to describe downhole vibrations:

- Top drive or kelly bounce: an indication of axial vibration
- Cyclic torque or rpm: an indication of torsional vibration (stick-slip)
- Sudden increase of torque: possible BHA whirl
- Low ROP: possible severe downhole vibration
- Fluctuating standpipe pressure: possible torsional vibration (stick-slip)

Post-run inspection. Downhole vibration can be detected through rig site inspection after the run. Specifically, the signs to look for are

- BHA component failures (motors, rotary steerable system, MWD tools, etc.)
- Cracks or wear on tool joints
- Washout and twist-off
- Premature bit wear
- Top-drive axial vibration and torque oscillations
- Stabilizer and collar wear

Any of these problems is a strong sign that downhole vibrations are occurring and should be brought to the attention of the supervisor and the rig crew. Proper actions, such as BHA/bit redesign and/or adding a real-time downhole vibration sensor, should be considered to prevent the failure from happening in the next run.

BHA Design for Directional Drilling

Design principles: Bit side force and bit tilt

The BHA is a portion of the drill string that affects the trajectory of the bit and, consequently, the trajectory of the wellbore. In general, the factors that determine the drilling tendency of a BHA are bit side force, bit tilt, hydraulics, and formation dip. The objective of the BHA design for directional control is to provide the directional tendency that will match the planned trajectory of the well.

The bit side force is the most important factor that affects the drilling tendency. The direction and magnitude of the bit side force determine the build, drop, and turn tendencies. A drop assembly is when the bit side force is acting toward the low side of the hole; a build assembly is when the bit side force is acting toward the high side of the hole; and a hold assembly is when the inclination side force at the bit is zero.

The bit tilt angle is the angle between the bit axis and the hole axis. Bit tilt affects the direction of drilling because a drill bit is designed to drill parallel to its axis.

Rotary assemblies

Rotary assemblies are designed to build, drop, or hold angle. The behavior of any rotary assembly is governed by the size and placement of stabilizers within the first 120 ft from the bit. Additional stabilizers run higher on the drill string will have a limited effect on the assembly's performance. Rotary assemblies are not steerable, for two reasons. First, the azimuth behavior (right/left turn) of a rotary assembly is nearly uncontrollable. Second, each rotary assembly has its own unique build/drop tendency, which cannot be adjusted from the surface. Thus, tripping for the assembly change is required in order to correct the wellbore course.

Commonly used stabilizer types are sleeve stabilizers, welded blade stabilizers, and integral blade stabilizers. Geology is the most important consideration for long wear life when selecting one type of stabilizer over another. Sleeve stabilizers (fig. 9–16) are the most economical, but ruggedness often becomes an issue. Welded blade stabilizers (fig. 9–17) are best suited to large holes in soft formations. Integral blade stabilizers (fig. 9–18) are most expensive but very rugged, making them the best choice in hard and abrasive formations. Sometimes, roller reamers are used along with stabilizers to open the hole to full gauge, extend the bit life, and prevent possible sticking problems.

Fig. 9–16. Sleeve stabilizer

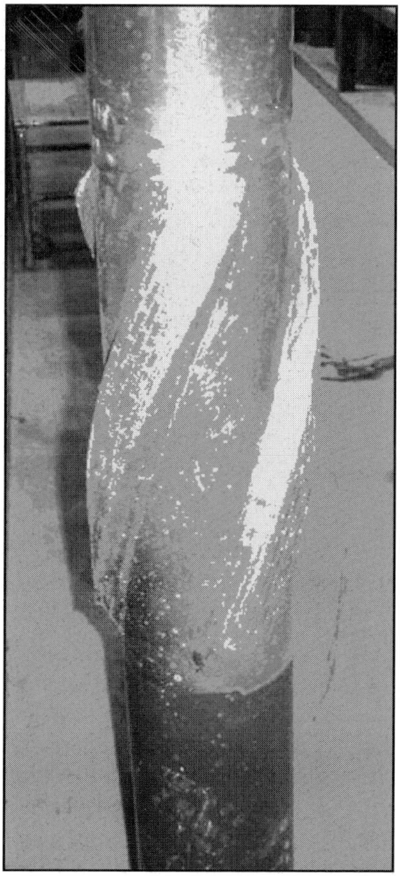

Fig. 9–17. Welded blade stabilizer

Fig. 9–18. Integral blade stabilizers

Building assemblies: The fulcrum principle

Building assemblies use the fulcrum principle; that is, a near-bit stabilizer, closely placed above the bit, creates a pivot point wherein the bending drill collars force the near-bit stabilizer to the low side of the hole and create a lateral force at the bit to the high side of the hole. Experience has shown that the more limber the portion of the assembly just above the fulcrum is, the faster will be the increase of angle.

A typical build assembly uses two to three stabilizers. The first (near-bit) stabilizer usually connects directly to the bit. If the direct connection is not possible, then the distance between the bit and the first stabilizer should be less than six feet, to ensure that the assembly remains an angle-building assembly. The second stabilizer is added to increase the control of side force and alleviate other problems.

Build rates can be increased by increasing the distance between the first and second stabilizers. When the distance between the stabilizers increases enough to cause the drill collar sag to touch the low side of the hole, the bit side force and bit tilt reach their maximum build rate for the assembly. In general, the drill collars will sag to touch the borehole wall when the distance between the stabilizers is greater than 60 ft. The amount of sag will also depend on the hole and collar sizes, inclination, stabilizer gauge, and WOB, among other parameters.

Other important factors for the fulcrum assemblies are inclination, WOB, and rotary speed. The build rate of a fulcrum assembly increases as inclination increases, because there is a larger component of the collar's own weight causing the drill collars to bend. Increasing the WOB will bend the drill collars behind the near-bit stabilizer more, so that the build rate will increase. A higher rotary speed tends to straighten out the drill collars, reducing the build rate. Therefore, low rotary speeds (70–100 rpm) are generally used with fulcrum assemblies. Sometimes, in soft formation, a high flow rate can lead to formation washout, resulting in decreased stabilizer contacts and, thus, a reduced build tendency.

Holding assemblies: Packed hole

Packed-hole assemblies contain three to five properly spaced stabilizers to maintain angle. The increased stiffness on the BHA from the added stabilizers keeps the drill string from bending or bowing and forces the bit to drill straight ahead. The assembly may be designed to have a slight build or drop tendency, to counteract formation tendencies.

Dropping assemblies: The pendulum principle

The pendulum effect is produced by removing the stabilizer just above the bit while retaining the upper stabilizers. While the remaining stabilizers hold the bottom drill collar away from the low side of the wall, gravity acts on the bit and the bottom drill collar and tends to pull them to the low side of the hole, decreasing the hole angle. Pendulum assemblies sometimes can be run slick (without stabilizer). Although a slick assembly is simple and economical, it is difficult to control and maintain the drop tendency.

A dropping assembly usually contains two stabilizers. As the distance between the bit and the first stabilizer increases, gravity pulls the bit to the low side of the hole, increasing the downward bit tilt and bit side force. If the distance between the bit and the first stabilizer is too large, the bit will begin to tilt upward, and the drop rate will reach a maximum. With a higher WOB, the drop assembly could even start building angle. In general, the distance between the bit and the first stabilizer will be about 30 ft. The second stabilizer is added to increase the control of side force.

Initially, low WOB should be used to avoid bending the pendulum toward the low side of the hole. Once a dropping trend has been established, moderate WOB can be used to achieve a higher ROP.

Deviation Tools

The most common deviation tools for directional drilling are steerable motor assemblies, which incorporate positive-displacement motors (hence, they are often called PDMs), and rotary steerable systems (RSS). Adjustable-gauge stabilizers, also known as two-dimensional rotary systems, to control inclination, have become quite popular to run in conjunction with rotary and PDM assemblies. Whipstocks—especially, casing whipstocks—are routinely used to sidetrack out of cased wellbores. Other tools such as turbines are mainly used in Russia, and jetting bits are seldom used today.

Steerable motor assemblies (or PDMs)

The most important advancement in trajectory control is the steerable motor assemblies, which contain PDMs with bent sub or bent housing. The PDM is based on the Moineau principle. The first commercial PDM was introduced to the petroleum industry in the late 1960s. Since then, their use has been accelerated greatly for directional drilling applications. Steerable motor assemblies are versatile and are used in all sections of directional wells, from kicking off and building angle to drilling tangent sections and providing accurate trajectory control. Among the PDM assemblies, the most commonly used today as a deviation tool is the bent-housing mud motor.

The bent sub and the bent housing use the bit tilt (i.e., misalignment of bit face away from the axis of the drill string) and bit side force to change the hole direction and inclination. The bent housing is more effective than the bent sub because of a shorter bit-to-bend distance, which reduces the bit offset and creates a higher build rate for a given size of bend. A shorter bit-to-bend distance also reduces the moment arm, which in turn reduces the bending stress at the bend. As a result, the bent-housing PDM is easier to orient and allows for long period of rotation. The only application for a bent sub is in larger hole sizes (22–26 in.). Owing in particular to the introduction of the adjustable bent housing, bent subs have become obsolete in most applications.

Before the personal computer became widely available, the simple *three-point curvature* calculation was used to predict the build rates of the motor assemblies, as follows:

$$BR = \frac{200\theta_b}{L_1 + L_2} \tag{9.15}$$

where

BR is the build rate, in degrees per 100 ft
θ_b is the bend angle, in degrees
L_1 is the distance from the first contact point (bit) to the second point (bend), in ft
L_2 is the distance from the second contact point to the third point (motor top stabilizer), in ft

For more accurate results, a BHA analysis program is often used to calculate the build/drop/turn rates of the motor assemblies. Note that the most popular BHA programs are usually based on the finite-element method, but they have been shown to not accurate enough to model the steerable motor system or the RSS. BHA software using the semianalytical method, though less flexible to use, is recommended when modeling PDMs and RSS, because of the better results.

Bent-housing motor components. A typical bent-housing motor contains four sections:

- Dump sub
- Power unit
- Transmission/bent-housing unit
- Bearing section

Dump sub. The dump sub, located on top of the motor assembly, contains a valve that is ported to allow fluid flow between the drill strings and the annulus. This allows the drill strings to fill when tripping in and empty when tripping out of the hole. The dump sub also permits low-rate circulation, bypassing the motor if required.

Power unit. Most motor assemblies use the Moineau pump principle to convert hydraulic energy into mechanical energy. A rotor/stator pair converts the hydraulic energy of the pressurized circulating fluid into the mechanical energy of a rotating shaft. The rotor and stator are of lobed design. The rotor and stator lobe profiles are similar, with the steel rotor having one less lobe than the elastomeric stator. The rotor and stator lobes are helical in nature, with one stage equating to the linear distance of a full *wrap* of the stator helix.

Power units may be categorized with respect to the number of lobes and effective stages. The speed and torque of a power section is directly linked to the number of lobes on the rotor and the stator. The higher the number of lobes is, the higher will be the torque and the lower will be the rotary speed. Typical rotor/stator configurations range from 1:2 lobes to 9:10 lobes. For best performance, the power section should be matched to the bit and the formation being drilled.

Transmission/bent-housing unit. The universal couplings inside the transmission/bent-housing unit eliminate all eccentric rotor motion and accommodate the misalignment motion of the bent housing while transmitting torque and downward thrust to the driveshaft, which is held concentrically by the bearing assembly.

Bearing section. The bearing assembly consists of multiple thrust-bearing cartridges, radial bearings, a flow restrictor, and a driveshaft. The thrust bearings support the downward thrust of the rotor, the hydraulic downward thrust owing to bit pressure loss, and the reactive upward thrust from the applied WOB. For larger-diameter motors, the thrust bearings are usually of a multi-stack ball-and-track design. Small-diameter motors utilize carbide- and/or diamond-enhanced friction bearings. Metallic and nonmetallic radial bearings are employed above and below the thrust bearings to absorb lateral side loading of the driveshaft.

Approximately 5–8% of the circulating fluid is allowed to flow through the bearing section by the flow restrictor to cool and lubricate the bearing assembly. The type of flow restrictor used is preselected and set in the motor shop, based on planned bit hydraulics. It cannot be changed at rig site. The driveshaft transmits both axial load and torque to the bit. The driveshaft is a forged component that is designed such that fatigue strength, axial strength, and torque strength are maximized. It has a threaded connection at the bottom end to facilitate connection to the drill bit.

Power delivered to the bit. In PDMs, the power delivered to the bit is expressed by the following relation:

$$\text{HP} = \frac{TN}{5{,}252} \tag{9.16}$$

where

HP is horsepower

T is the torque, in ft-lbs

N is the rotational speed, in rpm

Applications of PDMs in directional drilling. There are several applications of PDMs, the most important of which are described in the following sections.

Sidetracking. The most common method of sidetracking out of casing, especially when considerable drilling is to follow, is to mill a length of casing with a section mill and then to divert the trajectory with a bent-housing motor assembly. The assembly usually contains a stabilizer on the motor and may contain another one above the motor.

Steerable drilling and kicking off. The essential requirement for a steerable drilling system is that it can make both the inclination and azimuth changes. This is the most commonly used configuration because

- An average planned curvature can be adhered to by a combination of orienting and rotating
- After completing the buildup, the assembly can be rotated ahead to hold angle with minor corrections being made to inclination and azimuth as necessary
- Extended intervals can be drilled through different formations without tripping for assembly changes
- Drilling performance is maximized by efficiently delivering torque and horsepower at the bit

This system usually consists of a bent-housing motor and a stabilizer on the bearing housing. To enhance the sliding capability of the motor, the stabilizer has wide, straight blades tapered at either end and is undergauge relative to hole size (typically ⅛ in. to ½ in.). Depending on the application, additional stabilizers may be used above the motor. Although these stabilizers are generally spiraled, the blades should be tapered and undergauge.

The overall design of a steerable assembly will depend on the intended application, and the most important considerations are as follows:

- The expected build rate in oriented mode should be slightly greater (typically 1–2°/100 ft) than required, so as to guarantee the planned build rate
- The number of stabilizers used should be kept to a minimum to reduce drag in the oriented mode
- If the drill string is rotated in a curved section, bending stresses around the bent housing should be checked to ensure that they are below the endurance limit

Medium-radius applications (6–15°/100 ft DLS). For build rates in the range of 6–15°/100 ft, the vast majority of medium-radius drilling is undertaken in hole sizes of 12¼ in. and under, using motors of 8 in. diameter and less. A number of motor configurations are used to drill medium-radius wells, each having their own merits, including the following:

- Single-bent-housing motor
- Single bent housing with offset pad
- Double-bend motor: bent-housing motor with bent sub positioned on top of the motor and aligned with the bend
- Double-bent-housing motor

Intermediate- and short-radius applications. Intermediate-radius drilling systems are used to achieve build rates in the range of 15–65°/100 ft. The build and lateral sections are drilled with a short bearing pack motor. When the build rate exceeds 45°/100 ft, an articulated motor and a flexed MWD tool should be used. Both system types can be utilized for new or reentry wells.

Two types of motors are used to drill short-radius wells, with build rates in the range of 65–125°/100 ft: an articulated *build* motor (used to drill the build section) and a *hybrid lateral* motor (used to drill the horizontal lateral section). The articulated MWD tool is used on both the build and the lateral sections.

Rotary steerable systems

RSS represents an evolutionary step forward in directional drilling technology, overcoming the drawbacks associated with steerable motors and conventional rotary assemblies. In steerable motors, to initiate a change in the wellbore trajectory, the rotation of the drilling is halted in such a position that the bend in the motor points in the direction of the new trajectory. This mode, known as the sliding mode, typically creates higher frictional forces on the drill string. In extreme extended-reach drilling, the frictional force builds to the point at which there is no axial weight available to overcome the drag of the drill string against the wellbore; thus, further drilling is not possible. To overcome this limit of the steerable motor assemblies, RSS were developed in the early 1990s, responding to the needs of extended-reach drilling. The first RSS was used in BP's Wytch Farm (UK) extended-reach wells.

RSS allow continuous rotation of the drill string while steering the bit. Thus, they generally have a better ROP than do the conventional steerable motor assemblies. Other benefits include better hole cleaning, lower torque and drag, and better hole quality. RSS are much more complex mechanically and electronically and are therefore more expensive to run, as compared to conventional steerable motor systems. This economic penalty tends to limit their use to highly demanding extended-reach wells or to the very complex profiles associated with "designer" wells. Additionally, the technology is still very new. As a result, the present generation of systems (manufactured in 2002) are climbing a very steep learning curve with regards to run length, performance, and mechanical reliability.

Two steering concepts are used in RSS:

- Point the bit
- Push the bit

Point-the-bit systems use the same principle as is employed in bent-housing motor systems. In RSS, the bent housing is contained inside the collar, so that it can be oriented to the desired direction during drill string rotation.[12] Point-the-bit systems claim to allow the use of long-gauge bit to reduce hole spiraling and drill a straighter wellbore.[13] Push-the-bit systems use the principle of applying side force to the bit, pushing it against the borehole wall to achieve the desired trajectory. The force can be hydraulic pressure[14] or mechanical.[15] For either a point-the-bit or a push-the-bit RSS, a maximum build rate of about 6–8°/100 ft can generally be expected for an 8½ in. hole size tool.

Adjustable-gauge stabilizers

In the late 1980s, the industry developed adjustable-gauge stabilizers, whose effective blade OD could be changed while the tool was in downhole. With adjustable-gauge stabilizers, drillers could change stabilizer OD without having to make time-consuming and costly trips out of the hole. Adjustable stabilizers run in rotary assemblies were often placed near the bit or positioned about 15–30 ft from the bit. In these positions, changes in their gauge could effectively control the build/drop tendency of the assembly. Because they could control inclination while in the rotary mode, these assemblies became known as two-dimensional rotary systems. Adjustable-gauge stabilizers can also be run with steerable motor systems, making it possible to control inclination with the stabilizer while drilling in the rotary mode. If the wellbore requires a change in azimuth, one would have to revert to sliding mode.

Adjustable-gauge stabilizers have been widely used recently—in particular, in drilling into a horizontal section by using a geological steering or pay zone steering device, usually consisting of a logging-while-drilling tool. A resistivity sensor, with its deep investigation depth, can detect a geological change many feet before the bit penetrates that boundary. This ability may allow one to hold the drilling assembly in the reservoir and to steer away from either an upper or a lower boundary.[16]

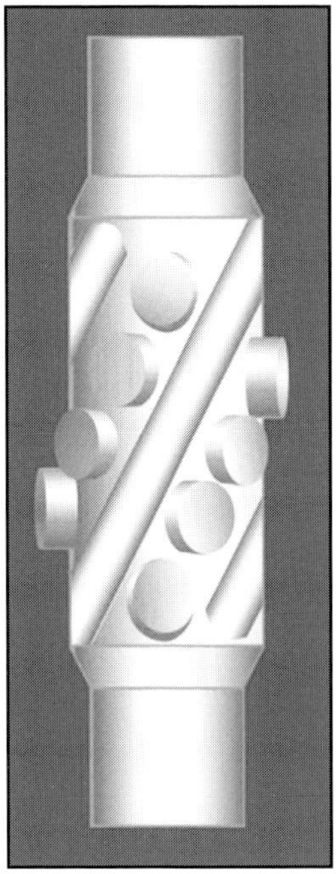

Fig. 9–19. Adjustable-gauge stabilizer

Whipstocks

Open hole whipstocks were the first type of deflection tool used for changing the wellbore trajectory but are seldom used today. Bent-housing motors have replaced open hole whipstocks as the most commonly used deviation tool in open hole sidetracking. Casing whipstocks, however, are still used routinely to sidetrack out of cased wellbores.

Whipstocks can be either retrievable or nonretrievable. The retrievable ones are ideal for drilling multiple laterals from a single wellbore.

A typical casing sidetracking operation involves multiple trips—to set the cement plug, set the whipstock and start the window mill, complete the mill and clean up, and so forth. To save time, recently developed systems can do all of this in one trip. Note that gyro survey is usually required for whipstock setting and for initial tool orientations.

Turbines

Turbines, also commonly known as turbodrills, are powered by the turbine motor, which has a series of rotors/stators (stages) connected to a shaft. As the drilling fluid is pumped through the turbine, the stators deflect the fluid against the rotors, forcing the rotors to rotate the driveshaft to which it is connected. Turbines are designed to run at high speed and low torque and thus are suited for running with diamond or PDC bits. Not only are turbines less flexible and less efficient than PDMs, but they are also more expensive; thus, turbines are not as widely used as PDMs, except in Russia.

Jetting bits

Using the hydraulic energy of the drilling fluid to erode a pocket out of the bottom of the borehole, jetting bits can change the trajectory of a borehole. A tri-cone bit with one large nozzle is oriented in the desired hole direction, to create a pocket. The drilling assembly is forced into the jetted pocket for a short distance. The procedure is continued until the desired trajectory change is achieved. Jetting is seldom used today owing to slow ROP and limitations in soft formations.

Supplementary Problems

Problem 9–1. Describe, in brief, the use of drill collars and list their functions.

Problem 9–2. Calculate the length of drill collars required in order to apply 25 kips WOB to drill an 8½ in. hole. Use a 6-in., 72 lbs/ft drill collar. You are given the following data:

- Mud weight = 12 lbs/gal
- Hole inclination = 15°
- Design factor = 1.1

Problem 9–3. Design a drill string given the following data:

- Hole size = 8½ in.
- Mud density = 13 lbs/gal
- Drill collars to be used: 6 in. × 2¼ in.; 83 lbs/ft
- WOB required = 50 kips
- Design factor = 1.1
- HWDP: 270 ft, 4½ in. × 2¾ in.; 41 lbs/ft
- MOP = 80 kips

Problem 9–4. Calculate the torque that can be applied when a tensile force pull of 80,000 lbs is applied to the string.

Problem 9–5. Calculate the curvature produced by buckling for an axial load of 250 kips given the following data:

- Drill pipe: 5 in.; 19.5 lbs/ft; NC 50 connection with tool joint OD = 6⅝ in.
- Hole size = 8½ in.

Nomenclature

BF	Buoyancy factor
BR	Build rate, in degrees per 100 ft
D	Drill pipe outer diameter (OD)
DF	Design factor
E	Young's modulus of elasticity
f	Coefficient of friction
F_F	Friction force; also, sliding friction force acting on the element
F_{hel}	Helical buckling load
F_N	Net normal force
F_{sin}	Sinusoidal buckling load
HP	Horsepower
L_1	Distance from the first contact point (bit) to the second contact point (bend), in ft
L_2	Distance from the second contact point to the third contact point (motor top stabilizer), in ft

L_c Depth at which collapse pressure is to be checked, in ft
M Torsion at the lower end of the element
N Rotational speed, in rpm
P_c Collapse pressure, in psi
r Radial clearance between the drill string and the wellbore
R Characteristic radius of the element
R_c Radius of curvature
T Torque, in ft-lbs
T_a Allowable axial tension, in lbs
T_h Calculated hook load, in lbs
T_{axial} Axial tension at the lower end of the element
w Buoyed unit weight of the drill string
W Buoyed weight of the drill string element
w_{dc} Buoyed weight of the drill collar, in lbs/ft
WOB Weight on bit, in lbs
x Inside fluid depth, in ft
θ Inclination angle at the lower end of the element
θ_b Bend angle, in degrees
ρ_m Mud density, in lbs/gal
ρ_{mi} Inside fluid density, in lbs/gal
σ_b Bending stress
ϕ Azimuth angle at the lower end of the element

References

1 *Recommended Practice for Drill Stem Design and Operating Limits.* 6th ed. 1998. API 7G.

2 Ibid.

3 Ibid.

4 Johancsik, C. A., D. B. Friesen, and R. Dawson. 1984. "Torque and Drag in Directional Wells—Prediction and Measurement." *Journal of Petroleum Technology,* June. 987–992.

5 Ibid.

6 Johancsik, Friesen, and Dawson, *Journal of Petroleum Technology,* June 1984. Rasmussen, B., J. O. Sorheim, E. Seiffert, O. Angeltvadt, T. Gjedrem. 1991. "World Record in Extended Reach Drilling, Well 33/9-C10, Statfjord Field, Norway." Paper presented at the SPE/IADC Drilling Conference, Amsterdam.

7 Ibid.

8 Dellinger, T., W. Gravley, and G. C. Tolle. 1980. "Directional Technology Will Extend Drilling Reach." *Oil and Gas Journal,* September 15, 153–69.

9 Dawson, R., and P. R. Paslay. 1984. "Drillpipe Buckling in Inclined Holes." *Journal of Petroleum Technology,* October.

10 Paslay, P. R., J. Yin-Min, J. E. E. Kingman, and J. D. Macpherson. 1992. "Detection of BHA Lateral Resonances while Drilling with Surface Longitudinal and Tortional Sensors." Paper SPE 24583, presented at the 67th SPE Annual Technical Conference and Exhibition, Washington, D.C.

11 Dykstra, M. W. 1996. "Nonlinear Drill String Dynamics." Ph.D. dissertation, University of Tulsa.

12 Vandiver, J. K., J. W. Nicholson, and R.-J. Shyu. 1990. "Case Studies of the Bending Vibration and Whirling Motion of Drill Collars." *SPE Drilling Engineering,* December, 282–90.

13 Dunayevsky, V. A., A. Judzis, and F. Abbassian. 1993. "Dynamic Stability of Drillstrings under Fluctuating Weights-on-Bit." *SPE Drilling and Completion,* June, 84–91. Field, D. J., A. J. Awarbrick, and G. A. Haduch. 1993. "Techniques for Successful Application of Dynamic Analysis in the Prevention of Field-Induced Vibration Damage in MWD Tools." Paper presented at the SPE/IADC Drilling Conference.

14 Field, D.J., A.J. Awarbrick, and G.A. Haduch, SPE/IADC Drilling Conference presentation.

15 Schaaf, S., D. Pafites, and E. Guichemerre. 2000. "Application of a Point the Bit Rotary Steerable System in Directional Drilling Prototype Well-Bore Profile." Paper SPE 62519, presented at the 2000 SPE/AAPG Western Regional Meeting, Long Beach, Calif.

16 Yonezawa, T., et al. 2002. "Robotic Controlled Drilling: A New Rotary Steerable Drilling System for the Oil and Gas Industry." Paper IADC/SPE 74458, presented at the IADC/SPE Drilling Conference, Dallas.

17 Barr, J. D., J.M. Clegg, and M.K. Russell. 1995 "Steerable Rotary Drilling with an Experimental System." Paper SPE/IADC 29382, presented at the SPE/IADC Drilling Conference, Amsterdam.

18 Gruenhagen, H., V. Hahne, and G. Alvord. 2002. "Application of New Generation Rotary Steerable System for Reservoir Drilling in Remote Areas." Paper IADC/SPE 74457, presented at the IADC/SPE Drilling Conference, Dallas.

19 Lawrence, L., J. Stymiest, and R. Russell. 2002. "Adjustable-Gauge Stabilizer in Motor Provides Greater Inclination Control." *Oil and Gas Journal,* February 18, 37–41.

10

Drilling Problems and Solutions

Introduction

During drilling operations, some type of a drilling problem will almost certainly occur, even in very carefully planned wells. The reason is that geological conditions for two wells that are near each other may differ (nonhomogeneous formation); therefore, different problems can be encountered.

The key to success in achieving well objectives is to design drilling programs based on anticipation of potential hole problems, rather than on containment and caution. Drilling problems, when encountered, can be very costly, the most prevalent of which are

- Pipe sticking
- Lost circulation
- Hole deviation
- Pipe failures
- Borehole instability
- Mud contamination
- Formation damage
- Hole cleaning
- Hydrogen sulfide–bearing formations and shallow gas
- Equipment- and personnel-related problems

An understanding of these problems, their causes, their anticipation and planning for solutions is essential to control overall well cost control and succed in reaching the intended target zone. This chapter addresses these problems, possible solutions, and when applicable, preventive measures.

Pipe Sticking

Definition

In drilling operations, a pipe is considered stuck if it cannot be freed (pulled out of the hole) without damage to the pipe and without exceeding the maximum allowed hook load of the drilling rig. There are two types of pipe sticking problems:

- Differential pressure pipe sticking
- Mechanical pipe sticking

Differential pressure pipe sticking

Differential pressure pipe sticking occurs when a portion of the drill string becomes embedded into a mud cake (an impermeable film of fine solids) that forms on the wall of a permeable formation during the drilling process. If the mud pressure, P_m, that acts on the outside wall of the pipe is greater than the formation fluid pressure, P_{ff}, as is generally the case except in underbalanced drilling, then the pipe is said to be differentially stuck, as shown in figure 10–1.

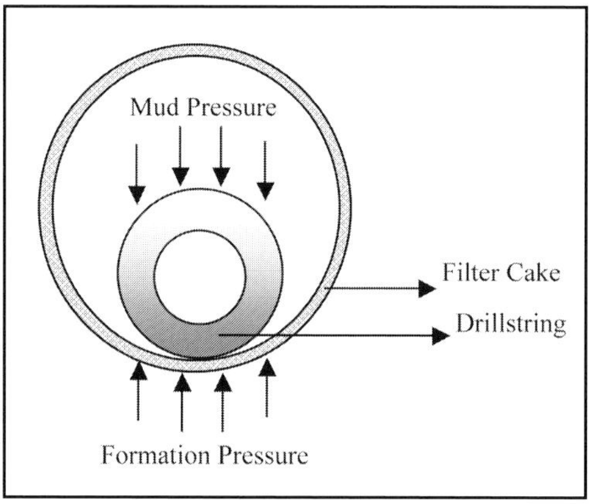

Fig. 10–1. Differential pressure pipe sticking

The differential pressure acting on the portion of the drill pipe that is embedded into the mud cake can be expressed as follows: $\Delta P = P_m - P_{ff}$. The *pull force,* F_{pull}, that is required in order to free the stuck pipe is a function of the differential pressure, ΔP; the coefficient of friction, μ; and the total contact area, A_c, between the pipe and mud cake surfaces:

$$F_{pull} = \mu A_c \Delta P \tag{10.1}$$

The *coefficient of friction,* μ, can vary from less than 0.04, for oil-based muds, to as much as 0.35, for weighted water-based muds with no lubricants added. The *contact area,* A_c, can be expressed in terms of the arc length, Ψ_{arc}, and length of the pipe body portion, L_{ep} (fig. 10–2) that are embedded in mud cake. The *arc length* is given as follows:[1]

$$\psi_{arc} = 2\sqrt{\left(\frac{D_h}{2} - t_{mc}\right)^2 - \left(\frac{D_h}{2} - t_{mc}\frac{D_h - t_{mc}}{D_h - D_{op}}\right)^2} \tag{10.2}$$

D_{op} must be equal to or greater than $2t_{mc}$ and equal to or less than $(D_h - t_{mc})$ where

D_h is the hole diameter
D_{op} is the outer pipe diameter
t_{mc} is the mud cake thickness

In equation (10.2), D_{op} must be equal to or greater than $2t_{mc}$ and equal to or less than $D_h - t_{mc}$. Thus, the contact area can be expressed as follows:

$$A_c = \Psi_{arc} L_{ep} \tag{10.3}$$

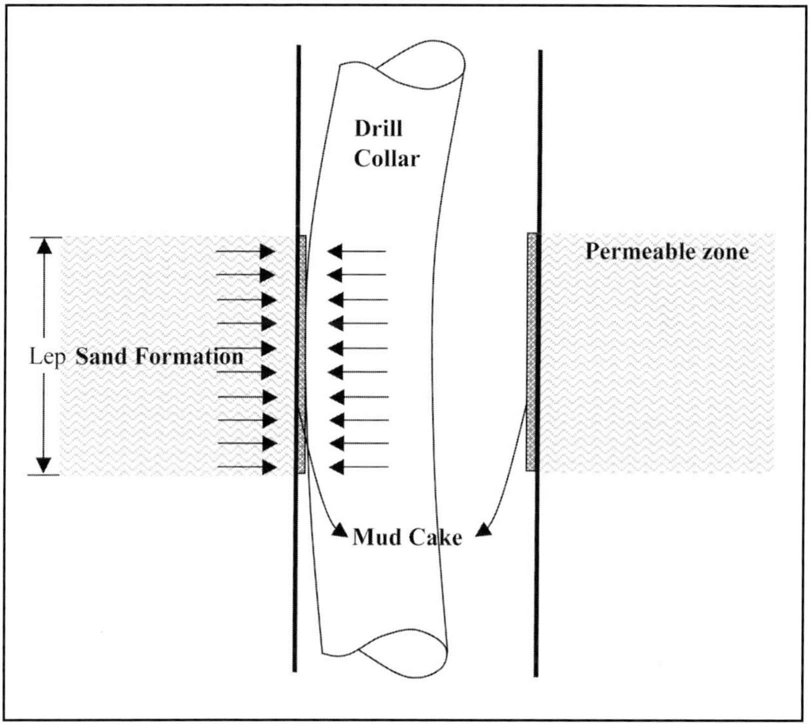

Fig. 10–2. Differential pressure pipe sticking with embedded pipe length

Equations (10.1)–(10.3) show controllable parameters that will cause higher pipe sticking force and the potential inability of freeing the stuck pipe. These are:

- Unnecessary high differential pressure
- Thick mud cake (high continuous fluid loss to formation)
- Low-lubricity mud cake (high coefficient of friction)
- Excessive embedded pipe length in mud cake (time delay in freeing operations)

Although hole and pipe diameters and hole angle play a role in the pipe-sticking force, they are uncontrollable variables. Once they are selected to meet well design objectives, they cannot be altered. However, the shape of drill collars, as in square drill collars and drill collars with spiral grooves and external upset tool joints, can minimize the sticking force.

Some of the indicators of differential pressure stuck pipe while drilling permeable zones or known depleted-pressure zones are

- Increase in torque and drag
- Inability to reciprocate the drill string and, in some cases, difficulty in rotating it
- Drilling-fluid circulation is not interrupted

Differential pressure pipe sticking can be prevented—or its occurrence can at least be mitigated—if some or all of the following precautions are taken:

- Maintain the lowest continuous fluid loss while adhering to the project economic objectives (e.g., controlling the mud cake thickness)
- Maintain the lowest amount of drilled solids in the mud system or, if economical, remove all drilled solids
- Use lowest differential pressure, with allowance for swab and surge pressures during tripping operations
- Select a mud system that will yield smooth mud cake (low coefficient of friction)
- Maintain drill string rotation at all times, if possible

As stated earlier, differential pressure pipe sticking problems may not be totally prevented. Therefore, when the problem does occur, common field practices for freeing the stuck pipe include

- Mud hydrostatic pressure reduction in the annulus
- Oil spotting around the stuck portion of the drill string
- Washing over the stuck pipe

Some of the methods used to reduce the hydrostatic pressure in the annulus are

- Reduction of mud weight by dilution
- Reduction of mud weight by gasifying with nitrogen
- Placement of a packer in the hole above the stuck point

Example 10–1

You are given the following well data:

- Drill collar outer diameter: $D_{op} = 6.0$ in.
- Hole diameter: $D_h = 9.0$ in.
- Mud cake thickness: $t_{mc} = 2/32$ in.
- Coefficients of friction: 0.15, for oil-based mud; 0.25, for water-based mud
- Length of drill collar embedded in mud cake: $L_{ep} = 20$ ft $= 240$ in.
- Differential pressure: $\Delta P = 500$ psi
- Determine the required pull force to possibly free the stuck pipe.

Solution: By use of equation (10.2), the arc length is found: $\psi_{arc} = 2.1$ in. By use of equation (10.1), the pull force is then calculated:

for the coefficient of friction 0.15,

$F_{pull} = 0.15 \times 500 \times 2.1 \times 240 = 37{,}530$ lbs; for the coefficient of friction 0.25
$F_{pull} = 0.25 \times 500 \times 2.1 \times 240 = 62{,}550$ lbs

Mechanical pipe sticking

The causes of mechanical pipe sticking include:

- High accumulation of drilled cuttings in the annulus
- Borehole instabilities—such as hole caving, sloughing, or collapse and plastic shale or salt sections squeezing (narrowing)
- Key seating

Drilled cuttings. Excessive drilled-cuttings accumulation in the annular space owing to inadequate hole cleaning can cause mechanical pipe sticking—particularly in directional well drilling, in which a stationary cuttings bed may form on the low side of the borehole. If this condition exists while tripping out, pipe sticking is very likely imminent. This is why it is a common field practice to circulate bottom up several times, with the drill bit off bottom, to flush out any cuttings bed that may be present prior to making a trip. A trend of increasing torque and drag and sometimes circulating drill pipe pressure is an indication of large accumulations of cuttings in the annulus and potential mechanical pipe-sticking problems. Some preventives measures include:

- Proper hydraulics and/or mud viscosity to maintain turbulent flow in annulus
- Drill string rotation at high enough speed to agitate any stationary cuttings bed that may have formed on the low side of the borehole

Borehole instability. This topic is addressed in more detail later; however, it is important at this time to mention, in brief, the pipe-sticking issues that are associated with the borehole instability problems. The most troublesome problem is posed by drilling into shale. Depending on mud composition and mud weight, shale can slough or plastically flow inward and therefore causes mechanical pipe sticking. In all types of formations, using too low of a mud weight can lead to the collapse of the hole, possibly causing mechanical pipe sticking. Also, drilling through salt that is known to exhibit plastic behavior under the overburden pressure can lead to problems. If mud weight is not high enough, the salt has the tendency to flow inward, causing mechanical pipe sticking. Indications of potential pipe-sticking problems due to borehole instability are a rise in circulating drill pipe pressure, an increase in torque, and in some cases, no fluid return to surface. (See fig. 10–3.)

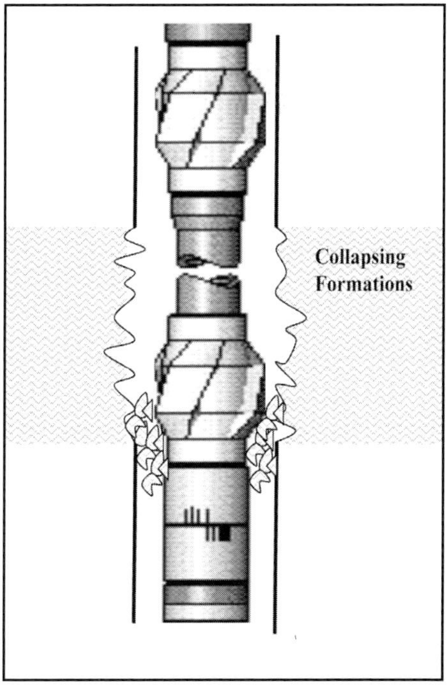

Fig. 10–3. Pipe sticking due to wellbore instability

Key seating. It is well known that *key seating* is a major cause of mechanical pipe sticking (fig. 10–4). The mechanics of key seating involve wearing a small hole (groove) into the side of a full-gauge hole. This groove is caused by drill string rotation with side force acting on it. This condition is created in either doglegs or undetected ledges in the vicinity of washouts.

The lateral force that tends to push the pipe against the wall that causes mechanical erosion and thus creates a key seat is given by $F_L = T\sin\theta_{dl}$, where F_L is the lateral force, T is the tension in the drill string just above the key seat area, and θ_{dl} is the abrupt change in hole angle (commonly referred to as the dogleg angle).

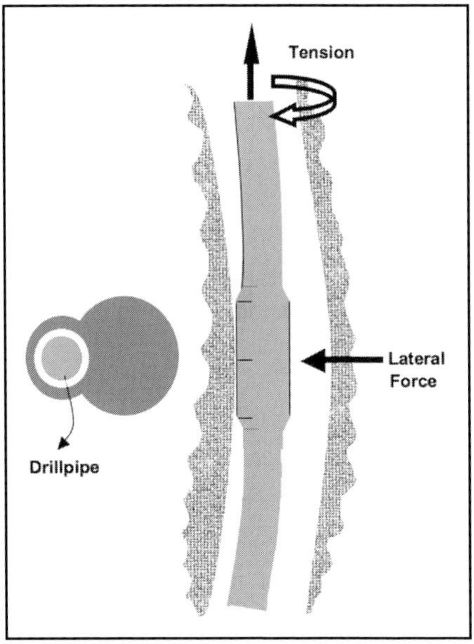

Fig. 10–4. Pipe sticking due to key seating

In general, long bit runs can cause key seats; therefore, it is common practice to make what is called wiper trips. Also, use of stiffer BHAs tends to minimize severe dogleg occurrences. During tripping out of the hole, a key-seating pipe-sticking problem is evident when several stands of pipe have been pulled out and then, all of a sudden, the pipe becomes stuck.

Freeing mechanically stuck pipe is achieved in many ways, depending on the suspected cause of sticking. For example, if cuttings accumulation is suspected to be the cause, then rotating and reciprocating the drill string and increasing the flow rate without exceeding the maximum allowed ECD are possible remedies for freeing the pipe. If hole narrowing owing to plastic shale is the cause, then an increase in mud weight may free the pipe, whereas hole narrowing owing to salt is the cause, then circulation of freshwater may free the pipe.

By contrast, if pipe is stuck in a key seat area, the most successful solution is backing off below the key seat and going back into the hole with an opener, to drill out the key section. Consequently, this will leads to a fishing operation, to retrieve the fish. The decision on how long to continue attempting to free stuck pipe versus back off, plug back, and then sidetrack is an economic issue that is generally addressed by the operating company.

Loss of Circulation

Definition

Lost circulation is defined as the uncontrolled flow of whole mud into a formation, sometimes referred to as a thief zone. The loss can be partial or total, as shown in figure 10–5. In partial lost circulation, mud continues to flow to surface with some loss to the formation. Total loss circulation, however, is when all the mud flows into a formation with no return to surface. In total loss circulation, if drilling is allowed to continue, it is referred to as blind drilling. This is not a common practice in the field, unless the formation above the thief zone is mechanically stable (i.e., there is no production and the fluid is clear water) and it is feasible, in terms of economics and safety, to go ahead with the drilling operation.

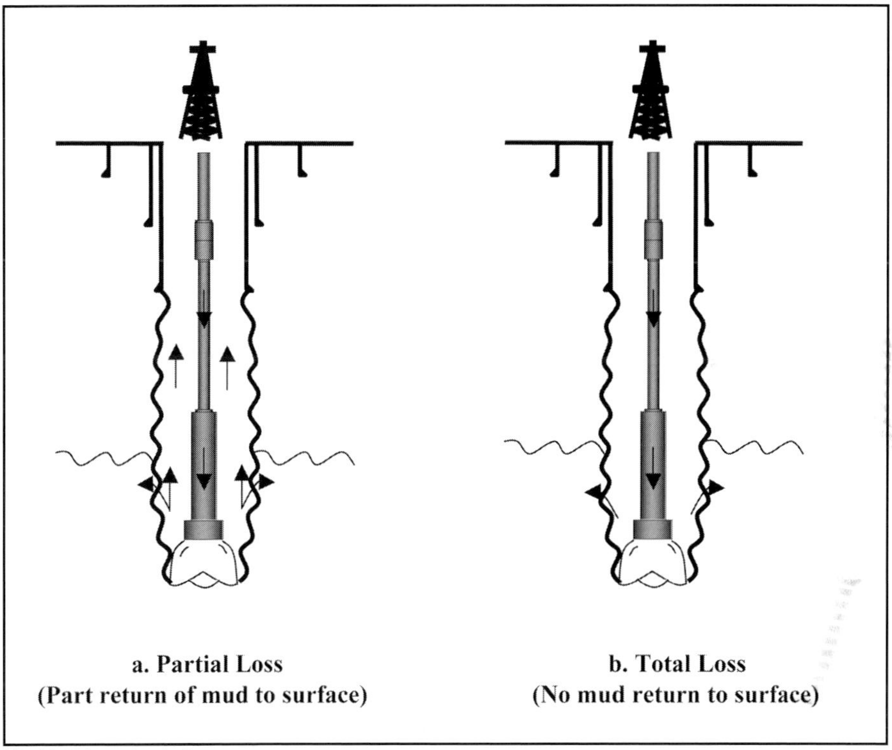

Fig. 10–5. Lost circulation

Lost circulation zones and causes

Formations that are inherently fractured, depleted, or cavernous or have a high permeability, on the one hand, are potential zones of lost circulation. On the other hand, under certain improper drilling conditions, induced fractures become potential zones of lost circulation. The major causes of induced fractures are excessive downhole pressures and improper intermediate casing-setting depth, especially in the transition zone.

Induced or inherent fractures may be horizontal at shallow depth or vertical at depths greater than approximately 2,500 ft. Excessive wellbore pressures are caused by high flow rates (high annular friction pressure loss) or tripping in too fast (high surge pressure), which can lead to high ECD. In addition, as a result of improper annular hole cleaning, excessive mud weight, or shutting a well in high pressure, shallow gas can also induce fractures and therefore cause lost circulation. Equations (10.4) and (10.5) show the conditions that must be maintained to avoid fracturing the formation during drilling and tripping in, respectively:

$$(ECD)_{drilling} = \rho_{eq} = \rho_m + \Delta\rho_f < \rho_{frac} \tag{10.4}$$

$$(ECD)_{tripping\ in} = \rho_{eq} = \rho_m + \Delta\rho_{srg} < \rho_{frac} \tag{10.5}$$

where

ρ_m is the static mud weight
$\Delta\rho_f$ is the additional mud weight owing to friction pressure loss in annulus
$\Delta\rho_{srg}$ is the additional mud owing to surge pressure
ρ_{frac} is the formation pressure fracture gradient in equivalent mud weight
ρ_{eq} is the equivalent circulating mud weight

Cavernous formations are limestone with inherently large caverns. This type of lost circulation is quick and total and is the most difficult to seal. High-permeability formations that are potential lost circulation zones are those of shallow sand of permeability in excess of 10 darcies. Generally, deep sand has a low permeability and therefore presents no loss-of-circulation problems. In noncavernous thief zones, mud level in mud tanks decreases gradually, and if drilling continues, total loss of circulation may occur.

Prevention of lost circulation

Complete prevention of lost circulation is not possible. This is because some formations—such as inherently fractured, low-pressure, cavernous, or high-permeability zones—are not avoidable when encountered during the drilling operation if the target zone is to be reached. However, mitigating the problem of lost circulation is possible if certain precautions, especially those related to induced fractures, are taken:

- Maintain proper mud weight
- Minimize annular friction pressure losses during drilling and tripping in
- Maintain adequate hole cleaning and avoid restrictions in the annular space
- Set casing to protect upper weaker formations during a transition zone
- Update formation pore pressure and fracture gradients for more accurate log and drilling data
- If lost circulation zones are anticipated, work out preventive measures prior to drilling that zone by treating the mud with lost circulation materials

Remedial measures

When lost circulation occurs, sealing the zone is necessary, unless the geological conditions allow blind drilling (which is unlikely in most cases). The common lost circulation materials that are mixed with the mud to seal loss zones may be grouped as fibrous, flaked, granular, and/or a combination of the three.

These materials are available in course, medium, and fine grades for an attempt to seal low- to moderate-loss circulation zones. In the case of severe-loss circulation zones, the use of various plugs to seal the zone becomes mandatory. It is important, however, to know the location of the lost circulation zone prior to setting a plug. Various types of plugs used throughout the industry include bentonite–diesel oil squeeze, cement–bentonite–diesel oil squeeze, cement, and barite.

Hole Deviation

Definition

Hole deviation is the unintentional departure of the drill bit from drilling along a preselected borehole trajectory. Whether drilling a straight or a curved hole section, the tendency of the bit to walk away from the desired path can lead to a higher drilling cost and legal problems with regard to the lease boundary. (See fig. 10–6.)

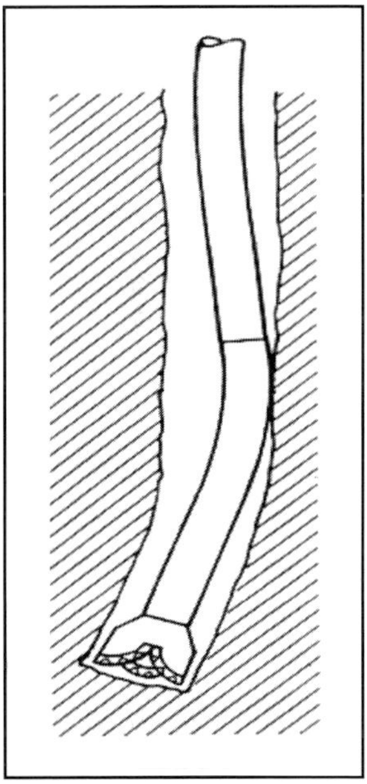

Fig. 10–6. Wellbore deviation

Causes

It is not exactly known what causes a drill bit to deviate from its intended path. It is, however, generally agreed that one or a combination of several of the following factors may be responsible for hole deviation:

- Heterogeneous nature of formation and dip angle
- Drill string characteristics and dynamic behavior
- Applied WOB
- Hole inclination angle from vertical
- Drill bit type and basic mechanical and hydraulic design
- Hydraulics at the bit
- Improper hole cleaning

A resultant force acting on a drill bit causes hole deviation to occur. The mechanics of this resultant force is complex and is mainly governed by the mechanics of the BHA, the rock-bit interaction, bit operating conditions, and to a lesser extent, the drilling-fluid hydraulics. The forces imparted to the drill bit owing to the BHA are directly related to the makeup of the BHA, (stiffness, stabilizers, reamers, etc.).

The BHA is a flexible elastic structural member that can buckle under compressive loads. The buckled shape of a given design of BHA depends on the amount of applied WOB. The significance of the BHA buckling is that it causes the axis of the drill bit to misalign with the axis of the intended hole path, causing the deviation. Pipe stiffness and length and number of stabilizers (their location and clearances from the wall of the wellbore) are two major parameters that govern the BHA-buckling behavior. The buckling tendency of the BHA is minimized by reduction of WOB and use of stabilizers with outside diameters that are almost in gauge with the wall of the borehole.

The contribution of the rock-bit interaction to bit deviating forces is governed by the following factors:

- Rock properties (cohesive strength, bedding or dip angle, and internal friction angle)
- Drill bit design features (tooth angle, bit size, bit type, bit offset [in the case of roller cone bits], teeth location and number, bit profile, and bit hydraulic features)
- Drilling parameters (tooth penetration into the rock and its cutting mechanism)

The mechanics of the rock-bit interaction is a very complex subject and is the least understood problem contributing to hole deviation. Fortunately, the advent of downhole measuring-while-drilling (MWD) tools that allow the monitoring of the advance of the drill bit along the desired path has made our lack of understanding with regard to the mechanics of hole deviation more acceptable.

Drill Pipe Failures

Drill pipe failures can be categorized as follows:

- Twist-off—due to excessive torque
- Parting—due to excessive tension
- Burst or collapse—due to excessive internal pressure or external pressure, respectively
- Fatigue—due to mechanical cyclic loads, with or without corrosion

Twist-off

Pipe failure due to twist-off occurs when the induced shearing stress owing to high torque exceeds the pipe material's ultimate shear stress. In vertical well drilling, excessive torques are not generally encountered under normal drilling practices. In directional and extended-reach drilling, however, torques in excess of 80,000 ft-lbs are common and can easily cause twist-off of improperly selected drill string components.

Parting

Pipe failure due to parting occurs when the induced tensile stress exceeds the pipe material's ultimate tensile stress. This condition may arise when pipe sticking occurs and an overpull is applied in addition to the effective weight of suspended pipe in the hole above the stuck point.

Collapse and burst

Pipe failure due to collapse or burst is not common. However, under extreme conditions of high mud weight and complete loss of circulation, pipe burst may occur.

Fatigue

Fatigue is a dynamic phenomenon that may be defined as the initiation and propagation of micro-cracks into macro-cracks as a result of repeated applications of stresses. It is a process of localized progressive structural fractures in material under the action of dynamic stresses.

It is well established that a structural member that may not fail under a single application of static load may very easily fail under the same load if it is applied repeatedly. Failure under cyclic (repeated) loads is referred to as fatigue failure.

Drill string fatigue is the most common and costly form of drill pipe failure in oil/gas and geothermal drilling operations. The combined action of cyclic stresses and corrosion can shorten the life expectancy of a drill pipe by a factor of thousands. Cyclic stresses are induced by dynamic loads owing to drill string vibrations and bending load reversals in curved sections of hole and doglegs owing to rotation. Pipe corrosion occurs during the presence of oxygen, carbon dioxide, chlorides, and/or hydrogen sulfide (H_2S). H_2S is the most severely corrosive element to steel pipe and is deadly to humans. Regardless of what may have caused pipe failure, the cost of fishing operations—and, sometimes, unsuccessful attempts to retrieve the fish out of the hole—can lead to the loss of multiple millions of dollars owing to rig downtime, the loss of expensive tools down hole, or the abandonment of the already-drilled section below the fish.

In spite of the vast amount of work that has been dedicated to pipe fatigue failure, it is still the least understood in terms of accurate analysis. This lack of understanding is attributed to the wide variations of statistical data in determining

- Type of service and environment of the drill string
- Magnitude of operating loads and frequency of occurrence (i.e., load history)
- Accuracy of methods to quantify the stresses
- Quality at time of manufacture (i.e., quality control)
- Applicability of material fatigue data

Pipe failure prevention

Although pipe failures cannot be totally eliminated, certain measures can be taken to minimize it. Fatigue failures can be mitigated by the minimization of induced cyclic stresses and by ensuring a noncorrosive environment during drilling operations. Cyclic stresses can be minimized by controlling dogleg severity and drill string vibrations. Corrosion can be mitigated by corrosive scavengers and by controlling the mud pH in the presence of H_2S. Proper handling and inspection of the drill string on a routine basis are the best measures for the prevention of failures.

Borehole Instability

Definition and causes

Borehole instability is an undesirable condition of an open hole interval that does not maintain its gauge size and shape and/or its structural integrity. The causes may grouped into the following categories:

- Mechanical—due to *in situ* stresses
- Erosion—due to fluid circulation
- Chemical—due to interaction of borehole fluid with the formation

Problems associated with borehole instabilities

There are four different types of borehole instabilities, as shown in figure 10–7, including

- Hole closure or narrowing
- Hole enlargement or washouts
- Fracturing
- Collapse

The problems associated with each type are addressed in the following sections.

Hole closure. Hole closure is a time-dependent process of borehole narrowing, sometimes referred to as creep, under the overburden pressure. It generally occurs in plastic flowing shale and salt sections. The associated problems are

- Increase in torque and drag
- Increase in potential for pipe sticking
- Increase in difficulty of casings landing

Hole enlargement. Hole enlargements are commonly termed washouts; that is, the hole becomes undesirably larger than intended. Hole enlargements are generally caused by hydraulic erosion and mechanical abrasion owing to the drill string and in inherently sloughing shale. The associated problems are

- Increase in difficulty of cementing
- Increase in potential for hole deviation
- Increase in hydraulics requirements for effective hole cleaning
- Increase in potential for problems during logging operations

Fracturing. Fracturing occurs when the wellbore drilling-fluid pressure exceeds the formation fracture pressure. The associated problems are

- Lost circulation
- Possible kick occurrence

Collapse. Borehole collapse occurs when the drilling-fluid pressure is too low to maintain the structural integrity of the drilled hole. The associated problems are

- Pipe sticking
- Possible loss of well

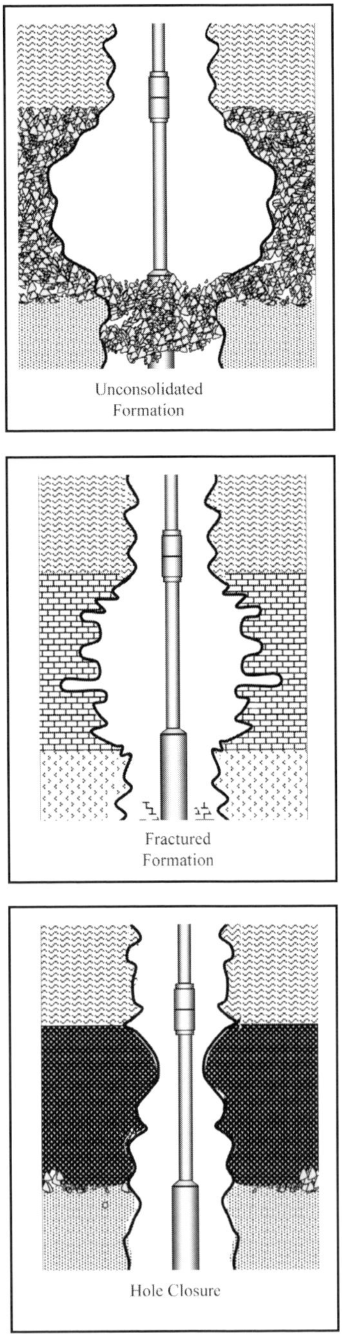

Fig. 10–7. Borehole instability–associated problems

Principles of borehole instability

Before drilling, the rock strength at some depth is in equilibrium with the *in situ* rock stresses (effective overburden stress and effective horizontal confining stresses). While a hole is being drilled, however, the balance between the rock strength and the *in situ* stresses is disturbed. In addition, foreign fluids are introduced, and an interaction process begins between the formation and borehole fluids. The result is the potential for a hole instability problem to arise. Although a vast amount of research has resulted in many borehole stability simulation models, all share the same shortcoming, of uncertainty in input data needed to run the analysis. Such data include *in situ* stresses, pore pressure, formation fluid chemistry (in the case of shale), and rock mechanical properties.

Mechanical rock failure mechanisms

Mechanical borehole failure occurs when the stresses acting on the rock exceed the compressive strength or the tensile strength of the rock. Compressive failure is caused by shear stresses as a result of low mud weight, whereas tensile failure is caused by normal stresses as a result of excessive mud weight.

The following failure criteria are used to predict hole instability problems:

- Maximum normal stress—for tensile failure
- Maximum strain energy of distortion—for compressive failure

In the maximum normal stress criterion, failure is said to occur when, under the action of combined stresses, one of the acting principal stresses reaches the failure value of the rock tensile strength. In the maximum of energy of distortion criterion, failure is said to occur when, under the action of combined stresses, the energy of distortion reaches the same energy of failure of the rock under pure tension.

Shale instability. Over 75% of drilled formations worldwide are shale formations. The drilling cost attributed to shale instability problems has been reported to be in excess of half a billion U.S. dollars per year. The causes of shale instability are twofold:

- Mechanical (stress change vs. shale strength environment)
- Chemical (shale/fluid interaction: capillary pressure, osmotic pressure, pressure diffusion, and borehole fluid invasion into shale)

Mechanical instability. As stated previously, mechanical rock instability can occur because the *in situ* stress state of equilibrium has been disturbed after drilling. The density of the mud in use may not bring the altered stresses to the original state; consequently, shale may become mechanically unstable.

Chemical instability. Chemical-induced shale instability due to the drilling-fluid/shale interaction alters shale strength, as well as the pore pressure in the vicinity of the borehole walls. This is caused by several mechanisms, including

- Capillary pressure
- Osmotic pressure
- Pressure diffusion in the vicinity of the borehole walls
- Borehole fluid invasion into the shale in overbalanced drilling

Capillary pressure. During drilling, the mud in the borehole contacts the native pore fluid in the shale through the pore throat interface, resulting in the development of capillary pressure, P_{cap}. which is expressed as

$$P_{cap} = \frac{2\lambda \cos\theta}{r} \tag{10.6}$$

where

λ is the interfacial tension
θ is the contact angle between the two fluids
r is the pore throat radius

To prevent borehole fluids from entering the shale and hence to stabilize the borehole, an increase in capillary pressure is required. This can be achieved by using oil-based or other organic low-polarity mud systems.

Osmotic pressure. When the energy level (referred to as the activity), A_s, in shale pore fluid is different than the energy level, A_m, in drilling mud, movement of water can occur in either direction across a semipermeable membrane, as a result of the development of osmotic pressure, p_{os}, or chemical potential, P_c. To prevent or reduce water movement across this semipermeable membrane that has a certain efficiency, E_m, one needs to equalize the activities or at least minimize their differential. If A_m is lower than A_s, it is suggested to increase E_m, and vice versa. The mud activity can be reduced by the addition of electrolytes that can be brought about through the use of mud systems such as seawater, saturated-salt polymer, KCl/NaCl polymer, and lime/gypsum.

Pressure diffusion. Pressure diffusion is a phenomenon of pressure change near the borehole walls that occurs over time. This pressure change is due to the compression of the native pore fluid by the borehole fluid pressure, p_{wfl}, and the osmotic pressure, p_{os}.

Borehole fluid invasion into shale. In conventional drilling, a positive differential pressure (difference between the borehole fluid pressure and the pore fluid pressure) is always maintained. As a result, borehole fluid is forced to flow into the formation (fluid-loss phenomenon) and may cause a chemical interaction that can lead to shale instabilities. To mitigate this problem, an increase of mud viscosity is generally used, although, in extreme cases, gilsonite may be used to seal off micro-fractures.

Wellbore stability analysis

There are several models in the literature that address wellbore stability analysis. These include very simple to very complex models, such as

- Linear elastic
- Nonlinear
- Elasto-plastic
- Purely mechanical
- Physico-chemical

Regardless of the model, the data needed include:

- Rock properties (Poisson ratio, strength, modulus of elasticity, etc.)
- *In situ* stresses (overburden and horizontal)
- Pore fluid pressure and chemistry
- Mud properties and chemistry

Analysis of the data (other than those data pertaining to the mud) is often confounded by problems of availability and/or uncertainties. However, sensitivity analysis can be conducted by assuming data for the many variables, to establish safety windows for mud selection and design.

Borehole instability prevention

Total prevention of borehole instability is unrealistic because restoring the physical and chemical *in situ* conditions of the rock are impossible. However, the drilling engineer can mitigate the problems of borehole instabilities by adhering to good field practices, including

- Proper selection and maintenance of mud weight
- Proper hydraulics to control the ECD
- Proper hole trajectory selection
- Compatible borehole fluid with the formation being drilled
- Minimum time spent in open hole
- Use of offset well data (utilization of the learning curve)
- Monitoring of trend changes (torque, circulating pressure, drag, fill in during tripping, etc.)
- Collaboration and sharing of information

Mud Contamination

Definition

A mud is said to be contaminated when a foreign material enters the mud system and causes undesirable changes in mud properties such as the density, the viscosity, and/or filtration. In general, water-based mud systems are the most susceptible to contamination of various types. Mud contamination can result from overtreatment of the mud system with additives or from material entering the mud during drilling.

Common contaminants—sources and treatments

The most common contaminants of water-based mud systems are

- Solids (added, drilled, active, and inert)
- Gypsum/anhydrite (Ca^{2+})
- Cement/lime (Ca^{2+})
- Makeup water (Ca^{2+} and Mg^{2+})
- Soluble bicarbonates and carbonates (HCO_3^- and CO_3^{2-})
- Soluble sulfides (HS^- and S^{2-})
- Salt/saltwater flow (Na^+ and Cl^-)

Solids contamination. Solids are those that are added to make up a mud system (e.g., bentonite and barite) and those that are drilled (all types—active and inert). Excess solids of any type are the most undesirable contaminant of drilling fluids, because they affect all mud properties. It has been shown that fine solids, of micron and submicron size, are the most detrimental to the overall drilling efficiency and must be removed if they are not a necessary part of the mud makeup.

The removal of drilled solids is achieved through the use of mechanical separating equipment (shakers, desanders, desilters, and centrifuges). Shakers remove solids of the size of cuttings (approximately 140 microns or larger). Desanders remove solids of the size of sand (down to 50 microns). Desilters remove solids in the size of silt (down to 20 microns). When solids become smaller than the cutoff point of desilters, centrifuges may have to be used. Chemical flocculants are sometimes used to flocculate fine solids into bigger size, so that they can be removed the by solids removal equipment. Total flocculants are those that do not discriminate between various types of solids, and selective flocculants are those that will flocculate drilled solids but not added barite solids. As a last resort, dilution is sometimes used to lower solids concentration.

Calcium ions contamination. The sources of calcium ions are gypsum, anhydrite, cement, lime, seawater, and hard/brackish makeup water. The calcium ion is a major contaminant to freshwater-based sodium/clay-treated mud systems. The calcium ion tends to replace the sodium ions on the clay surface through a base exchange, causing undesirable changes in mud properties such as rheology and filtration. It also causes thinners added to mud system to become ineffective.

The treatment chosen depends on the source of the calcium ions. For example, sodium carbonate (soda ash) is used if the source is gypsum or anhydrite. Sodium bicarbonate is preferred if the source is lime or cement. If treatment becomes economically unacceptable, break over to a mud system that can tolerate the contaminant, such as gypsum mud or lime mud.

Bicarbonate and carbonate contamination. The contaminant ions (CO_3^{2-} and HCO_3^-) come from drilling into carbon dioxide–bearing formations, thermal degradation of organics in mud, or overtreatment with soda ash and bicarbonate. These contaminants cause the mud to have high yield and gel strength and decreased pH. Treatment of the mud system with gypsum or lime is recommended.

Hydrogen sulfide contamination. The contaminant ions (HS^-, S^{2-}) generally come from drilling into H_2S-bearing formations. Hydrogen sulfide is the most deadly contaminant ion to humans and the most corrosive ion to steel in drilling operations (causing severe embrittlement to drill pipe). Scavenging of H_2S is done through the use of zinc, copper, or iron.

Salt/saltwater flows. The ions (Na^+ and Cl^-) that enter the mud system as a result of drilling into salt sections or as a result of formation saltwater flow cause a mud to have high yield strength, high fluid loss, and decreased pH. Some actions for treatment are dilution with freshwater, use of dispersants and fluid loss chemicals, and if the cost of treatment becomes excessive, ultimate conversion to a mud that tolerates the problem.

Producing-Formation Damage

Definition

Producing-formation damage has been defined by Brent Bennion as "the impairment of the unseen by the inevitable, causing an unknown reduction in the un-quantifiable." In a different context, formation damage is defined as the impairment to a reservoir (reduced production), caused by wellbore fluids used during drilling/completion and workover operations. It is a zone of reduced permeability within the vicinity of the wellbore (skin) that results from the invasion of foreign fluid into the reservoir rock. (See fig. 10–8.)

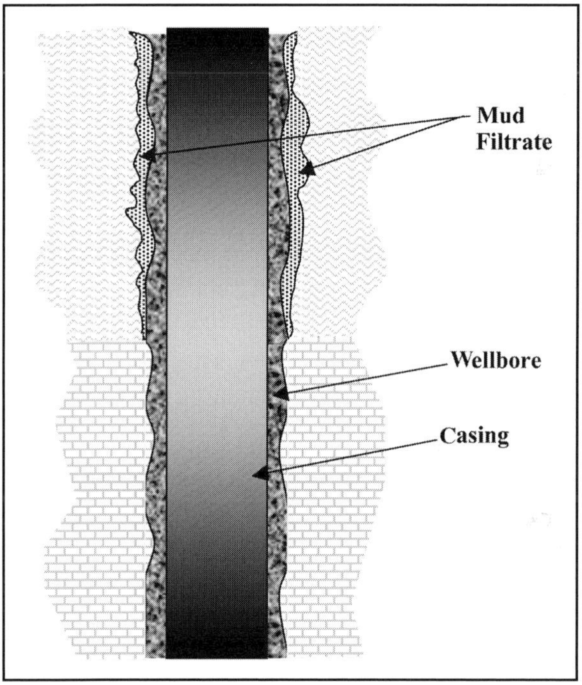

Fig. 10–8. Formation damage

Borehole fluids

Borehole fluids are classified as follows:

- Drilling fluids
- Completion fluids
- Workover fluids

Drilling fluids are categorized as mud, gas, or gasified mud. There are two types of mud: water based (pure polymer, pure bentonite, or bentonite/polymer) and oil based (invert emulsion or oil). Completion and workover fluids are mostly brines and solids free.

Damage mechanisms

The vast amount of literature identifies formation damage as a combination of several mechanisms:

- *Solids plugging.* Plugging of the reservoir rock pore spaces can be caused by either fine solids in the mud filtrate or dislodged solids within the rock matrix, as shown in figure 10–9. To minimize this form of damage is to minimize the amount of fine solids in the mud system and its fluid loss.
- *Clay particle swelling or dispersion.* This is a problem inherent in sandstone that contains water-sensitive clays. When a freshwater filtrate invades the reservoir rock, it causes the clay to swell and thus reduces or totally blocks the throat areas.
- *Saturation change.* Production is predicated on the amount of saturation within the reservoir rock. When a mud system filtrate enters the reservoir, this causes some change in water saturation and, therefore, potential reduction in production. High fluid loss causes increased water saturation, resulting in decreased rock relative permeability.
- *Wettability reversal.* Reservoir rocks are water wet in nature. It has been demonstrated that while drilling with oil-based mud systems, excess surfactants in the mud filtrate that enter the rock can cause wettability reversal. On the basis of field experience and laboratory tests, as much as 90% production loss can be caused by this mechanism. Therefore, to guard against this problem, the amount of excess surfactants used in oil-based mud systems should be kept to a minimum.
- *Emulsion blockage.* Inherent in oil-based mud systems is the use of excess surfactants. These surfactants, on entering the rock, can form an emulsion within the pore spaces, thus hindering production.
- *Aqueous filtrate blockage.* In water-based mud systems, the aqueous filtrate that enters the reservoir can cause blockage that reduces the production potential of the reservoir.
- *Mutual precipitation of soluble salts.* Any precipitation of soluble salts in the wellbore fluid filtrate, whether from the use of salt mud systems or from formation water or both, can cause solid blockage and thus hinder production.

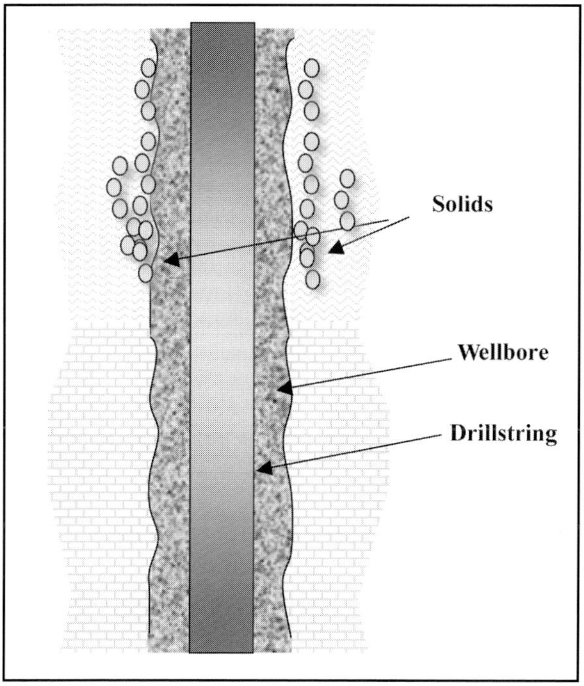

Fig. 10–9. Solids plugging

Hole Cleaning

Introduction to hole cleaning

Throughout the past decades, many studies have been conducted to increase our understanding of hole cleaning in directional well drilling. It has been demonstrated in laboratory work that drilling at inclination angle greater than approximately 30° from the vertical poses problems in cuttings removal that are not encountered in vertical wells. Formation of a moving or a stationary cuttings bed becomes an apparent problem if the flow rate for a given mud rheology is below a certain critical value, as illustrated in figure 10–10.

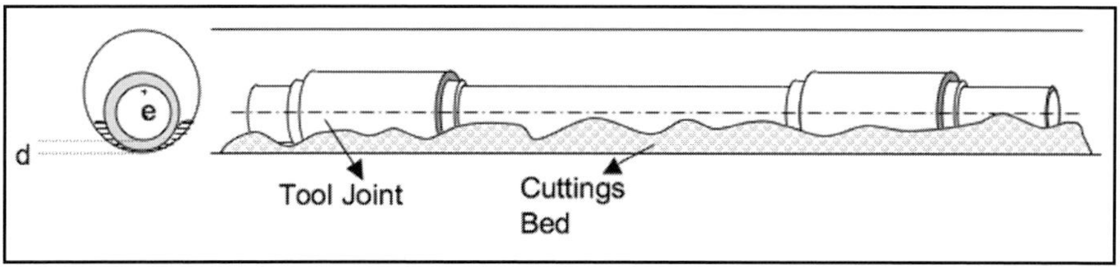

Fig. 10–10. Bed height formation owing to inadequate hole cleaning

Inadequate hole cleaning can lead to costly drilling problems, such as

- Mechanical pipe sticking
- Premature bit wear
- Slow drilling
- Formation fracturing
- Excessive torque and drag on drill string
- Difficulties in logging and cementing
- Difficulties in casings landing

The most prevalent of these is excessive torque and drag, which often makes it impossible to reach the target in high-angle/extended-reach drilling.

Factors affecting hole cleaning

The factors that have an impact on hole cleaning (annular cuttings removal) include

- Annular fluid velocity
- Hole inclination angle
- Drill string rotation
- Annulus/pipe eccentricity
- ROP
- Drilling-fluid properties
- Drilled-cuttings characteristics

A brief discussion on each of these factors and their limitations in field situations is presented in the following sections.

Annular fluid velocity. Irrespective of all other factors that affect annular hole cleaning, the flow rate is the dominant factor in cuttings removal while drilling directional wells. An increase in flow rate will result in more efficient cuttings removal under all conditions. However, the extent to which the flow rate can be increased may be limited by the following:

- The maximum allowed ECD
- Susceptibility of the open hole section to hydraulic erosion
- Rig hydraulic power availability

Hole inclination angle. It has been demonstrated through laboratory work that as hole angle increases from 0° to about 67° from the vertical, hole cleaning becomes more difficult and, therefore, the flow rate requirement increases. The flow rate requirements reach a maximum around 65–67° and then slightly decrease toward the horizontal. Also, it has been shown that around 25–45°, a sudden pump shutdown can cause cuttings sloughing to the bottom and may result in mechanical pipe sticking.

Although hole inclination can lead to cleaning problems, it is mandated by the needs of drilling into an inaccessible reservoir, offshore drilling, avoiding troublesome formations, side tracking, and drilling horizontally into the reservoir. Company objectives in total field development (primary and secondary production), environmental concerns, and economics are some of the factors influencing hole angle selection.

Drill string rotation. It has been shown in laboratory studies and reported from field cases that drill string rotation has moderate to significant effects that enhance hole cleaning. The level of enhancement is a combined effect of pipe rotation, mud rheology, cuttings size, flow rate, and very importantly, the string dynamic behavior. It has been proved that the whirling motion of the string around the wall of the borehole as it rotates is the major contributor to hole-cleaning enhancement. Also, mechanical agitation of the cuttings bed on the low side of hole and exposure to higher fluid velocities as the pipe moves to the high side of the hole are benefits of pipe whirling action.

Although pipe rotation definitely enhances hole cleaning, there are certain limitations to its implementation. For example, during angle building while using a downhole motor (i.e., in sliding mode), rotation cannot be induced. With the new rotary steering systems, this is no longer a problem. However, pipe rotation causes cyclic stresses that can cause acceleration of pipe failures owing to fatigue, casing wear, and in some cases, mechanical destruction to open hole sections. In addition, in slim hole drilling, high pipe rotation can cause high ECDs owing to the increase in annular friction pressure losses.

Hole/pipe eccentricity. In the inclined section of hole, the pipe has the tendency to rest on the low side of the borehole, owing to gravity. This creates a very narrow gap in the annulus section below the pipe; consequently, the fluid velocity becomes extremely low, leading to an inability to transport cuttings to the surface. As illustrated in figure 10–11, as eccentricity increases, particle fluid velocity decreases in the narrow gap, especially for high-viscosity fluid. However, since eccentricity is governed by the selected well trajectory, its adverse impact on hole cleaning may not be avoidable.

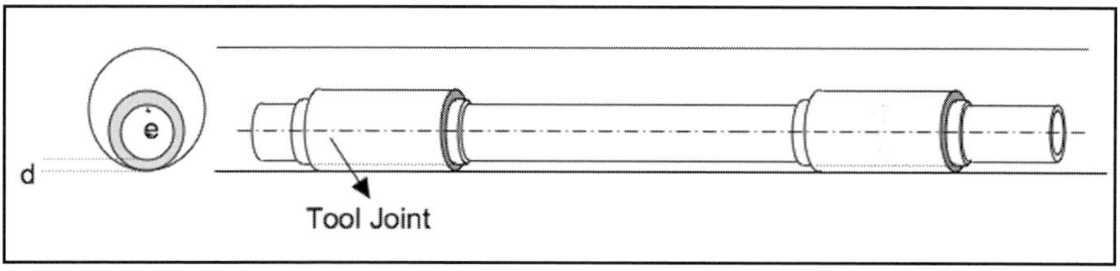

Fig. 10–11. Hole/pipe eccentricity

ROP. Under similar conditions, an increase in the drilling rate always results in an increase in the amount of cuttings in the annulus. To ensure good hole cleaning during high ROP drilling, the flow rate and/or pipe rotation have to be adjusted. If the limits of these two variables are exceeded, the only alternative is to reduce the ROP. Although a decrease in the ROP will have a detrimental impact on the drilling cost, the benefit of avoiding other drilling problems, such as mechanical pipe sticking and excessive torque and drag, can outweigh the loss in ROP.

Mud properties. The functions of drilling fluids are many and can have unique, competing influences. The two mud properties that have a direct impact on hole cleaning are the viscosity and the density.

The main functions of density are mechanical borehole stabilization and prevention of the intrusion of formation fluids into the annulus. Any unnecessary increase in mud density, beyond fulfilling the aforementioned two functions, will have an adverse effect on the ROP and, under the given *in situ* stresses, may cause fracturing of the formation. Importantly, mud density should not be used as a criterion to enhance hole cleaning.

Viscosity has as its primary function the suspension of added desired weighting materials, such as barite. Only in vertical well drilling and high-viscous pill sweep is viscosity used as a remedy in hole cleaning.

Cuttings characteristics. The size and distribution, the shape, and the specific gravity of cuttings affect their dynamic behavior in a flowing media. The specific gravity of most rocks is around 2.6 and therefore can be assumed as a nonvarying factor in cuttings transport. The cuttings size and shape are functions of the bit types (roller cone, PDC, or diamond matrix), of the regrinding that takes place after they are generated, and further of breakage by their own bombardment and bombardment with the rotating drill string. Thus, it is impossible to control the size and shape of cuttings, even if a specific bit group has been selected to generate them. Note that smaller cuttings are more difficult to transport in directional well drilling; however, with some viscosity and pipe rotation, fine particles seem to stay in suspension and therefore are easier to transport.

Hydrogen Sulfide–Bearing Zones and Shallow Gas

Hydrogen sulfide–bearing formation drilling poses one of the most difficult and dangerous problems to humans and equipment. If the presence of hydrogen sulfide is known or anticipated, there are very specific requirements to abide by, in accordance with IADC rules and regulations. The reader is advised to consult sources specific to this topic.

Shallow gas may be encountered at any time in any region of the world. The only way to combat this problem if it is encountered is to never shut in the well but rather divert the gas flow through a diverter system, which will hopefully be in place. Shallow gas of high pressure can be encountered at depths as low as a few hundred feet, where the formation fracture gradient is very low. The danger is that if the well is shut in, formation fracturing is more likely to occur, resulting in underground blow, the most severe blowout problem.

Equipment- and Personnel-Related Problems

Equipment

The integrity of drilling equipment and its maintenance are major factors in the minimization of drilling problems. Proper rig hydraulics (pump power) are needed for efficient bottom and annular hole cleaning; proper hoisting power is required for efficient tripping out; proper derrick design loads and drilling-line tension load will allow safe overpull in case of sticking problems; a proper well control system (ram preventers, annular preventers, internal preventers) will allow kick control under any kick situation; a proper monitoring and recording system will allow the monitoring of trend changes in all drilling parameters and will facilitate the accessibility of drilling data for retrieval at a future date for learning; proper tubular hardware should be specifically suited to accommodate all anticipated drilling conditions; and finally, effective mud handling and maintenance equipment will ensure the mud properties fit their intended functions.

Personnel

Under the assumption that all conditions are equal during drilling/completion operations, personnel are the key to the success or failure of those operations. Overall well cost owing to any drilling/completion problem can become extremely high. Therefore, providing continuing education and training to personnel who are involved, directly or indirectly, is essential for successful and economic drilling/completion practices.

Supplementary Problems

Problem 10–1. Calculate the pull-out force, given the following well data:

- Depth of the stuck point = 10,000 ft
- Drill collar OD = 6.25 in.
- Hole size = 8.5 in.
- Mud cake thickness = ⅛ in.
- Coefficient of friction = 0.20
- Mud weight = 10 lbs/gal
- Length of the embedded portion = 50 ft
- Formation pressure at the stuck point = 4,950 psi

Problem 10–2. Determine the volume and pressures during spotting, given the following well data:

- Drill collar OD = 6.25 in. × 3 in.
- Hole size = 8.5 in.
- Hole depth = 10,000 ft
- Mud weight = 10 lbs/gal
- Length of embedded drill collar = 100 ft
- Spotting fluid weight = 7.5 lbs/gal

Assume 3° hole inclination, 5% washout, and inside diameter of the string uniform with 3 in.

Problem 10–3. Calculate the allowable mud weight at 5,000 ft, given the following data:

- Cake thickness = ⅛ in.
- Diameter of the hole = 12¼ in.
- Pipe diameter = 6¼ in.
- Coefficient of friction = 0.25
- Pore pressure at 5,000 ft = 2,340 psi
- MOP = 75 kips

Problem 10–4. An 8½ in. hole is drilled through a depleted sand section of 100 ft. The wellbore pressure is 1,500 psi greater than the formation pressure. If a 6¼ in. drill collar is stuck over the entire section of the sand, calculate and plot the pull-out forces for cake thicknesses of 1/32 in., 1/16 in., 3/32 in., and ⅛ in., using coefficients of friction 0.1, 0.15, 0.2, 0.25, and 0.3. State your conclusions.

Problem 10–5. Estimate the approximate filter cake thickness for the following conditions: 100 ft of 6½ in. drill collar got stuck at 8,000 ft from the bottom; formation pressure equivalent at 8,000 ft is 12 lbs/gal; and 75 kips was applied to release the differential pressure stuck pipe. Assume a coefficient of friction of 0.35 and a mud weight of 14 lbs/gal. Repeat the calculation for coefficients of frictions 0.1, 0.15, 0.2, 0.25, and 0.3. State your conclusions.

Problem 10–6. Calculate the optimum fishing time, given the following data:

- Well depth = 3,000 m
- Estimated stuck depth = 2,500 m
- Daily operating rate of the rig = $1,500/day
- Value of drill string below stuck point = $100,000
- Time taken to drill original hole below stuck point = 30 h

Nomenclature

A_c	Contact area
C_R	Hourly rig operating rate, in $
C_s	Value of drill string below stuck point, in $
D_h	Diameter of the hole, in in.
D_{ip}	Pipe inside diameter (ID), in in.
D_{op}	Pipe outer diameter (OD), in in.
D_s	Stuck point, in ft
D_{sm}	Estimated depth of stuck point, in m
e	Stretch length, in in.
E	Elastic modulus
L_{ep}	Length of embedded portion, in ft
L_p	Length of spotting fluid inside pipe, ft
ΔF	Differential force, in lbs
F_L	Lateral force, in lbs
F_{pull}	Pull-out force, in lbs
ΔP	Differential pressure, in psi
P_{arc}	Arc length
P_m	Mud pressure
P_{ff}	Formation pressure, in psi
P_{gm}	Mud pressure gradient, in psi/ft
P_{gsf}	Spotting-fluid pressure gradient, in psi/ft
r	Pore throat radius

T	*T*ension in the drill string just above the key seat area
T_{ds}	Tension in the drill string at the point just above dogleg, in degrees per 100 ft
t_{mc}	Thickness of the filter cake or the mud cake, in in.
TVD	True vertical depth, in ft
V_{sf}	Volume of spotting fluid inside pipe, in in.^{3}
w	Weight of the pipe, in lbs/ft
α	Angle of contact between the BHA section and the mud cake, in radians
δ	Strain
ε	Deformation of mud cake at the midpoint of contact, in in.
θ	Contact angle between the two fluids
θ_{dl}	Dogleg angle
λ	Interfacial tension
μ	Coefficient of friction
ρ_{eq}	Equivalent circulating mud weight
$\Delta\rho_f$	Additional mud weight owing to friction pressure loss in the annulus
ρ_{frac}	Formation pressure fracture gradient in equivalent mud weight
ρ_m	Static mud weight, in lbs/gal
$\Delta\rho_{srg}$	Additional mud weight owing to surge pressure
σ	Tensile stress, in psi

References

1 Outmans, H. D. 1958. "Mechanics of Differential Pressure Sticking of Drill Collars," *Petroleum Transactions*, 213, 8.

11

Casing and Cementing Design

Drilling the Pay Zone: Selecting the Interval and the Initial Design

The completion begins when the drill bit first penetrates the pay zone. Drilling the pay zone is one of the most important parts of the drilling operation. The drilling mud that is adequate for drilling other intervals of the well may not be acceptable for the pay zone. Whereas formation damage created by the mud is acceptable in a nonproductive interval, it cannot be tolerated in the zone intended for production. A mud is needed that can control fluid loss into the formation without creating permanent damage. The mud may require special treatment and may occasionally need to be changed out to a nondamaging fluid. There are several goals in drilling that are of interest to the completion engineer:

- *Drilling a usable hole.* A hole through the pay should accept the design size of casing.
- *Minimizing formation permeability damage.* High drilling-mud overbalance pressure, uncontrolled particle size, a mud filtrate that causes clay swelling, and poor fluid-loss control may mask the response of a productive formation to a drill stem test (DST) and may lead to the bypassing of a producing zone.
- *Controlling washouts.* Wellbore instability problems may cause hole enlargements that make perforation and formation breakdown much more difficult.

The most important aspects of a drilling mud, from a formation damage standpoint, are to prevent loss of the drilling-mud filtrate and to make sure that any filtrate that is lost will not react with the formation to reduce permeability. Fluid reactivity is usually controlled by using potassium chloride or other salts to stabilize the clay in the formation.[1] Potassium chloride may not always control clay reactions or may require as much as 4% or more salt where smectite clay is present in the larger pore passages.

Fluid-loss control is accomplished by rapidly sealing off the permeable sections of the formation.[2] The mud accomplishes this fluid-loss control by creating an almost impermeable mud cake of particles on the surface of the formation. By carefully controlling the size range of solids that could invade the pores of the formation, invasion damage from particles can be stopped.[3] In some drilling and workover fluids, fine particles and at least parts of the solids in the fluids will be designed to be acid soluble.[4]

In cases where the formation matrix permeability is between approximately 0.5 md and 100 md and the pressure differential is small ($\Delta P < 100$ psi), the filtrate of even a damaging mud will not likely extend into the formation beyond a depth of a few inches, provided that the filter cake is successful in controlling fluid loss. Higher-permeability formations pose special problems if the mud cake cannot be formed quickly. During tripping, scraping of mud cake can occur; therefore, the cake must re-form quickly, to prevent the high fluid loss into the formation.

The importance of drilling a usable hole through the pay and its importance with regard to running and cementing pipe cannot be overstated. Failure to get a casing string or a liner to the bottom can be very costly, in terms of cost of an additional string or liner and the reduction of working space where pumps and other equipment needs to be set. Simply drilling a hole with a certain diameter of drill bit through a formation does not lead to a hole that will accept a string of pipe of an outside diameter (OD) just smaller than that of the drill bit.

In instances when casing cannot be run in a freshly drilled hole, the drift diameter of the hole is not equal to the bit diameter. Normally, casing strings are run with 1½ to 2 in. minimum clearance between the hole diameter and the OD of the pipe. In a straight hole, this clearance is adequate; however, improper BHA can cause problems for running casing. For example, using a slick BHA (drill collars in the BHA without stabilizers) usually leads to a hole with a diameter less than the diameter of the drill bit. If the hole has been drilled with the intention of running a liner, the problem may be even more pronounced. Liners are usually characterized by close tolerances between the hole diameter and the liner OD.

Thick mud cake can also cause problems in landing casing to the bottom, because of the potential for differential pressure pipe sticking.[5] Differential pressure sticking caused by thick, spongy mud cake is also an important problem. The pipe is held against the mud cake by the differential between the wellbore and the formation pressures. Diagnostics of stuck casing are often performed after examining the drilling record and trying different types of pipe movement and circulation. (For more details on these drilling problems, see chap. 10.)

Initial Completion Design

Selecting the pay zone

Selecting the pay and deciding where to place the wellbore are two of the most important pieces of engineering that must occur in the completion process. Many rocks, from shales to fractured granites, contain hydrocarbons, but not every rock type or reservoir can qualify as a pay zone. Selection of the pay hinges on several basic considerations:

- Prospect development economics
- Porosity and permeability requirements
- Hydrocarbon type and saturation requirements
- Recoverable hydrocarbon volumes (by primary, secondary, and tertiary methods)
- Pressure support
- Reservoir stability
- Recognition of compartmentalization
- Availability of technology for cost-effective production of the reserves
- Ability to plug and abandon the reservoir
- Environmental and other risks

The economics of a project depend simply on whether enough money can be made from the sale of the productive hydrocarbons in a limited amount of time, to offset the total cost of the project. This includes lease, development, and operation; royalties; interest on the money used; profit risk; plug and abandonment; and contingency funds for problems such as blowouts and cleanup operations. Every project—from the shallowest stripper well, at 2 bbl/d, to 100 millions of standard cubic feet per day (mmscf/d) or 30,000 bbl/d oil wells—must be judged in terms of risk versus cost recovery and profit.

Porosity and permeability represent the reservoir storage and the pathway, respectively, of flowing fluids. Porosity is the void space between the grains in which fluids can be stored. Permeability is a measurement of the ability of fluids to flow through the formation. Rock such as shale and chalk, on the one hand, may have extremely high porosities, approaching 30–40%; however, because the porosity is not linked, the permeability is very low. On the other hand, naturally fractured formations may have extremely high permeabilities, approaching tens of darcies in some cases, but have very low porosity, often only 4–6%. The amount of porosity and permeability necessary for a project depends on the production rate needs, although operations such as hydraulic fracturing can increase the production rate of a well by a factor of 2–10 or more. Fracturing alone may not make the project economic. The economics of a project are such that every factor must be weighed individually in terms of the economic justification and critical factors, such as hydrocarbon storage and the permeable pathway, which must be available before even a huge reservoir with billions of barrels of oil can be made productive. In reservoir selection, often a porosity or permeability cutoff is used for pay versus no-pay identification. Recognition of this level by use of porosity logs and flow tests is often critical to the establishment of minimum pay requirements.

Hydrocarbon type and saturation determine the amount of hydrocarbons that may occupy the pore space of a reservoir. Many factors, such as moveable versus irreducible saturations, and changing factors, such as relative permeability, can make the saturation and permeability values "moving targets." There are no set minimum values for hydrocarbon saturation; however, the best parts of the reservoir will usually have higher values of hydrocarbon saturation. Saturation of water may also be a key factor in pay identification. Extremely high saturations of water may indicate hydrocarbon depletion or movement of an aquifer into the pay.

The recoverable hydrocarbon volumes are usually calculated from the measured values of porosity and saturation. Oil-in-place quantities do not indicate that all of that oil can be recovered. The porosity of a formation varies, from very large pores to very small pores; the oil in very small pores often will not flow from the small capillaries, even under very high depletion pressures. How much oil will flow from a rock is dependent on the size of the pore spaces, the oil saturation and type, and the amount of energy available to push the oil toward the wellbore. Recoverable-hydrocarbon estimates may vary by many percentage points from what reality shows later on. The differences are in how well the pressure supports the drive mechanism in producing the fluids.

The pressures in the reservoir dictate how much fluid will ultimately be recovered. Many different types of pressure supports are available. Typical pressure support mechanisms include bottom- and edgewater drives, gas cap drives, volumetric depletion, and other pressure sources (e.g., reservoir compaction), among other factors. Each of these pressure support mechanisms has advantages and disadvantages with respect to recovery in a reservoir.

Among the most effective types of reservoir pressure support are the bottom- and edgewater drives. These systems may maintain pressure at initial values until the end of the project. The problem is that they may produce large amounts of water along with the oil. Volumetric depletion is usually found in a sealed reservoir. Such a reservoir may deplete without producing any water. The recovery, however, from such reservoirs is extremely low, since reservoir energy bleeds off very quickly. Pressure support can be added, in some cases, by the use of water floods, gas repressurization, or other types of pressure maintenance, such as tertiary floods. When factors such as bottom- and edgewater drive are recognized early, the location of the wellbores can be selected so as to take advantage of flow paths of the drive fluids and thereby enhance recoveries.

Reservoir stability is an issue that may affect the initial completion or repairs or recompletions throughout the life of the reservoir. Many geologically young formations lack sufficient strength for formation coherency during all phases of production. Recognition of this stability issue is usually easy because of rapid drilling rates, sand strength issues in the wellbore, or other factors. The decision to add a stabilizing completion is usually made after consideration of initial flow tests and other factors. The most common methods of completion in low-stability reservoirs are frac packs and gravel packs.

Recognition of compartmentalization is probably one of the most important factors in the initial design of well completions for a project. Compartmentalization is the division of a reservoir into partially or fully pressure-isolated compartments by faults, permeability or porosity pinchouts, folding, shale streaks, barriers, or other factors. When compartmentalization is recognized, the location and type of wellbores can be selected to efficiently drain the compartments and to take advantage of fluid flow patterns within the reservoir. Many failures, even of large fields, can be traced to a failure to recognize compartmentalization during the early development steps in the reservoir.

The availability of technology to produce the reserves is an area that keeps the oil industry active in research and development. Technologies such as water flooding, hydraulic fracturing, artificial lift, cold flow of heavy oils, and coal degasification have increased the world's recoverable hydrocarbons and continue to be critical in meeting the world's energy needs. When the reservoir flow patterns and other factors are understood, technology can often be developed within a moderate time frame to meet needs in specialized reservoirs.

The ability to produce hydrocarbons should never outstrip the ability to control the flow or the ability to plug an abandoned reservoir. Plug and abandonment intentions must take into account that the reservoir should be left in as good a condition as possible for potential tertiary operations that may recover even more fluids. Plug and abandonment costs can be a significant amount of the project cost. The cost of offshore plug and abandonment of fields may exceed $100 million. There are many associated risks, both political and environmental, in developing and producing a hydrocarbon depository. These risks must be taken into account during the economic justification for the reservoir and should offer as good a solution as is possible to the legitimate concerns posed in any situation. Once the values are known, selection of the pay can begin. The selection process uses a number of pieces of information gathered by electronics and other factors.

Completion design is a function of reservoir characteristics. The problem is that the reservoir data—in particular, design-sensitive data such as permeability, porosity, saturation, pressure, barriers, and longevity—are fully available only after most of the wells in the field have been drilled, completed, and tested. In many cases, after initial drilling and completion, reservoir barriers are finally recognized, and extreme redrilling or stimulations are needed to process the reservoir. The key to a good initial completion is to collect and assess the data at the earliest possible time, to allow the best early choice of completion.

Successful completions recognize the flow characteristics of the reservoir. There are a number of completion possibilities, as shown in table 11–1; each has a limited fit with regard to the reservoir properties. The following sections describe the completion types in general use, along with a few of the reservoir variables; the numbers for most variables, though typical, are only estimates.

Table 11–1. Well completion types

Vertical well Natural completion	• High permeability ($K_h \geq 10$ md for oil, ≥1 md for gas) open hole stable formation (no movement or spalling) • No bottom or edge water drives • Low K_v ($K_v < K_h$) (or deviated wells not considered possible) • No fracture planned/possible, no limits on surface/reservoir access • Laminations not "frequent."
Vertical well Open hole Frac planned	• No limits on permeability • Stable formation (no movement or spalling) • No bottom or edge water drive control needs low K_v ($K_v < 0.1K_h$) (or deviated wells not considered/possible) • No limits on surface/reservoir access • Multiple frac not planned • Laminations not frequent in zones not fractured • Bottom/edge water not penetrated by frac.
Vertical well Cased hole Frac or frac pack planned	• No limits on permeability • Cased hole 180° perforating and screenless pack frac for sand control 120°, 90° or 60° phased perfs for other fracs • Low K_v ($K_v < 0.1K_h$) (or deviated wells not considered/possible) • No limits on surface/reservoir access • Multiple fracs planned (all heavily laminated zones fractured) • Bottom/edge water not penetrated by frac.
Vertical well Open hole Gravel pack	• High permeability ($K_h \geq 10$ md for oil, ≥1 md for gas) • Laminations not "frequent" (h < 2 ft) • No bottom or edge water drive, control needed low K_v ($K_v < 0.5K_h$) (or deviated wells not considered/possible) • No limits on surface/reservoir access • Very high production rates possible • Gravel packs only where sand control needed.
Vertical well Cased hole Gravel pack	• High permeability ($K_h \geq 10$ md for oil, $K_h \geq 1$ md for gas) • Laminations not "frequent" (h < 2 ft) • Limited bottom or edge water control needed • Low K_v ($K_v < 0.5K_h$)or deviated wells not considered/possible) • No limits on surface/reservoir access • Gravel packs only where sand control needed.

Deviated path approach Vertical well in pay	• Surface/reservoir access limited • Deviated wellbore in pay not practical/possible • Laminated zones • Zones with barriers.
Multi-lateral well Vertical or horizontal	• Surface/reservoir access limited • Thick layered pay zones • Multiple well types needed • Compartmentalized reservoirs • Wellbore placed mostly for water control; wellbore placement for sweep/drainage • Very limited need for reentry (unless mechanical system used) • No pressure isolation needed.
Horizontal well Open hole	• $K_v > 0.5K_h$ or plan to frac open hole • No interbed barriers • No sealing lamination unless plan to frac • Stable formation (no movement or spalling or plan to gravel pack) • Good bottom water control possible • Surface/reservoir access restricted.
Horizontal well liner	• $K_v > 0.5K_h$ (unless plan to frac) • No interbed barriers • No sealing laminations, (unless plan to frac) • Some spalling control • No sand control problems • No multiple fracs planned (unless isolation packers set) • Limited bottom water drive control • Production logs/isolation not needed.
Horizontal well cased	• $K_v > 0.5K_h$ (unless plan to frac) • No interbed barriers • No sealing laminations (unless plan to frac) • No vugs or natural fractures (severe cement damage), unless plan to frac).

Special considerations

Steeply tilting pay. Examine the hydrocarbon/water fluid flow path to the wellbore, including the effects of K_v and K_h. Also investigate fracture growth and path. An uphill horizontal wellbore may be chosen to go after reserves that are above vertical well contact, known as attic or up-dip reserves.

High-permeability streaks. The size and permeability contrast to the reservoir location with respect to oil/water contact can significantly affect production or water breakthrough. The orientation of the well path or the decision to frac may be affected.

Salt or tectonic forces. Salt flow may produce extreme loads on casing. The normal approach requires concentric dual casing strings with annular spaces cemented. Tectonic forces and some horizontal collapse forces may create point loads on the casing, which are better handled by extremely-heavy-wall casing strings.

Sweep/floods. Well placement to process a reservoir uses the permeability pathways for best advantage. Wellbore location, orientation, and deviation may be influenced.

Fluid requirements. Heavy oil, scaling, organic precipitation, chronic emulsions, bubble and dew points, and other special requirements may make completion compromises or redesigns necessary.

Multiple zones. Multiple-zone completions and independent completions may be required by fluid pressure or royalty owners.

Completion design flexibility. The initial design is the starting place for the completion; however, it should never be construed to be unchangable. Flexibility is required for any completion to take advantage of information that can be obtained from drilling or other sources.

Completion types

There are several different types of well completions. Broadly, they can be classified as either open hole completions or cased hole completions.

Open hole completions. Open hole completions represent the simplest type of completions, and an open hole (or barefoot) completion is usually made by drilling to the top of the pay, then running and cementing casing. After these operations, the pay is drilled with a nondamaging fluid. Open hole completions have the largest possible formation contact with the wellbore, allowing injection or production with every part of the contacted interval.

Cased hole completions. A casing string is run to prevent the collapse of the wellbore and to act in concert with the cement sheath to isolate and separate the productive formations. The size of the casing is optimized on the basis of the expected productivity of the well and must be designed to withstand the internal and external pressures associated with completion, any corrosive influences, and the forces associated with running the casing.

An optimum casing string is one designed from the inside out—that is, the design is based on supplying a stable casing string of a size such that the total fluid production over the life of the well is optimized (including the possibility of secondary or tertiary floods). The effective design of a casing string for any well consists of four principal steps:

1. Determine the length and size of all casing strings that are needed to produce the well to its maximum potential.
2. Calculate the pressure and loads from predicted production and operations such as stimulation, thermal application, and secondary recovery.
3. Determine any corrosive atmosphere that the casing string will be subjected to and either select alloys that can resist corrosion or design an alternate corrosion control system.
4. Determine the weight and grade of casing that will satisfactorily resist all of the mechanical, hydraulic, and chemical forces applied.

The sizing of a casing string must be completed before finalizing the bit program, during the planning of the well. A typical casing string can be visualized as a very long telescope tube, with the surface casing or conductor pipe as the first segment and the deepest production string or liner as the smallest, most extended section. Each successive (deeper) segment of the casing string must pass through the last section with enough clearance to avoid sticking. Figure 11–1 illustrates the way the casing string fits together. The drill bits used for each section are usually 1½–3 in. or larger than the casing OD to be run. When one section is cased and cemented, a bit just small enough to pass through the casing drift inner diameter (ID) is run to drill to the next casing seat (casing-shoe set depth).

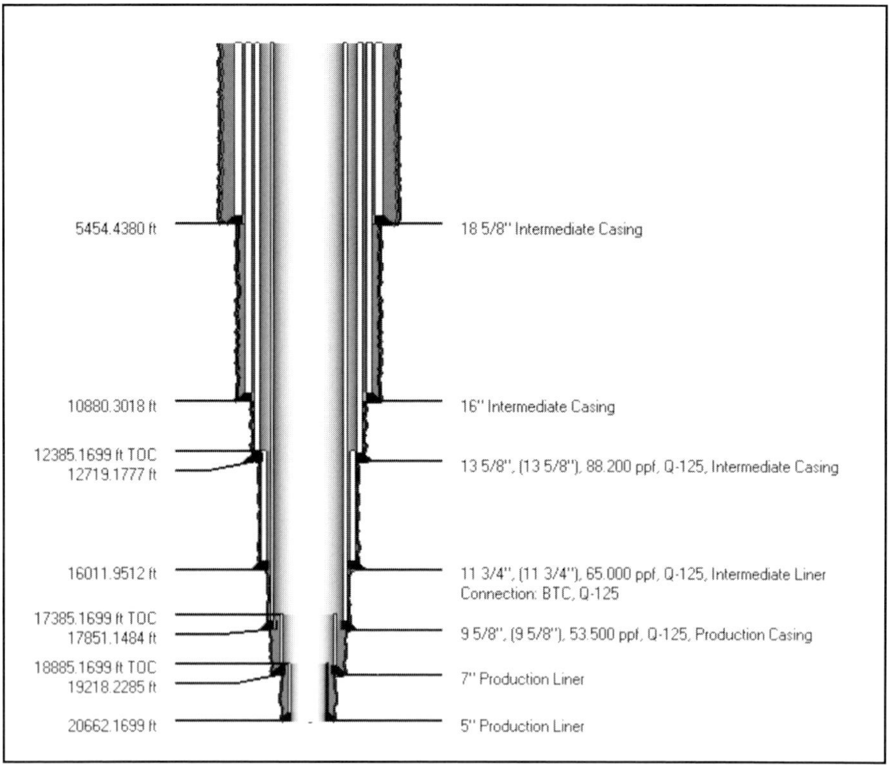

Figure 11–1. Typical casing strings for normal and high-pressure wells

During drilling, departing from the bit program is often required, especially in a wildcat well, where the fluid pressures in the formations cannot be controlled with a single mud weight without either breaking down some formations by hydraulic fracturing with the mud or allowing input of fluid from other formations because of low hydrostatic drilling-mud pressure (i.e., a kick). Ideally, just before this noncontrollable point is reached, the casing point or casing seat is designated and a casing string is run. The economics of drilling and cementing dictate that these casing points be as far apart as formation pressures and hole stability will allow. Use of as few casing strings as possible also permits larger casing to be used across the production zone without using extremely large-diameter surface strings.

Use of small casing severely restricts the opportunities for deepening the well or using larger pumps. Use of small casing to save on drilling costs is usually a poor choice in any area in which high production rates (including water floods) are expected.

Casing Design

Description of casing strings

Different casing strings, which vary in design, material of construction, and purpose, are run during the completion of a well. A brief description of each is provided in the following sections.

Conductor pipe. The conductor pipe is the first casing that is run in the well. This casing is usually large diameter and may be set with the spudding arm on the rig. (The spudding arm drives in the casing.) The primary purpose of the conductor casing is to serve as a flow line, allowing mud to return to the pits and stabilizing the upper part of a hole, which may be composed of loose soil. The depth of the conductor pipe is usually in the range of 50–250 ft and is set by the behavior of surface rocks and soil. It also provides a point for the installation of a blowout preventer (BOP) or another type of diverter system. This allows any shallow fluid flows to be diverted away from the rig and is a necessary safety factor in almost all areas. In areas with very soft and unconsolidated sediments, a temporary outer string, called a stovepipe, may be driven into place to hold the sediment near the surface.

Surface casing. The well is drilled out from the conductor pipe to a depth below the shallow freshwater sands. The surface casing string is run through the conductor pipe and has three basic functions:

- Protecting shallow freshwater sands from contamination by drilling fluids
- Preventing mud from being cut with brines or other water that may flow into the wellbore during drilling
- Providing sufficient protection of the zone to avoid fracturing of the upper hole, so that the drilling may proceed to the next casing point

The surface casing is cemented in place over the full length of the string and, from a safety standpoint, represents the second line of defense, sealing the well and handling any high-pressure flow.

Intermediate casing. The intermediate string is the next string of casing, and it is usually in place and cemented before higher mud weights are used. It allows control of the well in case of subsurface pressure higher than the mud weight and inflow of fluids. An inflow of well fluids encountered during the drilling or completion of the well is called a kick and may be extremely hazardous if the flowing fluids are flammable or contain hydrogen sulfide (sour gas). The intermediate casing may or may not be cemented in place and, if not cemented, may be removed from the well if an open hole completion is desired.

Liner. If a casing string is not hung from the surface, but rather is hung from some point down hole, it is called a liner. In most wells, the top of the liner is cemented in place to provide sealing. The top of the liner is set inside an upper casing string. The section where the liner runs inside another string is the overlap section.

Production liners. Production liners are permanent liners that are run through the productive interval. On some occasions, they may be run back to surface in a liner tieback operation. The tieback consists of a downhole mechanical sealing assembly in a hanger, into which a linear string or the tieback string is stabbed, to complete the seal. A cement job seals the liner into place in the casing and prevents leakage from the formation into the casing. The lower part of the casing string, into which the liner is cemented, is called the overlap section. Overlap length is usually only enough to ensure a good seal, typically 300–500 ft. Overlap length may be longer where water or gas channeling would create a severe problem.

Liners are run for a variety of reasons. If the operator wants to test a lower zone of dubious commercial quality, a liner can be set at less expense than a full casing string. Also, in lower-pressure areas where multiple strings of pipe up to the surface are not necessary to control corrosion or pressure, the liner can be an expense-saving item. In wells that are to be pumped by electric submersible pumps (ESPs), running the liner through the production section leaves the full hole diameter in the casing string above the pay for setting large pumps and equipment.

Production casing. The final casing run into the well is a string across the producing zone that is hung from the surface and may be completely cemented to the surface. This string must be able to withstand the full wellhead shut-in pressure if the tubing or the packer fails. Also, it must contain the full bottom-hole pressure and any mud or workover fluid kill weight when the tubing or packer is removed or replaced during workovers. The decision on whether to cement the full string is based on pressure control, economics, corrosion problems, pollution possibilities, and government regulations

Casing clearance

The necessary clearance between the outside of the casing and the drilled hole depends on the hole and the mud condition. In cases where mud conditioning is good or the mud is lightweight and the formations are competent, a clearance of 1½ in. total diameter difference is acceptable. Primary cementing operations may not be successful given this clearance, and cementing backpressures will be high. A better clearance for general-purpose well completions is 2–3 in. For higher mud weights, poorer mud conditioning, poor quality hole, and higher formation pressures, the clearance should be increased. Excessive clearances should also be avoided. If the annular area is too large, the cement cannot effectively displace the drilling mud. Nowadays, very-low-clearance wells are being drilled in narrow pore and fracture margins.

A reference for hole size and casing size for single- or multiple-string operations is provided by the chart in figure 11–2. To use the chart, select the casing or liner size to be employed and enter the chart at that size. Follow the solid lines to select the applications. The hole/casing combinations linked by the dashed lines represent close clearance and require special techniques; they should be avoided. The deepest casing is chosen first, and the bit/casing program is built in reverse (toward the surface from the maximum depth).

The solid lines indicate the common bit/casing combinations with adequate clearance for most operations. The dashed lines indicate less common (tighter) hole sizes or bit/casing combinations. Long runs of casing through close-clearance holes usually leads to problems. Tight clearances should be avoided if possible.

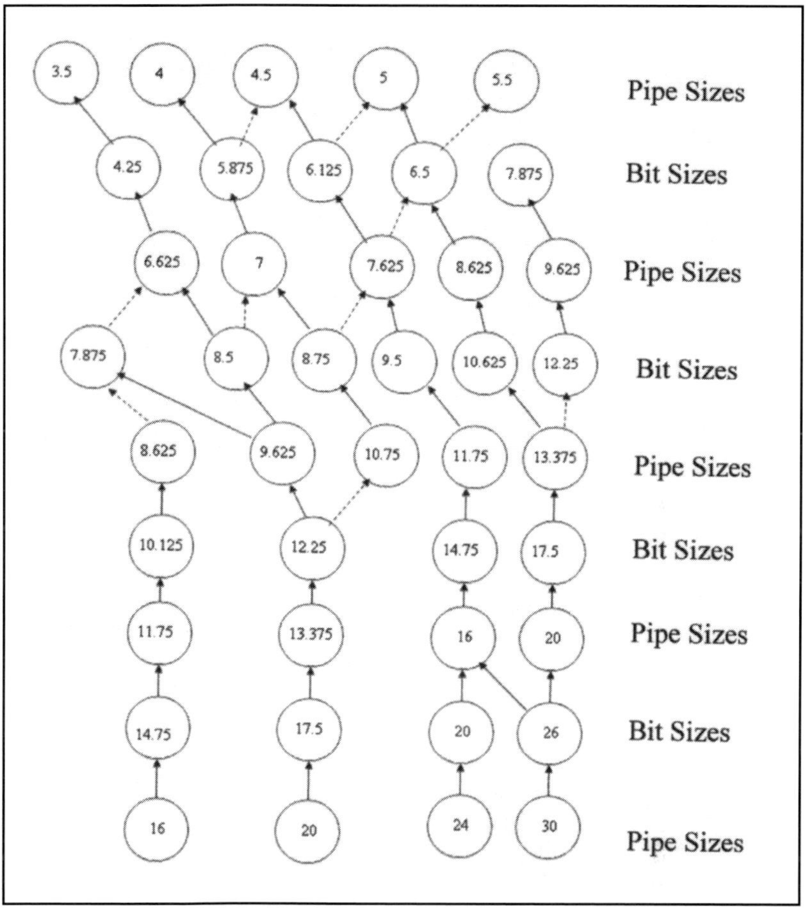

Fig. 11–2. Chart to select the bit/casing sizes required in most drilling programs

Casing connections

The threaded connection of casing or tubing is important because of strength and sealing considerations. The connections are isolated pressure vessels that contain threads, seals, and stop shoulders.[6] The fluid seal produced by a connection may be created in the threads either by a pipe dope fluid or by a metal or elastomer seal within the connection. Strength of the connection may range from less than pipe body strength to tensile efficiencies of over 115% of pipe body strength. Threads are tapered and designed to fit a matching thread in a particular collar. In the API round thread series, the connection may be either short thread and coupling (ST&C) or long thread and coupling (LT&C). As tensile loads increase toward the limit of the pipe, the connection will normally fail by shearing off the threads from the pipe or by thread jump-out caused by pipe deformation under severe loads.

To make a stronger joint in tubing, a thicker (larger OD) section is left at the end of the pipe, so that threads can be cut without making the wall thickness of the pipe thinner than in the body. This form of connection is called external upset (EUE) and is shown in figure 11–3.

Its ID is the same as that of the pipe. The OD of the EUE joint is larger than the nonupset (NU) connection, and the coupling or collar is normally manufactured on the pipe. Another method of increasing the strength of the threaded connection is by upsetting the connection to the inside of the pipe. This internal upset restricts the ID of the pipe at every joint and is used in drill pipe only when a constant OD is necessary.

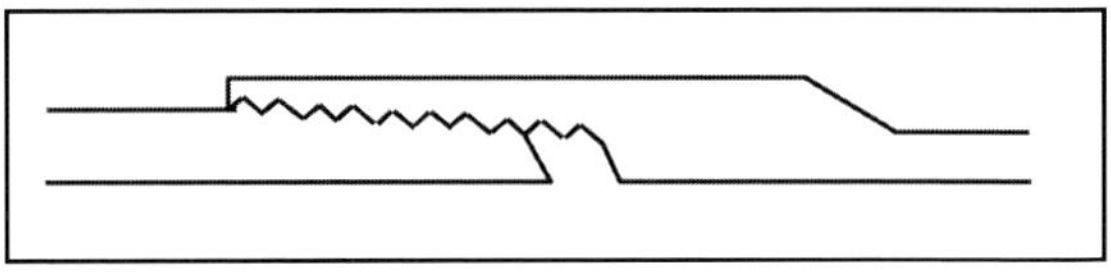

Fig. 11–3. Cutaway drawing of an EUE connection

Other sealing surfaces are available in special connections and have found popularity in applications where rapid, leak-free sealing is important. The two-step thread connection uses two sets of threads, with a metal sealing surface between them. In other connections, a groove at the base of the box may contain an elastomer seal. The disadvantage to the numerous thread and sealing combinations is that the connections cannot be mixed in a string without crossovers (adaptors).

Weights and grades of casing

Common casing diameters range from less than 4½ in. to more than 20 in., and common tubing sizes are from ¾ in. to more than 4½ in. In some high-rate wells, tubing may be 5½ in. to more than 7 in., and 2⅞ in. casing is run in tubingless completions.

After the size has been determined, the grade of the steel must be selected. The grades, along with weights, are based on pressure and corrosion requirements. The grades of steel used in oil industry tubulars are shown in table 11–2, for API and non-API sizes. The letters are assigned only to avoid confusion. Grades N80 and stronger are considered high-strength steels. Use of the higher-strength steels increases cost and decreases resistance to some forms of corrosion. Use of the very hard alloys, such as V150, makes packer setting difficult, since the slips have difficulty setting in the very hard steel of the casing.

The yield and burst strength values for each casing size and weight are used for selection of casing once the necessary strength calculations have been made. Often, because of economics or efforts to lighten the casing string, lower-grade or lighter-weight casing may be considered. Because the weight and pressure loads on a casing string change from top to bottom, a well-designed string may incorporate several weights and grades of casing.

Safety factors in casing design

The design criteria for casing strings depend on the intended use and the anticipated stresses. Because variance exists in both pipe manufacture and formation properties, safety factors for dynamic load must be incorporated into a design. Following are the common ranges for safety factors in normal completion:[7]

- Tension, 1.6–1.8
- Burst, 1.25–1.30
- Collapse, 1.0–1.25

Factors such as salt flows,[8] very high pressures, sour service,[9] reservoir compaction loads,[10] and thermal cycling[11] may change the safety factors.

Earth-shift forces, such as salt movement or other faulting and folding events caused by tectonic movement, are very difficult to address with a traditional safety factor. These forces have been successfully offset in some cases by designs using very-heavy-wall pipe or concentric pipe (casing cemented inside casing) over the affected zone. These types of casing designs are rare, and most are generated by trial and error.

Load description

The casing string must be designed for any load encountered from mud or reservoir fluids in placement or during any phase of stimulation or production. The common forces are tension, during casing running; internal pressures, during drilling, completion, or production; and external pressures, caused by drawdown, mechanical loads, and zone pressures. These loads are tension, burst, and collapse, respectively. The loads are often applied simultaneously in different parts of the string, and the forces may interact.

Collapse. Collapse loading is force applied from outside the casing by either fluid pressure in a zone or earth-shift forces. Forces from fluid pressure are collapse loadings, whereas earth-shift forces produce mechanical crush loadings. The largest collapse load from fluid pressure will usually be exhibited at the bottom of the string, where hydrostatic pressure is greatest. The exception is an isolated, very-high-pressure zone. These zones are usually noted on the drilling record as places where kicks are taken. Earth-shift zones due to faults or salt movement are much harder to locate, especially on wildcat wells, but may show up on the drilling record as sticking points (not associated with mud cake buildup) or zones that have to be reamed or redrilled to get back to gauge hole size. Salt zones are a very important indicator of potential casing problems.[12]

In the collapse design for fluid pressure, the worst-case loading occurs with the unlikely combination of an empty casing string in a hole full of mud. A proper design, for effects of collapse only, would be a casing string that is strongest at the bottom and weakest at the top. Collapse is also affected by the effects of tension, which reduces the collapse rating (or the set depth limit) of the casing. The point at which the effect of tension-induced narrowing of the string is at maximum is the surface, where effects of collapse pressure from hydrostatic pressure are the lowest.

Burst. Burst pressure is a force applied from inside the casing by produced-fluid pressures, the hydrostatic mud load, or the addition of surface pressure during stimulation or workover operations. Since there is usually mud hydrostatic pressure along the outside of the casing before and during cementing, the net pressure (or the difference between the pressure inside the casing and outside the casing) will be used in the design of the casing. Unlike collapse, however, the shallower casing section is also important in the burst calculations from a safety standpoint. Burst pressures exerted by produced fluids are maximum at the surface (i.e., there is no offsetting hydrostatic load), whereas those exerted by mud are maximum at the bottom of the well. During fracturing, high burst loads may be exerted all along the string. Because collapse loads offset the burst load at the bottom of the string, the burst calculation is usually important above the buoyancy neutral point.

Table 11–2. Commonly available casing types and grades

Size OD in	Nominal Weight lbs/ft	Grade	Collapse Resistance psi	Internal Yield Pressure at min. Yield P.E. psi	ST&C psi	LT&C psi
7	26	K55	4,320	4,980	4,980	4,980
7	26	C	5,250	6,790		6,790
7	26	NBC	5,410	7,240		7,240
7	26	180*	5,410	7,240		7,240
7	26	SS95*	7,800	7,240		7,240
7	26	S95	7,800	8,600		8,600
7	26	CYYS95	7,800	8,600		8,600
7	26	C95*	5,870	8,600		8,600
7	26	P110	6,210	9,950		9,960
7	29	C75	6,760	7,850		7,650
7	29	N8O	7,020	8,160		8,160
7	29	L80*	7,020	8,160		8,160
7	29	SS95*	9,200	8,160		8,160
7	29	595*	9,200	9,690		9,690
7	29	CYS95*	9,200	9,690		9,690
7	29	C95*	7,920	9,890		9,650
7	29	S105*	9,780	9,690		9,690
7	29	P110	8,510	11,220		11,220
7	25	V15O	9,800	15,300		15,300
7	32	C75	8,230	8,490		8,490
7	32	N80*	8,600	9,060		9,060
7	32	L80*	8,600	9,060		9,060
7	32	SS95*	10,400	9,060		9,060
7	32	S95*	10,400	10,760		10,780
7	32	CYS95*	10,400	10,760		10,760
7	32	C95*	9,730	10,760		10,780
7	32	S105*	11,340	10,760		10,760
7	32	P110	10,760	12,460		12,460
7	32	V150	13,020	16,990		16,990
7	35	C75	9,710	9,340		8,660
7	35	N80	10,180	9,960		9,240
7	35	180*	10,180	9,960		9,240
7	35	SS9S*	11,600	9,960		9,240
7	35	595*	11,600	11,830		10,970
7	35	CYS95*	11,600	11,830		10,970
7	35	C95*	11,640	11,830		10,970
7	35	S105*	12,780	11,830		10,970
7	35	P110	13,010	13,690		12,700
7	35	V150	16,230	18,670		17,320
7	38	C75	10,680	10,120		8,660
7	38	N80	11,390	10,800		9,240
7	38	180*	11,390	10,800		9,240
7	38	SS95*	12,700	10,800		9,240
7	38	595*	12,700	12,820		10,970
7	38	CYS95	12,700	12,620		10,970
7	38	C95	13,420	12,820		10,970
7	38	S105*	14,040	12,820		10,970
7	38	P110	15,110	14,550		12,700
7	38	V150	19,240	20,250		17,320

	Joint Strength			Body Yield 1,000 lbs	Wall Thickness in	Inside Diameter in	Drift Diameter in
BUTT psi	ST&C psi	LT&C psi	BUTT psi				
4,980	364	401	592	415	0.362	6.276	6.151
6,790		489	631	566	0.362	6.276	6.151
7,240		519	667	604	0.362	6.276	6.151
7,240		511	641	604	0.362	6.276	6.151
7,240		570	696	604	0.362	6.276	6.151
8,600		802	747	717	0.362	6.276	6.151
5,600		602	741	717	0.362	6.276	6.151
8,600		593	722	717	0.362	6.276	6.151
9,960		693	853	830	0.362	6.276	6.151
7,650		562	707	634	0.408	6.184	6.059
8,160		597	746	676	0.408	6.184	6.059
8,160		587	718	676	0.405	6.184	6.059
8,160		655	779	676	0.408	6.184	8.059
9,690		692	836	803	0.408	6.184	6.059
9,690		652	836	803	0.408	6.184	6.059
9,690		853	808	803	0.408	6.184	6.059
5,690		721	859	803	0.405	6.184	6.059
11,220		797	955	929	0.409	6.184	6.059
15,300		1,049	1,243	1,267	0.409	6.164	6.059
7,530		633	779	699	0.453	6.094	5.969
8,460		672	823	745	0.453	6.094	5.969
8,460		661	791	745	0.453	6.094	5.969
8,460		738	860	745	0.453	6.094	5.969
10,050		779	922	885	0.453	6.094	5.969
10,050		779	922	855	0.453	6.094	5.965
10,050		768	891	665	0.453	6.094	5.969
10,050		812	947	865	0.453	8.094	5.969
11,640		897	1,053	1,025	0.453	6.094	5.969
15,870		1,180	1,370	1,398	0.453	6.094	5.969
7,930		703	833	763	0.498	6.004	5.879
8,460		748	876	514	0.498	6.004	5.679
8,460		734	833	814	0.496	6.004	5.879
8,460		819	876	814	0.498	6.004	5.879
10,050		865	964	966	0.498	6.004	5.679
10,050		565	964	966	0.498	6.004	5.679
10,050		853	920	966	0.498	6.004	5.879
10,050		901	964	966	0.498	6.004	5.879
11,640		996	1,096	1,119	0.498	6.004	5.879
15,870		1,311	1,402	1,526	0.498	6.004	5.579
7,930		767	833	822	0.54	5.92	5.795
8,460		914	876	877	0.54	5.92	5.795
8,480		801	833	817	0.54	5.92	5.195
8,460		594	876	877	0.54	5.92	5.795
10,050		944	964	1,041	0.54	5.92	5.795
10,050		944	964	1,041	0.54	5.92	5.795
10,050		931	920	1,041	0.54	5.97	5.795
10,050		983	964	1,041	0.34	5.92	5.795
11,640		1,081	1,098	1,205	0.54	5.92	5.755
15,870		1,430	1,402	1,844	0.54	5.92	5.195

Tension. Tension is a force produced by the weight of the casing, the pressure differential, and the mud weights inside and outside the casing. It is largest at the top of the string and decreases, with depth, toward the bottom of the string. The tension load is partially offset by the buoyancy of the string in mud and is affected by pressure. When the pressure inside the tube rises, the pipe expands in diameter and shortens in length; alternatively, by anchoring of a pipe, the tension is increased to prevent upward movement. When the pressure outside a tube rises, the tube is elongated, or the compression is increased if the ends are fixed.

Buckling failure in casing usually results from axial compression (lengthwise) overloading. The load produces ridges in the casing walls or corkscrewing of the tube. Either of these actions relieve compression, but the pipe is usually permanently yielded.

The effects of buckling are a critical consideration in the design of the casing string. The neutral point is the dividing line between where buckling may occur and where it cannot occur in a tube that is evenly loaded around its radius. Above the neutral point, the tube is in tension and will not buckle; below the neutral point, the upward buoyancy of the mud and other forces, including pressure and mechanical loading, place the tube in compression. Buckling can occur if the compressive load is more than the pipe can tolerate in the wellbore surroundings.

The neutral point in the casing or tubing string can be described by the formula[13]

$$F_Z = P_i A_i - P_0 A_0 \tag{11.1}$$

where

F_Z is the true axial force
P_i is the pressure inside the tubular, in psi
A_i is the cross-sectional area of the ID, in in.
P_0 is the pressure outside the tubular, in psi
A_0 is the cross-sectional area of the OD, in in.

The right-hand side of equation (11.1) may also be referred to as the stability force. When F_z is algebraically greater than the stability force, the pipe tends to be straight; when F_z is less than the stability force, the pipe tends to buckle; and when F_z is equal to the stability force, the neutral point has been reached.

The true axial force, F_z, will vary from point to point in the string, and will also vary over the life of the well. Typical considerations necessary to compute F_z include the conditions at the time of cementing the casing or setting the packer in the case of tubing, as well as changes in the environment (temperature and pressure) to which the tubular is exposed. For casing, buckling primarily affects wear, particularly for intermediate strings through which additional drilling will be done. In extreme cases, splitting may also be common. For tubing, the radial clearance between tubing and casing is usually sufficient to allow corkscrewing, often producing permanent deformation of the buckled portion of the string.

A special-case expression for F_z uses Poisson's ratio:

$$F_Z = 2\mu P_i A_i \tag{11.2}$$

where μ denotes Poisson's ratio, which is about 0.3 for this application. Equation (11.2) is only appropriate for an open-ended tube, clamped at both ends, with $P_0 = 0$ and $\Delta T = 0$, ignoring weight; under the unlikely conditions of a weightless string with no outside pressure, buckling in tension is possible. To illustrate the impact of equation (11.2), consider a weightless tube that is open ended and subject to internal pressure only. For this loading, the only axial force is that due to ballooning, as given by equation (11.2). As the inside pressure is increased, F_z increases as $2\pi P$, but the right-hand side of equation (11.1)—that is, the stability force—increases as P_o. The tube will not only buckle immediately but also buckle in tension.

The stability equation has also been reported as[14]

$$F_s = F_a + P_i A_i - P_o A_o \tag{11.3}$$

where F_s is the stability force and F_a is the axial load. In relating equation (11.3) to equation (11.1), $F_s = -[F_z - P_i A_i - P_o A_o]$, and $F_z = -F_a$

Each zone or section of the casing string is checked for tensile requirements, following collapse and burst calculations. When corrections are made to a string design to compensate for tension load requirements, the order of selection is usually (1) stronger connection, (2) higher grade (stronger steel), and (3) higher weight. Increasing the connection strength and the steel grade is preferred, since they increase the total string strength without adding significant weight. For the sole purpose of derating the casing collapse strength owing to the effects of tension, a practical buoyancy neutral point can be estimated:

$$\text{NP} = D_t\left(1 - \frac{\rho_m}{65.4}\right) \tag{11.4}$$

where

NP is the neutral point for tension effects on casing collapse, in ft

D_t is the setting depth of the casing string, in ft

ρ_m is the mud density, in lbs/gal

65.4 is the weight of steel, in lbs/gal (based on a density of 489.5 lbs/ft)

The collapse resistance values given in the manufacturers' tables for casing are not affected by axial load. In a well, the casing will be stressed by fluid pressures, mechanical bending forces, and tensile forces produced by the hanging weight of the casing. See figure 11–4. In collapse calculations, axial tension produces a reduction in collapse resistance. The axial tension is assumed to be from tension loads on a straight, freely hanging pipe and not from bending loads. For an example using a curved section, see figure 11–5.

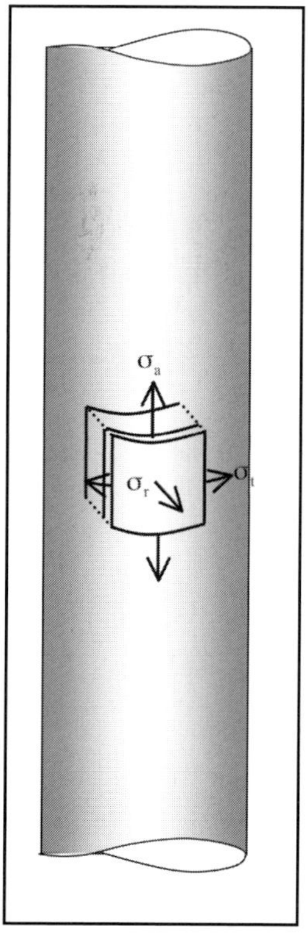

Fig. 11–4. Schematic of the location of axial (σ_a), tangential (σ_t), and radial (σ_r) stresses on a vertical suspended tube exposed to fluid pressure

The equivalent yield strength, Y_{pa}, is[15]

$$Y_{pa} = \left[\sqrt{1 - 0.75 \left(\frac{\sigma_a}{Y_p} \right)^2} - 0.5 \left(\frac{\sigma_a}{Y_p} \right) \right] Y_p \tag{11.5}$$

where
σ_a is the axial stress, in psi
Y_p is the minimum rated yield strength of the casing, in psi

The percent of yield stress, Y_t, may be calculated as follows:

$$Y_t = \frac{S_t}{Y_m} \tag{11.6}$$

where
S_t is the unit tensile stress, in psi
Y_m is the minimum yield stress of the casing body, in psi

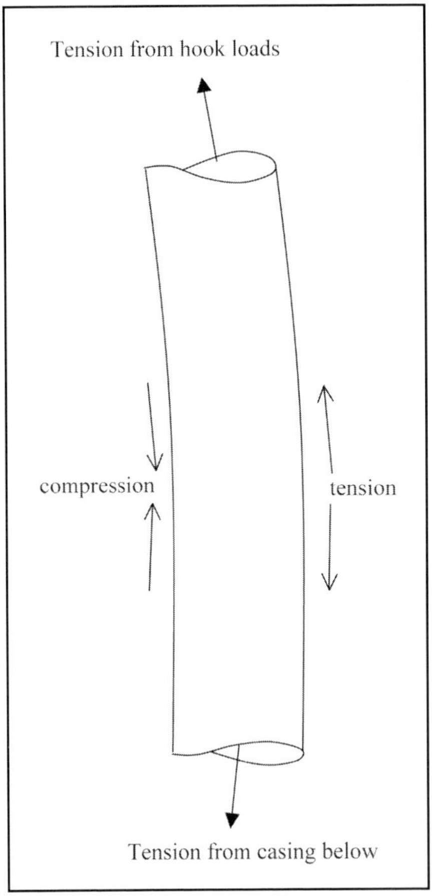

Fig. 11–5. Loads on a casing string during running and on casing landed in a curved section of the borehole

The percent yield stress is related to percent of full collapse pressure by the curve in figure 11–6. The product of the percent of collapse and the collapse rating is the effective collapse strength (derated for effects of tension). Because of buoyancy effects, tension exists only above the buoyancy neutral point, calculated by equation (11.4). The effects of compression below the buoyancy neutral point increase the casing collapse resistance, although the increase is not accounted for in this design.

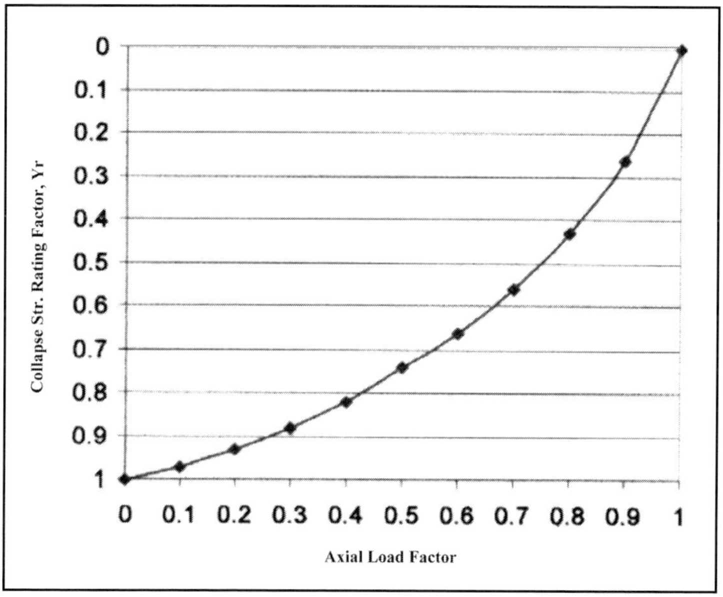

Fig. 11–6. Effect of axial tension on collapse strength of casing. Y_r is used in the collapse-derating calculation.

Design of the casing string—API method

A complex solution requiring a computer program is discussed first, in this section. The procedure illustrated here for determining the collapse strength is defined in API Bulletin 5C3.[16] This is intended merely as an introduction to the process. Subsequently, a practical, hand-calculated method will be given.

Several methods for designing casing strings produce an acceptable casing design. Most methods use an X-diagram graphical approach or a calculated design based on a single strength concern in each area of design. The API procedure recognizes the changes in steel behavior in elastic, transition, plastic, and ultimate yield.

When exposed to external pressure from mud or reservoir fluids and the effects of axial tension from the weight of the casing below any point (plus other loads), a piece of casing may fail in one of three possible collapse mechanisms: elastic collapse, plastic collapse, and exceeding the ultimate strength of the material. Each failure mode is bounded by limits of the ratio of casing diameter to thickness, *D/t*; a transition collapse formula was added arbitrarily, since the API minimum elastic and minimum plastic curves do not intersect. The transition formula covers this area.

The equations of the API casing-string design method are as follows:

1. Elastic:

$$P_c = \frac{46.95 \times 10^6}{(D/t)\left[(D/t)-1\right]^2} \tag{11.7a}$$

$$\text{applicable range } \frac{D}{t} \geq \frac{2+(B/A)}{3B/A} \tag{11.7b}$$

2. Transition:

$$P_{crt} = \text{YP}\left(\frac{F}{D/t} - G\right) \tag{11.8a}$$

$$\text{applicable range } \frac{\text{YP}(A-F)}{C+\text{YP}(B-G)} \leq \frac{D}{t} \leq \frac{2+(B/A)}{3B/A} \tag{11.8b}$$

3. Plastic:

$$P_{crp} = \text{YP}\left(\frac{A}{D/t} - B\right) - C \tag{11.9a}$$

$$\text{applicable range } \frac{\sqrt{(A-2)^2 + 8\left[B+(C/\text{YP})\right]} + (A-2)}{2\left[B+(C/\text{YP})\right]} \leq \frac{D}{t} \leq \frac{\text{YP}(A-F)}{C+\text{YP}(B-G)} \tag{11.9b}$$

4. Ultimate yield:

$$P_{cry} = 2\text{YP}\left[\frac{(D/t)-1}{(D/t)^2}\right] \tag{11.10a}$$

$$\text{applicable range } \frac{D}{t} \leq \frac{\sqrt{(A-2)^2 + 8\left[B+(C/\text{YP})\right]} + (A-2)}{2\left[B+(C/\text{YP})\right]} \tag{11.10b}$$

In equations (11.7)–(11.10), YP is the minimum yield strength of the pipe, in lbs; Y_{pa} is the yield strength of the axial stress equivalent, in psi, as given by equation (11.5); and other variables are as follows:

$$A = 2.8762 + (0.10679 \times 10^{-5})\text{YP} + (0.21301 \times 10^{-10})\text{YP} - (0.53132 \times 10^{-16})\text{YP} \qquad (11.11)$$

$$B = 0.026233 + (0.50609 \times 10^{-6})\text{YP} \qquad (11.12)$$

$$C = -465.93 + 0.030867\text{YP} - (0.10483 \times 10^{-7})\text{YP} + (0.36989 \times 10^{-13})\text{YP} \qquad (11.13)$$

$$F = \frac{\left\{\left(46.95 \times 10^{6}\right)\left[\dfrac{3B/A}{2+(B/A)}\right]\right\}^{3}}{\text{YP}\left[\dfrac{3B/A}{2+(B/A)} - \dfrac{B}{A}\right]\left[\dfrac{3B/A}{2+(B/A)}\right]^{2}} \qquad (11.14)$$

$$G = FB/A \qquad (11.15)$$

The addition of axial tension to a casing string has a net effect of reducing the yield stress of the pipe. The reduction may be severe enough to cause a derating of the pipe by a full grade, (e.g., from N80 to C75). Axial loads generally result from two forces:

- Hanging weight of the casing string
- Temperature-induced forces in thermal wells and in nonthermal wells where operating temperatures may change by over 100°F.

Buoyancy

When the hole is filled with cement or mud, there is a buoyant force exerted on the casing by the fluid in the hole and opposed by the fluid in the casing. Buoyant force acts on the entire casing string and results in a reduction in hanging weight. The amount of buoyant force exerted by the mud is equal to the mud weight displaced by the submerged casing.

The weight of steel, at 489.5 lb/ft or 65.4 lb/gal, is several times the heaviest mud weight; yet, the buoyant contribution of the mud is a significant factor in the hook load during running and cementing of the casing. Hook loads change dramatically during running and cementing operations, and conditions do exist (e.g., running closed-end casing and dry) under which hook load could diminish to zero (i.e., the casing floats).

Buoyant weight, W_b, for an open-ended casing string of air weight W_a, filled and surrounded by one fluid, is

$$W_b = W_a\left(1 - \frac{\rho_f}{\rho_s}\right) \qquad (11.16)$$

where

ρ_f is the density of the fluid

ρ_s is the density of the steel (489.5 lbs/ft or 65.4 lbs/gal)

When the fluid in the casing is different from the fluid outside the casing, the volumes contained in the casing and displaced by the casing must be calculated, and the weights must be summed against the air weight of the casing.

For the special case of an additional surface pressure, such as the holding pressure on the mud in the casing while cement sets in the annulus, the surface pressure is incorporated with the load produced by the mud. The buoyant force, F_b, applied to the air weight of the casing becomes

$$F_b = (P_{is} - 0.052\rho_i d)A_i - (P_{os} + 0.052\rho_o d)A_o \quad (11.17)$$

where

F_b is the buoyant force, in lbs
P_{is} is the surface pressure on the fluid inside the casing, in psi (usually zero)
P_{os} is the surface pressure on the fluid outside the casing, in psi (usually zero)
ρ_o is the density of the fluid inside the casing, in lbs/gal
ρ_i is the density of the fluid outside the casing, in lbs/gal
A_i is the inside area of the casing, in in.
A_o is the outside area of the casing, in in.
d is the depth, in ft

The pressure terms affect buoyancy much the same way as pressure affects tubing forces.

Example 11–1

A 10,000 ft string of 7 in., 26 lbs/ft N80 casing is run in a hole filled with 12 lbs/gal mud. The string is equipped with a float shoe. Calculate the following:

(a) Hook load of an empty closed-end casing string in mud. (Air weight = 10,000 ft × 26 lbs/ft = 260,000 lbs.)

(b) Hook load of a mud-filled string in mud (same mud weights inside and outside).

(c) Hook load of 16 lbs/gal cement–filled string in mud.

Solution:

(a) The displaced volume of mud is $(\pi/4)d^2 = (\pi/4)(7/12)^2(10{,}000) = 2{,}673$ ft, and the displaced weight of mud is $2{,}673_3 \times 12 \times 7.48 = 239{,}928$ lbs.

The displaced weight of the mud is a force acting upward on the casing, trying to float it out of the well. It has the effect of trying to lighten the total load. Thus, hook load = 260,000 – 239,928 = 20,062 lbs. (This ignores the weight of air.)

(b) Since the only action is on the cross section of the steel body, the buoyancy factor difference formula is adequate:

$$b_f = 1 - \frac{\rho_f}{\rho_s}$$

$$= 1 - \frac{12 \times 7.48}{489.5} = 0.817$$

$$W_b = W_a b_f$$

$$= (260{,}000 \text{ lbs})(0.817) = 212{,}420 \text{ lbs}$$

(c) Total buoyant weight = weight of casing in air + weight of cement – buoyancy of displaced mud. Weight of cement is

$(\pi/4)(6.276 \text{ in.})(16 \text{ lbs/gal})(0.052 \text{ gal/ft} \cdot \text{in}^2)(10{,}000 \text{ ft}) = 257{,}382 \text{ lbs.}$

From part (a), buoyant force is 239,928 lbs. Thus, the total buoyant weight is 260,000 + 257,382 – 239,928 = 277,455 lbs.

Collapse design—non-API method

A practical method that considers burst, collapse, and tensile forces is also available. This method may be worked by equations or by graphical methods. The design is conservative in collapse resistance because of the basic assumption of an empty string in a hole full of mud. In practice, most casing strings are filled with mud as they are run. The design begins at the bottom of the string.

The collapse force produced by fluid pressure from a homogeneous fluid in the well and an empty casing string is:

$$P_{cx} = 0.052\rho_f D_x \quad (11.18)$$

where

P_{cx} is the collapse pressure, in psi, at depth D

ρ_f is the density of the fluid on the outside of the pipe, in lbs/gal

D_x is the depth at the point of calculation, in ft

The forces can be visualized graphically by the methods illustrated in figures 11–7 through 11–12.[17]

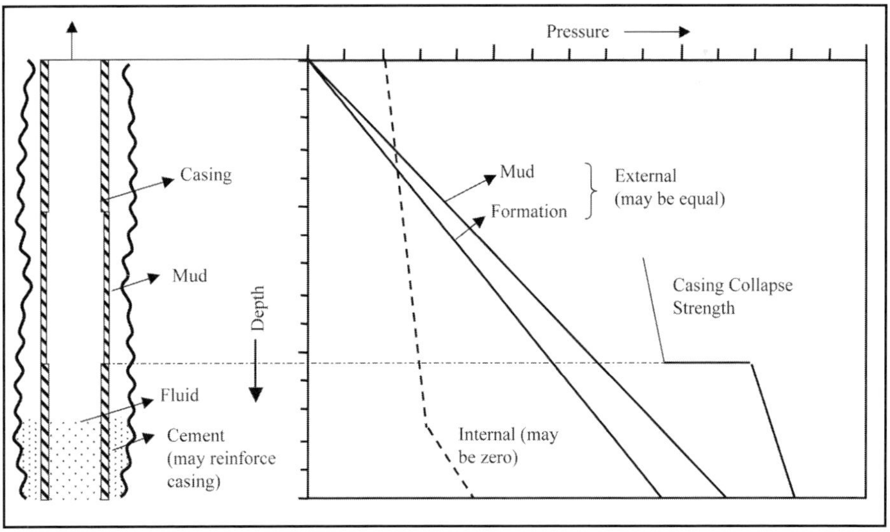

Fig. 11–7. Graphical analysis of collapse force

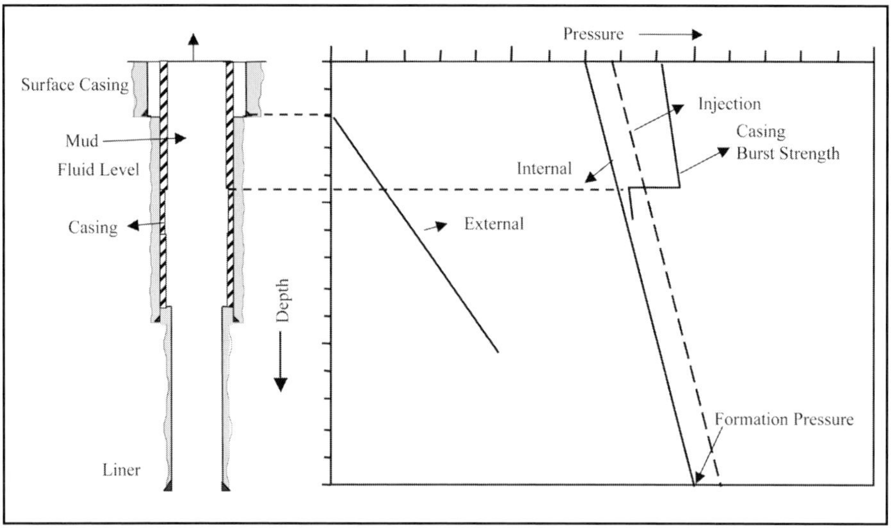

Fig. 11–8. Graphical analysis of burst force

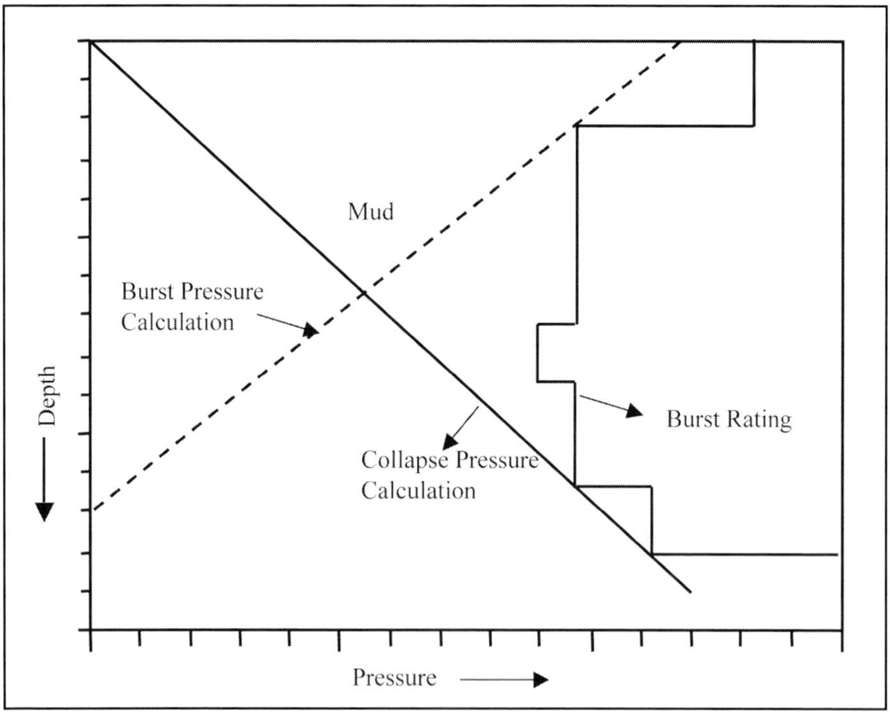

Fig. 11–9. X-diagram method—alternative casing design

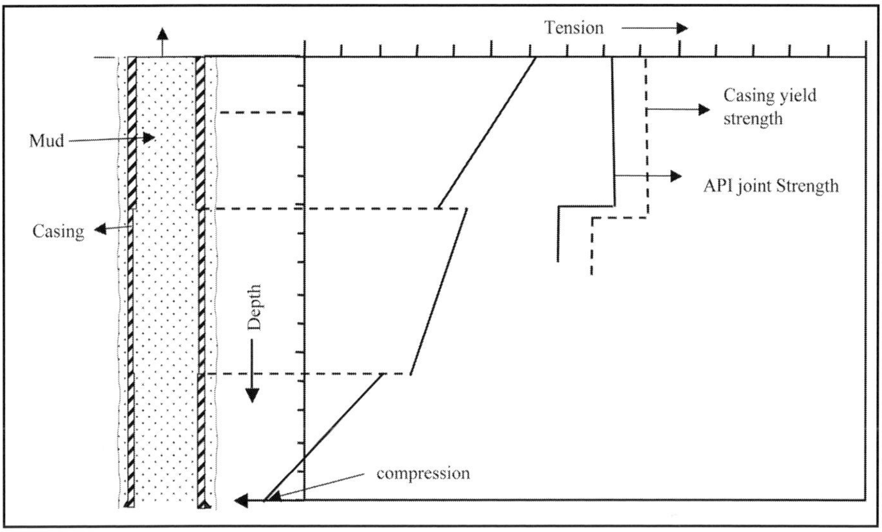

Fig. 11–10. Graphical analysis of tension

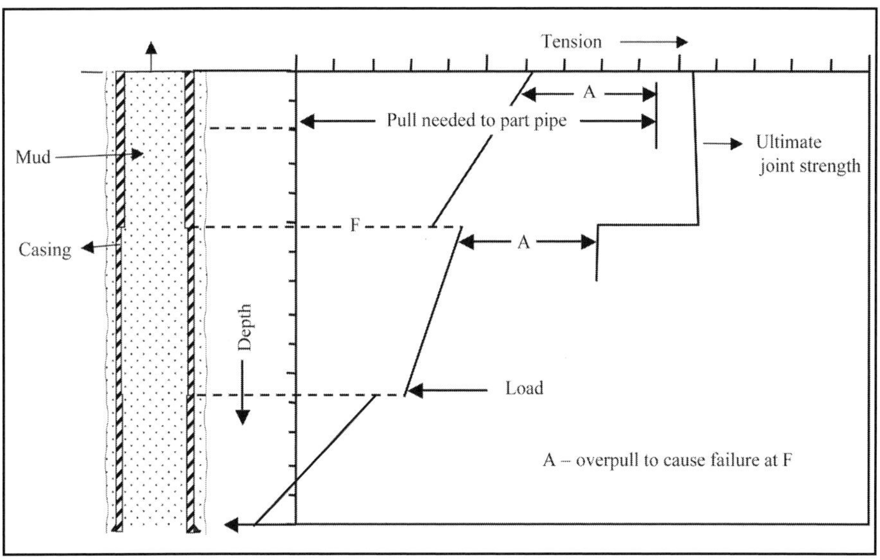

Fig. 11–11. Graphical analysis of overpull

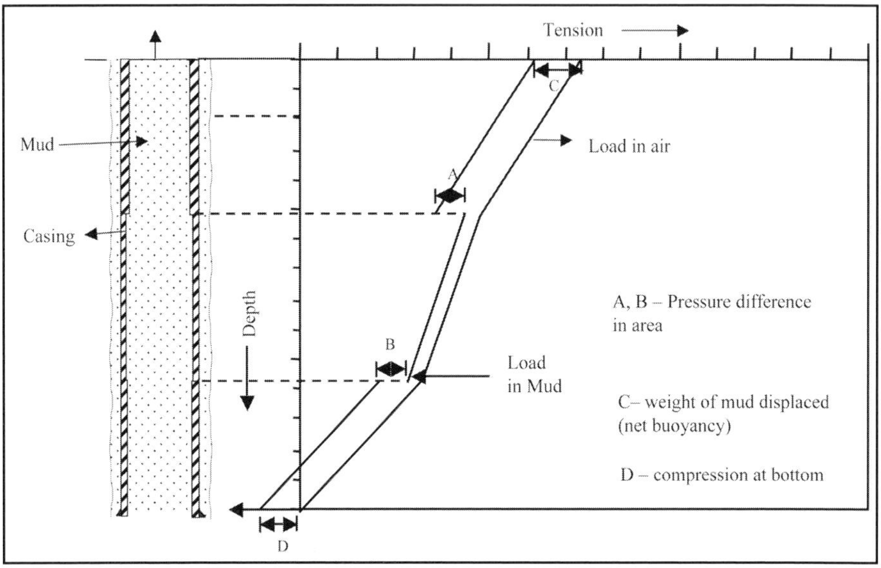

Fig. 11–12. Graphical analysis of buoyancy

When the effects of a different fluid, such as cement, inside the pipe is considered, the equation becomes[18]

$$P_{\mathrm{cx}} = (P_{\mathrm{o}} + D_{\mathrm{x}}G_{\mathrm{o}}) - (P_{\mathrm{i}} + D_{\mathrm{x}}G_{\mathrm{i}}) \tag{11.19}$$

where

P_{cx} is the collapse pressure, in psi, at depth D_x
P_{o} is the external surface pressure, in psi (usually zero)
D_{x} is the depth at the point of calculation, in ft
G_{o} is the pressure gradient exerted by the fluid outside the pipe, in psi/ft
P_{i} is the internal surface pressure, in psi
G_{i} is the pressure gradient exerted by fluid inside the pipe, in psi/ft

The external surface pressure, P_{o}, is used when there is an outer casing string and an annular pressure will be exerted. This technique is commonly used to back up the tubing and increase burst resistance of the inner string. It may also be used in some casing designs. Because outside surface pressure is rare, the term is generally dropped.

It is customary to design the primary strings for the worst possible case. Since the worst possible case will be when the pipe is empty, the equation reduces to $P_{\mathrm{cx}} = D_{\mathrm{x}}G_{\mathrm{o}}$. The outside surface pressure is assumed to be zero.

The design based on an empty pipe string may seem excessive, but it is done to eliminate consideration of triaxial forces produced by hole irregularities and other factors.[19] The worst-possible-case design therefore covers a multitude of other forces. Because of buoyancy produced by changes in axial load following setting of the cement, portions of the casing string may be in compression, rather than tension. Casing above the point of zero axial tension has less collapse resistance; casing below the point of zero axial tension has more collapse resistance, since it is in compression. The collapse forces on a casing are usually visualized as being applied by the pressure of the mud in which the string is run. The effect of tension in reducing the collapse strength of the casing is generally considered, but the effects of ballooning, ovality, and temperature changes during circulating are often neglected. These effects can be severe, especially in high-collapse-resistance casing, such as some 95 grades.

For the burst calculations, one of two API-approved formulas may be used. For plain-end (nonthreaded) pipe and pipe with premium couplings (couplings stronger than the pipe body), Barlow's formula is used:

$$P_y = 0.875 \frac{2Y_m t}{D_o} \tag{11.20}$$

where

P_y is the minimum internal yield pressure, in psi
Y_m is the minimum yield strength, in psi
D_o is the nominal OD, in in.
t is the nominal wall thickness, in in.

The 0.875 factor allows for a 12.5% variation in wall thickness, as permitted by API manufacturing specifications.

For threaded and coupled pipe, the internal yield is about the same as for plain-end pipe, except when low-clearance couplings are used or the coupling has lower test strength than the tube. In the case of lower strength couplings,

$$P_s = Y_m \frac{(D_o - D_i)}{D_o} \tag{11.21}$$

where

Y_m is the minimum yield of the coupling material, in psi
D_o is the nominal outside diameter of the coupling, in in.
D_i is the diameter at the root of the coupling when the pipe has been fully tightened, in in.

The design for the upper part of the casing string is dominated by the burst consideration. The safety aspect—controlling kicks and blowouts—places a large degree of importance on controlling full well pressure in the upper sections of the string. The worst case is when a blowout occurs and the hole contains only gas.

The maximum surface pressure, then, is the maximum formation pressure minus the hydrostatic load created by gas. As a simplification, the mud hydrostatic pressure is taken as the reservoir pressure (unless the reservoir pressure is known):

$$P_s = 0.52 \rho m D_t - G_g D_t \tag{11.22}$$

where

P_s is the maximum well shut-in pressure, in psi
ρ_m= density of mud, in lbs/gal
D_t is the depth of kick, in ft
G_g is the gas gradient, in psi/ft (0.1–0.15 psi/ft, varying based on pressure)

In most cases, the potential problem area will be at the casing shoe. The pressure at the shoe can be calculated as follows:

$$P_{bx} = G_{bd}D_c - G_g(D_c - D_x) \tag{11.23}$$

where
P_{bx} is the burst pressure, in psi, exerted at depth D_x
G_{bd} is the formation breakdown gradient, in psi/ft (usually 0.7–1.0 psi/ft)
D_c is the casing-setting depth, in ft
D_x is the depth at which P_{bx} is being determined, in ft

To determine the worst-case burst pressure exerted during a blowout with gas in the hole, the external pressure is subtracted from equation (11.23):

$$P_{bx} = G_{bd}D_c - G_g(D_c - D_x) - 0.052\rho_o D_x \tag{11.24}$$

where ρ_o is the external fluid density, in lbs/gal.

Burst force design may also be considered graphically. (See fig. 11–8.[20])

Equation (11.24) can be used to make the start of the X diagram of figure 11–9. The X diagram is constructed by use of collapse and burst calculations. The maximum burst line is drawn between the calculated burst at the surface and the calculated burst at the casing shoe. The collapse line is drawn between zero and the maximum collapse pressure at the casing shoe, as calculated by equation (11.22) or (11.23).

Tension design is the last step for each section of the casing string. The top of each section should be checked to see that the tensile ratings are not exceeded. The common safety factor is 1.6–1.8. When the tensile limits are exceeded, a change to a higher-strength joint should be made. Tension limits may be gathered from a table of casing properties or calculated by dividing the value for joint strength from API Bulletin 5C3 by the safety factor.

Example 11–2 (Casing design—single string and burst)

Design an economical casing string that will satisfy the pressure requirements of collapse and burst and the load requirements of tensile force. Strings increase in cost with increasing grade and weight. A packer will be set within 100 ft of the bottom of the well. The string design must allow packer running and setting, as well as efficient well operations. Assume an annulus full of mud and an empty string (the worst possible normal conditions).

Casing safety factors are as follows: collapse, 1.1; burst, 1.25; tensile, 1.7. You are given the following well data for a 10¾ in. surface string that has been landed and cemented at 450 ft:

- Depth = 7,600 ft
- Mud weight = 12 lbs/gal
- Bit size = 8¾ in.
- Fracture gradient = 0.77 psi/ft
- Straight gauge hole
- Reservoir pressure = 0.56 psi/ft
- Water-based mud

The buoyancy neutral point for collapse is Np = (7,600 ft)[1 – (12/65.4)] = 6,206 ft. The bottom-hole pressure is (12)(0.052)(7,600 ft) = 4,742 psi. The maximum collapse pressure (from the mud column) is (12 lbs/gal)(0.052 gal/in.^{2}/ft)(7,600 ft) = 4,742 psi. Pick a casing grade and weight that has a collapse rating in excess of 4,742 × 1.1 = 5,216 psi.

Single string design. The casing data on size, weight, grade, burst, collapse, body tensile and joint tensile are obtained from casing and tubing tables. The largest casing that can easily be run in an 8¾ in. hole is 7 in.

To satisfy collapse pressure requirements, pick 26 lbs/ft 0-75 casing as a first choice (collapse resistance = 5,250 psi). The first casing will extend from the bottom of the well upward to a point where a weaker (lower grade or lighter weight) and cheaper string can be run. That point depends on the pressure rating of the second string. Several small jumps in weight or grade or just a few large changes can be made. The route depends more on economics and inventory available than on a scientific selection process.

For the second casing, pick 23 lbs/ft 0-75 casing (collapse resistance = 3,770 psi; derating for the safety factor, 3,770/1.1 = 3,427 psi). To determine how deep the 23 lbs/ft casing can be set,

$$D_x = \frac{3,427}{12 \times 0.052} = 5,492 \text{ ft}$$

The bottom string (26 lbs/ft C75) has an initial design length of 5,492–7,600 ft; however, the collapse rating of the upper string (23 lbs/ft C75) must be derated for the effects of tension, since it is above the neutral point. The 26 lbs/ft casing is not derated for collapse, since maximum collapse pressure is exerted at the bottom of the well, where the casing is below the neutral point and in axial compression (which opposes collapse forces).

To derate the collapse rating of the upper string for effects of tension,

- The set depth of 5,492 ft should be derated slightly. Typically, 2–4% suffices. Hence, 5,400 ft will be used here.
- The unit tensile stress is calculated: (6206 ft – 5400 ft)(26 lbs/ft) = 20,956 lbs.
- The axial load factor, X, and derating factor, Y_r, are determined:
  X = 20,956/499,000 = 0.042
  Y_r = 0.99 (from fig. 11–11)
- The derated collapse (for effects of tension) of the 23 lbs/ft casing is 3,427 × 0.99 = 3,392.
- Collapse pressure at 5,400 ft is (5,400)(12)(0.052) = 3,370 psi.

The 3,370 psi collapse pressure is less than the 3,392 psi collapse rating (derated for the safety factor) on the 23 lbs/ft pipe, so the choice of 5,400 ft as the maximum set depth is a good one. The bottom of the 23 lbs/ft string sets the top of the 26 lbs/ft casing. The set points of the 26 lbs/ft casing are 7,600 ft to 5,400 ft.

The lesser of the two will almost always be the difference of the pore pressure and the gas gradient. In special cases where high stimulation pressures must be handled, the stimulation pressure may be selected as the design criteria. Since the top of the 26 lbs/ft casing string is above the neutral point, the burst design begins to govern. The 23 lbs/ft casing string will also be checked for burst to see how shallow it can be safely set. For a burst design, a surface pressure is needed. Use the lesser of the two possible formulations for the maximum pressure at the surface, $P_{s\,max}$:

$$\begin{aligned} P_{smax} &= \text{fracture pressure at bottom} - \text{gas gradient} \\ &= (7{,}600 \text{ ft} \times 0.77 \text{ psi/ft}) - (7{,}600 \text{ ft} \times 0.1 \text{ psi/ft}) \\ &= 5{,}852 - 760 = 5{,}092 \text{ psi} \end{aligned}$$

or

$$\begin{aligned} P_{smax} &= \text{pre pressure} - \text{gas gradient} \\ &= (7{,}600 \text{ ft} \times 0.56 \text{ psi/ft}) - (7{,}600 \text{ ft} \times 0.1 \text{ psi/ft}) \\ &= 4{,}256 - 760 = 3{,}496 \text{ psi} \end{aligned}$$

Since 3,496 psi is the highest likely pressure while running pipe (before cementing), it will be used for this design. There will be many cases for which the pore pressure is not known. In these cases, the mud weight is used to approximate the pore pressure.

Burst design. The burst rating of 23 lbs/ft C75 casing is 5,940 psi. Thus, the shallowest that the 23 lbs/ft casing string can be set is

$$D_x = \frac{3{,}496 - (5{,}940/1.25)}{0.624 - 0.1} = \frac{3{,}496 - 4{,}752}{0.524} < 0$$

Since the value of D_X is less than zero, the 23 lbs/ft 0-75 casing can be set to reach the surface. The 23 lbs/ft 0-75 casing runs from 5400 ft to 0 ft.

If the fracture breakdown pressure (with gas column), $P_{s\,max}$ = 5,092 psi, had been used, then the top of the 23 lbs/ft 0-75 casing would have been 649 ft, whereas the top of a 23 lbs/ft C95 casing would have bee needed to reach to surface.

Notice that a higher-grade pipe was used rather than a heavier-wall pipe. This allows a packer to be run from the surface that would set in the 23 lbs/ft pipe. For some close-clearance packers, the packer for 23 lbs/ft may not be able to pass through even one joint of heavier-wall pipe set at the top of the well.

Tensile (uncorrected for buoyancy) is calculated as follows:

- Feet of run: 6,206 ft to 5,400 ft
  - Casing: 26 lbs C75
  - Load = 20,956 lbs
  - Cumulative load = 20,956 lbs
  - Derated strength = 287,647 lbs
- Feet of run: 5,400 ft to 0 ft
  - Casing: 26 lbs/ft C75
  - Load = 124,200 lbs
  - Cumulative load = 145,156 lbs
  - Derated strength = 244,706 lbs

Thus, the design will be as follows:

- 0–5,400 ft, 23 lbs/ft 0-75
- 5,400–7,600 ft, 26 lbs/ft 0-75

Notice that the tensile calculations on the bottommost string extend only to the neutral point.

An intermediate string is slightly different from a single production string since it may experience the burst pressures from deeper zones. Collapse forces may be felt as well if there is zone communication between the interval below the intermediate string and any zone that contacts the outside of the intermediate string. This problem is more fully addressed later, in the section on liners. Depending on the job conditions and assumptions, a higher pore pressure in a lower zone may have to be used for burst calculations. It is reasonable in most intermediate-string casing designs to verify that the burst rating (with safety factor) is adequate for the mud from the lower zone. If the cement job extends along the full length of all strings, collapse or burst forces have much less effect.

Example 11–3 (Casing design—intermediate string)

Design a 9⅝ in. intermediate string, given the following information:

- 13⅜ in. surface casing at 2,500 ft
- 12¼ in. hole drilled to 9,500 ft
- Mud weight = 12 lbs/gal
- Future depth = 12,000 ft
- Future mud weight = 14.5 lbs/gal
- Fracture gradient at 9,500 ft = 0.884 psi/ft

An intermediate string is slightly different from a production string in that it will experience the collapse and burst forces encountered during the next deeper segment. If it is fully cemented, it is much less likely to fail, regardless of the pressure of the produced fluid or the mud. Depending on conditions, the job design may require that intermediate string withstand the pressures of the zone below, particularly the burst. Additionally, the mud weights inside and outside the string can be considered with the formulas.

First, determine the maximum bottom hole pressure (BHP):

$$\begin{aligned} \text{BHP} &= 0.052\ \rho_f D_t \\ &= 0.052 \times 1.45 \times 12{,}000 = 9{,}048 \text{ psi} \end{aligned}$$

A depth of 12,000 ft is used as the worst possible case.

Second, determine the collapse resistance design and pressures. The maximum collapse pressure generated by the mud weight outside the pipe is

$$P_{cx} = 0.052\ \rho_o D_x - 0.052\ \rho_i D_x + (P_o - P_i)$$

where ρ_o, ρ_i are the respective mud densities outside and inside the pipe and P_o, P_i are the respective external and internal surface pressures. Since the casing is empty, $\rho_i = 0$, and since the top of the string is at the surface and the casing is empty, $P_o = P_i$. Thus,

$$\begin{aligned} P_{cx} &= 0.052\ \rho_o D_x \\ &= 0.052 \times 12 \times 9{,}500 = 5{,}928 \text{ psi} \end{aligned}$$

The neutral point (i.e., the point at which load goes from compression to tension) is now calculated:

$$\text{NP} = P_{cx} = D_t(1 - \frac{\rho}{65.4})$$

$$= (9{,}500)(1 - \frac{12}{65.4}) = (9{,}500)(1 - 0.183) = 7{,}756 \text{ ft}$$

Note that this formula is usable for calculation of casing collapse only. To make calculations easier, 7,756 can be rounded to 7,760.

At this point, casing is selected, favoring the lightest, lowest grade. Use a safety factor of 1.125 for collapse. First, pick 47 lbs/ft S95 casing (collapse = 7,100 psi; burst = 8,150 psi; tension [joint] = 1,053,000 lbs). Thus, on incorporation of the safety factor, collapse = 7,100/1.125 = 6,311 psi (derated); this satisfies P.

Second, pick the next lower strength of casing, 43½ lbs/ft (collapse = 5,600; burst = 7,510; tension = 960,000 lbs). Thus, on incorporation of the safety factor, collapse = 5,600/1.125 = 4,977 psi.

Now calculate the maximum set depth for the 43½ lbs/ft section. This depth sets the top of the 47 lbs/ft section:

$$D_x = \frac{P_c}{0.052 \times 12}$$

$$= \frac{4{,}977}{0.052 \times 12} = 7{,}975 \text{ ft}$$

The 47 lbs/ft casing runs from 9,500 ft to 7,975 ft. Since 7,975 is below the neutral point of 7,760, no correction is needed.

To determine the top of the 43½ lbs/ft casing string, the base of the next string must be selected. Select 40 lbs/ft S95 casing (collapse = 4,230 psi; burst = 6,820 psi; tension = 858,000 psi). Thus, derated collapse = 4,230/1.125 = 3,760 psi, and derated burst = 6,820/1.125 = 6,062 psi (both collapse and burst are derated by the safety factor of 1.125).

The maximum set depth is

$$D_x = \frac{P_c}{0.052\ P_o}$$

$$= \frac{3{,}760}{0.052 \times 12} = 6{,}026 \text{ ft}$$

Since this section will be above the neutral point, the effect that tension has on collapse rating must be calculated. Hence, several further steps must be taken.

First, the 6,025 ft value is derated slightly. As a first guess, 5,900 ft is used.

Second, a unit tensile stress is calculated:

$S_t = (7{,}760 - 5{,}900) \text{ x } 43.5 = 80{,}910 \text{ lbs}$

Note that the weight calculated is the 43½ lbs/ft pipe, above the neutral point. This is the weight that the 40 lbs/ft pipe bears.

Third,

$$Y_R = \frac{80{,}910}{\text{tensile body strength}} = \frac{80{,}910}{1{,}088{,}000} = 0.074$$

(Note that tensile body strength is used here.)

Fourth, the derating factor is determined graphically. From figure 11–11, the derating factor is found to be 0.975.

Fifth, the collapse, derated for tension, is $P = 3{,}760 \times 0.975 = 3{,}666$ psi. Calculation of collapse pressure at 5,900 ft gives

$$P_c = 0.052\rho_o D_X$$
$$= 0.052 \times 12 \times 5{,}900 = 3{,}681 \text{ psi}$$

The guess of 5, 900 ft was a little too deep, but since 15 psi difference is not excessive, the 5,900 ft depth can be kept. The 43½ lbs/ft casing runs from 7,975 ft to 5,900 ft.

Sixth, since it is above the neutral point, the burst design will dominate the casing selection. To start the burst design, calculate a surface pressure that is the lesser of the fracture gradient at the 9⅝ in. casing shoe, less a gas gradient, and the formation pressure in the future pay (at 12,000 ft), less a gas gradient. The maximum surface pressure, *P*, that the fracture gradient will support is

$$P_s = D_t f_g - D_t G_g$$
$$= (9{,}500 \times 0.884) - (9{,}500 \times 0.1) = 7{,}448 \text{ psi}$$

For a calculation of the effect that formation (pore) pressure has on the surface pressure, the pressure on the zone at 12,000 ft is used. Since all that was given is the mud weight, that may be used as the formation pressure gradient. The gas gradient (multipled by the depth) is subtracted to get maximum surface pressure:

$$P_s = 0.052\ \rho_f D_t - 0.01 D_f$$
$$= (0.052 \times 14.5 \times 12{,}000) - (0.1 \times 12{,}000) = 7{,}484 \text{ psi}$$

If this maximum pressure at the surface ever occurred, the pressure at the shoe of the 9⅝ in. casing would be calculated as follows:

$$P_{D = 9{,}500 \text{ ft}} = 0.052\ \rho_f D_f - G_g(D_f - D_t)$$
$$= (0.052 \times 14.5 \text{ x } 12{,}000) - 0.1(12{,}000 - 9{,}500) = 8{,}798 \text{ psi}$$

or a gradient of 0.926 psi/ft. Since 8,798 psi is above the pressure necessary to fracture the formation at the shoe of the 9⅝ in. casing, the surface pressure of 7,848 psi would never occur.

To proceed with burst design, use the surface pressure of 7,448 psi from the fracture gradient. The shallowest depth that a casing string can be set (limited by burst) comes from

$$\begin{aligned} P_{bx} &= P_S + G_g D_X - 0.052\rho_o D_X \\ &= 7{,}448 + 0.1D_X(0.052 \times 12)D_X \\ &= 7{,}448 + D_X(0.1 - 0.6224) \\ &= 7{,}448 - 0.524D_X \end{aligned}$$

Hence, $D_X = (7{,}448 - P_{bx})/0.524$. Thus, the shallowest depth that the 40 lbs/ft pipe can be set is $D_X = (7{,}448 - 6{,}062)/0.524 = 2{,}645$ ft. (The derated burst of the 40 lbs/ft casing was 6,062.)

As a check, the collapse pressure is $P_c = 0.052 \times 12 \times 2{,}645 = 1{,}650$ psi. (The derated collapse was 3,760 psi for the 40 lbs/ft casing.) The 40 lbs/ft pipe runs from 5,900 ft to 2,645 ft.

Since the 40 lbs/ft pipe cannot be run above 2,645 ft, a stronger pipe is needed. Although 43½ lbs/ft pipe would handle the pressure, the smaller ID would restrict the packer size that could be run below it; hence, a stronger grade of pipe (at 40 lbs/ft) is needed. Since the casing tables show that a 40 lbs/ft P110 pipe is unavailable, the lower string design must be changed to 43½ lbs/ft strings to allow a packer to pass any section of the string and still be set in the bottom section. Select 43½ lbs/ft P110 pipe for the upper section (collapse = 4,430 psi; burst = 8,700 psi; tension = 1,106,000 lbs). This practical overriding of the design theory is very common, especially when higher grades are more available or, occasionally, more economical.

For the stronger pipe, burst = 8,700/1.125 = 7,733 psi, and $D_X = (7{,}448 - 7{,}733)/0.524 < 0$. Thus, the 43½ lbs/ft P110 pipe can be run to surface.

Check for tension design:

Pipe Length (ft)	Weight (lbs/ft)	Grade	Load	Cumulative	Rating
7,760 to 5,900	43.5	S95	80,910	80,910	960,000
5,900 to 2,645	43.5	S95	141,592	222,502	960,000
2,645 to 1,475	43.5	S95	50,895	273,398	960,000
1,475 to 427	43.5	P110	64,162	337,561	1,006,000

Final design:

Interval	Weight (lbs/ft)	Grade
0–1,475 ft	43.5	P110 (LT&C)
1,475–2,645 ft	43.5	S95 (LT&C)
2,645–5,900 ft	43.5	S95 (LT&C)
5,900–7,975 ft	43.5	S95 (LT&C)
7,975–9,500 ft	47	S95 (LT&C)

From 2,645 ft to 5,900 ft, the 40 lbs/ft string would have been used, except that upper sections had to be heavier weight with smaller diameter to handle the burst load. For packer access, the ID of the string must be at least as large at the top as it is in the sections below.

Dynamic loads

Dynamic loads from running casing into the well can be estimated with the following formula from Vreeland:

$$V_c = \frac{F_m - W}{\rho C_o A} \tag{11.25}$$

where

V_c is the critical velocity
F_m is the force required in order to yield the casing
W is the weight of casing hanging below the slips
ρ is the mass per unit volume
C_0 is the speed of sound in the material: $C_0 = \sqrt{E/\rho}$
A is the cross-sectional area

For steel, $\rho C_0 = \sqrt{\rho E}\, 150 \text{ lbs} \cdot \text{s/in.}^3$

For example, the load increment produced when a 29 lbs/ft casing with a diameter of 7 in. is lowered at 3 ft/s (12 s per 40 ft joint) and stopped suddenly can be calculated by rearranging equations (11.2) and (11.3) to solve for a force, F, applied at any casing running velocity, V: $F = V\rho C_0 A = (36 \text{ in./s})(150 \text{ lbs} \cdot \text{s/in}^3)(8.449 \text{ in.}^2) = 46{,}000$ lbs. The force would be in addition to string tensile weights.

Once a usable hole has been drilled, the casing string or liner is run to depth and cemented in place. Actual running of the string involves several operations to properly place the pipe and to avoid damaging the well. Like any pipe or drill string run in the well, casing is susceptible to differential sticking. To minimize the problem, mud conditioning and centralizers are used. In the case of differential sticking, the pull required in order to free a differentially stuck pipe is[21]

$$F = A_c C_f \Delta P \tag{11.26}$$

where

F is the pulling force, in lbs
ΔP is the differential pressure into the zone from the hole, in psi
A_c is the contact area of pipe and mud cake, in in.^{2}
C_f is the coefficient of friction (an average of 0.25 is normal)

As seen from equation (11.26), differential sticking force (equal in magnitude to F) is dependent on the area of contact between the pipe and the mud cake and on the mud weight. In a situation with 1.5–2 in. of diameter difference between pipe and hole, there is about 2 in. contact between the casing and the mud cake. When the mud cake is thick and spongy or when clearances are very tight, contact cross-sectional area may double.

The force involved in differential sticking is illustrated in the following example. For a casing string run across a series of permeable zones with a total zone thickness of 100 ft, the mud overbalance is 4,000 psi, at a depth of 8,000 ft, with a mud weight of 11 lbs/gal, against a pore pressure of 10.0 lbs/gal. $\Delta P = 8{,}000 \times (11 - 10) \times 0.052 = 416$ psi, and $A_c = 100 \times 12 \times 2 = 2{,}400$ in.^{2}; thus, $F = 416 \times 2{,}400 \times 0.25 = 249{,}600$ lbs. Increased mud weight and increased contact area can significantly increase the differential sticking force.

The differential sticking force can be considerably lessened by the use of centralizers. Although centralizers cause an increase in drag in the casing and are often difficult to run in the well near the surface, Bowman and Sherer[22] have estimated that the increase in drag from running centralizers will be less than the weight of the casing over the zone where differential sticking is a problem.

In the case of the previous example, the air weight of a 100 ft section of the 8⅝ in. casing (40 lbs/ft) is 4,000 lb. Thus, if the drag caused by centralizers over a 100 ft section were equal to the weight of the casing, the increase in drag would be 4,000 lbs, or 1.6% of the differential sticking force. Such an analysis and an associated estimate of cost savings is dependent on the zone of differential sticking being located near the bottom of the well. This would mean that only the section of casing to be landed across the zone would need to be centralized.

In reality, the simple example provides a very low estimate of centralizer drag, since centralizers need to be run over the entire string and more densely on casing that passes intervals prone to differential sticking. Centralizers, however, are a necessary design item on most strings, and every effort should be made to run them. It is nearly impossible to differentially stick a casing string with centralizers on every joint,[23] but field experience dictates that it may be difficult to run that many in a hole with clearance of two inches or less. Centralizers aid in getting the pipe to bottom, being able to move the pipe, and improving the chances of a successful primary cement job.

The number of centralizers needed depends on pipe and mud weight, hole deviation, and hole condition. Calculation programs are available for centralizer spacing. The number of centralizers per joint depends on the lateral forces exerted by the weight of both the casing and the casing contents. It is this weight that forces the casing into the side of the hole. The lateral force increases with casing weight, hole deviation angle, and weight differential between cement and mud while the cement is in the annulus. Doglegs and other hole problems also influence the force on casing. The use of centralizers in soft or unconsolidated formations becomes very difficult, since the centralizer blades or springs will tend to embed in the formation. The number of centralizers must be sufficient to offset the lateral force and position the casing near the middle of the hole.

For the lowest possible drag while running pipe, it is important that the right number and type of centralizers be run. Slim-hole centralizers should be used in close-clearance holes. Necessary support equipment includes stop devices to keep the centralizers in place and avoid bunching them on the pipe by freely sliding along the pipe at a bend or a point of close clearance. Casing collars are usually used for stop devices when only one centralizer is needed per joint. When more than one centralizer per joint is required or where a flush joint liner is used, stop devices must be added to position the centralizers. With the use of very hard grades of flush joint liners (e.g., V150), slip-on stop devices, even those with set screws or slips, may not work.

Centralizers should not hinder the ability to move pipe during cementing; rather, the centralizers prevent the sticking that hinders pipe movement. The inability to rotate some liners is often caused by insufficient starting torque or increased torque as cement travels up the annulus. The torque to start rotation may approach the maximum but should decrease as the string is rotated.[24]

The most successful methods of rotation are power swivels, rotary tables, and casing power tongs. Power swivels and rotary tables are the most reliable. Casing power tongs are normally only used for joint makeup.

It is a high-risk operation to run centralizers in a hole drilled without stabilizers. Bowman and Sherer have given four reasons why liners (and other casing strings) do not make it all the way to bottom:[25]

1. A usable hole was not drilled (the drift diameter is not equal to the bit diameter)
2. Differential sticking occurs because centralizers were not used
3. The wrong type centralizers and support equipment were used
4. A dirty hole was drilled (one full of bridges or fill)

Speed of running the string is a critical factor in formation damage, since the casing is like a piston in a cylinder, with mud as the pressure seal. The pressures created by running casing can easily be enough to fracture the formation and force mud into pay zones. Loss of mud in this manner will also reduce hydrostatic load and can cause a blowout. Safe running speeds on wells with normal clearances and low-viscosity mud are 40–60 s per joint (equal or less than 1 ft/s). Close-clearance strings or heavily gelled muds require longer run times.

A procedure for actual running of a casing string, originally reported by Bowman and Sherer,[22] is put forth in API Bulletin D-17. Prior to running the casing, the hole should be cleaned by mud circulation with the bit on the bottom, and a conditioned mud with low solids content and low viscosity should be circulated into the hole. After the casing is run, circulation of mud from the bottom up at least twice is recommended. The mud circulation with the casing in the well is necessary, since velocity of the mud in the close-clearance annulus of the casing and hole will be much higher than with the drill string in the hole and more cuttings will be moved. If these cuttings are not removed, they may form bridges in the annulus, blocking or restricting cement flow.

Design of the casing string—deviated wells

The basic difference in casing design for a deviated well is the magnitude of the loads encountered in the deviated or angle build section. This requires bending and torque considerations. The major points for consideration for deviated wells casing design are

1. Axial and torque loads can be estimated by analyzing separate sections
2. Relationships can be developed for various parameters in casing-string design
3. The pickup, setdown, and neutral states produce sufficiently different loads and magnitudes to analyze each separately

Bending force. Bending force needs to be considered for casing design in deviated wells owing to the effect of wellbore curvature and inclination angle. Equivalent axial force due to bending can be found by considering a beam subjected to pure bending:

$$F_b = E\frac{D}{2L}\left(\frac{\pi}{180}\right)\theta \tag{11.27}$$

where

θ is the inclination angle from vertical
D is the diameter of the casing
E is the elastic modulus of the pipe
L is the length considered

The dogleg severity (DLS) angle, α, is the change in angle, in degrees per 100 ft of borehole length. Expressing equation (11.27) in terms of DLS angle and considering steel,

$$F_b = 218\alpha A_s D \tag{11.28}$$

where

F_b is the effective axial force due to bending, in lbs/ft
A_s is the cross-sectional area of the casing, in in.^{2}

A_s can be approximately proportional to the nominal weight of the casing, and the constant of proportionality is given as 3.40 with an acceptable accuracy for most of the casing pipes. Thus, equation (11.28) can be expressed as

$$F_b = 64\alpha D w_n \tag{11.29}$$

where w_n is the unit weight of pipe, in lbs/ft.

Usually, the curvature of the casing with couplings is different when it is in contact with the wellbore wall. Consequently, the axial force calculated using equation (11.29) may be an underprediction. Lubinski has used the classical beam deflection theory to derive the maximum axial stress as

$$(\sigma_z)_{max} = 218\alpha d_n \frac{6KL_j}{\tan h(6KL_j)} \quad K = \sqrt{F_a / EI} \tag{11.30}$$

where *K* and L_j have units of in.$^{-1}$ and ft, respectively.

If the increased axial stress caused by bending is expressed as an equivalent axial force, F_{ab}, then

$$F_b = 64\alpha d_n w \frac{6KL_j}{\tanh\,(6KL_j)} \tag{11.31}$$

where F_b would be the axial force required in order to create the same maximum stress level in a straight section of pipe.

In the previous discussion, the effect of bending on casing failure was handled by consideration of the maximum stress present under the combined loading situation experienced. In this analysis, a possibility of failure is indicated when the maximum stress level exceeds the yield strength of the steel. An alternative approach sometimes used is to express the axial strength of the material in terms of tension and bending. This approach is used most commonly in rating the tensional joint strength of a coupling subjected to bending.

API formulas have been developed for the joint strength of round thread casing subjected to bending. When the axial tension strength divided by the cross-sectional area of the pipe wall under the last perfect thread is greater than the minimum yield strength, the joint strength is given by

$$F_{\text{cr}} = 0.95 A_{jp} \left\{ \sigma_{ult} - \left[\frac{140.5 \alpha d_{\text{n}}}{(\sigma_{\text{ult}} - \sigma_{\text{yield}})^{0.8}} \right]^5 \right\} \tag{11.32}$$

where $F_{\text{cr}}/A_{\text{jp}} \geq \sigma_{\text{yield}}$ and $A_{\text{jp}} = (\pi/4)[(d_{\text{n}} - 0.1425)^2 - (d_{\text{n}} - 2t)^2]$. When the axial tension strength divided by the cross-sectional area of the pipe wall under the last perfect thread is less than the minimum yield strength,

$$F_{\text{cr}} = 0.95 A_{jp} \left(\frac{\sigma_{\text{ult}} - \sigma_{\text{yield}}}{0.644} + \sigma_{\text{yield}} - 218.15 \alpha d_{\text{n}} \right) \tag{11.33}$$

For the previous calculations, the axial force in deviated portion of the well is calculated as follows:

$$F_{\text{T}} = (w_{\text{n}} L - F_{\text{buoy}}) \cos\theta \tag{11.34}$$

where

L is the length of the portion of pipe

F_{buoy} is the buoyant force

θ is the inclination of the deviated portion from vertical

Liner design

A liner is a casing string that does not extend back to surface. Liners may be permanent or temporary and are run for a variety of reasons:[27]

- Wear protection of the permanent casing string from rotating drill pipe or frequent tubing movement, especially in a deviated section
- Correction of hole difficulty (e.g., doglegs, key seats, or ledges)
- Zone isolation of a high-pressure zone or a lost circulation zone
- As an economical alternative to a full casing string
- To allow adequate room above the liner top for a large-diameter pump
- To reduce casing tension loads in deep wells or to allow use of an available, low-strength connection
- For repair of leaks or buckled casing
- For casing strengthening in earth-shift zones

Regardless of the reasons, most liner designs incorporate the cost reduction factor of not running casing back to surface.

Liners have drawbacks as well, including several problems related to liner cementing. From a design standpoint, perhaps the biggest concern is that the intermediate string or the last full string becomes the top part of the liner and must handle the burst force generated by the zones crossed by the liner. Before a decision is made on use of a liner, the strength of the top part of the last full string must be checked against the new maximum expected surface pressure. Unless proven otherwise by cement bond evaluation tools, the upper pipe strength alone (with no assistance from the cement sheath) must be capable of withstanding the pressure. If the upper section will not withstand the forces, it should be protected by running a full string; otherwise, the string should be run before perforating the well.

The first steps of liner design are the same as for full casing strings. The collapse, burst, and tension safety factors and design equations are the same. The basic differences between liner and casing designs become more profound for deep wells, hot environments, and very-high-pressure zones.

Example 11–4 (Liner design)

A well is being drilled through the formations and pressures shown in figure 11–13. A conventional design for the primary 9⅝ in. casing with safety factors of 1.125 for collapse, 1.125 for burst, and 1.8 for tension is $P = 0.052\rho_0 D_X - 0.052\ \rho_i D_X + (P_o + P_i)$. For $P_o = P_i$, $\rho_i = 8.3$ lbs/gal, and $\rho_i = 18.5$ lbs/gal, $P_{CX} = (0.052 \times 18.5 \times 13{,}250) - (0.052 \times 8.3 \times 13{,}250) = 6{,}821$ psi. (Note that ρ_i is normally assumed to be equal to zero for a conservative design; however, in this case, the high-pressure zone will be isolated by the casing and the liner during production, so a freshwater weight can be used in calculations.)

Further, the neutral point is

$$NP_1 = D_t\left(1 - \frac{\rho_m}{65.4}\right) = 13{,}250\left(1 - \frac{18.2}{65.4}\right) = 9{,}563 \text{ ft}$$

The casing for the lowest part of full string is chosen as follows: collapse minimum = 6,821 × 1.125 = 7,674 psi. This is satisfied by 53.5 lbs/ft P110 casing. (Note that although standard casing design can be used to design the full string to surface, it will not be done here.)

The first liner will be 6 5/8 in. Mud weight is 9.2 lbs/gal. For $P_{CX} = 7{,}080$ psi, collapse minimum = 7,080 × 1.125 = 7,965 psi. This is satisfied by 28 lbs/ft N80 casing. The neutral point is

$$NP_2 = 14{,}800\left(1 - \frac{9.2}{65.4}\right) = 12{,}718 \text{ ft (above the liner top)}$$

The second liner will be 3½ in. Mud weight is 15 lbs/gal. For $P_{CX} = 0.052 \times 15 \times 15{,}900 = 12{,}402$ psi, collapse minimum = 12,402 × 1.125 = 13,952 psi. This is satisfied by a 9.2 lbs/gal mud. The neutral point is

$$NP_2 = 15{,}900\left(1 - \frac{15}{65.4}\right) = 12{,}253 \text{ ft (above the liner top)}$$

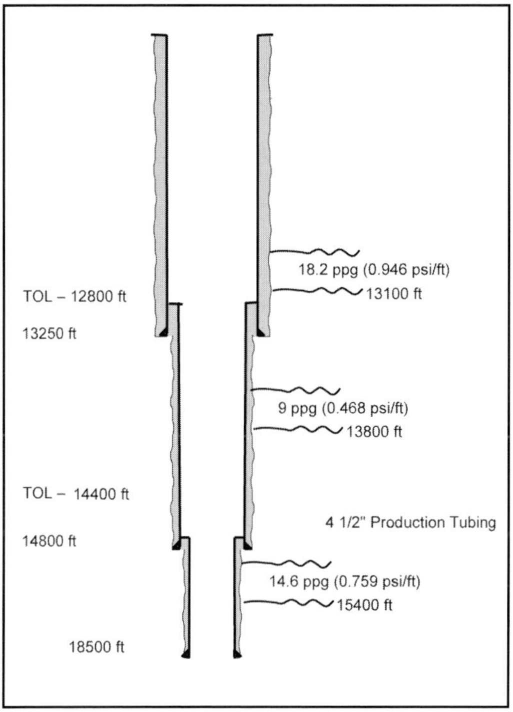

Fig. 11–13. Two-liner design in example 11–4

Burst failure of liners is usually caused by kicks while drilling below a liner,[28] pressure while fracture stimulating, or a tubing leak while producing a deep, high-pressure zone. All these factors are made worse by a poor cement job. Since filling the casing/open hole annulus with cement significantly strengthens and supports the pipe, a good cement job over high pressure or earth-shift zones is a requirement in any deep well.

Several complications may be caused by a poor cement job.[29] After a primary cement job on the 9⅝ in. string, the 6⅝ in. liner is run, and a circulation cement job is initiated. The low-pressure zone at 13,800 ft breaks down while cementing, returns are lost, and the cement top peaks at 13,400 ft. A top squeeze with cement seals the top of the liner but leaves an uncemented zone from 13,120 ft to 13,400 ft. While drilling near TD to test a deep structure, the temperature of the mud goes from 270°F (i.e., the temperature of the zone at 15,400 ft) to 345°F (i.e., the bottom-hole temperature at 18,500 ft). Since both ends of the liner are cemented, stress will be induced by the temperature rise.

The amount of stress at fixed end casing is

$$\sigma E\alpha(DT) = 200T$$

where

σ is the stress, in psi

α is the coefficient of thermal expansion for steel (6.9×10^{-7}in./in.°F)

E is the Young's modulus for steel (30×106 psi)

ΔT is the increase in temperature, in °F

The stress is the compressional axial load that can cause pipe splitting or deformation. Repair of the collapse may be made by running the 3½ in. liner from above the first liner to TD.

Liner tiebacks. Although the liner completion described in figure 11–13 may be used only as a liner, there are some conditions under which the liner may be run back to the surface. Such a liner tieback may be temporary, to protect permanent casing strings from wear during drilling,[30] or permanent, by use of cement, to strengthen the casing string through a high-pressure zone or isolate and stop a casing[31] or a liner top[32] from leaking.

To allow a tieback, most liners are equipped with a liner tieback receptacle. The receptacle is an internally polished receiver that will accept the seal assembly of the tieback liner. The tieback liner usually incorporates some type of cement port, or sliding sleeve, and a seat that will allow cement to be circulated to the annulus and upward through the annulus to the surface. The seat, float equipment, and/or plugs used in the cementing operation are of drillable material.

Cementing

Cementing is one of the most critical steps in well completion. Sadly, because it comes at the end of drilling, in the haste to put a well on production, the time and commitment to get a good cement job done is not always taken. Consequently, significantly more time and money is spent in correcting the cement job than would have been required to do it correctly (i.e., the initial cost).

Primary cementing comprises the main isolation of the casing and liner strings, where cement may extend from the shoe (or bottom) of the string to the surface or to a design height that covers and isolates all flowable productive and corrosive zones.[33] Squeeze cementing is the operation of repairing a cement job that failed to effectively protect or isolate the upper zones. In primary cementing, the object is to place a continuous sheath or band of cement around the pipe—extending, without channels or voids, from the casing wall outward, to the formation face, with a height that matches the design requirements of the well. When a primary cement job fails to isolate completely the section of interest, repair of the cement job must be done before drilling can proceed. These repair steps are covered under the collective label of squeeze cementing. In a squeeze job, cement is forced into the zone through perforations, ports in tools, a hole produced by corrosion, or the clearance between casing overlap liners or strings.

Cement fills and seals the annulus between the casing string and the drilled hole. It has three general purposes:

- Zone isolation and segregation
- Corrosion control
- Formation stability and pipe strength improvement

Cement forms an extremely strong, nearly impermeable seal from a thin slurry. The properties and the behavior of the cement slurry depend on its components and additives. This section will focus on the basics of the cementing process.

Most cements used in the oil industry are a type of portland cement. The name portland was taken from an English Channel island with a limestone quarry that was used as the source of stone for the development of cement. Portland cement is produced from limestone and either clay or shale by roasting at 2,600–3,000°F. High temperature fuses the mixture into a material called clinker cement. After the roasting step, the rough clinker product is ground to a size specified by the grade of the cement. The final size of the cement particles has a direct relationship to how much water is required in order to make a slurry without producing an excess of water at the top of the cement or in pockets as the cement hardens. The crystals seen in set cement include tricalcium silicate (C_3S_i), dicalcium silicate (C_2S_i), tetracalcium aluminoferrite (C_4AlF), tricalcium aluminate (C_3Al), periclase or magnesium oxide (MgO), and free lime (CaO).[34]

Not all cements, even those made from the same components, will react in the same manner when mixed with water. Basically, the differences are the fineness of the grind of the cement, impurities in the water, and some minor additives added during the cement manufacturing process. Table 11–3 lists the API-designated classes of cements. These classifications were created in response to deeper and hotter downhole conditions. Note that the useful depths were derived from average pumping times of *neat* cement (containing no additives) for average temperatures involved at these depths. The actual well environment controls the limits of the cement. Also, additives such as accelerators and retarders can be used to modify the behavior of the cement. In this manner, a class H cement, for example, can be used at much greater depths than the 8,000 ft limit given in the table 11–3.

Depth estimates are based on conditions imposed by the casing-cementing, well-simulation tests (see API-RP10-B) and should be considered approximate values. The response of the cement to these conditions can be modified by additives.

There are a number of other cements that do not fall specifically into any general classification. These cements are special blends of portland and additives or cements based on other chemistry. They include pozzlin cement that incorporates organic resin technology; expanding cements that increase in volume as the cement sets; silica and lime cement, for hot wells; and low-heat–generating cements, for permafrost applications. These cements are rarely used in general completions because they are more expensive than portland or have other traits that are less desirable than those of portland.

Environmental conditions and available completion equipment may significantly affect the performance of the cement or may place special requirements on the cement. The unique problems that result from the effect that temperature has on cement setting and on long-term strength of cements have led to development of special cements, for both steam wells and wells in arctic environments. High temperatures sharply reduce cement strength and durability, necessitating the development of stabilizers. Silica additives and lime-based cements have proved effective in thermal wells. Permafrost cement was developed in response to a need to cement formations to depths of 2,000 ft without producing sufficient heat of hydration from setting the cement to melt and destabilize the permafrost.

Table 11–3. API-designated cement classes

Class Type	Details
Class A:	For use from surface to 6,000 ft (1,830 m) depth, when special properties are not required.
Class B:	For use from surface to 6,000 ft (1,830) depth, when conditions require moderate to high sulfate resistance.
Class C:	For use from surface to 6,000 ft (1,830 m) depth, when conditions require high early strength.
Class D:	For use from 6,000 ft to 10,000 ft depth (1,830 m to 3,050 m), under conditions of high temperatures and pressures.
Class E:	For use from 10,000 ft to 14,000 ft depth (3,050 m to 4,270 m), under conditions of high temperature and pressures.
Class F:	For use from 10,000 ft to 16,000 ft depth (3,050 m to 4,880 m), under conditions of extremely high temperatures and pressures.
Class G:	Intended for use as a basic cement from surface to 8,000 ft (2,440 m) depth. Can be used with accelerators and retarders to cover a wide range of well depths and temperatures.
Class H:	A basic cement for use from surface to 8,000 ft (2,440 m) depth as manufactured. Can be used with accelerators and retarders to cover a wider range of well depths and temperatures.
Class J:	Intended for use as manufactured from 12,000 ft to 16,000 ft (3,600 m to 4,880 m) depth under conditions of extremely high temperatures and pressures. It can be used with accelerators and retarders to cover a range of well depths and temperatures.

The most important aspect of cement blending is obtaining a consistent slurry by use of the proper amount of additives and mixing water. The optimum water-to-cement ratio for a cement slurry is a compromise. Maximum cement strength occurs at a water-to-cement ratio of about 2.8 gal/sack. This is the minimum amount of water necessary to fully hydrate and chemically react with the cement ground to a size that represents class G. However, a slurry mixed at this water rate has a very high viscosity and cannot be pumped. If too much water is used to aid in pumping and displacement, low strength and a very high free water quantity will result. The trade-off between cement strength and the volume of mixing water is evident from the data presented in figure 11–14.[35] Free water is defined as water that is not needed by the cement for reaction. When flow stops, free water separates out to the top of the cement column. Separation may occur at the top of a long column or in pockets in highly deviated wells.[36] These pockets contribute to annular gas leakage and other annular flow problems.

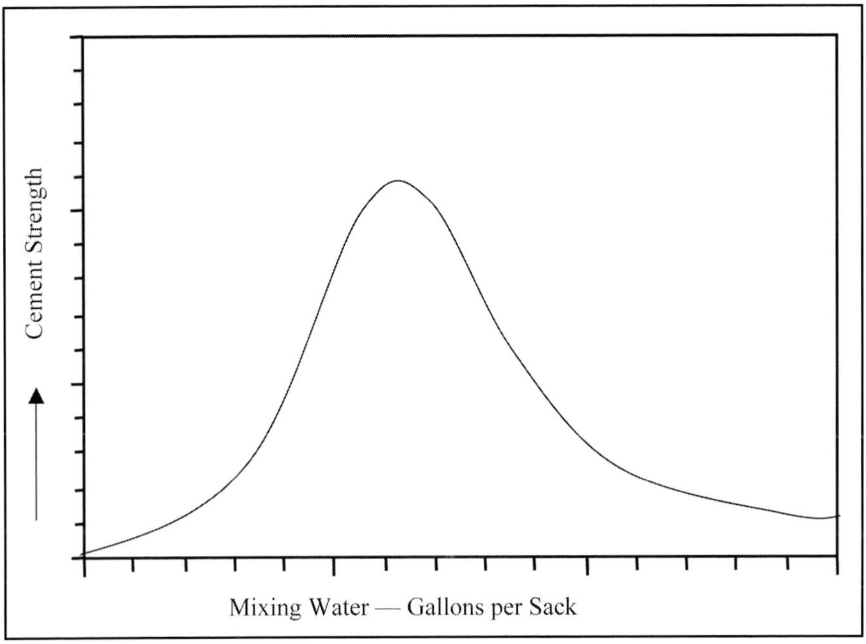

Fig. 11–14. Cement compressive strength versus volume of mixing water

Cement is mixed by jet mixers that combine cement and water in a single-pass operation or by the more precise batch mixers, which mix by circulating in a large tank but mix only a limited volume at a time. Although an acceptable slurry can be achieved when a jet mixer is used by an experienced operator, a batch mixer allows closer control in critical, small jobs. Jet mixers are used for almost all large jobs that require a constant supply of cement slurry at a high rate. The density of slurries mixed by these methods must be checked periodically with a pressurized mud balance to obtain consistent density, which is important to control the reservoir pressure and prevent formation fracture breakdown.

The quality of the water used to mix the cement varies widely, depending on the specifications required by the company involved. Freshwater, seawater, and some brackish waters are used to mix cement slurries. For any source of water, the behavior of the resultant cement in terms of setting time and pumpability must be known before mixing. Pumpability is measured by a laboratory instrument called a consistometer.[37] This device measures the setting time of a cement slurry by stirring the slurry (under pressure) until it thickens too much to stir. The output is as units of consistency and is time related. This test yields the time that a particular slurry can be pumped at a given temperature. Because of the development of offshore fields, seawater has become very widely used for cementing. Seawater, like most inorganic salt brines, slightly accelerates the setting time of cement. Use of brackish water (from bays, swamps, sewage, or produced waters) can cause problems. High salt contents, especially calcium chloride, may decrease the cement set time. Organic contaminates (e.g., oil-based mud) may slow the cement setting time—sometimes so severely that the slurry does not set.

Accelerators or retarders may be used in the cement to change the setting time from a few minutes to many hours. A retarder is used in deep or very hot wells to prevent the setting of the cement before the job is complete. Accelerators are used in shallow or cool wells to speed up the setting of cement, so that less rig time is lost waiting for the cement to set. Values such as filtrate-loss control and cement expansion can also be directly affected.

Cement additives may be divided into two general classifications based on their reaction type: chemical and nonchemical. Nonchemical additives are usually materials that affect the cement by altering density or controlling fluid loss. Chemical additives modify the hydration (water intake).

Cement density

Controlling the cement slurry density is critical for placing a column of cement where the formation may be fractured by a heavy slurry or would allow the well to flow if the cement slurry was lighter than the pore pressure. For a lighter-weight cement as compared to the normal of 15–16 lbs/gal, bentonite clay may be added to absorb water to yield a lighter cement with a higher volume of bound water. A cement density of 10–12 lbs/gal can be achieved in this manner. Grinding the cement to a very small size will also require more water in order to satisfy the high surface area and lighten the slurry to the 10–12 lbs/gal range. Ultralightweight cements,[38] using hollow ceramic or glass beads, can reduce the overall weight of the cement slurry to less than 9 lbs/gal. Even lower densities can be achieved by foaming the cement with a compressed gas, such as nitrogen.[39] Foamed cements can create densities of 4–7 lbs/gal but require careful control of annulus surface pressures in order to avoid gas channels and voids.

All of these lightweight cements, although strong enough to support the pipe, have less strength than regular portland cement. Heavyweight materials are added to the cement to increase the cement density, usually to control the pressure in the formation during the pumping of the cement. Iron ore, barite (barium sulfate), and sand can create slurries of up to 25 lbs/gal. Other methods of preparing heavyweight slurries include the use of dispersants, which allow less water to be used in a cement while still maintaining pumpability. For a chart of cement density for various methods of density control, see table 11–4.

Table 11–4. Density ranges for various types of cement slurries

Cement Slurry Type	Weight Range lb/gal
Densified and weighted	16 to 22
Neat slurry	14 to 18
High water ratio slurries	11 to 15
Ceramic bead extended slurry	9.5 to 12+
Glass bubble extended slurry	7.5 to 12+
Foam cement	6 to 12+

In some treatments in which lightweight cements are not used—whether based on preference, economics, or reasons of strength—staging tools can be used to control the pressures on a zone by running a multistage cement job. A staged job separates the cement job into small cement jobs that only support a portion of the total column and weight. The tools prevent the cement columns from contacting each other until set. An example of a cement-staging tool is seen in figure 11–15. The simplest tool uses a drillable plug to seat the pipe below the tool and to open a set of ports, allowing the next cement stage to turn the corner and start up the annulus. Some tools are equipped with a seal device that, by creating channels or by exerting more pressure on the lower zones, prevents cement from falling down the annulus and ruining the job.

With these tools, even a deep well with several zones can be cemented by turning the job into several consecutive jobs. The staged concept can also be done by cementing with a small volume and perforating the pipe above the last cement top and repeating the process. However, the use of staging tools saves waiting-on-cement (WOC) time between jobs. The obvious drawback to the staging tools is the same as for all downhole tools—namely, reliability.

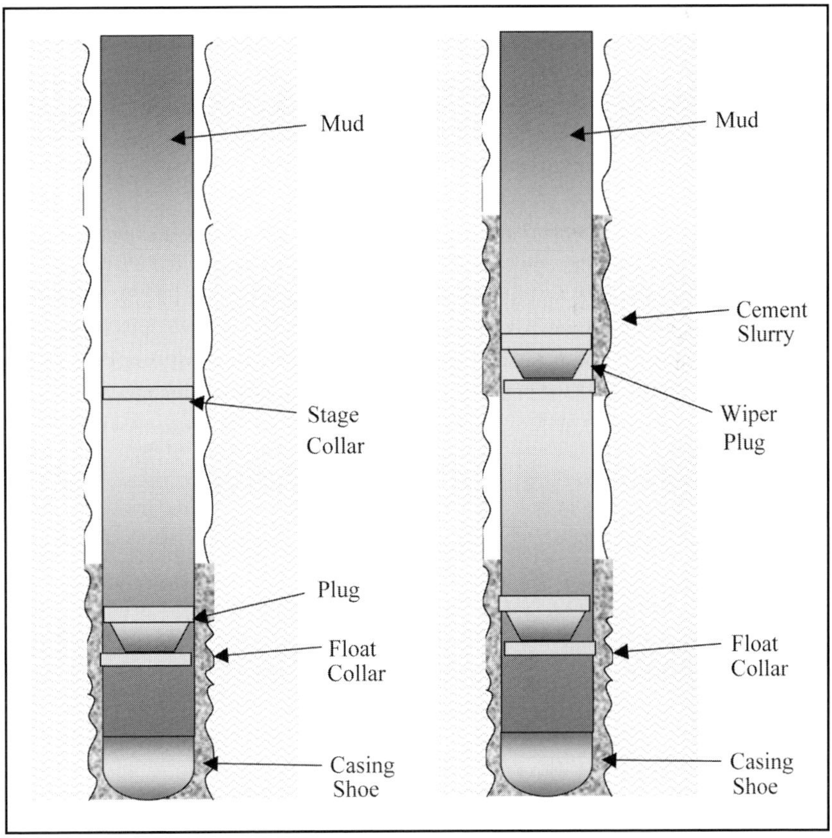

Fig. 11–15. Schematic of a cement-staging tool with inflatable seal protection

Fluid loss

Lost circulation materials control the flow of whole cement into natural fractures or extremely large vugs. The control materials come in three basic types: granular, lamellated, and fibrous. Granular materials, such as sand and other products, set a secondary matrix by filling cracks and vugs in the formation. They may range in size from ¼ in. diameter to fine powder to achieve control. Lamellated or flaked products, such as shredded cellophane, stop at the formation face and create a blockage on which cement will form a filter cake. Fibrous materials, such as paper, nylon, or polypropylene, are best suited to the bridging of small fractures.

Loss of cement filtrate in permeable sections creates dehydrated masses of cement and impedes flow. Any decrease in cement flow rate may impair the ability of the cements.[40] Restrictions in the annular space also increase the circulating pressure and the risk of breaking down the formation. When fluid-loss control cannot be achieved by traditional methods, a plug of cement is often set across the zone, and the cement is drilled out, to continue with well operations. The thin sheath of cement on the formation and the cement that invades the natural fractures are often adequate to halt fluid loss.

Chemical additives react with cement and usually alter the chemical process of hydration. Accelerators are usually inorganic salts that speed up the hydration of cement by increasing the temperature of the slurry.

Factors affecting cement

Increasing temperature accelerates the setting time of cement. Both the bottom-hole temperature and the heat produced by a large mass of cement during hydration may affect the set time. To correctly design a cementing job, the bottom-hole temperature of the formation and an estimate of circulating temperature should be known. A temperature gradient is calculated from this information, to give the temperature exposure the cement will face during the pumping operation. Heat transfer from the cement rising in the annulus to the cement coming down the inside of the casing is also important. Cementing computer simulations that model the temperature profile commonly provide a temperature schedule.

Pressure has a tendency to accelerate the thickening time and the setting time of the cement, but the effect of pressure is not nearly so pronounced as that of temperature. Testing a cementing design in the laboratory requires simulated downhole pressure and temperature be applied to the test design.

The strength requirements of oil well cements are dependent on several factors. The cement must be strong enough to secure the pipe in the hole, to isolate the zone, and to withstand the nominal shock of drilling, perforating, and fracturing. For drilling ahead, the minimum WOC times are usually based on the time required for the cement to develop 50 psi tensile strength. The strength of cement has always been of interest, since strength develops over a long period of time and rig time can be lost waiting for cement to set. The WOC time can be shortened by the use of accelerators. Cement requires very little strength in order to physically support the casing. More strength is required in order to withstand loading from drill bits and pressure. In the design of the cementing operation, it is imperative that high-strength cements be used around the casing shoe (the bottom end of the pipe) and across potential pay and thief zones (i.e., areas of fluid loss) and water-producing zones. Filling the annulus behind pipe and zone separation requires very little strength, and more economical cements or cement extenders may be used.

While the cement slurry is liquid, the hydrostatic force from the weight of the slurry exerts force to prevent entry of gas into the wellbore annulus. When pumping stops, the cement starts to gel and set, and it begins to support itself by the initial bond to the formation. This initial attachment, coupled with fluid loss to the formation, reduces the applied hydrostatic load.[41] Fluids can then enter the annulus, creating voids and channels in the cement behind the pipe. Methods of control include reaction with the formation gas to plug the channels[42] and reduction in permeability to stop the gas from entering.[43] Use of an external inflatable casing packer (ECP) is also an option.[44] This tool operates like a hydraulic set packer between the casing and the open hole.

Excess volume

The necessary volume of cement is the volume of the open hole less the volume of the casing across the zone. An excess of 30–100% of the total is usually added to the cement volume to allow for washouts and mud contamination. The 30–100% range of excess cement volumes is large, even for the technology of the oil field. It reflects the variability of drillers' expertise and formation conditions.

Hole volume is calculated from the caliper log. The bit diameter should not be used for hole volume calculations, since it will not reflect washouts. In most operations, four-arm caliper tools that give two independent diameters are more accurate than three-arm calipers that give a maximum or averaged reading.

Cementing design

The first recorded use of cement in the oil industry was in 1903, in a water shut-off attempt in California. At first, cement was hand mixed and run in a dump bailer to spot a plug. The benefit of pumping the cement down a well was soon recognized, and a forerunner of the modern two-plug method was first used in 1910.[45] The plugs were employed to minimize contact between the mud and the cement. Although mechanical and chemical improvements have since been made in the cementing process, the original plug concept is still valid.

Cementing design includes the selection of additives and equipment to remove mud and properly place and evaluate the cement. The cementing design depends on the purpose of the cementing operation. The initial cement is usually to fill the annular space between the casing and the hole from the casing shoe to the surface or to a point several hundred feet above the zone that must be isolated. The first cement job is called primary cementing, and its success is absolutely critical to the success of subsequent well control and completion operations; hence, if success is not ensured, repair (squeeze cementing) must be done promptly. Although squeeze cementing has become commonplace, it is expensive; its use can be curtailed through improved primary-cementing procedures.

Primary cementing

In primary cementing, the object is to place a continuous sheath or band of cement around the pipe, extending outward, without channels or voids, to the formation face. Primary cementing is not an easy operation to do correctly. Many things can happen during this process to create problems or weak spots in the primary cement design.

Application

The mixing of cement and water is the first critical area of application of cementing technology. To prevent fracturing or loss of control, the water and the cement must be blended together at the proper slurry density. The weight of the slurry is equal to the weight of the set cement less any weight of free water.

One of the first questions that should be asked about a design is the volume of cement needed for a job. In a short string or a shallow string, complete cement fill of the annulus is needed, plus at least 30% excess to displace the lead cement that is in contact with the mud as the cement displaces the mud from the annulus. Cement contaminated with mud will not form an effective seal; mud channels may run through it, and it may not develop any strength. In cases where the mud has not been adequately conditioned before cementing, as much as 100% excess may be appropriate.

The volume of the hole should be measured with a caliper tool after removing the drilling string and before running casing. Calipers may be available in three-arm, four-arm or multi-arm styles. Three-arm calipers report an average round hole diameter on the basis of the smallest diameter reading of one of the arms. Four-arm calipers work as a pair of two-arm calipers. The data from this tool draw an average of the hole on the basis of two circles or ellipses. Both tools are capable of underestimating the hole volume.

Caliper tools report data on a log track that shows deviation from a theoretical line reflecting gauge hole or bit size. Washouts and irregular hole volumes must be calculated to give an accurate reading on hole size. The easiest way to calculate hole volume in a washout is to use an average washout diameter equal to at least 90% of the maximum caliper-measured diameter where the diameter is fluctuating widely and 100% of the maximum diameter where the hole diameter is more consistent. Calculating the volume of the hole in vertical segments of similar diameter yields usable results. The problems in cementing through a washout are that fluid velocity becomes very low in a washout; swept debris at the leading edge of the cement drops out or mixes in, and the cement slurry can no longer scour or clean the mud cake in the washout.

Cement mixing

There are two types of oil field cement-mixing equipment: on the fly and batch. Batch mixing is done in a large tank with circulation or paddle mixers. The cement and the water are measured into the tank, sometimes with an on-the-fly mixer, with small additions of cement or water to get the right slurry density. Although batch mixing is by far the most accurate method, the size of the cement job is limited by the volume of the tank at hand. Mixing on the fly involves moving steady streams of cement and water through a zone of turbulence produced by high-velocity flow. The cement slurry produced in this manner is highly dependent on the experience and attention of the mixer operator. Numerous problems with variances in slurry weight have led to averaging *pods,* or tanks downstream of the on-the-fly mixer. To minimize the damage produced from slurries that are lighter or heavier than designed, most cementing service companies have density-monitoring devices to report slurry density back to the mixer operator.

Incorrect cement density can cause gas migration, poor set strength, inadequate cement bond, blowouts, formation fracturing, and lack of mud displacement. Cement slurry density must be rigorously controlled for the subsequent well completion steps to be carried out successfully.

Once a consistent cement slurry blend has been achieved, the second critical step, displacement, begins. To effectively bond the pipe to the formation with cement, the drilling mud and the drilling-mud filter cake must be completely removed. Failure to remove the cake or the mud will lead to failure of the primary cement job by leaving mud channels in the cement. Failures necessitate squeeze cementing (repair operations).

Mud conditioning and displacement are the next critical steps in the cementing application. For cement to isolate zones, a sheath of cement must completely surround the pipe and bond the formation wall to the pipe. The mud cake must be removed, and the pipe must be centralized. Centralization is needed to provide sufficient standoff, or clearance between the casing and the borehole wall. Removal of the mud and the mud cake is done by a combination of chemical and physical actions that are well documented but often overlooked during application. The ease of mud removal depends on the physical condition of the mud and the access to the mud.

Mud displacement begins with decreasing the gel strength of the mud and removing cuttings. After casing is run in the well, the annular space open to flow is smaller than when drill pipe was present. The smaller annular area creates higher velocities that can disturb deposits of cuttings. Cuttings can accumulate in the lead portion of the cement, contaminating the cement and creating blockages that can create lost circulation. The presence of a mud cake will prevent bonding of the cement to the formation. An estimation of the volume of cement needed for removal of mud cake by turbulent flow is[46]

$$V_t = t_c q(5.616 \text{ ft}^3/\text{bbl}) \tag{11.35}$$

where

V_t is the volume of fluid (in turbulent flow), in ft^3
t_c is the contact time, in min
q is the displacement rate, in bbl/min

Studies have shown that a contact time (during pumping) of 10 min or longer provides better mud removal than shorter contact times. Equation (11.35) is valid as long as all the fluid passes the point of interest; however, equation (11.35) will not be valid for mud outside the path of the flowing fluid, such as where the casing is uncentralized and presses against the formation.

Movement of the pipe during cementing is one of the best methods of improving the mud displacement and reducing the number of mud channels remaining after cementing.[47] Reciprocation (up and down) and rotation of casing help force the mud from the pipe/formation contact areas and ensure a more even distribution of cement. Rotation of the pipe requires special rotating heads in order to allow pumping while turning. Reciprocation, or moving the casing up and down a few feet while cementing, can be done more easily but does not force the mud from the contact area in the same manner as rotation. Addition of scrapers to the casing can help remove hard mud cake.[48] Use of centralizers minimizes contact area and may make pipe movement easier.

Displacement of the mud and the mud cake cannot always be accomplished by flowing cement. Heavily gelled muds and tightly compressed filter cakes are very resistant to removal by any flowing fluid. Special removal procedures are necessary. The basic mud removal step is to pump the cement in turbulent flow; the high velocity, the high viscosity, and the abrasive nature of cement work in unison to scour the formation and the casing. During scouring, much of the mud and cake materials are mixed in with the first cement pumped. This contaminated cement must be removed from the well. In the cement volume design, an allowance for contaminated cement is provided, in part, through the 30–100% excess cement normally designed into most jobs.

If muds and mud cakes cannot be removed by cement flow, special preflush fluids and mechanical devices are available to improve displacement. To improve mud and mud cake displacement, the binding agent in the mud must be broken down. In most cases, the mud binders are clay, polymers, or surfactants. Chemical flushes of acids, solvents, or surfactants are useful but must be selected for action on specific muds. These flushes are pumped ahead of the cement or spotted in the annulus before the cement job.

Mechanical devices for mud and mud cake removal include casing centralizers, scratchers for cake removal, and turbulence-inducing devices to improve mud displacement. Wire or wire rope scratchers break up the mud cake during the running of the casing. Complete removal of the cake is not necessary; the action of the cement will often be sufficient to remove the cake fragments once the integrity of the cake has been disrupted.

The alignment of the casing in the borehole is an often-neglected factor that has a tremendous impact on mud conditioning, cementing, perforating, and production, particularly in highly deviated or horizontal holes.[49] Uncentralized casing always lays on the low side of the hole. In soft formations, the casing may even embed or bury into the wall of the formation. When casing contacts the wall, the drilling-mud cake and some whole mud is trapped between the casing and the rock. This mud cannot be removed. Mud removal attempts by flushes and turbulent cement flow will have little contact, as indicated in the velocity profile sketches of figure 11–16. Cement bypasses the mud, and channels are left behind the pipe. These channels may completely undermine the principles of zone separation by cement and usually require repair by squeeze cementing. Channels are the most common form of primary cement failure.

Centralizers and pipe movement can improve cementing.[50] Centralizers hold the pipe away from the wall of the hole so that cement may more evenly displace the mud and completely fill the annulus. The design of centralizers varies widely with the application. Centralizion of casing in nearly straight holes is relatively easy, but as holes become more deviated, centralization becomes more difficult. In the more deviated wells, the weight of the casing will flatten most spring centralizers and may deeply embed some of the solid fin body units. The actual number of centralizers needed for a well depends on the acceptable deflection of the pipe and the severity of doglegs in the well.

For examples of centralizers and their spacing, see table 11–5. Note that the centralizer spacing decreases (i.e., more centralizers are needed) as hole angle pipe size and clearance increase.[51] The spacing is usually calculated by a computer, using a model such as that by Lee et al.[52] These programs project spacing on the basis of the input depth, DLS, lateral load, tension, and deviation. Typical spacing is 30–60 ft between centralizers.

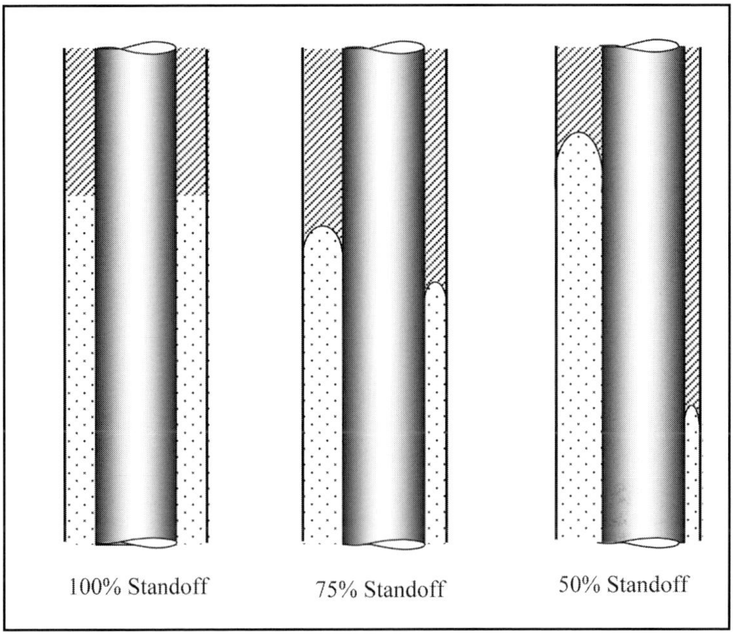

Fig. 11–16. Cement velocity schematic at different casing standoffs

Table 11–5. Maximum spacing (in ft) between centralizers (standoff = 1 in.)

Hole Angle Deg	5½ in 8¾ in	7 in 10¾ in	9⅝ in 12¼ in	10¾ in 12¼ in	10¾ in 14¾ in	13⅜ in 17½ in
10	63	62	65	66	73	70
25	57	53	55	56	60	53
20	53	49	48	48	50	42
25	50	45	40	43	40	36
30	48	42	38	39	37	29
35	46	38	33	36	33	26
40	44	35	30	33	30	22
45	42	33	28	29	27	20
50	40	30	26	27	25	18

The variance in casing weight can be illustrated by the following examples of buoyed weight of casing:

$$W_{cb} = W_{cd} + 0.0408(\rho_i d_i^2 - \rho_o d_o^2) \tag{11.36}$$

where
W_{cb} is the buoyant weight of the casing, in lbs/ft
W_{cd} is the dry weight of the casing, in lbs/ft
ρ_i is the density of fluid in the casing, in lbs/gal
ρ_o is the density of fluid in the annulus, in lbs/gal
d_i is the ID of the casing, in in.
d_o is the OD of the casing, in in.

For circulation with an 11.5 lbs/gal mud in a 7 in., 26 lbs/ft N80 casing, $W_{cb} = 26 + 0.0408\,[(11.5 \times 6.276^2) - (11.5 \times 7^2)] = 21.5$ lbs/ft. If freshwater (8.33 lbs/gal) is used to displace 16 lbs/gal cement to the float collar, the buoyed weight becomes $W_{cb} = 26 + 0.0408[(8.3 \times 6.276^2) - (16 \times 7^2)] = 7.4$ lbs/ft.

Mud retards (slows) the set of cement. Minimizing this effect requires mud removal and separation from the cement whenever possible. Most casing strings are run full of mud during casing placement for assistance in well control. Cement displaces the mud from the casing before it flows up the annulus. If the mud is lighter than the cement or the mud has high gel strength, the cement will tend to finger, or channel, through the mud during its flow down the casing, thus mixing cement with mud.

Mixture of mud and cement in the tubulars can be prevented by use of the two-plug system. Before the cement is circulated down the well, a hollow rubber plug (fig. 11–17) with a disk that can be ruptured at high pressure is placed in front of the cement. The cement pushes this plug down to the bottom of the well, wiping the inside of the casing and displacing the mud from inside the casing ahead of the cement. At the bottom of the well, the plug lands, or is *bumped,* and pressure builds up, rupturing the disk. Cement comes through the plug and can *turn* up the annulus. The second plug is dropped at the end of the calculated cement volume, and the cement is displaced down the well with mud or water. The second plug, or *top plug,* is solid and has the same set of wipers as the first plug. At the bottom of the hole, the top plug reaches the top of the first plug, and pressure rises, indicating that the plug has been bumped. The plugs are made of drillable material that can be easily removed if the well is deepened.

Correct loading of the plugs is critical. If the plug sequence is accidentally reversed and the top plug is dropped first, the job will end when this solid plug hits bottom, and the casing will be left filled with cement. The actual displacement in the wellbore is very different from what the surface pump rate may indicate, especially when the density of the mud is much less than the density of the cement. When a lighter mud is displaced, the cement is in a *free fall.* The cement density is enough to rapidly push the mud ahead and displace it from the well without the driving pressure of the pump. This is most noticeable in the later stages of the job, during displacement, when the casing contains more mud than cement.

Surface pressure can go to almost zero at low injection rates, and the well is said to go *on vacuum.* At this point, the well is taking fluid faster than it is being injected, and mud return rate from the well can be more than the cement injection rate. (This is referred to as a vacuum because void space is being created in the casing at the surface.) As the cement turns the corner at the bottom of the well and starts up the annulus, the injection pressures caused by the heavier cement density will climb. The well returns, which are monitored continuously at the surface, may go to zero as the cement fills the void volume in the pipe that was evacuated during free fall. It may appear that the well has lost returns by breaking down (fracturing) the formation. This rapid movement of fluids must be included in the design, to allow control of the mud.

The problems involved with free fall are rapidly increasing bottom-hole pressure, caused by resistance to faster than design mud flow rates around the shoe, and apparent loss of returns, as the cement fills the voids created during the initial free fall. Figure 11–17 is an example of a field job showing pump and return rates. If, for example, the low rate of returns after two hours prompted the operator to reduce the injection rate in an attempt to limit the apparent loss of cement, the cement would not be in turbulent flow, and the mud cake might not be cleaned off the formation.

The problem of free fall of a heavier cement in a lighter mud system is illustrated in figure 11–17. The upper schematic shows the placement of the 16 lbs/gal cement with the lighter mud. The graph at left is a model prediction with measured data from the field.[41]

After the plug has been bumped, the WOC time begins, and pressure is held until the cement has set. Pressure control is assisted by the float equipment. These devices are flapper or poppet valves near the bottom of the string that prevent the cement from returning to the casing. The one-way valves are of drillable material and are designed to withstand the high-velocity flow of large quantities of abrasive cement without damaging the sealing mechanism. Examples of float valves are shown in figure 11–18. If the float is at the end of the casing string, it is called a float shoe. If it is placed a joint or two off bottom, then it is called a float collar. The preferred location will depend on the operator, but for reasons of cement contamination control, float collars are usually preferred. The float collar results in a joint or two above the shoe being filled with the last cement pumped. This last cement may be contaminated with residual mud scraped from the casing wall by the top plug. Use of both a float collar and a float shoe is an accepted practice in some areas; such dual floats serve as an extra barrier against back pressure leak

After WOC, a drill bit just smaller than the casing ID is run if the well is to be deepened. The hole is drilled through the casing shoe and into the formation beneath this string. At this point, the casing shoe is generally tested to ensure

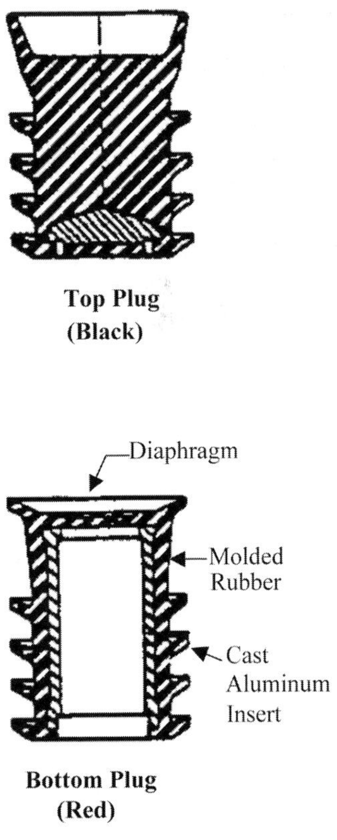

Fig. 11–17. Schematics of the two-plug system

that a good, leak-tight cement job has been obtained. If there are leaks evident during this pressure test, the well is squeezed with cement until a pressure-tight seal can be obtained. Since the casing shoe is the weak spot for blowout control, this step is a necessity.

In summary, proper placement of a good primary cement job is contingent on several factors: selection of the right cement blend, conditioning of the mud, removal of the mud cake, centralization and movement of the pipe to ensure full cement contact around the perimeter of the outside casing wall, and use of enough cement to isolate the full zone.

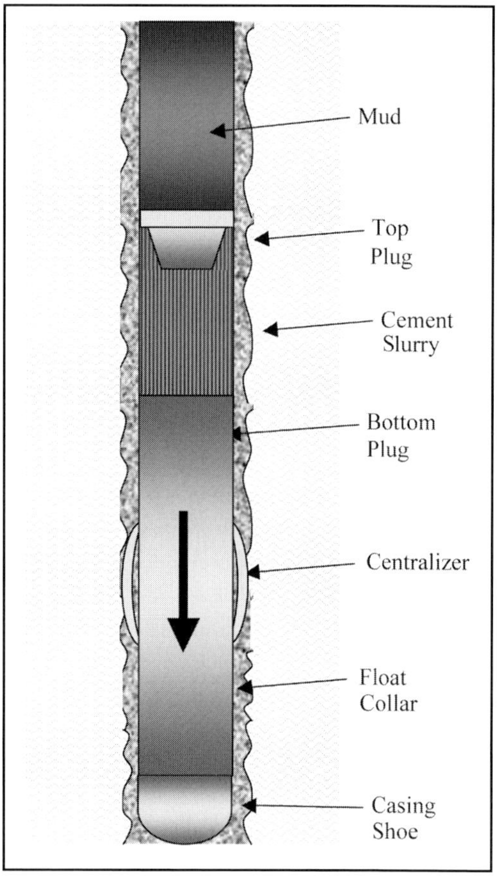

Fig. 11–18. Floats and float locations. Note that the float collar is located one joint (or more) above the end of the string.

Cementing calculations

Buoyant force on the casing by the fluid in the hole tries to float the casing. Hydrostatic pressure acts against the effective area of the casing, causing the upward force. The pressure acts on the full area of the closed-end casing if the float is in place and holding or on the material cross-sectional area created if the casing is open ended. The weight of the casing string minus the upward buoyant force gives the buoyed or true weight of the casing string in the hole.

For 13⅜ in., 61 lbs/ft K55 casing in a 17 in. hole filled with 10 lbs/gal mud, closed-end area is $\pi d_o^2/4$ = 141 in.², and effective area is $\pi(d_o^2 - d_i^2)/4$ = 1.75 in.². Hydrostatic pressure at 4,000 ft is 4,000 ft × 10 × 0.052 psi/ft = 2,080 psi. Thus, the hydrostatic effect on casing is 2,080 psi × 17.5 in. = 36,400 lbs, and casing-string air weight is 61 lbs/ft × 4,000 ft = 244,000 lbs. The buoyed weight of the casing in mud divided by the outside area of the casing gives the pressure needed to balance the string: 207,600 lbs/141 in. = 1,472 psi.

Thus, a bottom-hole kick or other pressure increase of over 1,472 psi (an additional 0.368 psi/ft or 7.1 lbs/gal) could start the casing moving upward. At shallower depths, especially with large-diameter casing, the additional pressure to lift the buoyed weight can be 100 psi or less.

The pressure to land the top plug when displacing 16 lbs/gal cement with freshwater to 4,000 ft (assuming complete annulus fill with cement) is determined in terms of the respective hydrostatic pressures of cement in the annulus and water in the casing. Cement hydrostatic pressure in the annulus is $P_{ch} = 0.052\gamma_c$ (TVD) × 16 × 4,000 = 3,328 psi, and water hydrostatic pressure in the casing is $P_{wh} = 0.052\gamma_c$ (TVD) = 0.052 × 8.33 × 4,000 = 1,733 psi. Thus, the pressure to land the plug is 3,328 – 1,733 = 1,595 psi.

In wells where all the exposed formations will not support the full weight of the cement while fracturing, the cement must be lightened, or the zone must be protected by filling the annulus with only a partial column of cement (i.e., by staged cementing). Assuming that the zone at 4,000 ft (bottom hole) has a fracture gradient of 0.72 psi/ft, the height of a 16 lbs/gal cement column that will be 200 psi below fracturing pressure is calculated as follows:

- Bottom-hole fracture pressure: 4,000 ft × 0.72 psi/ft = 2,880 psi
- Allowable bottom-hole pressure: 2,880 psi – 200 psi = 2,680 psi
- Cement gradient: 16 lbs/gal × 0.052 psi · gal/ft-lb = 0.832 psi/ft
- Full-column pressure: 4,000 ft × 0.832 psi/ft = 3,328 psi

If 16 lbs/gal cement is used, the maximum column height (within the allowable pressure) is 2,680 psi/0.832 psi/ft = 3,221 ft. If a full cement column is needed, the maximum cement density is 2,680 psi/4,000 ft = 0.67 psi/ft, or 12.9 lbs/gal. Cement densities are only part of the picture; the friction pressures developed by pumping the cement past restrictions adds to the hydrostatistic pressure of the cement.

Squeeze cementing

Squeeze cementing forces a cement slurry behind the pipe, to repair leaks or shut off fluid loss. Squeeze cementing is normally thought to be a repair step, but it is also used to seal off depleted zones or to stop unwanted fluid production.

Eight major uses of squeeze cementing for repair and recovery control purposes have been documented[33]:

1. *To control high gas/oil ratios (GORs).* By squeezing the top section of the perforations, gas production can be made to pass vertically through the top part of the formation matrix, slowing the gas production by the contrast of vertical versus horizontal permeabilities.
2. *To control excessive water.* Squeezing lower perfs can delay water production. Only if an impenetrable barrier separates the oil and water or if vertical permeability is very low will effective water reduction be achieved.
3. *To repair casing leaks.* Cement can be squeezed through holes in casing. This is best accomplished by very-small-particle cement.
4. *To seal off thief zones or lost circulation zones.* Cement slurry may penetrate natural fractures for only a centimeter or two but may develop sufficient blockage to help control leakage. However, sealing off natural fractures is often difficult.
5. *To stop fluid migration from a separate zone.* This is usually a block squeeze or channel repair operation.
6. *To isolate zones.* This entails the selective shutoff of depleted or abnormally low- or high-pressure zones.
7. *To repair the primary cement job.* Filling voids or channels and repairing liner tops are common squeeze-cementing operations.
8. *Abandonment squeezes.* These involve shutting off depleted reservoirs or protecting freshwater sands.

Squeeze cementing is separated into high-pressure squeezing and low-pressure squeezing.[54] High-pressure squeezing involves fracturing the formation with cement until a required surface pressure is reached. High pressures at the end of the job, although popular with many companies, are actually of little importance; in actual practice, final pressures should be well below 1 psi/ft.[55] The high-pressure squeeze uses neat cement with very high fluid loss. The best uses of the technique are to shut off depleted zones and to seal perforations.[56]

The low-pressure squeeze technique is probably more efficient in placing a controlled amount of cement in a problem area of the well. With this technique, formation fracturing is completely avoided. The pressure is achieved by applying increasingly larger pressures on the cement and allowing the cement to filter out on the formation, creating a block in the annulus. Once the cement slurry has hardened or dehydrated to a sufficient extent, no more fluid will be displaced. The excess cement that is still in the drill pipe or the annulus can be displaced from the well by opening the casing valve and flushing with a displacement fluid. The advantages of the low-pressure squeeze are less pressure exposure to tubing and casing and special cementing tools and a smaller quantity of cement.

For either squeeze-cementing process, relatively low water loss and strong cement are included as part of the design. Most operations use nonretarded API class A, class G, or class H cement, all of which are suitable for squeeze conditions down to 6,000 ft without additives. For deeper wells, class G or class H cement can be retarded to gain necessary pumping time. In hotter wells (above 230°F), additives should be considered to increase strength at high temperature.

Although squeeze cementing is often used to help repair primary cement failures and to protect the pipe, it is possible to collapse the casing during squeeze cementing. If a packer is set immediately above the zone to be squeezed and an open channel exists that links the back side of the casing above the packer to the interval being squeezed (as in fig. 11–19), then the outside of the casing above the packer may be exposed to the full pressure of the cement squeeze. When the inside of the casing is not be loaded or pressurized, casing failure can occur if the value of ΔP is above the pipe strength.

The thickening and setting times of cement used in squeeze operations are calculated in the same manner as those used in primary cementing. Squeeze pressure does affect the dehydration of the slurry, particularly across zones that are very permeable. Fluid-loss additives may be included if the slurry is designed to move any significant distance across a permeable formation. Normal dehydration of a cement on a permeable section is severe enough to seal off the flow channel before complete displacement is accomplished.

Cement squeeze tools. A drillable or retrievable cement retainer is a modified packer that helps control the placement of cement and protects other zones from pressure and excess cement. Retrievable tools can be set and released several times and can be used to accomplish several squeeze repairs in one trip. A drillable tool is a single-use tool that stays in place and is drilled out (if needed) after the cement has set. Drillable tools are modified packers and are available in compression set and tension set models. Compression set models are normally used below 3,000 ft, at which depth the weight of the string is adequate to completely engage the slips.

Drillable cement tools are more restricted in terms of setting and application than are retrievable tools, but they offer more control on the set cement. The drillable models are preferred where continued pressure must be maintained after squeezing.[57] When squeezing formations that are naturally fractured, it is more important to fill the fractures, rather than build up a filter cake.

A two-slurry system is successful in fractures[44]; this system employs a highly accelerated slurry and a moderate-fluid-loss slurry. Accelerated slurries are pumped into the zones of least resistance and allowed to take an initial set. After the first slurry has gelled, the moderate-fluid-loss slurry is forced into the narrower fractures. The first slurry used for this type of squeeze should take an initial set within 10–15 min of placement.

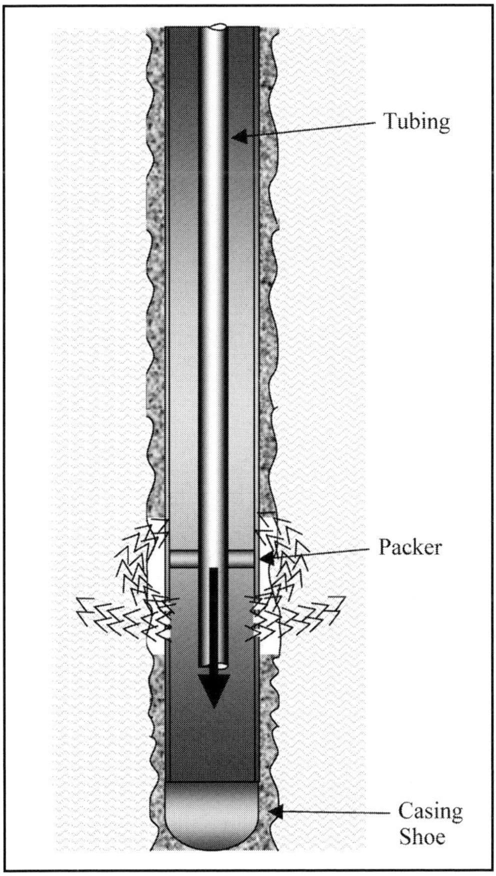

Fig. 11–19. Illustration of how cement squeezes below a packer can result in casing collapse

Liner cementing

Cementing a liner in place requires very closely controlled application of existing technology and a fair amount of risk. Calculating the volume of cement to be used in a liner-cementing job is extremely difficult and requires more information than is available from a simple caliper run. For maximum caliper information, a four-arm device capable of determining elliptical holes should be utilized for hole volume. Cement excesses of between 20% and 100% have been used on a number of liner jobs, with larger excesses being responsible for better bonding and less channels. There is also a direct correlation between absence of channels and pipe movement.

Three cementing methods are generally accepted for liners. In a single-stage cementing job, cement is circulated to the top of a liner. This is much like a primary cement job and may include pipe movement during cementing.

In a planned squeeze program, the lower part of the liner is cemented, and the top part of the liner is squeezed later. This technique does not have good middle support and should not be used to isolate high-pressure zones. This procedure is more widely followed in worldwide operations because of perceived problems of disengaging the running assembly from the liner and of flash setting of cement. Disengaging from the liner before cementing eliminates the ability to move the liner and almost universally results in poor cement jobs.

A third procedure, commonly reserved for short liners, is to fill part of the hole with cement and then slowly run the casing string into the cement, forcing the cement to flow upward around the pipe. While this method can be accomplished with the minimum amount of pumping, the lack of circulation can result in poor removal of drilling mud. The technique is also called a puddle job.

Most liner jobs do not include plans to move the liner during the primary cementing.[58] The reasons for this include the following:

- Detaching the drill pipe from the liner before cementing minimizes the risk of being unable to detach from the liner once the cement is in place.
- It may be necessary to change to a higher-strength drill string to allow pipe movement.
- Movement may cause the liner hanger to become tangled with the centralizers near the top of the string.
- Swab or surge pressures may be created during liner movement, especially in close-clearance wellbores.
- Movement of the liner during cementing may knock off debris from the borehole wall. The debris may cause bridges and reduce the possibility of circulating cement.

Despite the quoted disadvantages of staying attached during the cementing operation, several serious disadvantages with releasing the liner before cementing are as follows:

- If the liner is hung off, the small bypass area around the liner offers a greater restriction to flow and causes more lost circulation because of the backpressure on the flowing cement.
- If a downhole rotating liner hanger is used (rotation only), additional torque is required in order to initiate rotation to overcome bearing friction. Pipe often rotates easier when it is being raised or lowered; the difference in torque required is often substantial.
- The potential for sloughing shale and annulus bridging is lessened when the operator can alternate between rotation and reciprocation.

- Premature shearing of the pins in the liner-wiper plug is less likely because there is no relative movement between the liner and the setting tool. (These two pieces of equipment move together.)
- If there are cement channels, as well as a large hydrostatic pressure difference between the inside and the outside of the running tools, the cups or seals can give way before cementing of the liner is complete.
- The displacement efficiency of cement around the tubulars when pipe is not moved is lessened.

When liners are close clearance, then the density differences between mud and cement should be as close as possible. This negates the advantages of hole cleaning by use of higher-density cement. Reciprocation of the string is helpful because it produces lateral pipe movement that causes the pipe to change sides in the wellbore while it is alternately compressed and stretched (slacked off and picked up). Rotation helps by mixing the cement into wellbore irregularities and displacing the mud, owing to drag forces produced by the flowing cement.

Although liner movement should be a goal in any liner operation, well conditions may prevent any type of movement. In many cases, however, liner movement can be achieved in a well-conditioned hole. The following are two clear cases where a liner should not be moved:

- When a short or small liner ($3\frac{1}{2}$ in. or smaller) is run in a deep well, the liner should be hung off first, since it may be impossible to determine from the weight indicator whether the liner has been released from the drill pipe.
- In cases of hole deviation over 35°, reciprocation may be difficult owing to high drag forces.

Many of the problems in liner running can be lessened by drilling a usable hole. Problems with key seats, ledges, washouts, and other nongauge issues intensify when close-tolerance liners are run.

When cement is circulated from the liner bottom to over the liner top, the cement must remain fluid long enough to detach from the liner and to circulate the cement from the well or to pull up above the top of the cement with the drill string. If the cement flash sets, then the drill string will become cemented in place, and the hole most likely will be lost. Cement may prematurely set or thicken, or cement circulation may be lost for a number of reasons:

- Improper thickening or pump times caused by a poor design, ineffective field operations, or bad test results
- Poor density control on the cement or poor mixing of the cement at the surface
- Bridging in the annulus, caused by a buildup of cuttings
- Plugging from dehydration of cement caused by excessive water loss in open hole sections below the overlap
- Increased hole cleaning of the cement as compared to the drilling mud

One of the most troublesome problems in cementing design is inadequate hole cleaning prior to cementing. This is especially true when lightweight, low-viscosity muds are used and little attention is paid to cuttings removal. Heaving shales are also a problem in hole fill and may cause washouts. Under no circumstances should circulation be halted with the liner in the hole before all of the cement has been displaced. Because of the small clearances and the yield point of cement, it may be very difficult to start circulation again.

Liner-cementing techniques. Cementing of liners requires special equipment and techniques to obtain a seal in the close clearances found between the liner and the open hole or the casing string. Two cementing techniques are use for liner cementing: a modified circulation job (which looks much like a cement squeeze) and a puddle cement job.

In the former technique (circulation/squeeze), the liner and its associated equipment are run on the drill string with a liner-running tool and a retrievable packer assembly. After the base of the liner is squeezed (usually up to the shoe of the outer casing or slightly above), the liner-running tool is pulled out of the liner up to a point just above the liner top, and the top section of the liner is squeezed. After drill-out of the remaining cement, a liner packer may be run.

Cementing liners, especially deep liners at high pressures, is complicated since the liner is often isolated from the rest of the string by packers and close clearances. The result is that pressures are often trapped behind the pipe. Consequently, pipe collapse and deformation are common.[59] Liner-cementing technology is little different from full-string technology except that pipe movement (including rotation) is done on drill pipe,[60] and use of plugs requires two-part plugs. Liner tieback operations may require special circulating guidelines because of the narrow clearances.[61]

Liner hanger clearances near the top will be critical in minimizing backpressure if the cement is circulated around the top of the liner in a complete circulation job. Close clearances created by a large liner hanger can raise the backpressure and the ECD. In some cases, this increase in equivalent density is enough to fracture the well.

In a puddle job, the cement slurry is spotted by the drill pipe over the section in which the liner is to be run. The volume calculation for the puddle of cement must consider hole volume and liner volume. Undetected washouts in the hole can lead to lack of cement around the liner top. Although the procedure is much simpler than the circulation/squeeze technique, it is often less effective in providing a seal. This technique is used for short liner sections.

Supplementary Problems

Problem 11–1. Briefly explain the functions of the following casing strings:

(a) Drive pipe

(b) Conductor pipe

(c) Surface casing

Problem 11–2. Estimate the corrected collapse resistance for the following casing:

- Casing: size 7 in.; grade L-80.
- Nominal weight = 29 lbs/ft
- Axial load = 142,880 lbs
- ID = 6.184 in.

Problem 11–3. A 10,000 ft string of 7 in., 38 lbs/ft N-80 casing is equipped with a float shoe. Given a mud density of the fluid in the hole of 12 lbs/gal mud, calculate the hook load of

(a) An empty, closed-end casing string in mud

(b) A mud-filled string in mud

(c) 15 lbs/gal cement inside the string and mud outside

Problem 11–4. Design an economical casing string that will satisfy the pressure requirements of collapse and burst and the load requirements of tensile force, given the following information:

- Packer-setting depth = 7,450 ft
- Worst condition: Annulus full of mud and an empty string
- Casing safety factors: collapse = 1.1; burst = 1.20; tensile = 1.8

You are also given the following data from the well file:

- A 10¾ in. surface string has been landed and cemented at 450 ft.
- Depth = 7,500 ft
- Mud weight = 12 lbs/gal
- Bit size = 8½ in.
- Fracture gradient = 0.75 psi/ft
- Reservoir pressure = 0.55 psi/ft

Problem 11–5. Using the data from problem 11–4, redesign the casing with the following safety factors: tension = 1.5; collapse = 0.9; burst = 1.1. Summarize your conclusions.

Problem 11–6. You are given the following data:

- 13⅜ in. surface casing at 4,000 ft
- 12¼ in. hole drilled to 10,500 ft
- Mud weight = 12 lbs/gal
- Future depth = 12,000 ft
- Future mud weight = 14.5 lbs/gal
- Fracture gradient at 10,500 ft = 0.90 psi/ft
- Casing safety factors: collapse = 1.1; burst = 1.20; tensile = 1.8

Design a 9⅝ in. intermediate casing.

Problem 11–7. Design a 9⅝ in. protective casing, given the following data:

- Setting depth = 9,000 ft
- Mud weight = 13 lbs/gal, 15 lbs/gal
- Fracture gradient at 9,000 ft = 0.88 psi/ft
- Expected depth of 7 in. production casing = 13,000 ft

Use the following safety factors: tension = 1.8; collapse = 1.1; burst = 1.0.

Problem 11–8. Calculate the maximum compressive stress in the following casing string:

- Length = 10,000 ft
- Casing size = 9⅝ in.
- Weight= 47 lbs/ft
- Mud weight = 11 lbs/gal

Problem 11–9. Design a 7 in. production casing, given the following data:

- Surface casing depth = 1,500 ft
- Intermediate casing = 7,000 ft
- Production casing = 12,000 ft
- Mud density = 10 lbs/gal
- Expected reservoir pressure = 5,800 psi

Use the following safety factors: tension = 1.8; collapse = 1.0; burst = 1.0.

Problem 11–10. Using the data from problem 11–4, redesign the casing with the following safety factors: Tension = 1.5; Collapse = 0.9; Burst = 1.1.

Summarize your conclusions based on the cost and feasibility.

Problem 11–11. You are given the following well data:

- Surface casing
  - Casing-shoe depth = 1,000 ft
  - Mud density = 9 lbs/gal
- Intermediate casing
  - Casing-shoe depth = 8,000 ft
  - Mud density = 10.5 lbs/gal
- Production casing
  - Casing-shoe depth = 15,000 ft
  - Mud density = 13 lbs/gal
- Maximum reservoir pressure expected = 8,800 psi
- Safety factors: tension = 1.8; collapse = 1.0; burst = 1.1

  (a) Design the bit program for this entire well.

  (b) Design the casing program for the entire well.

Problem 11–12. Calculate the weight at which the following production casings would fail in tension by axial yielding rather than by joint failure:

- 7 in. casing; grade N-80
- 5½ in. casing; grade P-110

Use available options of different casing couplings.

Problem 11–13. For a 16,000 ft string of 5½ in., 20 lbs/ft 0-75 casing, calculate the final hook load at the end of the cement job when 16 lbs/gal cement is in the annulus to the surface and a 10.5 lbs/gal mud is inside the casing. Surface pressure on the mud in the casing will be 500 psi.

Problem 11–14. A 5,000 ft, 4 1/2 in., 10.5 lbs/ft K-55 liner is run on 3½ in., 13 lbs/ft drill pipe. The liner top is at 12,500 ft. Calculate the maximum hook load in a 15 lbs/gal mud if the drill pipe and the liner are filled with mud.

Problem 11–15. Calculate the weight of mud that will just balance the weight (float) of a 5,000 ft string of a 10¾ in., 40.5 lbs/ft K-55 casing string if the string is run empty. (Neglect air weight.)

Problem 11–16. Calculate the buoyant weight of a 9⅝ in., 47 lbs/ft N-80 casing when 16.4 lbs/gal cement just reaches the shoe and there is a 8.33 lbs/gal surfactant water flush in the annulus.

Problem 11–17. Derive a formula for the required density of a displacing flush that will give a neutral density with a flush ρ_0.

Problem 11–18. For a well with a fracture gradient of 0.6 psi/ft at 7,000 ft, calculate the maximum cement height above the zone when using 16.4 lbs/gal cement. (Use a maximum bottom-hole pressure of 200 psi below fracture pressure.)

Problem 11–19. For the casing string in problem 11–15, calculate the cement density that will allow a full cement column to the surface. (Use a maximum bottom-hole pressure of 200 psi below fracture pressure.)

Nomenclature

A Cross-sectional area
A_c Contact area of pipe and mud cake, in in.^{2}
A_i Cross-sectional area of inside diameter (ID), in in.
A_o Cross-sectional area of outside diameter (OD), in in.
C_f Coefficient of friction (average of 0.25 is normal)
C_o Speed of sound in the material ($C_o = \sqrt{E/\rho}$)
d Depth, in ft
D_c Casing-setting depth, in ft
d_i ID of casing, in in.
d_o OD of casing, in in.
D_o Nominal OD, in in.
D_t Settling depth of casing string, in ft
D_x Depth of calculation, in ft
E Young's modulus for steel (30 × 106 psi)

F Pulling force, in lbs
F_a Axial load
F_b Buoyant force, in lbs
F_m Force required in order to yield the casing
F_s Stability force
F_z True axial force; also, axial force rating of the tubular
G_{bd} Formation breakdown gradient (usually 0.7–1.0 psi/ft)
G_g Gas gradient, in psi/ft (varies from 0.1 to 0.15 psi/ft based on pressure)
G_i Pressure gradient exerted by fluid inside pipe, in psi/ft
G_o Pressure gradient exerted by the fluid outside pipe, in psi/ft
K_h Horizontal permeability
K_v Vertical permeability
NP Neutral point for tension effects on casing collapse, in ft
ΔP Differential pressure into the zone from the hole, in psi
P_{bx} Burst pressure exerted at depth D_x, in psi
P_{cx} Collapse pressure at depth D_x, in psi
P_i Pressure inside tubular, in psi
P_{is} Surface pressure on fluid inside casing, in psi (usually zero)
P_o Pressure outside tubular, in psi
P_{os} Surface pressure on fluid outside casing, in psi (usually zero)
P_s Maximum well shut-in pressure, in psi
P_y Minimum internal yield pressure, in psi
q Displacement rate, in bbl/min
S_t Unit tensile stress, in psi
t Nominal wall thickness, in in.
ΔT Increase in temperature, in degrees Fahrenheit
t_c Contact times, in min
V_c Critical velocity
V_t Volume of fluid (in turbulent flow), in ft^3
W Weight of casing hanging below the slips
W_{cb} Buoyant weight of casing, in lbs/ft
W_{cd} Dry weight of casing, in lbs/ft
Y_m Minimum yield stress of casing body, in psi
Y_p Minimum rated yield strength of casing, in psi
Y_{pa} Yield strength of axial stress equivalent, in psi
Y_t Percent of yield stress
YP Minimum body yield strength of pipe, in lbs
α Coefficient of thermal expansion for steel (6.9×10 in units of inverse degrees Fahrenheit)
μ Poisson's ratio (about 0.3 for this application)
ρ Mass per unit volume
ρ_f Fluid density
ρ_i Density of fluid inside casing, in lbs/gal
ρ_m Mud density, in lbs/gal
ρ_o Density of fluid outside casing, in lbs/gal

ρ_s Density of steel (489.5 lbs/ft or 65.4 lbs/gal)
σ Stress, in psi
σ_a Axial stress, in psi

References

1 Greeham, T., and A. McKee. "Drilling Mud: Monitoring and Managing It." *Oilfield Review,* 1 (2), 41. Jones, F. O., Jr. 1964. "Influence of Chemical Composition of Water on Clay Blocking of Permeability." *Journal of Petroleum Technology,* April, 441–46.

2 Peden, J. 1982. "Reducing Formation Damage by Better Filtration Control." *Offshore Services and Technology,* Vol. 15. January. 26–28. Arthur, K. G., and J. M. Peden. 1989. "The Evaluation of Drilling Fluid Filter Cake Properties and Their Influence on Fluid Loss." Paper SPE 17617, presented at the International Meeting on Petroleum Engineering, Tianjin, China.

3 Arthur and Peden, paper SPE 17617. Krueger, R. F., and L. C. Vogel. 1954. "Damage to Sandstone Cores by Particles from Drilling Fluids." *Drilling and Production Practices, API,* 158. Glenn, E. E., and M. L. Slusser. 1957. "Factors Affecting Well Productivity. II. Drilling Fluid Particle Invasion into Porous Media." *Journal of Petroleum Technology,* May, 132–39. Krueger, R. F., P. W. Fischer, and L. C. Vogel. 1967. "Effect of Pressure Drawdown on the Clean-up of Clay or Silt Blocked Sandstone." *Journal of Petroleum Technology,* March, 397–403.

4 Tuttle, R. N., and J. H. Barkman. 1974. "New Nondamaging and Acid-Degradable Drilling and Completion Fluids." *Journal of Petroleum Technology*, November issue. 1221–1276.

5 Wilson, G. E. 1976. "How to Drill a Usable Hole." *World Oil,* September, 47.

6 Bollfrass, C. A. 1985. "Sealing Tubular Connections." *Journal of Petroleum Technology,* June, 955–65.

7 Goins, W. C., Jr., B. J. Collings, and T. B. O'Brien. 1965. "A New Approach to Tubular String Design, Part 1." *World Oil,* November. Goins, W. C., Jr., B. J. Collings, and T. B. O'Brien. 1965. "A New Approach to Tubular String Design, Part 2." *World Oil,* December, 83–88. Klementich, E. F., and M. J. Jellison. 1986. "A Service Life Model for Casing Strings." *SPE Drilling Engineering,* April, 141–52.

8 Rike, E. A., G. A. Bryant, and S. D. Williams. 1986. "Success in Prevention of Casing Failures Opposite Salts, Little Knife Field, North Dakota." *SPE Drilling Engineering,* April, 131–40.

9 Greer, J. B., and W. E. Holland. 1981. "High-Strength Heavy-Wall Casing for Deep, Sour Gas Wells." *Journal of Petroleum Technology,* December, 2389–98.

10 Wooley, G. R., and W. Prachner. 1988. "Reservoir Compaction Loads on Casing and Liners." *SPE Production Engineering,* February, 96–102.

11 Holliday, G. H. 1969. "Calculation of Allowable Maximum Casing Temperature to Prevent Tension Failures in Thermal Wells." Paper presented at ASME Conference, 1969, Tulsa.

12 Rike, Bryant, and Williams, *SPE Drilling Engineering,* April 1986, 131–40.

13 Pattillo, P. D., and N. C. Huang. 1982. "The Effect of Axial Load on Casing Collapse." *Journal of Petroleum Technology,* January 159–64.

14 Ibid.

15 API Bulletin 5C3. 1983. "Formulas and Calculations for Casing, Tubing, Drillpipe, and Line-Pipe Properties." American Petroleum Institute.

16 Ibid.

17 Lindsey, H. E., Jr. 1979. "Techniques for Liner Tie-Back Cementing." *Production Operations,* 37–39.

18 API Specification 5A. 1982. "Specifications for Casing, Tubing and Drillpipe." American Petroleum Institute. Halliburton Modern Well Completion Course.

19 Bollfrass, *Journal of Petroleum Technology,* June 1985, 955–65.

20 Hills, J. O. 1951. "A Review of Casing String Design Principles and Practice." *API Production Practices.*

21 Wojtanowics, A. K., and E. E. Maidla. 1987. "Minimum Cost Casing Design for Vertical and Directional Wells." *Journal of Petroleum Technology,* October, 1269–82.

22 Bowman, G. R., and B. Sherer. 1988. "How to Run and Cement Liners, Part 3." *World Oil,* May, 58–66. Bowman, G. R., and B. Sherer. 1988. "How to Run and Cement Liners, Part 4." *World Oil,* July, 84–88.

23 Bowman, *World Oil,* May 1988, 58–66.

24 Bowman, *World Oil,* May 1988, 58–66. Lindsey, H. E., Jr. 1986. "Rotate Liners for a Successful Cement Job." *World Oil,* October 1986 , 39–43.

25 Bowman, *World Oil,* July 1988, 84–88.

26 Lindsey, *World Oil,* October 1986, 39–43.

27 Linsey, *Production Operations,* 1979, 37–39.

28 Durham, K. S. 1987. "How to Prevent Deep-Well Liner Failure (Part 1)." *World Oil,* October, 46–50. Durham, K. S. 1987. "How to Prevent Deep-Well Liner Failure (Part 2)." *World Oil,* November, 47–49.

29 Ibid.

30 Manley, D. D. 1988. "Installation of Retrievable Liners." Paper SPE 17523, presented at SPE Rocky Mountain Regional Meeting, Casper, Wyo.

31 Ibid.

32 Agnen, J. W., and R. S. Klein. 1984. "The Leaking Liner Top." Paper SPE 12614, presented at the SPE Deep Drilling and Production Symposium, Amarillo, TX.

33 Smith, D. K. 1987. *Cementing.* SPE monograph.

34 Ibid.

35 Smith, R. S. Amoco internal report on cementing.

36 Keller, S. R., R. J. Crook, R. C. Haut, and D. Kulakofsky. 1987. "Deviated Wellbore Cementing: Part 1—Problems." *Journal of Petroleum Technology,* August, 955–60.

37 API Bulletin RP-10B. 1975. "Deviated Wellbore Cementing." *Journal of Petroleum Technology,* June, 759–94.

38 Harms, W. M., and J. T. Ligenfelter. 1981. "Microspheres Cut Density of Cement Slurry." *Oil and Gas Journal,* February 2, 59–66.

39 Harms, W. H. 1985. "Cementing of Fragile Formation Wells with Foamed Cement Slurries." *Journal of Petroleum Technology,* June, 1049–57.

40 Davies, D. R., J. J. Hartog, and J. S. Cobbett. 1981. "Foamed Cement—a Cement with Many Applications." Paper SPE 9598, presented at the Middle East Oil Technology Conference, Manama, Bahrain.

41 Smith, R. C., and D. G. Calvert, 1975. "The Use of Sea Water in Wet Cementing." *Journal of Petroleum Technology,* June, 759–64.

42 Smith, A. C., C. A. Powers, and T. A. Dobkins. 1980. "A New Ultralightweight Cement with Super Strength." *Journal of Petroleum Technology,* August, 1438–44.

43 Benge, O. G., and L. B. Spangle. 1982. "Foamed Cement—Solving Old Problems with a New Technique." Paper SPE 11204, presented at the SPE Annual Technology Conference, New Orleans.

44 Bannister, C. E. 1978. "Evaluation of Cement Fluid-Loss Behavior under Dynamic Conditions." Paper SPE 7592 presented at the SPE Annual Fall Technical Conference, Houston.

45 Bannister, paper SPE 7592. Brice, J. W., and B. C. Holmes. 1964. "Engineered Casing Cementing Programs Using Turbulent Flow Techniques." *Journal of Petroleum Technology,* May, 503–8.

46 Cheung, P. A., and R. M. Beirute. 1982. "Gas Flow in Cements." Paper SPE 11207, presented at the SPE Annual Technology Meeting, New Orleans.

47 Bannister, C. E., G. E. Shuster, L. A. Woolridge, and M.J. Jones. 1983. "Critical Design Parameters to Prevent Gas Invasion During Cementing Operations." Paper SPE 11982, presented at the SPE Annual Technology Meeting, San Francisco.

48 Cheung, P. A., and B. D. Myrick. 1984. "Field Evaluation of an Impermeable Cement System for Controlling Gas Migration." Paper SPE 13045, presented at the SPE Annual Technology Meeting, Houston.

49 Smith, R. C. 1984. "Successful Primary Cementing Can Be a Reality." *Journal of Petroleum Technology,* November, 1851–58.

50 R. S. Smith, internal report.

51 D. K. Smith, SPE monograph.

52 Sauer, C. W. 1987. "Mud Displacement during Cementing: A State of the Art." *Journal of Petroleum Technology,* September, 1091–1101.

53 Smirk, D. E., D. P. Kundert, and H. L. Vacca. 1987. "Application of Primary Cementing Principles, Rocky Mountains and Texas." Paper SPE 16208, presented at the Production Operations Symposium, Oklahoma City.

54 McLean, R. H., C. W. Manrey, and W. W. Whittaker. 1967. "Displacement Mechanics in Primary Cementing." *Journal of Petroleum Technology*, February, 251–60.

55 Parker, P. N., B. J. Ladd, W. M. Ross, and W. W. Wahl. 1965. "An Evaluation of a Primary Cementing Technique Using Low Displacement Rates." Paper SPE 1234, presented at the 40th Annual Meeting of the SPE, Denver.

56 Clark, C. R., and L. G. Carter. 1973. "Mud Displacement with Cement Slurries." *Journal of Petroleum Technology,* July, 3–19.

57 Garcia, J. A. 1985. "Rotating Liner Hanger Helps Solve Cementing Problems." *Petroleum Engineering,* September, 38–48.

58 Bowman, G. S., and B. Sherer. 1988. "How to Run and Cement Liners, Part 1." *World Oil,* March, 38–46.

59 Crook, R. J., S. R. Keller, and M. A. Wilson. 1987. "Deviated Wellbore Cementing: Part 2—Solutions." *Journal of Petroleum Technology,* August, 961–66.

60 Crook, Keller, and Wilson, *Journal of Petroleum Technology,* August 1987, 961–66.

61 Reiley, R. H., J. W. Black, T. O. Stagg, D. A. Walter, and G. S. Atol. 1987. "Cementing of Liners in Horizontal and High-Angle Wells at Prudhoe Bay, Alaska." Paper SPE 16682, presented at the 62nd Annual Meeting of the SPE, Dallas.

62 Wilson, M. A., and F. L. Sabins. 1987. "A Laboratory Investigation of Cementing Horizontal Wells." Paper SPE 16928, presented at the 62nd Annual Meeting of the SPE, Dallas.

63 Lee, H. K., R. C. Smith, and R. E. Tighe. 1986. "Optimal Spacing for Casing Centralizers." *SPE Drilling Engineering,* April, 122–30. API Specification 10D. 1986. "Casing Centralizers." 3rd ed. American Petroleum Institute.

64 Baret, J. R. 1988. "Why Cement Fluid Loss Additives Are Necessary." Paper SPE 17630, presented at the SPE International Meeting, Tianjin, China.

65 Lee, Smith, and Tighe, *SPE Drilling Engineering,* April 1986, 122–30.

66 Beirute, R. M. Field data.

67 Baret, paper SPE 17630. Patton, L. Douglas. 1987. "Squeeze Cementing Made Easy." *Petroleum Engineer International,* October, 46–52.

68 Douglas, *Petroleum Engineer International,* October 1987, 46–52. Rike, J. L. 1973. "Obtaining Successful Squeeze-Cementing Results." Paper SPE 4608, presented at the fall meeting of the SPE, Las Vegas.

69 Rike, paper SPE 4608.

70 Goolsby, J.L. "A Proven Squeeze—Cementing Technique in a Dalomite Reservoir." Reprinted from *Journal of Petroleum Technology.* Oct 1969. 1341–1346.

71 Lindsey, *World Oil,* October 1986, 39–43.

72 Durham, *World Oil,* October 1987, 46–50. Durham, *World Oil,* November 1987, 47–49.

73 Lindsey, *World Oil,* October 1986, 39–43. Gust, D. A., and R. S. McDonald. 1989. "Rotation of a Long Reach Liner in a Shallow Long-Reach Well." *Journal of Petroleum Technology,* April, 401–4.

74 Lindsey, H. E., Jr. "Field Results of Liner Rotation During Cementing." *SPE Production Engineering,* February 1987. 9–14.

75 Mahoney, B.J. and J.R. Barrios.1962. "Cementing Liners Through Deep, High Pressure Zones." *Engineering Essentials of Mud Drilling.* HBJ Publishers.

12

Well Planning in Drilling Engineering

Introduction

Well planning is the key to being able to safely and economically drill a usable hole for oil and gas production. The economic feasibility of drilling a well is established from cost estimation of drilling, completion, and production operations, while safety and cost control are achieved through the proper planning of all the relevant programs that have an impact on the proposed prospect well.

Planning for drilling an oil/gas well requires many detailed studies evaluating every aspect that directly or indirectly influences the successful economic outcome of the project. Planning requires intuitive, common sense judgments controlling decision making, along with thorough analyses representing the coordinated efforts of many individuals, each of whom contributes specific skills to the task.

A good well plan requires the coordination of a team that includes the drilling engineer, the drilling supervisor, the drilling superintendent, the geologist, the production engineer, and the reservoir engineer, as well as individuals responsible for safety, environmental, and government regulations.

Objectives

The objectives of well planning in the oil patch are twofold:

- To identify and address all significant engineering parameters, events, regulations, and other situations that are likely to have a direct or indirect impact on the proposed drilling venture
- To prepare a well plan that addresses all issues and that will enhance the success of the prospect well by drilling in a safe, efficient, and economical manner, in compliance with all federal, state, and local government rules and regulations

Information Needed for Effective Well Planning

In development well drilling, the best source of information is offset well data. This includes the following information from daily drilling report summaries: hydraulics, tubular, mud, BHAs, directional surveys, drill bits, logging, casing and cementing, geological information, reservoir characteristics, logistics, weather, service companies' and product suppliers' recommendations, government regulations, personnel and type of contractual agreement, and above all, problems encountered and the success or failure of attempted solutions. By contrast, in exploratory well drilling, a meager supply of information is available from surface geology information and seismic data (formation tops and pore/fracture pressure information).

Responsibilities of the Drilling Engineer

The drilling engineer is the well architect and planning coordinator for the project, whose responsibilities include

- Estimating the overall well cost—that is, *authorization for expenditures* (AFE)
- Collecting and reviewing the available data on all offset wells in the area
- Designing all the drilling programs (drilling fluid, drill bit, wellbore hydraulics, casing/cementing, directional plans, tubulars and BHAs, and well control)
- Anticipating drilling problems likely to be encountered and working out contingencies
- Selecting the drilling rig and its specifications
- Preparing drilling-cost and drilling-time curves
- Coordinating bid requests and evaluating contractors to ensure optimum rig selection and rig personnel efficiency and safety records
- Coordinating the activities of purchasing, environmental, regulatory, and other engineering groups to ensure that all phases of the well program are completed economically, safely, and on schedule.

Considerations in Well Planning

Following is a list of some of the more important considerations in well planning and a brief discussion of each:

- *Area geology.* This includes identification of formation tops to be penetrated, problem zones (e.g., loss circulation zones), shales, abnormal conditions (pressures, temperatures, etc.), and possible production intervals.
- *Formation pore pressure and fracture gradients.* An accurate knowledge of formation pressure and fracture gradients is one of the most important pieces

of geological information for the drilling engineer to have. It allows proper selection of casing-setting depths in conjunction with the optimum type of drilling fluid to be used in each interval of the wellbore.

- *Logging program.* This program should be coordinated with the geologist in charge. Details as to type of log to be run and data to be derived should be worked out well in advance of drilling.
- *Casing program.* This section contains a casing schematic that represents well construction details. In addition, each casing string should have complete, detailed design data.
- *Mud program.* This is a detailed discussion, by interval, of desired drilling-fluid type, properties, and maintenance. Many downhole problems are the direct results of improper use of drilling fluids.
- *Cementing program.* This section should include amounts of additives, probable bottom-hole temperatures, estimated amounts and types of cement to be used, setting time allowed, curing time required, and types of shoe, floats, centralizers, and scratchers required for cementing each string of casing. In addition, casing running and handling must be included.
- *Well control.* Specifications and drawings of the BOP stack and all other supporting auxiliary equipment (choke manifold, kill and fill lines, etc.) should be included. The engineering design of the well control program should be established, along with the procedures to be followed by the drilling crew, the tool pusher, and the drilling supervisor.
- *BHAs.* Reamers, centralizers, shock absorber subs, short drill collars, large-diameter drill collars, and other types of equipment that lend themselves to general improvement of drilling should be specified here. Proper use of BHAs can eliminate very costly hole deviation problems.
- *Hydraulics program.* The hydraulics program dictates the rig hydraulic power requirements. It consists of the calculations of all pressures (friction and dynamic) in the rig circulating system. Optimum utilization of hydraulic horsepower at the bit improves ROP and increases bit life. Proper design of hydraulics can also ensure effective drilled-cuttings removal and safe equivalent circulation density (ECD) of mud.
- *Drill bit program* The types of bits to be used must be specified, and optimum operating conditions (WOB, rpm, optimum flow rates, and corresponding optimum nozzle sizes) should be selected for each bit. Post-analysis of previous bit records is an essential part of this program.
- *Routine drilling.* This include the operator's desires as to routine daily procedures that the drilling contractor should follow. For example, it may include the requirements of blowout drills, mud maintenance, tripping practices, equipment inspection, and so forth.

- *Drilling-time curve.* A drilling-time curve (a plot of time vs. depth out) should be included in the program plan and compared daily with actual progress. When differences occur between predicted performance and actual performance, the reasons should be found, and appropriate adjustments should be made to the drilling parameters.
- *Drilling rig.* A thorough evaluation of all rig systems (power, circulating, hoisting, rotary, well control, and data acquisition/monitoring) must be conducted if proper implementation of optimized drilling programs is to be achieved.

The Drilling Program

Planning the drilling program

The purpose of planning the drilling program is to provide a basis for the safe and successful drilling and completion of a well at the lowest overall well cost. The amount and details of planning required vary greatly with the complexity of the well. For example, when development wells are drilled, most of the information is taken from offset wells. However, with the rapid changes in drilling technologies and products, it is important to take a fresh look at each successive well, making certain that no issues critical to drilling efficiency have been overlooked.

Although certain circumstances pertaining to a given well will dictate the final form and complexity of the drilling-program plans, the following outline should serve as a guide:

- Well summary (drilling and geological prognosis, pore/fracture pressures, drilling time and cost curves, casings, drilling procedures, well construction schematics, completion, abandonment)
- Drilling programs
  - Mud program (mud type, properties, and maintenance)
  - Drill bit program (drill bit type, WOB, rpm, and hydraulics)
  - Drill string program (drill pipe, drill collars, and BHA)
  - Hydraulic program (ECD and rig hydraulic power requirement)
  - Casing/cementing (equipment, materials, and accessories)
  - Well control program (equipment, control methods, pump pressure schedule, and pressure profile in the annulus during well control)
- Wellhead equipment
- Rig specification
- Evaluation phase (sampling, coring, logging, and testing)
- Emergency (contingency plans for all encountered problems)
- Miscellaneous (AFE, reporting, permits, offset records, contracts, rentals, etc.)

Updating the drilling program

All drilling programs should be designed with some flexibility, to accommodate changes in response to unexpected encountered problems. Initial plans are based on assumptions that will be confirmed or corrected as the well is drilled. However, major changes in well plans require thorough review, analysis, and redesign. Even though the plans are flexible to some degree and safety factors are built into the critical design parameters, the mechanical and economical impact of major proposed changes must be reviewed, documented, and approved in the same manner as the original plans. Anticipating or knowing potential problems—and designing contingency plans to combat them—is a major issue in well planning.

Post-Well Analysis

An important consideration to improve on the next effort to drill a well is to conduct post-well analysis of all completed efforts for the previous well. Both successful and failed aspects of the drilling operation should be documented immediately following project completion. An objective, critical internal review of the planning, engineering, operations, and third-party services can lead to more successful and cost-effective future drilling operations. The post-well analysis report should include the following:

- Well description (well name, field, lease, API/government identification number, and elevation)
- AFE (number, amount, supplements, and budget)
- Results (TD reached, economically producing, dry hole, abandoned, etc.)
- Dates (rig contract date, rig relocation date, spud date, completion date, and plug and abandonment date)
- Geology (purpose of well, formation tops, zones of interest, cores and interval taken, logs and interval taken, and drill stem testing if conducted)
- Casing/cementing (description by hole section, well construction, and wellhead schematics)
- Mud (type by hole interval, cost, and loss circulation zones if any)
- Contractor (rig name, contract type, cost, and service companies used)
- Drill bits (types and performance)
- Drilling time curves (estimated and actual)
- Directional plans (preselected, actual, surveys, and BHA summary)
- Well cost (proposed and actual)
- Recommendations (of drilling engineer, supervisor, and superintendent)

Well Cost Estimation

Overall well cost, excluding production, may be expressed as follows:

$$C_{\mathrm{wdo}} = C_{\mathrm{d}} + C_{\mathrm{o}} \tag{12.1}$$

In equation (12.1), C_{d} is the cost, including bit and rig rental only, of drilling a foot of hole. The other cost, C_{o}, in equation (12.1) is related to all other costs of drilling a foot of hole (e.g., casings, mud, cementing services, logging services, coring services, site preparation, fuel, transportation, and completion; because this cost can be easily accounted for, no further discussion will be given here.

Bit run performance

The drilling cost, C_{d}, or hole-making cost, needs to be synthesized on the basis of individual bit run performances. Bit run cost may be expressed as given previously, in equation (8.1).

Example 12.1

Using the following data, as well as the data listed in Table 12.1 for different bits with their respective drilling performance determine the cost analysis and select the bit that will result in lowest drilling cost.

Operating cost of the rig is $12,000/day,
Trip time is 10 hours
Connection time is 1 minute per connection.

Table 12.1 Bit Performance

Bit	Bit Cost ($)	Rotating Time (hours)	Connection Time (hours)	ROP (ft/hr)
A	1,000	15	0.1	14
B	3,000	35	0.2	13
C	4,000	45	0.3	10
D	4,500	65	0.3	11

Solution:

The cost per foot for each of the bits can be calculated using eq. (12.16).

For bit A

$$C_f = \frac{1{,}000 + \frac{12{,}000}{24}(15 + 0.1 + 10)}{14 \times 15} = \$65/\text{ft}.$$

For Bit B

$$C_f = \frac{2{,}000 + \frac{12{,}000}{24}(35 + 0.2 + 10)}{13 \times 35} = \$54/\text{ft.}$$

For bit C

$$C_f = \frac{4{,}000 + \frac{12{,}000}{24}(45 + 0.3 + 10)}{10 \times 45} = \$70/\text{ft.}$$

For bit D

$$C_f = \frac{4{,}500 + \frac{12{,}000}{24}(55 + 0.3 + 10)}{11 \times 65} = \$52/\text{ft.}$$

It can be clearly seen that the lowest drilling cost was obtained using bit D. This example provides the necessity of the performing the cost analysis rather than merely selecting the bit based on the rate of penetration alone.

Time estimates

Since all of the variables in the right-hand side are known, equation (8.1) can be directly used in post-well analysis of bit run cost; however, for estimating drilling cost prior to drilling, the drilling time ($T_{d\,i}$), the trip time ($T_{t\,i}$), and the connection time ($T_{c\,i}$) must be estimated first. The trip time and the connection time may be determined from past experience with regard to rig crew pipe-handling efficiency.

The trip time may be determined from the following equation:

$$T_{ti} = 2\frac{t_s}{L_s}D_i \tag{12.2}$$

where

$T_{t\,i}$ is the trip time required in order to change a bit and resume drilling operations, in h
t_s is the average time required in order to handle one stand of drill string, in h
L_s is the average length of one stand of drill string, in ft
D_i is the depth at which the trip was made, in ft

Thus, the time required per trip increases linearly with depth. In addition, the footage drilled by a single bit tends to decrease with depth, causing the number of trips required in order to drill a given depth increment to also increase with depth.

Trip depths may be estimated from the well-known drilling rate versus depth relationship, as follows, if the approximate bit life is known:

$$\frac{dD}{dT_r} = k\varepsilon^{\alpha D} \tag{12.3}$$

where k and α are constants and must be determined from past field data. Assuming that the average bit life for a given bit group in a given bit group in a given hole interval is T_b and integrating equation (12.3) between D_i, the depth of the last trip, and D_{i+1}, the depth of the next trip, gives the following:

$$D_{i+1} = \frac{1}{\alpha} ln(\alpha k \bar{T}_b + e^{\alpha D_i}) \tag{12.4}$$

The drilling time it takes for a given bit to drill from depth D_i to depth $D_i + L$ may also be obtained by integrating equation (12.3)—that is,

$$T_{di} = \frac{1}{k\alpha}\left(e^{\alpha D_{i+1}} - e^{\alpha D_i}\right) \tag{12.5}$$

Thus, with α and k known, the drilling and tripping times can be predicted; therefore, drilling cost, C_{di}, for a given bit run interval can be calculated. Total drilling cost can then be obtained by simple algebraic summation:

$$C_d = \sum_{i=1}^{N} C_{di} \tag{12.6}$$

The constants α and k can be approximately determined if ROP versus depth data are available from past well records. This is accomplished, with reference to equation (12.3), by plotting the data on semilog paper, fitting a straight line through the plot, and extrapolating the line to zero depth. The intercept with the abscissa (ROP) is k, and the slope of the line is α.

Learning Curve

It has been shown by Ikoku[1] that drilling cost can be predicted using the concept of the *learning curve*. The success of any commercial venture depends on the ability to make a profit. Profit might be simply taken as the difference between selling price and all well costs. Thus, if one has the ability to forecast prices and costs and has some degree of certainty about volume, profits may be readily calculated, and business decision making will be quite easy. The learning curve is consistently used in many industries as a basis for projecting costs, forecasting manpower requirements, rating personnel, scheduling production, and negotiating multimillion-dollar contracts.

It is important that individuals engaged in repetitive tasks improve on their performances. Quantifying this improvement gives rise to the progressive function that is commonly is referred to as the learning curve. The most common factors that give rise to performance improvement as cumulative output increases include the following:

- Job familiarization by the individual
- Improvement in tool coordination, shop organization, and engineering liaison
- Development of better methods
- Development of better drilling technology and more efficient tools

In the petroleum industry, the first well drilled in a new field (a wildcat well) generally will have the highest cost. As the operator becomes more familiar with the area, optimized drilling implementation will continue to decrease the cost of each subsequent well drilled until a point is reached at which there is no significant improvement.

Cost Control in Well Drilling

Overall *cost control* in drilling ventures is achieved through the proper design, maintenance, and implementation of all drilling programs, including

- Mud program
- Hydraulic program
- Drill bit program
- Drill string program
- Casing program
- Well control program

Other programs of importance in the control of overall well costs are

- Completion
- Stimulation
- Production
- Separation, storage, and transportation

Optimum drilling programs collectively allow the drilling of a usable hole at the lowest overall cost ($/ft). The mud program is the preeminent program governing well cost control.

Almost every drilling problem can be related, directly or indirectly, to the particular drilling fluid in use. Once the fluid type is selected, its properties (viscosities, weight, fluid loss, solids, etc.) must be maintained at the levels that are essential to the performance of its specific functions.

Time Value of Money

The concept of *time value of money* enables the calculation of the future value of an investment made today and the calculation of the present value of past investments. It provides a way to compare and adjust the authorization of expenditure, to arrive at realistic budget estimates and calculate the return on investment. Time value of money is based on the concept that the present-day dollar is worth more than that in the future. It will have a time value as long as interest can be earned. In a basic sense, this helps to estimate the project cost and financial goals and to evaluate alternative courses of actions during the planning and the execution of projects.

For example, a company may have a bought a drill bit three years ago at a price of \$2,000 and may not have used it yet. The concept of time value of money lets us find the future or present value based on the company's average rate of return. Future value is given as

$$FV = PV\left(1+\frac{r}{n}\right)^{nm} \tag{12.7}$$

where

FV denotes the future value
PV denotes the present value
r is the periodic interest rate or growth rate, expressed as a fraction
n is the number of payments per year
m is the number of years

For continuous compounding of interest, equation (12.7) can be rewritten as

$$FV = PV \times e^{rn} \tag{12.8}$$

On the basis of equation (12.8), it is evident that the present value becomes less both as the number of period becomes larger and as the interest rate period becomes larger. Therefore, the future value is influenced by the rate of return the company is providing to the investors, as well as by the number of payments per year.

Example 12–2

A company bought an external casing packer (ECP) five years ago for \$10,000 and did not use it. What would its value be today to break even, assuming an average rate of return for the past five years of 14%.

Solution:

Past value = \$10,000
Interest rate = 14% per year
$n = 4$ periods per year

Thus, future value is

$$FV = 10{,}000\left(1+\frac{0.14}{4}\right)^{4\times 5} = \$19{,}897$$

Price Elasticity

Price elasticity is used to measure the effect that economic variables, such as demand or supply of rigs or wells drilled, have on change in the crude oil price. Elasticity allows quantification of how sensitive the one variable is with respect to the other one. Furthermore, it is independent of units of measurement, expressing the ratio of the percentage of change of wells and footage drilled to the percentage change in the crude price.

Price elasticity describes the degree of responsiveness, of the rig in demand or the rig in supply, to the change in the crude price. Price elasticity of drilling also measures the direction of change of the drilling rigs when the price of oil changes. That is,

$$\text{Drilling price elasticity } (\%) = \frac{\%\text{ age change in wells drilled}}{\%\text{ age change in crude oil price}} = \times\ 100\% \tag{12.9}$$

Elasticity, E, is given as

$$E = \frac{\%\Delta R}{\%\Delta P} = \frac{dR/R}{dP/P} = \frac{dR}{dP}\left(\frac{P}{R}\right) \tag{12.10}$$

For example, if the oil price increases from \$30–\$34.00 per barrel and if the number of drilling rigs increases from 1,220 to 1,240, then elasticity of drilling rigs would be calculated as follows:

$$E = \frac{\left[(1{,}120 - 1{,}140)/1{,}120\right] \times 100\%}{\left[(30 - 34)/30\right] \times 100\%} = \frac{1.79\,\%}{13.33\,\%} = 0.133$$

Midpoint formula is another method. This gives the answer regardless of the direction of change:

$$E = \frac{(R_2 - R_1)/[(R_2 + R_1)/2]}{(P_2 - P_1)/[(P_2 + P_1)/2]} \tag{12.11}$$

Applying this to the example in equation (12.11) yields

$$E = \frac{(1{,}140 - 1{,}120)/1{,}130}{(34 - 30)/32} = 0.14$$

To compare the calculated elastic values, the elasticity can be classified among five *ranges of elasticity:*

1. *Inelastic.* The number of drilling rigs does not respond strongly to the oil price (i.e., $E < 1$).
2. *Elastic.* The number of drilling rigs responds strongly to the oil price (i.e., $E > 1$).
3. *Perfectly inelastic.* The number of drilling rigs does not respond to the oil price change (i.e., $E = 0$).
4. *Perfectly elastic.* The number of drilling rigs responds infinitely to the oil price change (i.e., $E > \infty$).
5. *Unit elastic.* The number of drilling rigs responds by the same percentage as the oil price change (i.e., $E = 1$).

Supplementary Problems

Problem 12–1. An oil company is planning to drill wells where the prospective formations are classified as prospects 1, 2, and 3. The probabilities of striking oil in the three prospective formations are 0.30, 0.45, and 0.25, respectively.

On the basis of data obtained from offset wells earlier, the probability of striking oil is 40% in prospect 1 formation, 20% in prospect 2 formation, and 30% in prospect 3 formation. If oil is struck at this site, determine the probability that it is prospect 1 formation.

Problem 12–2. Find the cost (in \$/ft) to drill between 9,000 ft and 10,000 ft for a prospect well, given the following data:

- Well depth target = 17,000 ft
- Casing setting = 500 ft, 4,000 ft, 9,000 ft, 14,000 ft, and 17,000 ft
- Rig can handle triples, and the rental cost is \$12,000/d
- Tripping time per stand is 2.5 min

Bits used for the different casing settings and control well data for the region are as given in table 12–2. The formation constants for the third casing setting are $k = 300$ and $\alpha = 0.0003$.

Table 12–2. Bit details for different phases in problem 12–2

Phase	Size(in)	Average Life(hours)	Cost(\$)
1	22	40	4,000
2	17½	35	3,000
3	13½	24	2,400
4	8⅜	20	2,000
5	5¾	10	1,500

Problem 12–3. Assuming that striking oil is independent from one location to another and that the probability of striking oil at an individual location is 0.3,

(a) What is the expected value of success?

(b) What is the expected value of the number of locations having oil if drilling is carried out simultaneously at six different locations?

Problem 12–4. Assuming that striking oil is independent from one location to another and that the probability of striking oil at an individual location is 0.2,

(a) What is the probability of striking oil in exactly two wells if drilling is carried out simultaneously at six different locations?

(b) What is the probability that the fourth time striking oil will occur at the sixth drilling if drilling is carried out sequentially at one location after another?

Problem 12–5. The demand of workover rigs is given by $Q_{wor} = 450 + 16.5P$, and the crude oil price per barrel is \$35. If demand of the rig shifts outward such that the oil price rises by 10%, will the demand response be relatively price elastic or inelastic?

Nomenclature

C_0 Cost of services, in \$/h
$C_{b\,i}$ Cost of bit i, in \$
C_d Drilling cost, in \$/h
$C_{d\,i}$ Drilling cost, in \$/ft, for bit run i
C_r Rig cost, in \$/h
D_i Depth where trip was made, in ft
ΔD_i Formation interval drilled, in ft, by bit i
E Elasticity
FV Future value
k Constant
L_s Average length of one stand of drill string, in ft
m Number of years
n Number of payments per year
PV Present value
r Periodic interest rate or growth rate, expressed as a fraction
$T_{c\,i}$ Connection time, in h, for bit i
$T_{d\,i}$ Drilling time, in h, for bit i
t_s Average time required in order to handle one stand of drill string, in h
$T_{t\,i}$ Trip time, in h, for bit i
$T_{t\,i}$ Trip time required in order to change a bit and resume drilling operations, in h
α Constant

References

[1] Ikoku, C. U. 1978. "Application of Learning Curve Models to Oil and Gas Well Drilling." Paper SPE 7129, presented at the California Regional Meeting of SPE-AIME, San Francisco.

Bibliography

Aadnoy, B. S. 1987. "Modeling of the Stability of Highly Inclined Boreholes in Anisotropic Rock Formations." Paper SPE 16526, presented at SPE Offshore Europe '87, Aberdeen.

———. *Modern Well Design.* 1996. Houston: Gulf Publishing.

Abram, A. 1977. "Mud Design to Minimize Rock Impairment Due to Particle Invasion." *Journal of Petroleum Technology,* May, 586.

Adams, N. 1977. "How to Control Differential Pipe Sticking." *Petroleum Engineer,* September.

Anderson, E. E, and J. J. Azar. 1993. "PDC Bit Performance under Simulated Borehole Conditions." *SPE Drilling and Completion,* 8 (3).

Annis, M. R., and P. H. Monoghan. 1962. "Differential Pressure Sticking—Laboratory Studies of Friction between Steel and Mud Filter Cake." *Petroleum Transactions,* May.

Annual Statistical Review, Petroleum Industry Statistics, 1965–1974. 1975. American Petroleum Institute.

Azar, J. J. 1997. "Important Issues in Cuttings Transport for Drilling Directional Wells." Paper SPE 39020, presented at the 5th Latin American and Caribbean Petroleum Engineering Conference and Exhibition, Rio de Janeiro.

Azar, J. J., and J. L. Lummus. 1975. "The Effect of Drilling Fluid pH on Drill Pipe Corrosion Fatigue Performance." Paper SPE 5516, presented at the fall meeting of the SPE, Dallas.

———. 1976. "Corrosion Fatigue of Drill Pipe." *Petroleum Engineering International,* September.

Billingston, S. A. 1963. "Practical Approach to Circulation Problems." *Drilling Contractor,* July–August, 52.

Black, A. D., S.J. 1977. "Drillability of Sandstone and Dolomite at Simulated Depths." Paper presented at the ASME Petroleum Division Conference.

Bourgoyne, A. T., Jr., B. H. Caudle, and O. K. Kimbler. 1972. "The Effect of Interfacial Films on the Displacement of Oil by Water in Porous Media." *SPE Journal*, February 1972, 60–68.

Bowker, A. H. J., and G. J. Lieberman. 1972. *Engineering Statistics.* Englewood Cliffs, NJ: Prentice-Hall.

Bowman, G. R., and B. Sherer. 1988. "How to Run and Cement Liners, Part 1." *World Oil,* March, 38–45.

———. 1988. "How to Run and Cement Liners, Part 2." *World Oil,* May 58–66.

Bradley, W. B. 1974. "Deviation Forces from the Wedge Penetration Failure of Anisotropic Rock." *Transactions of the ASME,* 95 (Series B, No. 4), 1093.

———. 1974. "Factors Affecting the Control of Borehole Angle in Straight and Directional Wells." Paper SPE 5071, presented at the SPE-AIME 49th Annual Meeting, Houston.

———. 1979. "Failures of Inclined Boreholes." *Journal of Energy Resources Technology, Transactions of AIME,* 232.

———. 1990. "A Task Force Approach to Reducing Stuck Pipe Costs." Paper SPE/IADC 21999, presented at the SPE/IADC Drilling Conference, Amsterdam.

Buzarde, L. L., Jr., R. L. Kastor, W. T. Bell, and C. L. DePriester. 1972. "Production Operations Course. 1. Well Completions." SPE.

Cagle, W. S., and H. D. Mathews. 1977. "An Improved Lost Circulation Slurry Squeeze." *Petroleum Engineer,* July, 26.

Chambie Syndicate de la Recherche, Comitedes Technicians. 1981. *Blowout Prevention and Well Control.* Paris: Editions Technip.

Chenevert, M. E. 1969. "Shale Hydration Mechanics." *Journal of Petroleum Technology*, Vol. 22, No. 9, September 1970, 1141–1148.

Chien, S. F. 1971. "Annular Velocity for Rotary Drilling Operations." *Proceedings of the SPE Fifth Conference on Drilling and Rock Mechanics,* 5–16.

Civan, F., R. M. Knapp, and O. A. Henry. 1989. "Alteration of Permeability by Fine Particle Processes." Journal of Petroleum Science and Engineering, 2.

Clancy, L.W., and Boudreau, M., Jr. 1981. "High-Water Loss High-Solids Slurry Stops Lost Circulation with Oil Mud." *Oil and Gas Journal,* January, 99.

Clark, D. A. 1982. "An Experimental Investigation of the Mechanics of Bit-Teeth Rock Interaction." M.S. thesis, University of Tulsa.

Cunningham, R. A., and J. G. Eenink. 1959. "Laboratory Study of Effect of Overburden, Formation and Mud Column Pressure on Drilling Rates of Permeable Formations." *Transactions of the AIME,* 217, 9.

da Fontoura, S. A. B., and H. M. R. dos Santos. 1989. "In-Situ Stresses, Mud Weight and Modes of Failure around Oil Wells." *Proceedings of the 15th Canadian Rock Mechanics Symposium,* 235.

Deily, F. H., and A. J. Durelli. 1958. "Bottomhole Stress in a Wellbore." Paper SPE 1095-G. Houston.

Detournay, E., and A. H. Cheng. "Poroelastic Response of a Borehole in a Non-hydrostatic Stress Field." *International Journal of Rock Mechanics, Mining Sciences, and Geomechanics,* abstract no. 25, p. 171.

Dodge, N. A. 1963. "Friction Losses in Annular Flow." Paper ASME 63-WA-11.

Dunbar, M. E., T. M. Warren, and A. G. Kadaster. 1986. "Bit Sticking Caused by Borehole Deformation." *SPE Drilling Engineering,* 1 (4), 417.

Echel, J. R., and J. P. Nolley. 1950–51. "An Analysis of Hydraulic Factors Affecting the Rate of Penetration of Drag-Type Rotary Bits." *API Drilling and Production Practice.*

Edwards, J. W. 1964. "Engineering Design of Drilling Operations." API Drilling and Production Practice, 39–55.

Fabrycky, W. J., P. M. Ghare, and P. E. Torgersen. 1972. *Industrial Operations Research.* Englewood Cliffs: Prentice-Hall, pp. 172–99.

Feenstra, R., and D. H. Zijsling. 1984. *The Effect of Bit Hydraulics on Bit Performance in Relation to the Rock Destruction Mechanism at Depth.* SPE 13205 SPE Annual Technical Conference. Houston.

Fontenot, J. E., and F. E. Clark. 1974. "An Improved Method for Calculating Swab and Surge Pressures and Circulating Pressures in a Drilling Well." *SPE Journal,* October, 451–62.

Freeland, T. Jr. 1961. "Dynamic Stresses in Long Drill Pipe Strings." *Petroleum Engineer,* May.

Galle, E. M., and H. B. Woods. 1960. "Variable Weight and Rotary Speed for Lowest Cost Drilling." Paper presented at the annual meeting of AAODC, New Orleans.

Garcia-Gavito, D., and J. J. Azar. 1994. "Proper Nozzle Location, Bit Profile, and Cutter Arrangement Affect PDC Bit Performance Significantly." *SPE Drilling and Completion,* 9 (3). 167–175.

Garnier, A. J., and N. H. Van Lingen. 1959. "Phenomena Affecting Drilling Rate at Depth." *Transactions of the AIME,* 217, 232.

Gill, J. A. 1989. "How Borehole Ballooning Alters Drilling Responses." *Oil and Gas Journal,* March, 43.

Glowka, D. 1981. "Optimization of Bit Hydraulic Configurations." Paper SPE 10240, presented at the 56th Annual Fall Technology Conference, San Antonio.

Gnirk, P. F. 1972. "The Mechanical Behavior of Uncased Wellbores in Elastic/Plastic Media under Hydrostatic Stress." *SPE Journal,* February, 49.

Goins, W. C., Jr. 1952. "How to Combat Circulation Loss." *Oil and Gas Journal,* June 71.

Goins, W. C. Jr., and R. Sheffield. 1983. *Blowout Prevention.* 2nd ed. Houston: Gulf Publishing.

Govier, G. W., and K. Aziz. 1972. *The Flow of Complex Mixtures in Pipes.* New York: Van Nostrand–Reinhold.

Gray, G. R., and H. C. H. Darley. 1981. *Composition and Properties of Oil Well Drilling Fluids.* 4th ed. Houston: Gulf Publishing.

Haden, E. L., and G. R. Welch. 1961. "Techniques for Preventing Sticking of Drill Pipe." *Drilling and Production Practices.*

Hale, A. H., F. K. Mody, and D. P. Salisbury. 1993. "The Influence of Chemical Potential on Wellbore Stability." *SPE Drilling and Completion,* 8 (3), 207.

Hanshford, J. E., and A. Lubinski. 1966. "Cumulative Fatigue Damage of Drill Pipe in Dog-Legs." *Journal of Petroleum Technology,* March.

Hariharan, P. R., and J. J. Azar. 1994. "PDC Bit Hydraulics Design and Profile Are Key to Reducing Bit Balling." *Oil and Gas Journal,* 94 (50).

Helmick, W. E., and Longley, A. J. 1957. "Pressure Differential Sticking of Drill Pipe." *Oil and Gas Journal,* June, 132.

Hempkins, W. B., et al. 1987. "Multivariate Statistical Analysis of Stuck Drill Pipe Situations." *SPE Drilling Engineering,* 2 (3), 237–244.

Hirschmann, W. B. 1964. "Profit from the Learning Curve." *Harvard Business Review,* January–February, 125–39.

Historical Statistics of the United States—Colonial Times to 1957, a Statistical Abstract Supplement. 1961. Washington, DC: U.S. Department of Commerce.

Howard, G. C., and P. P. Scott, Jr. 1951. "An Analysis of Lost Circulation." *Transactions of the AIME,* 171.

Hsiao, C. 1987. "A Study of Horizontal Wellbore Failure." Paper SPE 16927, presented at the SPE annual conference, Dallas.

Hussaini, S. M., and J. J. Azar. 1983. "Experimental Study of Drilled Cuttings Transport Using Common Drilling Fluids." *SPE Journal,* February, 11.

Jenkins, P. B., and A. L. Crockford. 1975. "Drilling Costs." Paper SPE 5266, presented at the SPE-European Spring Meeting, London.

Joint Association Survey of the Oil and Gas Producing Industry, API, 1955 through 1977. American Petroleum Institute.

Kendal, H. A. 1977–78. "Fundamentals of Pressure Control." 6-Part Series. *Petroleum Engineer International,* October, November, and December 1977, and January, February, and March 1978.

Kendal, W. A., and W. C. Goins. 1960. "Design and Operation of Jet-Bit Programs for Maximum Hydraulic Horsepower, Impact Force, or Jet Velocity." *Transactions of the ASME,* 219, 238–47.

Krueger, R. F. 1986. "An Overview of Formation Damage and Well Productivity in Oil Field Operations." *Journal of Petroleum Technology,* February, 131.

Lubinski, A. 1961. "Maximum Permissible Dog-legs in Rotary Boreholes." *Journal of Petroleum Technology,* February, 194.

———. 1977. "Fatigue of Range 3 Drill Pipe." *IFP Review,* 2.

Lubinski, A. and H. B. Woods. 1953. "Factors Affecting the Angle of Inclination and Doglegging in Rotary Boreholes." *Drilling and Production Practices,* p. 222.

Lummus, J. L., and J. J. Azar. 1986. *Drilling Fluids Optimization: A Practical Field Approach.* Tulsa: PennWell.

Ma, D., and J. J. Azar. 1985. "Dynamic of Roller Cone Bits." *ASME Journal of Energy Resources Technology,* 107 (4), 356.

———. 1986. A Study of Rock-Bit Interaction and Wellbore Deviation." *ASME Journal of Energy Resources Technology,* 1 (1).

———. 1989. "A New Way to Characterize Gauging Scraping Action of Roller Cone Bits." Paper SPE 19448.

Maggie, L. 2003. "Geared To Maximize Drilling Performance." *American Oil and Gas Reporter,* April.

Mahony, B. J., and J. R. Barrios. 1976. "20,000 Feet in 100 Days." Paper SPE 5172, presented at the SPE-AIME Deep Drilling and Production Symposium, Amarillo, TX.

Brown, D.M. and D.A. Gledhill. 2003. "Expandable Drill Bit Provides New Method of Drilling Increased Diameter Hole." Paper SPE 79793. SPE/IADC Drilling Conference, Amsterdam.

McKinney, L. K. 1986. "Formation Damage Due to Oil-Base Drilling Fluids at Elevated Temperature and Pressure." M.S. thesis, University of Tulsa.

McLamore, R. T. 1971. "The Role of Rock Strength Anisotropy in Natural Hole Deviation." *Journal of Petroleum Technology,* November, 1313.

McLean, M. R., and M. A. Addis. 1990. "Wellbore Stability Analysis: A Review of Current Methods and their Field Applications." Paper SPE/IADC 19941, p. 261. SPE/IADC Conference, Houston.

McNeill, T. F., and D. S. Clark. 1966. *Cost Estimating and Contract Pricing.* New York: Elsevier.

Messenger, J. U. 1981. *Lost Circulation.* Tulsa: PennWell Publishing.

Miller, T. W. and J. B. Cheatham. 1971. "Rock/Bit Tooth Interaction for Conical Bit Teeth." *SPE Journal,* June, 162.

Mitchell, M. J., and J. M. Campbell. 1963. "Effect of Tooth Geometry on Tooth Wear Rate of Rotary Rock Bits." Paper 851-37-I, presented at the spring meeting of the Mid-Continent District, API, Oklahoma City.

Moore, P. L. *Drilling Practices Manual.* 2nd ed. Tulsa: PennWell Publishing.

Morita, N., and K. E. Gray. 1980. "A Consecutive Equations for Non-linear Stress-Strain Curves in Rocks and Its Application to Stress Analysis around a Borehole during Drilling." Paper SPE 9328, presented at the SPE Annual Conference and Exhibition, Dallas.

Muecki, T. W. 1979. "Formation Fines and Factors Controlling Their Movement in Porous Media." *Journal of Petroleum Technology,* February, 144.

Murray, A. S., and R. A. Cunningham. 1955. "Effect of Mud Column Pressure on Drilling Rates." *AIME Petroleum Transactions,* 204

Nicholson, R. W. 1974. "Acceptable Dogleg Severity Limits." *Oil and Gas Journal,* April, 73.

———. 1974. "Minimize Drill Pipe Damage and Hole Problems." *Transactions of IADC.*

O'Brien, D. E., and M. E. Chenevert. 1973. "Stabilizing Sensitive Shales with Inhibited, Potassium Based Drilling Fluids." *Journal of Petroleum Technology,* September, 255.

Ooutmans, H. D. 1963. "The Effect of Some Drilling Variables on the Instantaneous Rate of Penetration." *AIME Petroleum Reprint Series,* No. 6, *Drilling,* 102–14

Paccaloni, G., M. Tambini, and M. Galoppini. 1988. "Key Factors for Enhanced Results of Matrix Stimulation." Paper SPE 17154, presented at Formation Damage Symposium, Bakersfield, CA.

Peret, J. 1967. "A Review of Modern Drilling Costs." *API Drilling and Production Practices,* 130–42.

Peterson, C. R. 1970. "Roller Cutter Forces." *SPE Journal,* March, 57.

Pistle, H., and W. H. Thompson. 1963. "Optimization—the Key to Reduced Well Costs." *Proceedings of the 6th World Petroleum Congress,* 343–57.

Placido, J. C., J. J. Azar, J. R. Sorem, and F. Kessler. 1994. "Drill Pipe Fatigue Life Prediction Model Based on Critical Plane Approach." Paper OTC 7569, presented at the Offshore Technology Conference, Houston.

Pratt, C. A. "Increased Penetration Rates Achieved with New Extended Nozzle Bits." *Journal of Petroleum Technology,* August 1978. 1191–1198.

"Program to Determine Factors Affecting Rate of Penetration in Slow Drilling Formations." 1980. Proposal presented by Drilling Research Laboratory, Salt Lake City.

Rabia, H. *Oil Well Drilling Engineering.* 1985. London: Graham & Trotman.

Reed, D., and A. Pilehvari. 1993. "A New Model for Laminar, Transitional and Turbulent Flow of Drilling Muds." Paper SPE 25456, presented at the SPE/IADC Drilling Conference, Amsterdam.

Robinson, L. H., Jr. "Effects of Core and Confining Pressure on Failure Characteristics of Sedimentary Rocks." *Petroleum Transactions of AIME,* 216, 1959.

Roegiers, J. C., and E. Detournay. 1988. "Consideration on Failure Initiation in Inclined Boreholes." In *Key Questions in Rock Mechanics,* proceedings of the 29th U.S. Symposium, September 1988. Edited by Cundall. Rotterdam: Balkema, p. 461.

Rollins, H. M. 1966. "Drill Stem Failures Due to H_2S." *Oil and Gas Journal,* January.

Rowley, D. S., R.J. Howe, and F.H. Deily. 1960. "Laboratory Drilling Performance of the Full Scale Rock Bit." Paper SPE 1545-G, presented at the 35th annual meeting of the SPE.

Sample, K. J., and A. T. Bourgoyne. 1978. "Development of Improved Laboratory and Field Procedures for Determining and Carrying Capacity of Drilling Fluids." Paper SPE 7497, presented at the SPE Annual Technological Conference and Exhibition, Houston.

Sanner, D. 1989. "Effect of Drilling Fluid Filtrates on Flow Properties of Various Rocks." M.S. thesis, University of Tulsa.

Savins, J. G. 1958. "Generalized Newtonian (Pseudoplastic) Flow in Stationary Pipes and Annuli." *Transactions of the AIME,* 213, 325.

Schuh, F. J. 1961. "Computer Makes Surge Pressure Calculations Useful." *Oil and Gas Journal,* August 3, 96.

———. 1989. "Horizontal Well Planning-Build Curve Design." Paper presented at the Centennial Symposium Petroleum Technology into the Century, New Mexico Tech, Albuquerque, NM.

Sharma, M. M., and R. W. Wunderlich. 1987. "The Alteration of Rock Properties Due to Interactions with Drilling Fluid Components." Journal of Petroleum Science and Engineering, 1, 127.

Simon, R. 1963. "Energy Balance in Rock Drilling." Paper SPE 499, presented at the SPE Conference on Drilling and Rock Mechanics, Austin, TX.

Smith, D. E. 1975. "Prior Review of Drilling-Related Costs Saves $250,000 in Texas Infill-Drilling Program." *Oil & Gas Journal,* June 16, 62–66.

Stenevik, B. C. 1991. "Design and Construction of a Large-Scale Wellbore Simulator and Investigation of Hole Size Effects on Critical Cuttings Transport Velocity in Highly Inclined Wells." M.S. thesis, University of Tulsa.

Storli, C. 1987. "Formation Damage of Primary and Secondary Emulsifiers and Amine Compounds at Elevated Temperatures Using Various Concentrations." M.S. thesis, University of Tulsa.

Subramanian, R., and J. J. Azar. 2000. "Experimental Study of Friction Pressure Drop for Non-Newtonian Drilling Fluids in Pipe and Annular Flow." Paper presented at the SPE International Oil and Gas Conference, Beijing, China.

Sutko, A. A. 1973. "Drilling Hydraulics—a Study of Chip Removal Force under a Full-Size Jet Bit." *SPE Journal,* August, 233–38.

Sutko, A. A., and G. M. Myers. 1971. "The Effect of Nozzle Size, Number, and Extension on the Pressure Distribution under a Tricone Bit." *Journal of Petroleum Technology.*1299–1304.

Thomas, R. P., J. J. Azar, and T. E. Becker. 1982. "Drill Pipe Eccentricity Effect on Drilled Cuttings Behavior in Vertical Wellbores." *AIME Petroleum Transactions.* September, 273.

Tomren, P. H., A. W. Iyoho, and J. J. Azar. 1986. "An Experimental Study of Cuttings Transport in Directional Wells." *SPE Drilling Engineering,* 1 (1), 43.

Tosaya, C. A. 1982. "Acoustical Properties of Clay-Bearing Rocks." Ph.D. dissertation, Stanford University.

Tovar, J. 1990. "Formation Damage Studies Using Whole Drilling Muds in Simulated Boreholes." M.S. thesis, University of Tulsa.

Van Der Knapp, W. 1959. "Nonlinear Behavior of Porous Media." *Transactions of AIME,* 126, 179–87.

Veeken, C. A. M., et al. 1989. "Use of Plasticity Models for Predicting Borehole Stability." *Proceedings of the International Symposium of the IRSM-SPE,* 835.

Vidick, B., J. A. Yearwood, and H. Perthuis. 1989. "How to Solve Lost Circulation Problems." Paper SPE 17811, presented at the International Meeting on Petroleum Engineering, Tianjin, China.

Vidrine, D. J., and E. J. Benit. 1968. "Field Verification of the Effect of Differential Pressure on Drilling Rate." *Journal of Petroleum Technology,* July, 676–82.

Walker, B. H., et al. 1980. "Field and Laboratory Applications of Bits with Synthetic Formed Diamond Cutters." Paper presented at the ASME Energy Sources Conference, New Orleans.

Walker, R. E., and T. M. Mayes. 1975. "Design of Muds for Carrying Capacity." *Journal of Petroleum Technology,* July, 893–900, *AIME Petroleum Transactions.*

Warren, J. E. 1940. "Causes, Preventions, and Recovery of Stuck Drill Pipe." *Drilling and Production Practices.* API. p. 30.

Warren, T. M. 1981. "Drilling Model for Soft Formations Bits." *Journal of Petroleum Technology,* December, 963–70.

———. 1984. "Factors Affecting Torque for a Roller Cone Bit." *Journal of Petroleum Technology,* September, 1500–1508.

Warren, T. M., and W. J. Winters. 1984. "Effect of Nozzle Diameter on Jet Impact for a Tricone Bit." *SPE Journal,* February, 9–18.

Williams, C. E., and G. H. Bruce. 1951. "Carrying Capacity of Drilling Muds." *AIME Petroleum Transactions,* 111–120

Wilson, G. E. 1976. "How to Drill a Usable Hole." *World Oil,* August, 40.

Winters, W. J. 1987. "Roller-Bit Model with Rock Ductility and Cone Offset." Paper SPE 16696, presented at the annual meeting of the SPE, Dallas.

Winters, W. J., J. Senger, and M.S. Oliver. 1987. "Application of the 1987 IADC Roller Bit Classification System." Paper SPE/IADC 16143.

Wolfson, L. 1974. "Three Dimensional Analysis of Constrained Directional Drilling Assemblies in a Curved Hole." M.S. thesis, University of Tulsa.

Woods, H. B., and A. Lubinski. 1955. "Use of Stabilizers in Controlling Hole Deviation." *Drilling and Production Practices,* p. 165.

Yang, J. H., and K. E. Gray. 1965. "Single-Blow Bit-Tooth Impact Tests on Saturated Rocks under Confining Pressure and at Elevated Pore Pressure." *SPE Journal,* September, 389–408.

Young, F. S., Jr. 1968. "Computerized Drilling Control." Paper SPE 2241, presented at the fall meeting of the SPE. Journal of Petroleum Technology, April 1969. 483–496.

———. 1976. "Data for Drilling Optimization." Paper presented at the 1976 Rotary Drilling Conference in IADC, Houston.

Zoback, M. D., et al. 1985. "Well Bore Breakouts and In Situ Stress." *Journal of Geophysical Research,* 90, June, 5523–5530.

Index

A

B

D

E

F

G

H

I

N

O

P

R

S

T

U

V

W

Y

About the Authors

Dr. J. J. Azar, PhD, PE, SPE, NAE, is a professor emeritus of Petroleum Engineering at the University of Tulsa. He joined the university in 1965 as an assistant professor in Mechanical/Aerospace Engineering, and was promoted to associate professor in 1969. In 1975, he joined the Petroleum Engineering Department as full professor and Director of the University of Tulsa Drilling Research Projects, a joint oil and gas industry research program. In 1996, he stepped down from the directorship to devote full time to teaching and writing. He served as acting chairman of the Petroleum Engineering Department in 2001. Dr. Azar officially retired from the University May 31, 2002, to pursue his own activities in teaching and consulting. Professor Azar is a world-renowned lecturer and researcher in drilling engineering. As a director of drilling research, he is credited with building one of the most comprehensive and top drilling engineering curricula in the Department of Petroleum Engineering at the University of Tulsa, worldwide. He has led his graduate students to pioneering research work in areas of cuttings transport in directional well drilling, drilling fluids hydraulics, drill bit, and drill string dynamics. He is closely associated and worked with all major oil and gas companies across the globe. His contributions to the advancement of drilling technology through teaching, research, and publications are notable. He is the author and/or coauthor of four textbooks and more than 60 publications in refereed technical journals. He has given more than 100 presentations at technical meetings and professional functions. He has been invited as keynote speaker at several technical conferences. Professor Azar was a participant speaker at the "Advances in Drilling Technologies" workshop organized by the U.S. National Research Council and the National Academy of Sciences. In 2004, he was invited by the University of Chicago to be session leader and group facilitator on Materials science conference held in Doha, Qatar. He has served as a member of the SPE Education and Professionalism Technical Committee, SPE Student Chapter adviser, SPE Career Guidance and Student Development Committee, Technical reviewer, Committee Chairman/Member on the "SPE Distinguished Achievement Award for Petroleum Engineering Faculty," Committee Chairman/Member on the "AIME Mineral Industry Education Award" and Member of Editorial Advisory Board of Offshore.

Dr. Azar received the 1998 SPE Drilling Engineering Award and the 1997 SPE Distinguished Achievement Award for Petroleum Engineering Faculty. He was elected as SPE Distinguished Member for the year 2004. He was inducted into the U.S. National Academy of Engineering in Washington, D.C, in 2004. He is a registered professional engineer and a member of SPE. His areas of interest in teaching and research include: wellbore fluids hydraulics, wellbore mechanics, drill bit mechanics, drillstring mechanics, and well control mechanics in application to vertical, directional, horizontal, multilateral, coiled tubing, under-balanced, and slim hole drilling. Dr. Azar can be reached via email at adc.training@sbcglobal.net, or by phone at 918-492-6542.

Dr. G. Robello Samuel has been a Chief Technical Advisor for Drilling Engineering in the Drilling Evaluation division of Halliburton since 1998. He has more than 25 years of experience in domestic and international oil/gas drilling and completion operations, management, consulting and teaching. He has been an educator in the United States, Venezuela, Mexico, China, and India. Also, he has been an adjunct professor for the past seven years in the Petroleum Program of Chemical and Mechanical Engineering Technology Departments at the University of Houston where he teaches advanced drilling engineering, downhole tools and technologies, and complex well architecture courses.

Dr. Samuel has written and co-written more than 80 journal articles, conference papers, and technical articles. He has given several graduate seminars at various universities. Dr. Samuel has been the recipient of numerous awards including the "CEO for A Day" Halliburton award. He has also worked at the Oil and Natural Gas Corporation (ONGC), India, from 1983 to 1992 as a drilling engineer.

Dr. Samuel holds B.S. and M.S. (Mechanical Engineering) degrees from the University of Madurai (Madurai), the College of Engineering, Guindy, and Anna University (Chennai). He received his M.S. and Ph.D. (Petroleum Engineering) degrees from the University of Tulsa. He is also the author of *Advanced Drilling Engineering* (Gulf Publishing), *Formulas and Calculation for Drilling* (Wiley-Scrivener) and the co-author of *Drilling Engineering* (Pennwell Publishers). Dr. Samuel can be reached via e-mail at robellos@hotmail.com or on the web at www.sigmaquadrant.com or www.oilwelltechnology.com.